S0-BPZ-053

Symmetry Principles
and
Atomic Spectroscopy

Symmetry Principles and Atomic Spectroscopy

Brian G. Wybourne
Department of Physics
University of Canterbury
Christchurch, New Zealand

Including an Appendix of Tables
by P. H. Butler

Wiley-Interscience *A Division of John Wiley & Sons*
New York · London · Sydney · Toronto

10 9 8 7 6 5 4 3 2 1

Library of Congress Catalogue Card Number: 70-96043

SBN 471 96508 1

Printed in the United States of America

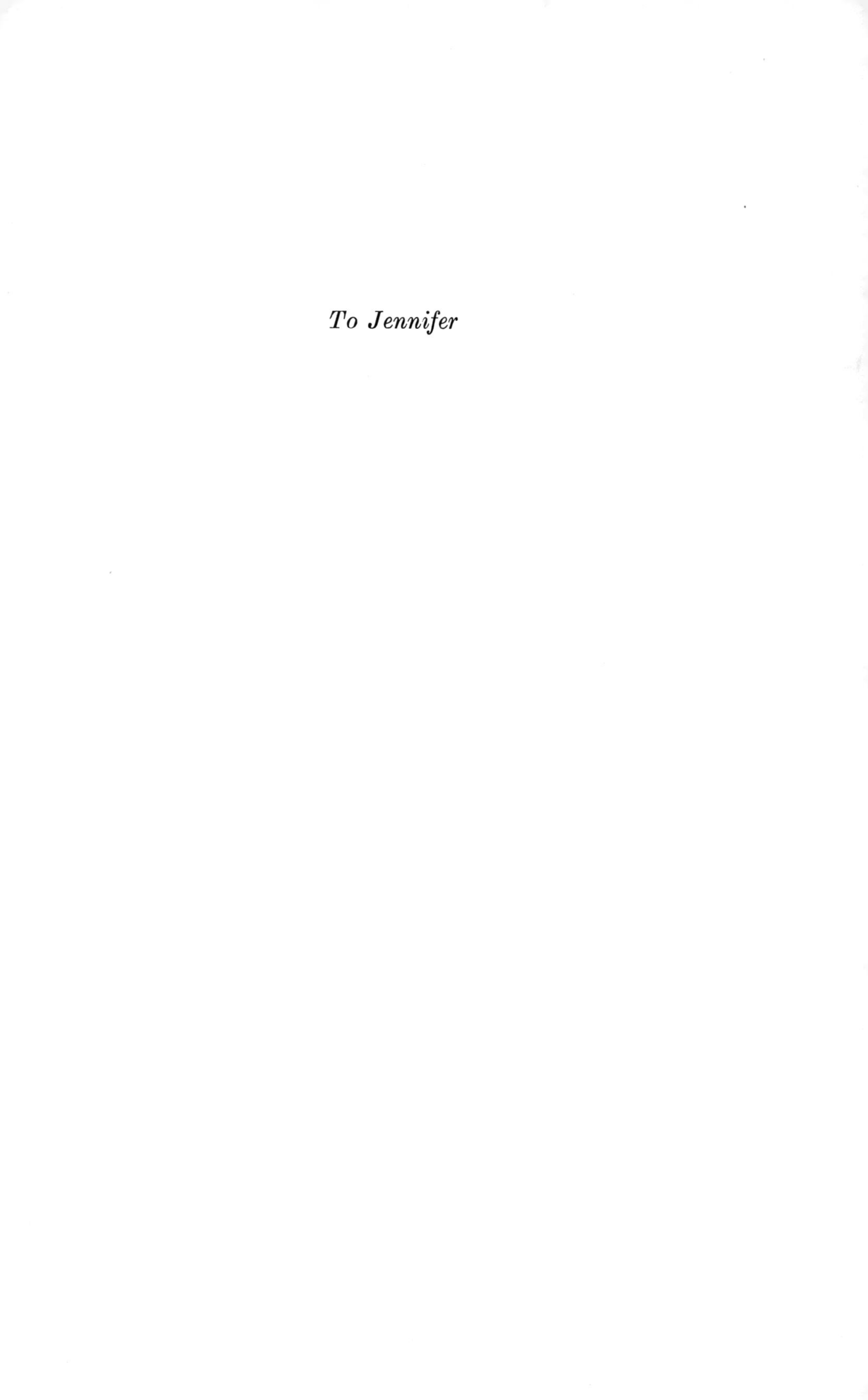

To Jennifer

Preface

The importance of symmetry principles was recognized very early in the development of quantum mechanics and was skillfully exploited in the works associated with the names of van der Waerden, Weyl, Wigner, and Yamanouchi. The application of these principles to the description of atomic wavefunctions is most closely associated with the name of the late Professor Guilio Racah. Atomic spectroscopists were surprisingly slow at realizing the full significance of Racah's work and much of the further development was made by nuclear physicists, notably Jahn, Elliott, and Flowers. More recently much effort has gone into the use of symmetry principles in the description of the properties of elementary particles.

Most of these applications of symmetry principles have drawn heavily on the theory of continuous groups developed by Lie, Cartan, Weyl, and others. In recent years there has been an increasing tendency to emphasize the duality that exists between the symmetric group and the full linear group making extensive use of Young's theory of substitutional analysis and D. E. Littlewood's development of the theory of Schur functions (or *S*-functions).

In this volume I propose to illustrate the application of these recent mathematical developments in group theory to problems of atomic spectroscopy. I do not intend to give a rigorous treatment of the mathematical developments but rather to state, and explain, results that are unfamiliar to many workers in atomic spectroscopy and other allied fields of physics. An extensive bibliography is given to which the reader may turn for further details.

Throughout the book I have endeavored to develop a systematic notation for labeling the various group representations and to include worked examples wherever appropriate. My principal concern has been with the essential development of the theory of the unitary, orthogonal, and symplectic groups and no attempt has been made to encompass the much wider subject of the noncompact groups.

One of my students, Mr. P. H. Butler, has programmed many of the results given in the text and constructed a comprehensive set of tables which appears in Appendix II. These tables are presented with the hope that they will assist the reader in making actual calculations as he masters the methods described. The members of the University of Canterbury's Computing Centre have been of considerable assistance in preparing these tables.

The text was developed from a series of lectures I gave at the Laboratoire Aimé Cotton, Centre National de la Recherche Scientifique, Orsay, in early 1968 and later to the Lithuanian Academy of Sciences at Vilnius. I am grateful to those who attended these lectures and whose remarks unwittingly influenced portions of the book. The critical remarks of my students and of Professor B. R. Judd have been of considerable value in its preparation. I am also grateful to my secretary, Mrs. M. A. Sewell, for her skillful preparation of a difficult manuscript.

This book could not have been written without the special leave generously granted by the Council of the University of Canterbury. The research effort that has resulted in this work has also been generously supported by a grant from the United States Air Force Office of Scientific Research, Office of Aerospace Research, to whom I take this opportunity of expressing my gratitude.

BRIAN G. WYBOURNE

Christchurch, New Zealand
September 1969

Contents

1 Introduction 1

2 The Symmetric Group S_n 3

2.1 Permutations 3
2.2 Cycles 4
2.3 Odd and Even Permutations 4
2.4 Regular Permutations 4
2.5 Cayley's Theorem 4
2.6 Classes 5
2.7 Partitions 5
2.8 Number of Permutations in a Class 6
2.9 Example of S_4 6
2.10 Young Diagrams 7
2.11 Conjugate Young Diagrams 8
2.12 Standard Young Tableaux 8
2.13 Dimensions of Irreducible Representations of S_n 8
2.14 Hook Graphs of Young Diagrams 9
2.15 Characters of the Symmetric Group 10
2.16 Character Tables 11
2.17 Orthogonality Properties of Characters 11
2.18 Compound Characters 11
2.19 Immanants of Matrices 12
2.20 Immanants and Characters 13
2.21 Characters and Partitions 13
2.22 Application to S_6 14

3 Schur Functions and the Symmetric Group 18

3.1 Symmetric Functions 18
3.2 Elementary Symmetric Functions 18
3.3 Homogeneous Product Sums 19

3.4 Power Sums 19
3.5 The Schur Functions 20
3.6 Frobenius's Formula for the Characters of the Symmetric Group 21
3.7 Inner and Outer Products for S_n 22
3.8 Inner Products 22
3.9 Outer Products 22
3.10 Outer Products and S-Functions 22
3.11 Inner Products of S-Functions 25

4 *The Full Linear Group* 28

4.1 S-Functions and the Full Linear Group 28
4.2 Compound Matrices 29
4.3 Induced Matrices 31
4.4 Invariant Matrices 32
4.5 Decomposition of the Direct Product for $GL(n)$ 32
4.6 Degree of Irreducible Representations of $GL(n)$ 33

5 *The Restricted Groups* 35

5.1 Restricted Groups 35
5.2 The Orthogonal Group 36
5.3 Characters of $O(n)$ 37
5.4 Frobenius's Notation for Partitions 38
5.5 Characters of $O(n)$ (Continued) 38
5.6 Branching Rule for $U(n) \to O(n)$ 39
5.7 Associated Characters of $O(n)$ 40
5.8 Characters of the Rotation Group $R(n)$ 40
5.9 The Symplectic Group 40
5.10 Branching Rule for $U(n) \to Sp(n)$ 42
5.11 Nonstandard Symbols 42
5.12 Branching Rules for the Group G_2 46
5.13 Products of Characters for $O(n)$ and $Sp(n)$ 46

6 *The Plethysm of S-Functions* 49

6.1 Basis Functions and the Full Linear Group $GL(n)$ 49
6.2 Plethysm of S-Functions 51
6.3 Rules of Plethysm 52
6.4 The Evaluation of Plethysms of S-Functions 52
6.5 The Degree of a Plethysm 55
6.6 Examples of Plethysms of S-Functions 55

7 *Plethysm and the Restricted Groups* 57

7.1 Intransitive Groups 59
7.2 Spin Representations 59

7.3 The Ternary Orthogonal Group 60
7.4 The Quaternary Orthogonal and Rotation Groups 61
7.5 The Rotation Group in Six Dimensions 63
7.6 Branching Rules and Plethysm 63
7.7 Examples of Branching Rules 64
7.8 Branching Rules and the Ternary Orthogonal Groups 66

8 Classification of Atomic States of Equivalent Electron Configurations 69

8.1 Classification of Atomic States 69
8.2 *LS* Coupling 70
8.3 The Seniority Classification 70
8.4 Additional Classificatory Symbols 71
8.5 Example of f^3 72
8.6 *jj*-Coupling 74
8.7 *LL*-Coupling 75

9 Classification of Atomic States for Mixed Configurations 77

9.1 Mixed Configurations 77
9.2 The Basic Group for $(l_1 + l_2 + l_3)^n$ Configurations 77
9.3 Reduction of the $R_{2(l_1+l_2+l_3+3/2)}$ Symmetry 78
9.4 Example of $(d + s + s')^n$ 79
9.5 Example of $(d + s)^n$ 80
9.6 Parity and Mixed Configurations 81

10 Symmetry Properties of Interactions in Equivalent Electron Configurations 84

10.1 Transformation Properties of Single-Particle Operators 84
10.2 Quasi-Spin Classification of Operators 86
10.3 Quasi-Spin Properties of the Matrix Elements of Single-Particle Operators 88
10.4 Selection Rules and the Matrix Elements of Single-Particle Operators 90
10.5 The Wigner-Eckart Theorem and Isoscalar Factors 93
10.6 The Coulomb Interaction 96
10.7 Symmetrization of the Orbit-Orbit Interaction 102
10.8 Spin-Dependent Two-Particle Interactions 105
10.9 Generalized Two-Particle Operators 108
10.10 Effective Two-Particle Operators 111
10.11 Effective Three-Particle Operators 112

11 Symmetry Properties of Interactions in Mixed Configurations 117

11.1 Transformation Properties of Single-Particle Operators 117
11.2 Coulomb Interaction in $(d + s)^n$ Configurations 119

11.3 Effective Two-Particle Operators in Mixed Configurations 121
11.4 Scalar Three-Particle Operators in $(d + s)^n$ Configurations 122

References 125

Appendix I *Spin and Difference Characters for the Rotation Groups* 133

Appendix II *Tables of Group Properties* 140

Index I-1

Symmetry Principles and Atomic Spectroscopy

1

Introduction

The analysis and interpretation of the spectra of atoms and nuclei has drawn heavily on parallel developments in the theory of complex spectra. Notable progress in the formulation of the theory of complex spectra came from Racah's[1,2] exploitation of the symmetry properties of atomic wavefunctions using the classical theory of continuous groups as developed by Lie,[3] Cartan,[4] Weyl,[5,6] and others. Later developments[7–9] have made extensive use of Young's theory of substitutional analysis[10,11] and Littlewood's development[12,13] of the theory of Schur functions[14] (or S-functions) emphasizing the close connection between the theory of the symmetric and continuous groups. These later developments have been summarized by Judd.[15]

Mathematical research over the last thirty years has contributed enormously to the simplification of the theory of groups. Much of this work has been almost entirely overlooked by physicists to the detriment of the development of applications of symmetry principles to physical problems. The aim of this book is to illustrate the application of recent mathematical developments in group theory to problems of atomic spectroscopy. It is not intended to give a rigorous treatment of these mathematical developments but rather to state results that are unfamiliar to many workers in atomic spectroscopy and to illustrate their applications to actual problems. By way of applications I propose to discuss three central problems in the theory of atomic spectra:

1. The symmetry classification of the atomic states of multielectron configurations.

2. The analysis and classification of the many-body operators that arise in the application of perturbation theory to atomic problems.

3. The application of selection rules in evaluating the matrix elements of interactions.

I have assumed that the reader is already acquainted with the elementary theory of continuous groups and the irreducible tensor algebra as outlined, for example, by Judd[15] and shall outline briefly some relevant parts of the theory of the symmetric group and the general properties of group characters. I propose then to discuss Littlewood's development of the theory of the symmetric and continuous groups in terms of the theory of S-functions with particular emphasis upon the algebra of plethysm. Finally, this body of knowledge is applied to some practical problems in the theory of atomic spectra.

2

The Symmetric Group S_n

2.1 PERMUTATIONS

The $n!$ permutations of the n letters $(\alpha_1, \alpha_2, \ldots, \alpha_n)$ form a group known as the symmetric group which we shall designate as the group S_n. We may write a typical permutation as

$$\begin{pmatrix} 1 & 2 & \cdots & n \\ \alpha_1 & \alpha_2 & \cdots & \alpha_n \end{pmatrix}$$

where the n letters $(\alpha_1, \alpha_2, \ldots, \alpha_n)$ denote the numbers $(1, 2, \ldots, n)$ in some particular order, e.g.,

$$\begin{pmatrix} 1 & 2 & 3 \\ 3 & 1 & 2 \end{pmatrix}$$

i.e., $1 \to 3$, $2 \to 1$, $3 \to 2$. The product ST of two permutations S and T is defined to be the permutation obtained by first operating with T and then with S from right to left, e.g.,

$$\begin{pmatrix} 1 & 2 & 3 & 4 & 5 \\ 2 & 4 & 5 & 3 & 1 \end{pmatrix}\begin{pmatrix} 1 & 2 & 3 & 4 & 5 \\ 5 & 4 & 2 & 3 & 1 \end{pmatrix} = \begin{pmatrix} 1 & 2 & 3 & 4 & 5 \\ 1 & 3 & 4 & 5 & 2 \end{pmatrix}$$

and

$$\begin{pmatrix} 1 & 2 & 3 & 4 & 5 \\ 5 & 4 & 2 & 3 & 1 \end{pmatrix}\begin{pmatrix} 1 & 2 & 3 & 4 & 5 \\ 2 & 4 & 5 & 3 & 1 \end{pmatrix} = \begin{pmatrix} 1 & 2 & 3 & 4 & 5 \\ 4 & 3 & 1 & 2 & 5 \end{pmatrix}$$

We note from these two examples that the multiplication of permutations is usually noncommutative.

2.2 CYCLES

Permutations may be resolved into cycles, e.g.,

$$\begin{pmatrix} 1 & 2 & 3 & 4 & 5 & 6 & 7 & 8 \\ 2 & 3 & 1 & 5 & 4 & 7 & 6 & 8 \end{pmatrix}$$

may be regarded as the product of four *closed* cycles

$$(1\ \ 2\ \ 3)(4\ \ 5)(6\ \ 7)(8)$$

The cycles have no common elements and can be said to commute with one another; hence the order in which the cycles are written is irrelevant. A cycle containing two symbols is called a *transposition.* Any cycle can be written as a product of transpositions having elements in common, e.g., $(1\ \ 2\ \ 3) = (1\ \ 3)(1\ \ 2)$. The cycle on n letters can be written as the product of $n - 1$ transpositions.

2.3 ODD AND EVEN PERMUTATIONS

The difference between the number of symbols and the number of independent cycles is defined as the *decrement* of the permutation. Permutations with an *even* (or *odd*) decrement are said to be *even* (or *odd*). For example, in S_6 the permutation $(1\ \ 2\ \ 3)(4\ \ 5)(6)$ is *odd* while $(1\ \ 2\ \ 3)(4\ \ 5\ \ 6)$ is *even.* The *even* permutations of degree n form a group, the *alternating group* A_n which has $n!/2$ elements and which is obviously a subgroup of S_n.

2.4 REGULAR PERMUTATIONS

Regular permutations have the property of leaving no symbol (apart from the identity element) unchanged and form a regular subgroup of S_n. They can at best be decomposed into cycles containing no common elements and of the same length. There are n possible regular permutations in S_n and they form a subgroup of S_n of order n. Since a regular permutation can be resolved only into cycles of equal length, the subgroup will be cyclic if n is a prime.

2.5 CAYLEY'S THEOREM

We shall not make a detailed investigation of the subgroups of S_n. The analysis of the subgroups of S_n assumes considerable importance in the theory of finite groups since Cayley's demonstration[16] that "*every group G of order n is isomorphic with a subgroup of S_n.*"

2.6 CLASSES

Two permutations α and β of S_n are said to belong to the same class if it is possible to find a permutation γ of S_n such that

$$\gamma\alpha\gamma^{-1} = \beta$$

The element β is said to be *conjugate* to α. Conjugate permutations have the same cycle structure so that the permutations of a given class are necessarily all even or all odd. Furthermore, each class contains the inverse of each of the permutations contained in it.

In designating permutations we shall normally omit cycles which involve unpermuted symbols, e.g.,

$$\begin{pmatrix} 1 & 2 & 3 & 4 & 5 \\ 2 & 3 & 1 & 4 & 5 \end{pmatrix} \rightarrow (1 \quad 2 \quad 3)(4)(5) \rightarrow (1 \quad 2 \quad 3)$$

In these cases it is important to remember the degree of the permutation. Thus for the group S_4 there are five distinct classes:

1. e

2. (1 2), (1 3), (1 4), (2 3), (2 4), (3 4)

3. (1 2)(3 4), (1 3)(2 4), (1 4)(2 3)

4. (1 2 3), (1 3 2), (1 2 4), (1 4 2), (1 3 4), (1 4 3), (2 3 4), (2 4 3)

5. (1 2 3 4), (1 2 4 3), (1 3 2 4), (1 4 2 3), (1 4 3 2), (1 3 4 2)

N.B. The elements e; (1 2 3 4); (1 3)(2 4); (1 4 3 2) are examples of regular permutations and form a cyclic group of order 4 which is the regular subgroup of S_4.

2.7 PARTITIONS

Any permutation can be resolved into independent cycles. If ν_1 = no. of one-cycles, ν_2 = no. of two-cycles, etc., then since there are just n symbols for a given permutation of S_n

$$1\nu_1 + 2\nu_2 + \cdots + n\nu_n = n \tag{1}$$

The cycle structure may be designated as $(1^{\nu_1} 2^{\nu_2} \cdots n^{\nu_n})$ or just (ν). All permutations of S_n which have the same cycle structure (ν) form a conjugate class in S_n. Thus each solution for positive integers $\nu_1, \nu_2, \ldots, \nu_n$ gives a class in S_n; hence the number of classes is just the number of such solutions. Let

$$\begin{aligned} \nu_1 + \nu_2 + \cdots + \nu_n &= \lambda_1 \\ \nu_2 + \cdots + \nu_n &= \lambda_2 \\ \cdot \quad & \cdot \\ \cdot \quad & \cdot \\ \cdot \quad & \cdot \\ \nu_n &= \lambda_n \end{aligned} \tag{2}$$

then

$$\lambda_1 + \lambda_2 + \cdots + \lambda_n = n \qquad \text{and} \qquad \lambda_1 \geq \lambda_2 \geq \cdots \geq \lambda_n \geq 0 \tag{3}$$

Thus each solution of Eq. 1 corresponds to a partition of n into $[\lambda_1, \lambda_2, \ldots, \lambda_n]$ and leads to an important result:

The number of classes in S_n is equal to the number of partitions of n into positive integers such that Eq. 3 is satisfied.

Given a partition as in Eq. 3 there is a corresponding cycle structure, namely,

$$\begin{aligned} \nu_1 &= \lambda_1 - \lambda_2 \\ \nu_2 &= \lambda_2 - \lambda_3 \\ \cdot \quad & \cdot \\ \cdot \quad & \cdot \\ \cdot \quad & \cdot \\ \nu_n &= \lambda_n \end{aligned} \tag{4}$$

2.8 NUMBER OF PERMUTATIONS IN A CLASS

A cycle (ν) of a particular class can be resolved into ν_1 one-cycles, ν_2 two-cycles, etc. We can distribute n symbols in $n!$ ways. This leads to duplication since, for example, if 1 and 2 appear in one-cycles there is no distinction between (1)(2) and (2)(1). All the ν_1 one-cycles can be permuted in $\nu_1!$ ways. Thus a given permutation will be repeated $\nu_1!\,\nu_2!, \ldots, \nu_n!$ times. Also a two-cycle like (1 2) can appear as (2 1); a three-cycle (1 2 3) can appear as (2 3 1) or (3 1 2), etc.; thus each permutation is repeated $1^{\nu_1} 2^{\nu_2} \cdots n^{\nu_n}$ times. Thus the number of *distinct* permutations of S_n having the cycle structure (ν) is

$$n(\nu) = \frac{n!}{\nu_1!\,1^{\nu_1}\nu_2!\,2^{\nu_2} \cdots \nu_n!\,n^{\nu}} \tag{5}$$

2.9 EXAMPLE OF S_4

We first determine the number of partitions of 4 subject to Eq. 3. This gives five partitions [4]; [3 1]; [2^2]; [2 1^2]; [1^4] and hence there are five classes in

S_4. We now determine the cycle structure of each class using Eq. 4, e.g., consider the partition [3 1], we have $\lambda_1 = 3$, $\lambda_2 = 1$ and from Eq. 4 we find $\nu_1 = 2$, $\nu_2 = 1$ and hence the cycle structure is $(1^2\,2^1)$. The number of elements in each class follows from Eq. 5 to give the following for S_4:

Partition	Cycle structure	No. elements in class
[4]	(1^4)	1
[3 1]	$(1^2\,2^1)$	6
$[2^2]$	(2^2)	3
$[2\,1^2]$	$(1^1\,3^1)$	8
$[1^4]$	(4^1)	6
	Total no. elements:	24

We note that the total number of elements is equal to 4! which is the order of the group. We recall from the theory of finite groups that the number of irreducible representations of a finite group is equal to the number of classes; hence S_4 has five irreducible representations. The number of irreducible representations for a symmetric group S_n is simply equal to the number of partitions of n. Each class may be labeled uniquely by specifying its corresponding partition. Later we show that the partitions may also be used to uniquely label the irreducible representations of S_n.

2.10 YOUNG DIAGRAMS

We may represent each class of conjugate elements of S_n by a partition $[\lambda] = [\lambda_1, \lambda_2, \ldots, \lambda_n]$, where

$$\sum_{i=1}^{n} \lambda_i = n \tag{6}$$

Since the cycles can be arranged in any order there is no restriction in assuming $\lambda_1 \geq \lambda_2 \geq \cdots \geq \lambda_n \geq 0$. Then, corresponding to each class of conjugates $[\lambda]$, Young has shown that there exists a uniquely determined irreducible representation defined in terms of the Young diagram

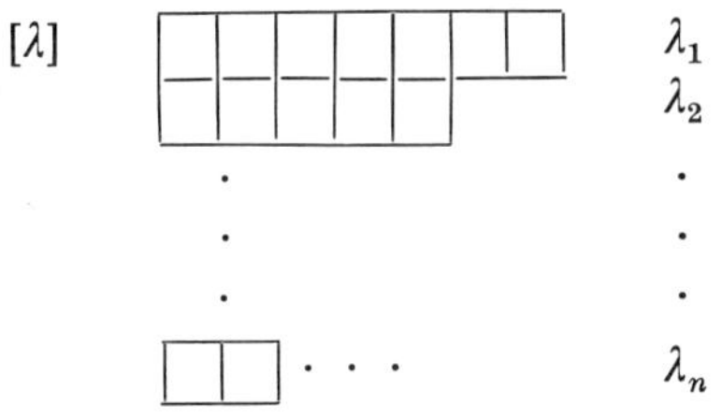

there being λ_i cells in the ith row. Thus for S_4 there are the five diagrams

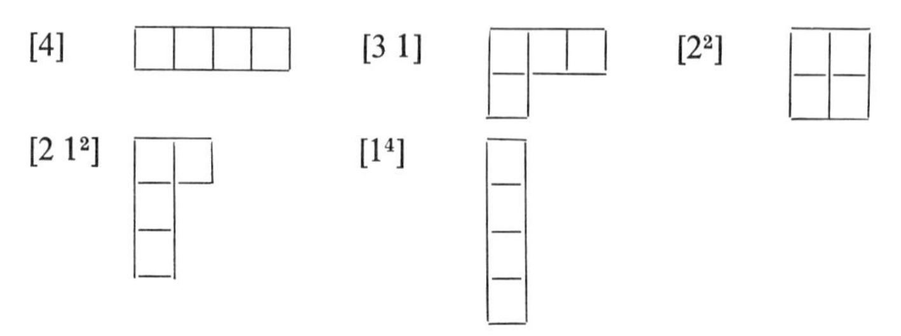

Associated with every partition of S_n is a *shape* known as the Young diagram.

2.11 CONJUGATE YOUNG DIAGRAMS

The shape formed by interchanging the rows and columns of $[\lambda]$ is known as the *conjugate* (or *adjoint*) of $[\lambda]$ and will be designated as $[\tilde{\lambda}]$; for example,

$[3\ 1] \equiv$ $\qquad$ $[\widetilde{3\ 1}] =$

$\therefore \quad [\widetilde{3\ 1}] \equiv [2\ 1^2]$

If $[\lambda] = [\tilde{\lambda}]$ the partition is said to be self-associated.

2.12 STANDARD YOUNG TABLEAUX

Of the $n!$ tableaux of shape λ there will be a certain number f^λ which have letters in each row and column in lexical order. These are said to form *standard Young tableaux*. For example, consider S_4. For [4] there is only one standard tableau 1|2|3|4; for [3 1] there are three standard tableaux

1	2	3
4		

1	2	4
3		

1	3	4
2		

etc.

2.13 DIMENSIONS OF IRREDUCIBLE REPRESENTATIONS OF S_n

Rutherford,[10] Robinson,[11] and Boerner[17] have presented a number of theorems relating to the properties of standard tableaux giving in particular the two fundamental theorems:

1. *The dimension f^λ of the irreducible representation associated with the partition λ of n symbols is equal to the number of distinct standard tableaux associated with the shape λ.*

2. *The number f^λ of standard tableaux belonging to the partition $[\lambda_1\ \lambda_2 \cdots \lambda_r]$ where $\lambda_1 + \lambda_2 + \cdots + \lambda_r = n$ is given by*

$$f^\lambda = n!\,\frac{\prod_{i<k} (l_i - l_k)}{l_1!\, l_2! \cdots l_r!} \tag{7}$$

where

$$l_1 = \lambda_1 + r - 1,\ l_2 = \lambda_2 + r - 2, \ldots, l_r = \lambda_r$$

2.14 HOOK GRAPHS OF YOUNG DIAGRAMS

Robinson[18] has introduced the concept of *hook graphs* of Young diagrams. The *hook length* of a cell in a Young diagram is equal to $(\alpha + \beta + 1)$ where β is the number of cells directly *below* the given cell in the same column and α is the number of cells to the *right* of the given cell in the same row. For the [4 2 1] partition of S_7 we have the hook graph

6	4	2	1
3	1		
1			

Let $l_0, l_1, \ldots, l_p$ be the hook lengths in the first column. Then the *product* of the hook lengths of the graph is defined as

$$H_{[\lambda]} = \frac{l_0!\, l_1! \cdots l_p!}{\prod_{i<j} (l_i - l_j)} \tag{8}$$

e.g.

$$H_{[421]} = \frac{6!\,3!\,1!}{5 \cdot 3 \cdot 2} = 144$$

Theorem

Let $H_{[\lambda]}$ be the product of hook lengths in the hook graph of an irreducible representation $[\lambda]$ of S_n. Then the degree f^λ of $[\lambda]$ is

$$f^\lambda = n!/H_{[\lambda]} \tag{9}$$

Equation 9 may thus be used to determine rapidly the degree of the irreducible representation corresponding to any partition $[\lambda]$.

Example of S_4

[4]	$\begin{array}{\|c\|c\|c\|c\|}\hline 4 & & & \\ \hline\end{array}$	$H_{[4]} = 4!$	$f^{[4]} = \dfrac{4!}{4!} = 1$
[31]	$\begin{array}{\|c\|c\|c\|c\|}\hline 4 & & & \\ \hline 1 \\ \cline{1-1}\end{array}$	$H_{[31]} = \dfrac{4!\,1!}{3}$	$f^{[31]} = \dfrac{4!}{4!/3} = 3$
$[2^2]$	$\begin{array}{\|c\|c\|}\hline 3 & \\ \hline 2 & \\ \hline\end{array}$	$H_{[2^2]} = \dfrac{3!\,2!}{1}$	$f^{[2^2]} = \dfrac{4!}{3!\,2!} = 2$
$[21^2]$	$\begin{array}{\|c\|c\|}\hline 4 & \\ \hline 2 \\ \cline{1-1} 1 \\ \cline{1-1}\end{array}$	$H_{[21^2]} = \dfrac{4!\,2!\,1!}{2\cdot 3\cdot 1}$	$f^{[21^2]} = \dfrac{4!}{4!\,2!/6} = 3$
$[1^4]$	$\begin{array}{\|c\|}\hline 4 \\ \hline 3 \\ \hline 2 \\ \hline 1 \\ \hline\end{array}$	$H_{[1^4]} = \dfrac{4!\,3!\,2!\,1!}{3\cdot 2\cdot 1\cdot 2}$	$f^{[1^4]} = \dfrac{4!}{4!} = 1$

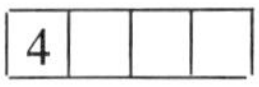

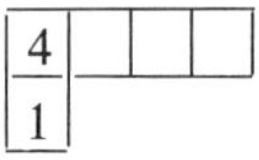

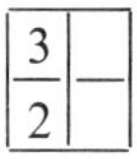

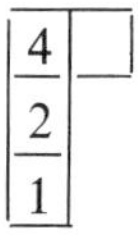

We may check our working by noting that

$$\sum_{[\lambda]} (f^{[\lambda]})^2 = n! \tag{10}$$

2.15 CHARACTERS OF THE SYMMETRIC GROUP

In most of what follows we shall be concerned with group *characters* rather than *representations*. We shall first outline our notation and then state, without proof,[19] a few basic theorems pertaining to the theory of group characters.

Notation

Consider a group H containing h elements $S_i (i = 1, 2, \ldots, h)$. These elements may be grouped into classes. We shall use the letter C_ρ to denote the h_ρ elements of the ρth class. The class containing the inverses of the elements of C_ρ will be designated as C'_ρ.

The *spur* or *trace* of the matrix representing the element S_i which belongs to the irreducible representation $\Gamma^{(j)}$ is called the *characteristic* of S_i in $\Gamma^{(j)}$ and is denoted by $\chi^{(j)}(S_i)$. The set of characteristics of all the elements S of the group as represented in $\Gamma^{(j)}$ is called a *group character* and will be written as $\chi^{(j)}$.

All the elements of the same class ρ have the same characteristic and this will be denoted by $\chi_\rho^{(j)}$. The characteristic of an inverse element is the complex conjugate of the characteristic of the element.

$$\chi^{(j)}(S_i^{-1}) = \chi^{(j)}(S_i^*) \tag{11}$$

In the case of the symmetric group the inverse of an element occurs in the same class as the element itself and hence the characteristics for the symmetric group are all *real*.

2.16 CHARACTER TABLES

If a group contains λ classes then each character will consist of λ characteristics and since the number of classes is equal to the number of representations there must be only λ^2 chracteristics. These characteristics may be arranged in a square array called the *table of characters*; for example, for the group S_4

Class	(1^4)	$(1^2\,2)$	(13)	(4)	(2^2)
Order	1	6	8	6	3
$\{4\}$	1	1	1	1	1
$\{31\}$	3	1	0	-1	-1
$\{2^2\}$	2	0	-1	0	2
$\{21^2\}$	3	-1	0	1	-1
$\{1^4\}$	1	-1	1	-1	1

we have the character table shown here, in which we have used the partitions of 4, enclosed in curly brackets, to label the five representations of S_4.

2.17 ORTHOGONALITY PROPERTIES OF CHARACTERS

The *rows* of a character table satisfy the orthogonality relation

$$\sum_\rho h_\rho \chi_\rho^{(i)} \chi_{\rho'}^{(j)} = h\delta_{ij} \tag{12}$$

while the *columns* satisfy the relation

$$\sum_i h_\rho \chi_\rho^{(i)} \chi_\sigma^{(i)} = h\delta_{\sigma\rho'} \tag{13}$$

2.18 COMPOUND CHARACTERS

A group character may be *simple* or *compound*. If the representation is *irreducible* the character is *simple*. A *reducible* representation has a *compound* character. A compound character φ with a set of characteristics φ_ρ may be expressed as a sum of simple characters, if their values are known, by making use of the orthogonality relations. Suppose

$$\varphi = \sum_i g_i \chi^{(i)} \tag{14}$$

then the coefficients g_i may be found as follows:

$$\begin{aligned} \sum_\rho h_\rho \varphi_\rho \chi_{\rho'}^{(i)} &= \sum_\rho g_i h_\rho \chi_\rho^{(i)} \chi_{\rho'}^{(i)} = h g_i \\ \therefore \quad g_i &= \frac{1}{h} \sum_\rho h_\rho \varphi_\rho \chi_{\rho'}^{(i)} \end{aligned} \tag{15}$$

If φ is a simple or compound character, then

$$\begin{aligned} \sum h_\rho \varphi_\rho \varphi_{\rho'} &= \sum g_i g_j h_\rho \chi_\rho^{(i)} \chi_{\rho'}^{(j)} \\ &= h \sum g_i^2 \end{aligned}$$

hence the condition that a character φ should be simple is that

$$\sum h_\rho \varphi_\rho \varphi_{\rho'} = h \tag{16}$$

2.19 IMMANANTS OF MATRICES

The concept of the *immanants* of a matrix was introduced by Littlewood and Richardson[12,20] and is a natural extension of the idea of the *determinant* and *permanent* of a matrix.[21,22] The immanants of a matrix play a key role in the development of the character theory of the symmetric and full linear groups.

The immanants of a matrix of order n^2,

$$[a_{st}]$$

where s is the index of the row and t the index of the column of the matrix, is defined as follows. Let S denote any permutation $e_1, e_2, \ldots, e_n$ of the set of numbers $1, 2, \ldots, n$, and let $\chi^{(\lambda)}(S)$ be the character of the symmetric group corresponding to the partition (λ). Then the immanant $|a_{st}|^{(\lambda)}$ is defined as

$$|a_{st}|^{(\lambda)} = \sum \chi^{(\lambda)}(S) P_S \tag{17}$$

where the summation is over the $n!$ permutations of the symmetric group and

$$P_S = a_{1e_1} a_{2e_2} \cdots a_{ne_n} \tag{18}$$

Consider the case of the immanants of the matrix $[a_{st}]$ of order 3^2. We have the character table for S_3 as

Class	(1^3)	$(1\ 2)$	(3)
Order	1	3	2
$\{3\}$	1	1	1
$\{2\ 1\}$	2	0	1
$\{1^3\}$	1	1	1

hence the three immanants

$$|a_{st}|^{(3)} = a_{11}a_{22}a_{33} + a_{11}a_{23}a_{32} + a_{12}a_{21}a_{33} + a_{13}a_{22}a_{31} + a_{12}a_{23}a_{31} + a_{13}a_{21}a_{32}$$

$$|a_{st}|^{(21)} = 2a_{11}a_{22}a_{33} - a_{12}a_{23}a_{31} - a_{13}a_{21}a_{32}$$

$$|a_{st}|^{(1^3)} = a_{11}a_{22}a_{23} - a_{11}a_{23}a_{32} - a_{12}a_{21}a_{33} - a_{13}a_{22}a_{31} + a_{12}a_{23}a_{31} + a_{31}a_{21}a_{32}$$

It is easily seen that $|a_{st}|^{(3)}$ is simply the permanent and $|a_{st}|^{(1^3)}$ the determinant of the matrix $[a_{st}]$ and generally

$$\overset{+\quad +}{|a_{st}|} = |a_{st}|^{(n)} \tag{19}$$

and

$$|a_{st}| = |a_{st}|^{(1^n)} \tag{20}$$

2.20 IMMANANTS AND CHARACTERS

Littlewood and Richardson[12] have made use of the theory of immanants to deduce the characters for the symmetric groups up to order 10!.

Let $\alpha_1, \alpha_2, \ldots, \alpha_n$ be n symbols which are permuted by the operations of the symmetric group of order $n!$. The matrix of transformation of the α's for any substitution S, is a permutation matrix, which will be represented by A_S. Littlewood and Richardson then demonstrated that in general the sum of the immanants of the principal r-rowed minors[21] of the matrix A_S is a compound character of the group. If the principal minor is a permutation matrix, its immanant may be read directly from the table of characters of the symmetric group of order $r!$. If it is not a permutation matrix, its immanant is zero.

2.21 CHARACTERS AND PARTITIONS

Since the number of characters is equal to the number of classes and is therefore equal to the number of partitions of n we shall associate each simple character with one of these partitions. The character defined by a partition will be unique if

$$\lambda_1 \geq \lambda_2 \geq \cdots \geq \lambda_i \geq 0$$

The partitions labeling the *characters* are enclosed in curly brackets, whereas those labeling *classes* are enclosed in parentheses.

If a character is obtained from the sum of the immanants of the principal minors of order $(n - \lambda_1)^2$ and the immanants correspond to the partition of $n - \lambda_1$

$$n - \lambda_1 = \lambda_2 + \lambda_3 + \cdots + \lambda_i \tag{21}$$

then the partition of n

$$n = \lambda_1 + \lambda_2 + \cdots + \lambda_i \tag{22}$$

is to be associated with the character. If the correspondence is obtained consecutively for the different degrees in their natural order the logical sequence is retained.

2.22 APPLICATION TO S_6

We now determine the characters of S_6. There is a class corresponding to every partition of 6. The 11 classes and their orders are

$$\begin{array}{ccccccccccc} 1^7, & 1^4 2, & 1^3 3, & 1^2 4, & 1^2 2^2, & 123, & 15, & 6, & 24, & 2^3, & 3^2 \\ 1, & 15, & 40, & 90, & 45, & 120, & 144, & 120, & 90, & 15, & 40 \end{array}$$

The first character, $\chi^{\{6\}}$ will be equal to unity for each class. Permutation matrices, A_S, typical of each class may now be written out as follows:

$$(1^6): \begin{bmatrix} 1 & 0 & 0 & 0 & 0 & 0 \\ 0 & 1 & 0 & 0 & 0 & 0 \\ 0 & 0 & 1 & 0 & 0 & 0 \\ 0 & 0 & 0 & 1 & 0 & 0 \\ 0 & 0 & 0 & 0 & 1 & 0 \\ 0 & 0 & 0 & 0 & 0 & 1 \end{bmatrix} \qquad (1^4 2): \begin{bmatrix} 1 & 0 & 0 & 0 & 0 & 0 \\ 0 & 1 & 0 & 0 & 0 & 0 \\ 0 & 0 & 1 & 0 & 0 & 0 \\ 0 & 0 & 0 & 1 & 0 & 0 \\ 0 & 0 & 0 & 0 & 0 & 1 \\ 0 & 0 & 0 & 0 & 1 & 0 \end{bmatrix}$$

$$(1^3 3): \begin{bmatrix} 1 & 0 & 0 & 0 & 0 & 0 \\ 0 & 1 & 0 & 0 & 0 & 0 \\ 0 & 0 & 1 & 0 & 0 & 0 \\ 0 & 0 & 0 & 0 & 1 & 0 \\ 0 & 0 & 0 & 0 & 0 & 1 \\ 0 & 0 & 0 & 1 & 0 & 0 \end{bmatrix} \qquad (1^2 4): \begin{bmatrix} 1 & 0 & 0 & 0 & 0 & 0 \\ 0 & 1 & 0 & 0 & 0 & 0 \\ 0 & 0 & 0 & 1 & 0 & 0 \\ 0 & 0 & 0 & 0 & 1 & 0 \\ 0 & 0 & 0 & 0 & 0 & 1 \\ 0 & 0 & 1 & 0 & 0 & 0 \end{bmatrix}$$

$(1^2 2^2)$: $\begin{bmatrix} 1 & 0 & 0 & 0 & 0 & 0 \\ 0 & 1 & 0 & 0 & 0 & 0 \\ 0 & 0 & 0 & 1 & 0 & 0 \\ 0 & 0 & 1 & 0 & 0 & 0 \\ 0 & 0 & 0 & 0 & 0 & 1 \\ 0 & 0 & 0 & 0 & 1 & 0 \end{bmatrix}$ (123): $\begin{bmatrix} 1 & 0 & 0 & 0 & 0 & 0 \\ 0 & 0 & 1 & 0 & 0 & 0 \\ 0 & 1 & 0 & 0 & 0 & 0 \\ 0 & 0 & 0 & 0 & 1 & 0 \\ 0 & 0 & 0 & 0 & 0 & 1 \\ 0 & 0 & 0 & 1 & 0 & 0 \end{bmatrix}$

(15): $\begin{bmatrix} 1 & 0 & 0 & 0 & 0 & 0 \\ 0 & 0 & 1 & 0 & 0 & 0 \\ 0 & 0 & 0 & 1 & 0 & 0 \\ 0 & 0 & 0 & 0 & 1 & 0 \\ 0 & 0 & 0 & 0 & 0 & 1 \\ 0 & 1 & 0 & 0 & 0 & 0 \end{bmatrix}$ (6): $\begin{bmatrix} 0 & 1 & 0 & 0 & 0 & 0 \\ 0 & 0 & 1 & 0 & 0 & 0 \\ 0 & 0 & 0 & 1 & 0 & 0 \\ 0 & 0 & 0 & 0 & 1 & 0 \\ 0 & 0 & 0 & 0 & 0 & 1 \\ 1 & 0 & 0 & 0 & 0 & 0 \end{bmatrix}$

(24): $\begin{bmatrix} 0 & 1 & 0 & 0 & 0 & 0 \\ 1 & 0 & 0 & 0 & 0 & 0 \\ 0 & 0 & 0 & 1 & 0 & 0 \\ 0 & 0 & 0 & 0 & 1 & 0 \\ 0 & 0 & 0 & 0 & 0 & 1 \\ 0 & 0 & 1 & 0 & 0 & 0 \end{bmatrix}$ (2^3): $\begin{bmatrix} 0 & 1 & 0 & 0 & 0 & 0 \\ 1 & 0 & 0 & 0 & 0 & 0 \\ 0 & 0 & 0 & 1 & 0 & 0 \\ 0 & 0 & 1 & 0 & 0 & 0 \\ 0 & 0 & 0 & 0 & 0 & 1 \\ 0 & 0 & 0 & 0 & 1 & 0 \end{bmatrix}$

(3^2): $\begin{bmatrix} 0 & 1 & 0 & 0 & 0 & 0 \\ 0 & 0 & 1 & 0 & 0 & 0 \\ 1 & 0 & 0 & 0 & 0 & 0 \\ 0 & 0 & 0 & 0 & 1 & 0 \\ 0 & 0 & 0 & 0 & 0 & 1 \\ 0 & 0 & 0 & 1 & 0 & 0 \end{bmatrix}$

Taking the permanents of the one-rowed principal minors of these matrices, which corresponds to taking their spurs we obtain the compound character $\varphi^{(1)} = 6, 4, 3, 2, 2, 1, 1, 0, 0, 0, 0$. It follows from Eqs. 21 and 22 that the new simple character will be labeled by the partition $\{51\}$. Removing from $\varphi^{(1)}$ the simple character $\chi^{\{6\}}$, we obtain the residue

$$\varphi^{(1)'} = 5, 3, 2, 1, 1, 0, 0, -1, -1, -1, -1$$

Since from Eq. 16

$$\sum h_\rho \varphi^{(1)'2} = 6!$$

we conclude that $\varphi^{(1)'}$ is indeed a simple character; hence

$$\varphi^{(1)} = \chi^{\{6\}} + \chi^{\{51\}}$$

and

$$\chi^{\{51\}} = 5, 3, 2, 1, 1, 0, 0, -1, -1, -1, -1$$

The next compound character is found by taking the sum of the permanents of the principal two-rowed minors of the matrices A_S to give

$$\varphi^{(2)} = 15, 7, 3, 1, 3, 1, 0, 0, 1, 3, 0$$

Using Eq. 15, we find that $\varphi^{(2)}$ contains the simple characters $\chi^{\{6\}}$ and $\chi^{\{51\}}$ once each and the residual is the simple character

$$\chi^{\{42\}} = 9, 3, 0, -1, 1, 0, -1, 0, 1, 3, 0$$

The third compound character $\varphi^{(1^2)}$ is found by taking the sum of the determinants of the principal two-rowed minors of the matrices A_S to give

$$\varphi^{(1^2)} = 15, 5, 3, 1, -1, -1, 0, 0, -1, -3, 0$$

This compound character contains the simple character $\chi^{\{51\}}$ once and the residue is the simple character

$$\chi^{\{41^2\}} = 10, 2, 1, 0, -2, -1, 0, 1, 0, -2, 1$$

The next compound character $\varphi^{(3)}$ is found by taking the sum of the permanents of the principal three-rowed minors of A_S to give

$$\varphi^{(3)} = 20, 8, 2, 0, 4, 2, 0, 0, 0, 0, 2$$

This compound character may be resolved into the simple characters $\chi^{\{3^2\}} + \chi^{\{42\}} + \chi^{\{51\}} + \chi^{\{6\}}$ to give

$$\chi^{\{3^2\}} = 5, 1, -1, -1, 1, 1, 0, 0, -1, -3, 2$$

A further compound character $\varphi^{(21)}$ may be found by summing the $|a_{st}|^{(21)}$ immanants of the principal three-rowed minors of the A_S to yield

$$\varphi^{(21)} = 40, 8, 1, 0, 0, -1, 0, 0, 0, 0, -2$$

Upon resolving $\varphi^{(21)}$ into the simple characters

$$\chi^{\{321\}} + \chi^{\{41^2\}} + \chi^{\{42\}} + \chi^{\{51\}}$$

we obtain

$$\chi^{\{321\}} = 16, 0, -2, 0, 0, 0, 1, 0, 0, 0, -2$$

We need go no further now since we can show that

$$\chi^{\{\lambda\}}\chi^{\{1^n\}} = \chi^{\{\tilde{\lambda}\}} \tag{23}$$

hence immediately write down the remaining characters to give the complete character table for S_6 as

Class	(1^6)	(1^42)	(1^33)	(1^24)	(1^22^2)	(123)	(15)	(6)	(24)	(2^3)	(3^2)
Order	1	15	40	90	45	120	144	120	90	15	40
$\{6\}$	1	1	1	1	1	1	1	1	1	1	1
$\{51\}$	5	3	2	1	1	0	0	−1	−1	−1	−1
$\{42\}$	9	3	0	−1	1	0	−1	0	1	3	0
$\{41^2\}$	10	2	1	0	−2	−1	0	1	0	−2	1
$\{3^2\}$	5	1	−1	−1	1	1	0	0	−1	−3	2
$\{321\}$	16	0	−2	0	0	0	1	0	0	0	−2
$\{2^3\}$	5	−1	−1	1	1	−1	0	0	−1	3	2
$\{31^3\}$	10	−2	1	0	−2	1	0	−1	0	2	1
$\{2^21^2\}$	9	−3	0	1	1	0	−1	0	1	−3	0
$\{21^4\}$	5	−3	2	−1	1	0	0	1	−1	1	−1
$\{1^6\}$	1	−1	1	−1	1	−1	1	−1	1	−1	1

We note that in deriving the characters for the group S_n we need only know the character tables up to those for S_r where $r \not> n/2$.

Character tables for the symmetric groups up to order 16! have been reported in the literature.[12,13,23–26]

3

Schur Functions and the Symmetric Group

3.1 SYMMETRIC FUNCTIONS

The theory of symmetric functions[13,27–29] plays a central role in all branches of physics where the methods of group theory are employed. Three distinct, but related symmetric functions on the n arguments $\alpha_1, \alpha_2, \ldots, \alpha_n$ may be considered.

3.2 ELEMENTARY SYMMETRIC FUNCTIONS

The *elementary symmetric functions* are defined as follows:[29]

$$\left.\begin{aligned} a_0 &= 1 \\ a_1 &= \sum \alpha_1 \\ a_2 &= \sum \alpha_1 \alpha_2 \\ a_3 &= \sum \alpha_1 \alpha_2 \alpha_3 \\ &\cdot \\ &\cdot \\ &\cdot \\ a_n &= \alpha_1 \alpha_2 \alpha_3 \cdots \alpha_n \end{aligned}\right\} \tag{24}$$

These functions are directly connected with the equation whose roots are $\alpha_1, \alpha_2, \ldots \alpha_n$. Thus

$$(x - \alpha_1)(x - \alpha_2) \cdots (x - \alpha_n) = x^n - a_1 x^{n-1} + a_2 x^{n-2} \cdots + (-1)^n a_n$$
$$= 0$$

i.e., the a_r's turn up as the coefficients of the expansion of the functions $g(x)$ and $f(x)$ defined as follows:

$$g(x) = \prod_{i=1}^{n} (x - \alpha_i) = \sum_{r=0}^{n} a_r(-1)^r x^{n-r} \tag{25}$$

and

$$f(x) = \prod_{i=1}^{n} (1 - \alpha_i x) = \sum_{r=0}^{n} a_r(-1)^r x^r \tag{26}$$

3.3 HOMOGENEOUS PRODUCT SUMS

The coefficients of the expansion of the function $1/f(x)$ provides a further set of symmetric functions, the *homogeneous product sums* denoted by h_r. The defining series is then

$$F(x) = 1/f(x) = \sum_{r=0}^{\infty} h_r x^r \qquad \text{with} \quad h_0 = 1 \tag{27}$$

but

$$F(x) = 1/\Pi(1 - \alpha_i x) = \Pi(1 + \alpha_i x + \alpha_i^2 x^2 + \alpha_i^3 x^3 + \cdots) \tag{28}$$

Comparing Eqs. 27 and 28 we conclude that the symmetric function h_r is the sum of all the monomial symmetric functions of degree r in the α_i's. Thus

$$\begin{aligned} h_1 &= \sum \alpha_1 \\ h_2 &= \sum \alpha_1^2 + \sum \alpha_1\alpha_2 \\ h_3 &= \sum \alpha_1^3 + \sum \alpha_1^2\alpha_2 + \sum \alpha_1\alpha_2\alpha_3 \end{aligned} \tag{29}$$

etc.

3.4 POWER SUMS

The third type of symmetric functions is the *power sum* S_r which is simply the sum of the rth powers of the α_i's, i.e.,

$$S_r = \sum \alpha_1^r \tag{30}$$

A generating function for the power sums may be obtained as follows:

$$f(x) = \Pi(1 - \alpha_i x)$$
$$\log f(x) = \sum \log (1 - \alpha_i x)$$

Differentiating leads to the result

$$\frac{f'(x)}{f(x)} = \sum \frac{-\alpha_i}{1 - \alpha_i x} = \sum - \alpha_i - \alpha_i^2 x - \alpha_i^3 x^2 - \cdots$$
$$= -S_1 - S_2 x - S_3 x^2 - \cdots - S_n x^{n-1} \tag{31}$$

3.5 THE SCHUR FUNCTIONS

Numerous relationships between the functions a_r, h_r, and S_r may be established by manipulation of the various generating formulas.[29] Two are of particular importance for our work. If we represent by $[Z_r]$ the matrix

$$[Z_r] = \begin{bmatrix} S_1, 1, 0, 0, \ldots & & 0 \\ S_2, S_1, 2, 0, \ldots & & 0 \\ S_3, S_2, S_1, 3, 0, \ldots & & 0 \\ \cdots\cdots\cdots\cdots\cdots & \cdots & \cdots \\ & & S_2, S_1, r-1 \\ S_r, S_{r-1} & & S_3, S_2, S_1 \end{bmatrix} \tag{32}$$

we have

$$r!\, a_r = |Z_r| \tag{33}$$

and

$$r!\, h_r = \overset{+}{|}Z_r\overset{+}{|} \tag{34}$$

Equations 33 and 34 are illustrative of two particular immanants of the matrix $[Z_r]$ and suggest the possibility of the other immanants being of interest. To this end we define the so-called *Schur functions* (abbreviated to *S-functions*).

If $(\lambda) = (\lambda_1, \lambda_2, \ldots, \lambda_p)$ is a partition of r with the parts in descending order, the S-function

$$\{\lambda\} \equiv \{\lambda_1, \lambda_2, \ldots, \lambda_p\}$$

is defined by the equation

$$r!\,\{\lambda\} = |Z_r|^{(\lambda)} \tag{35}$$

It follows from Eqs. 33 and 34 that

$$\{r\} = h_r \quad \text{and} \quad \{1^r\} = a_r \tag{36}$$

Let ρ denote the class $(1^\alpha 2^\beta \cdots)$ of order h_ρ of the symmetric group of order $r!$ and let $S_\rho = S_1{}^\alpha S_2{}^\beta \cdots$; then we may obtain an expression for the sum of products $\sum_\rho P_S$ obtained from the matrix $[Z_r]$ by a simple comparison of the known results,[29]

$$a_r = \sum \pm h_\rho S_\rho \tag{37}$$

and

$$a_r = \sum \pm P_S \tag{38}$$

to give

$$\sum_\rho P_S = h_\rho S_\rho \tag{39}$$

where of the alternative signs, (+) is taken for *even* permutations and (−) for *odd* permutations and the summation in Eq. 39 is over the permutations of the class ρ.

Upon comparing Eqs. 17, 35, and 39 we obtain the important result

$$r!\,\{\lambda\} = \sum \chi_\rho^{(\lambda)} h_\rho S_\rho \tag{40}$$

which provides the crucial link between S-functions and the characters of the symmetric group.

3.6 FROBENIUS' FORMULA FOR THE CHARACTERS OF THE SYMMETRIC GROUP

The celebrated formula of Frobenius for determining the characters of the symmetric group follows from Eq. 40 if we multiply both sides by $\chi_\sigma^{(\lambda)}$ and sum over all the partitions (λ) using the orthogonal properties of group characters to give

$$r! \sum \{\lambda\}\chi_\sigma^{(\lambda)} = \sum h_\rho \chi_\rho^{(\lambda)} \chi_\sigma^{(\lambda)} S_\rho = r!\, S_\sigma$$

and then replacing σ by ρ, and multiplying by the alternating function

$$\begin{aligned}\Delta(\alpha_1, \ldots, \alpha_n) &= \prod_{r<s} (\alpha_r - \alpha_s),\\ &= \sum \pm \alpha_1^{n-1}\alpha_2^{n-2} \cdots \alpha_{n-1}\end{aligned}$$

to give

$$S_1{}^\alpha S_2{}^\beta S_3{}^\gamma \cdots \Delta(\alpha_1, \ldots, \alpha_n) = \sum \pm \chi_\rho^{(\lambda)} \alpha_1^{\lambda_1+n-1} \alpha_2^{\lambda_2+n-2} \cdots \alpha_n^{\lambda_n} \tag{41}$$

which is Frobenius' result.[30] The right-hand side is summed for all permutations of the suffixes taking the minus sign for odd permutations.

The application of Frobenius' formula proceeds by equating the coefficients of $\alpha_1^{\lambda_1+n-1}\alpha_2^{\lambda_2+n-2} \cdots \alpha_n^{\lambda_n}$ on both sides of Eq. 41. The practical application of Frobenius' is formidable (though complete) in all but the simplest cases. A considerable literature has arisen on the computation of characters for the symmetric group[10–13,20,30–35] but we shall not need to pursue this subject further.

3.7 INNER AND OUTER PRODUCTS FOR S_n

The concept and analysis of the inner and outer products of the irreducible representations of the symmetric group is of vital importance in the development of both the theory of the symmetric group and of the full linear group. Due to the duality that exists between these two groups[6,11] a considerable part of the theory of the full linear group, and its subgroups, may be deduced directly from corresponding results obtained for the symmetric group. Furthermore, isomorphisms exist between the mutliplicative properties of S-functions and group characters that lead to immense simplifications of the character theory of these groups.

3.8 INNER PRODUCTS

If $(S_i)^\rho$ and $(S_i)^\sigma$ are matrices representing S_i in two irreducible representations ρ and σ of the group H, then the Kronecker products[16] of these matrices

$$(S_i)^\rho \times (S_i)^\sigma$$

yield a representation of H known as the *inner product* and designated as $\rho \circ \sigma$. The analysis of the inner products amounts to performing the character decomposition

$$\chi^\rho \chi^\sigma = \sum_\nu g_{\rho\sigma\nu} \chi^\nu \tag{42}$$

3.9 OUTER PRODUCTS

Let us now form the Kronecker product of the matrices $(S_i)^\sigma$ and $(S'_j)^\sigma$ where H and H' are distinct groups. These matrices yield a representation of the direct product $H \times H'$ which we shall denote by $\rho \times \sigma$. If the direct product $H \times H'$ is a subgroup of a finite group H'' then we shall call the representation of H'' induced by the irreducible representation $\rho \times \sigma$ of $H \times H'$ an outer product $\rho \cdot \sigma$. Note that we reserve the *open dot* ($\circ$) for indicating the *inner product* and the *full dot* ($\cdot$) for the *outer product*.

3.10 OUTER PRODUCTS AND S-FUNCTIONS

In the case of the symmetric group we may set $H = S_n$ and $H' = S_m$ and take $H'' = S_{m+n}$ in defining the outer product.

Littlewood and Richardson[12] have shown that the product of two S-functions $\{\rho\}$ and $\{\sigma\}$ of degrees m and n, respectively, may be expressed as the sum of integral multiples of the S-functions $\{\nu\}$ of degree $m + n$, i.e.,

$$\{\rho\}\{\sigma\} = \sum_{\nu} \Gamma_{\rho\sigma\nu}\{\nu\} \tag{43}$$

Furthermore, they have shown that there exists an isomorphism between the multiplication of S-functions and the formation of the outer products $\rho \cdot \sigma$ of the irreducible representations $\rho \times \sigma$ of the symmetric group. That is, there is a one-to-one correspondence between the partitions (ρ), (σ), and (ν) that characterize the S-functions of Eq. 43 and those that label the irreducible representations of the symmetric group that arise in the decomposition of the outer product $\rho \times \sigma$.

We illustrate the multiplication of S-functions by a few simple examples and then give the general rule.

Consider the product $\{1\}\{1\}$. From Eqs. 24, 29, and 36 we have

$$\{1\}\{1\} = (\sum \alpha_1)(\sum \alpha_1) = \sum \alpha_1^2 + 2 \sum \alpha_1\alpha_1$$

and since

$$\{2\} = \sum \alpha_1^2 + \sum \alpha_1\alpha_2 \quad \text{and} \quad \{1^2\} = \sum \alpha_1\alpha_2$$

we conclude that

$$\{1\}\{1\} = \{2\} + \{1^2\}$$

or diagrammatically

$$\square \cdot \square = \square\square + \begin{array}{c}\square\\\square\end{array}$$

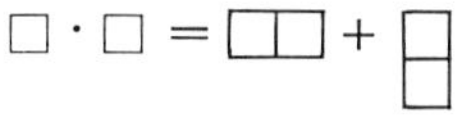

Now consider the product $\{2\}\{1^2\}$. We have

$$\{2\} = h_2 = \sum \alpha_1^2 + \sum \alpha_1\alpha_2 \quad \text{and} \quad \{1^2\} = a_2 = \sum \alpha_1\alpha_2$$

hence,

$$\begin{aligned}\{2\}\{1^2\} &= (\sum \alpha_1^2 + \sum \alpha_1\alpha_2)(\sum \alpha_1\alpha_2) \\ &= \sum \alpha_1^3\alpha_2 + \sum \alpha_1^2\alpha_2^2 + 3 \sum \alpha_1^2\alpha_2\alpha_3 + 6 \sum \alpha_1\alpha_2\alpha_3\alpha_4\end{aligned}$$

We expect this product to be expressible as the sum of integral multiples of S-functions of degree 4. Using Eq. 40 and the character table of S_4 we have

$$4!\,\{31\} = 3S_1^4 + 6S_1^2S_2 - 6S_4 - 3S_2^2$$

$$4!\,\{21^2\} = 3S_1^4 - 6S_1^2S_2 + 6S_2 + 6S_4 - 3S_2^2$$

hence

$$\{31\} + \{21^2\} = \frac{S_1^4 - S_2^2}{4}$$

Using Eq. 30, we find

$$\{2\}\{1^2\} = \{31\} + \{21^2\}$$

or diagrammatically

$$\square\square \cdot \begin{array}{c}\square\\\square\end{array} = \begin{array}{l}\square\square\square\\\square\end{array} + \begin{array}{l}\square\square\\\square\\\square\end{array}$$

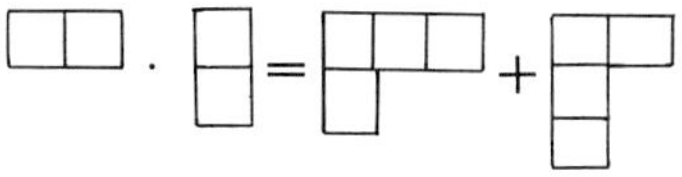

Similar examples lead to the general rule[11,12] for the multiplication of S-functions $\{\rho\}\{\sigma\}$:

The S-functions appearing in the product

$$\{\rho\}\{\sigma\} = \{\rho_1, \rho_2, \ldots, \rho_m\}\{\sigma_1, \sigma_2, \ldots, \sigma_n\}$$

are those corresponding to graphs which can be built up by adding to a graph corresponding to the partition (ρ), σ_1 identical symbols α, σ_2 symbols β, σ_3 symbols γ, etc., in this order and in all ways consistent with the following rules:

1. *After the addition of each symbol there are no two identical symbols in the same column of the compound graph.*
2. *If, after the addition of each symbol, we count the α's, β's, γ's, etc., from right to left by rows starting at the top then at all times while the count is being made the number of α's counted must be greater than, or equal to, the number of β's counted which in turn must be greater than, or equal to, the number of γ's and so on.*
3. *The graphs we obtain after addition of each symbol must be regular.*

As an example, consider the product $\{32\}\{2^21\}$; we have

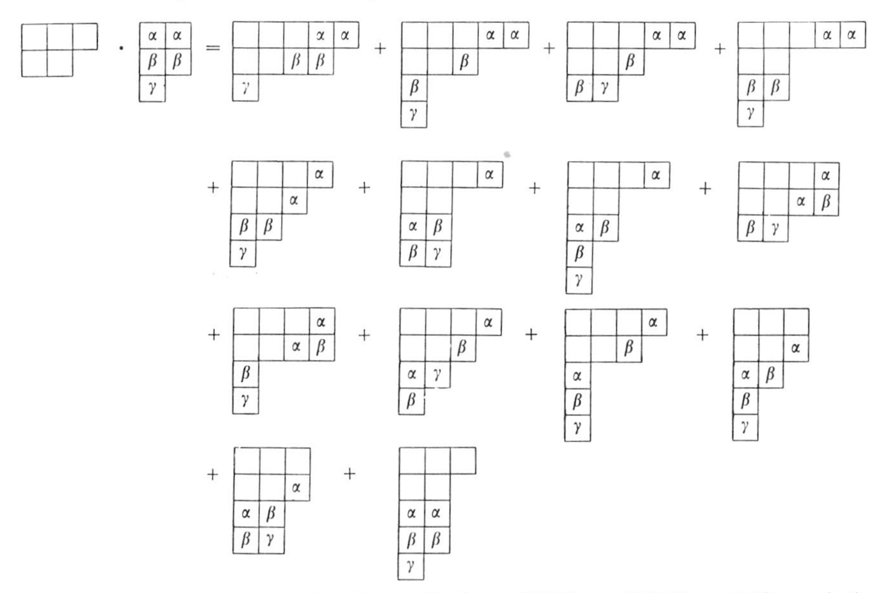

$$\therefore \quad \{32\}\{2^21\} = \{541\} + \{531^2\} + \{532\} + \{52^21\} + \{4321\} + \{42^3\} \quad (44)$$
$$+ \{42^21^2\} + \{4^22\} + \{4^21^2\} + \{4321\} + \{431^3\} + \{3^221^2\}$$
$$+ \{3^22^2\} + \{32^31\}$$

The reduction of the outer product $\{\rho\} \cdot \{\sigma\}$ may be dimensionally checked by noting that if

$$\{\rho\}\{\sigma\} = \sum_{\nu} \Gamma_{\rho\sigma\nu}\{\nu\}$$

then[11]

$$f^{\{\rho\}\{\sigma\}} = \frac{(m+n)!}{m!\,n!} f^{\{\rho\}} f^{\{\sigma\}} = \sum \Gamma_{\rho\sigma\nu} f^{\{\nu\}} \tag{45}$$

where $f^{\{\rho\}}$, $f^{\{\sigma\}}$, and $f^{\{\nu\}}$ are the degrees of the representations $\{\rho\}$, $\{\sigma\}$, and $\{\nu\}$, respectively. The degrees of the representations may be readily computed using Eq. 9 and the properties of hook graphs. Thus in the example of Eq. 44 we find

$$\frac{(5+5)!}{5!\,5!} 5 \cdot 5 = 288 + 567 + 450 + 525 + 768 + 300 + 567 + 252$$
$$+ 300 + 768 + 525 + 450 + 252 + 288$$

Both sides agree and thus the dimensional check holds. It should be noted that this check only checks the dimensions of the representations and there can still exist the possibility of accidental errors involving irreducible representations of the same degree or alternatively linear combinations of irreducible representations having identical sums.

3.11 INNER PRODUCTS OF *S*-FUNCTIONS

The reduction of the inner product $\{\rho\} \circ \{\sigma\}$ for the symmetric group is considerably more complex than for the outer product. Numerous attempts at simplifying the reduction of inner products have been made with varying degrees of success.[37–45]

Littlewood[42] has attempted to solve the problem by defining an inner product of *S*-functions as follows: If (ρ) and (σ) are partitions of the same integer n and

$$\chi_{\alpha}^{(\rho)}\chi_{\alpha}^{(\sigma)} = \sum g_{\rho\sigma\nu}\chi_{\alpha}^{(\nu)}$$

where $\chi_{\alpha}^{(\nu)}$, etc., denote symmetric group characteristics then the expression

$$\{\rho\} \circ \{\sigma\} = \sum g_{\rho\sigma\nu}\{\nu\} \tag{46}$$

is called the inner product of the *S*-functions $\{\rho\}$ and $\{\sigma\}$. The problem of expressing the Kronecker product of two irreducible representations of the symmetric group S_n as direct sums of irreducible representations is then equivalent to evaluating the inner product $\{\rho\} \circ \{\sigma\}$.

Littlewood has given a theorem which allows the inner product of *S*-functions to be evaluated without recourse to character tables. We state this theorem without proof.

Theorem[42]

$$(\{\lambda\}\{\mu\}) \circ \{\nu\} = \sum \Gamma_{\rho\sigma\nu}(\{\lambda\} \circ \{\rho\})(\{\mu\} \circ \{\sigma\}) \tag{47}$$

where

$$\{\rho\}\{\sigma\} = \sum \Gamma_{\rho\sigma\nu}\{\nu\}$$

To illustrate the application of this theorem consider the evaluation of the inner product $\{51\} \circ \{321\}$. We have from Eq. 47

$$(\{5\}\{1\}) \circ \{321\} = \sum \Gamma_{\{\rho\}\{\sigma\}\{321\}}(\{5\} \circ \{\rho\})(\{1\} \circ \{\sigma\})$$

But $\{1\} \circ \{\sigma\} = \{1\}$ since σ can only be a partition of 1 and $\{5\} \circ \{\rho\} = \{\rho\}$ since $\{5\}$ corresponds to the symmetric representation of S_5 and hence

$$\begin{aligned}(\{5\}\{1\}) \circ \{321\} &= \sum \Gamma_{\{\rho\}\{1\}\{321\}}\{\rho\}\{1\} \\ &= (\{32\} + \{31^2\} + \{2^21\})\{1\} \\ &= \{42\} + \{41^2\} + \{3^2\} + 3\{321\} + \{31^3\} \\ &\quad + \{2^3\} + \{2^21^2\}\end{aligned}$$

But

$$\begin{aligned}(\{5\}\{1\}) \circ \{321\} &= \{51\} \circ \{321\} + \{6\} \circ \{321\} \\ &= \{51\} \circ \{321\} + \{321\}\end{aligned}$$

$$\therefore \quad \{51\} \circ \{321\} = \{42\} + \{41^2\} + \{3^2\} + 2\{321\} + \{31^3\} + \{2^3\} + \{2^21^2\}$$

We may check our result by noting that if

$$\{\rho\} \circ \{\sigma\} = \sum g_{\rho\sigma\upsilon}\{\upsilon\}$$

then

$$f^{\{\rho\}\circ\{\sigma\}} = f^{\{\rho\}}f^{\{\sigma\}} = \sum g_{\rho\sigma\upsilon}f^{\{\upsilon\}} \tag{48}$$

A number of special formulas exist which can be of considerable value in simple cases. In particular,

$$\begin{aligned}\{n-1,1\} \circ \{n-1,1\} &= \{n\} + \{n-1,1\} + \{n-2,2\} + \{n-2,1^2\} \\ \{n-1,1\} \circ \{n-2,2\} &= \{n-1,1\} + \{n-2,2\} + \{n-2,1^2\} + \\ &\quad \{n-3,2,1\} \\ \{n-1,1\} \circ \{n-2,1^2\} &= \{n-1,1\} + \{n-2,2\} + \{n-2,1^2\} + \\ &\quad \{n-3,2,1\} + \{n-3,1^3\} \\ \{n-2,2\} \circ \{n-2,2\} &= \{n\} + \{n-1,1\} + 2\{n-2,2\} + \{n-2,1^2\} \\ &\quad + \{n-3,3\} + 2\{n-3,2,1\} + \{n-3,1^3\} \\ &\quad + \{n-4,4\} + \{n-4,3,1\} + \{n-4,2^2\}\end{aligned}$$

The above formulas sometimes yield S-functions with parts that are not in the standard descending form. These S-functions may be converted into the standard form by using Littlewood's three conversion rules,[13] which we now state:

I. In any S-function two consecutive parts may be interchanged provided that the *preceding* part is decreased by unity and the *succeeding* part increased by unity, the S-function being thereby changed in sign.

$$\{\lambda_1, \ldots, \lambda_i, \lambda_{i+1}, \lambda_{i+2}, \ldots, \lambda_p\} = -\{\lambda_1, \ldots, \lambda_{i+1} - 1, \lambda_i + 1, \lambda_{i+2}, \ldots, \lambda_p\} \quad (49)$$

II. In any S-function if any part exceed by unity the preceding part the value of the S-function is zero.

III. The value of any S-function is zero if the last part is a negative number.

As an example of the application of these rules we have

$$\{22\} \circ \{22\} = \{4\} + \{31\} + 2\{22\} + \{21^2\} + \{13\} + 2\{121\} + \{1^4\} + \{04\} + \{031\} + \{022\}$$

But

$$\{121\} = -\{121\} = 0 \qquad \text{(Rule I)}$$
$$\{022\} = -\{112\} = 0 \qquad \text{(Rule II)}$$
$$\{04\} = -\{31\} \qquad \text{(Rule I)}$$
$$\{031\} = -\{211\} \qquad \text{(Rule I)}$$
$$\{13\} = -\{22\} \qquad \text{(Rule I)}$$
$$\{22\} \circ \{22\} = \{4\} + \{22\} + \{1^4\}$$

If

$$\{\lambda\} \circ \{\mu\} = \sum g_{\lambda\mu\rho}\{\rho\}$$

then

$$\{\lambda\} \circ \{\tilde{\mu}\} = \sum g_{\lambda\mu\rho}\{\tilde{\rho}\}$$

Furthermore,

$$\{\tilde{\lambda}\} \circ \{\tilde{\mu}\} = \{\lambda\} \circ \{\mu\}$$

These general results for conjugates may be usefully applied in practical calculations. We note that the inner product $\{\rho\} \circ \{\sigma\}$ contains the identity representation if and only if $\{\rho\} \equiv \{\sigma\}$ and the alternating representation $\{1^n\}$ if and only if $\{\rho\} \equiv \{\tilde{\sigma}\}$. These two observations are of particular relevance to a number of problems in atomic spectroscopy.[46]

4

The Full Linear Group

4.1 S-FUNCTIONS AND THE FULL LINEAR GROUP

The infinite set of all nonsingular matrices A of order n^2 form a continuous group known as the *full linear group* $GL(n)$ on n variables. The theory of the group characters of $GL(n)$ is closely related to that of the symmetric group.[6,11,13] The study of the properties of the irreducible representations of $GL(n)$, and its subgroups, is of key importance in developing the applications of symmetry principles to problems in quantum mechanics. We shall generally only sketch the main results leaving the reader to consult the relevant literature for detailed derivations.

The set of matrices A themselves form a matrix representation of $GL(n)$ corresponding to the unary representation. The character of this representation will be given by the spurs of the matrices A which is equivalent to the elementary symmetric function a_1 of the latent roots of A. But from Eq. 36 we see that

$$a_1 = \{1\}$$

hence the simple character is just the S-function $\{1\}$. The identity representation will have a character given by the trivial S-function $\{0\}$. We seek then to see if all the characters of $GL(n)$ can be expressed in terms of the S-functions of the matrices A.

4.2 COMPOUND MATRICES

Let X and Y be two-column vectors with elements $x_1, x_2, \ldots, x_n$ and $y_1, y_2, \ldots y_n$, respectively. These two-column vectors may each be transformed into new sets X' and Y' by the same nonsingular matrix $A \equiv [a_{st}]$ so that

$$X' = AX \qquad \text{and} \qquad Y' = AY \tag{50}$$

We may form $n(n-1)/2$ alternating linear functions of the vectors X and Y and these will transform into alternating linear functions of X' and Y'. Putting

$$x^{ij} = x_i y_j - x_j y_i \tag{51}$$

the corresponding functions of X' and Y' are given by

$$x'^{ij} = \sum a_{ip} x_p \sum a_{jq} y_q - \sum a_{iq} x_q \sum a_{jp} y_p \tag{52}$$

The coefficient of $x^{pq} = x_p y_q - x_q y_p$ is then

$$a_{ip} a_{jq} - a_{iq} a_{jp} = \begin{vmatrix} a_{ip} & a_{iq} \\ a_{jp} & a_{jq} \end{vmatrix} \tag{53}$$

Thus the matrix of transformation of the $n(n-1)/2$ expressions x^{ij} is a matrix whose elements are the two-rowed minors of the matrix A and is known as the *second compound* matrix of A and designated as $A^{\{1^2\}}$.

As an example, consider the two vectors $X \equiv (x_1, x_2, x_3)$ and $Y \equiv (y_1, y_2, y_3)$. Then from Eq. 50

$$\begin{bmatrix} x_1' \\ x_2' \\ x_3' \end{bmatrix} = \begin{bmatrix} a_{11} & a_{12} & a_{13} \\ a_{21} & a_{22} & a_{23} \\ a_{31} & a_{32} & a_{33} \end{bmatrix} \begin{bmatrix} x_1 \\ x_2 \\ x_3 \end{bmatrix} = \begin{bmatrix} a_{11}x_1 + a_{12}x_2 + a_{13}x_3 \\ a_{21}x_1 + a_{22}x_2 + a_{23}x_3 \\ a_{31}x_1 + a_{32}x_2 + a_{33}x_3 \end{bmatrix}$$

and

$$\begin{bmatrix} y_1' \\ y_2' \\ y_3' \end{bmatrix} = \begin{bmatrix} a_{11} & a_{12} & a_{13} \\ a_{21} & a_{22} & a_{23} \\ a_{31} & a_{32} & a_{33} \end{bmatrix} \begin{bmatrix} y_1 \\ y_2 \\ y_3 \end{bmatrix} = \begin{bmatrix} a_{11}y_1 + a_{12}y_2 + a_{13}y_3 \\ a_{21}y_1 + a_{22}y_2 + a_{23}y_3 \\ a_{31}y_1 + a_{32}y_2 + a_{33}y_3 \end{bmatrix}$$

The alternating function

$$\begin{aligned}
x'^{12} &= x_1' y_2' - x_2' y_1' \\
&= (a_{11}x_1 + a_{12}x_2 + a_{13}x_3)(a_{21}y_1 + a_{22}y_2 + a_{23}y_3) \\
&\quad - (a_{21}x_1 + a_{22}x_2 + a_{23}x_3)(a_{11}y_1 + a_{12}y_2 + a_{13}y_3) \\
&= (a_{11}a_{22} - a_{21}a_{12})(x_1y_2 - x_2y_1) + (a_{11}a_{23} - a_{21}a_{13}) \\
&\quad \times (x_1y_3 - x_3y_1) + (a_{12}a_{23} - a_{22}a_{13})(x_2y_3 - x_3y_2) \\
&= \begin{vmatrix} a_{11} & a_{12} \\ a_{21} & a_{22} \end{vmatrix} x^{12} + \begin{vmatrix} a_{11} & a_{13} \\ a_{21} & a_{23} \end{vmatrix} x^{13} + \begin{vmatrix} a_{12} & a_{13} \\ a_{22} & a_{23} \end{vmatrix} x^{23}
\end{aligned}$$

Similarly,

$$x'^{13} = \begin{vmatrix} a_{11} & a_{12} \\ a_{31} & a_{32} \end{vmatrix} x^{12} + \begin{vmatrix} a_{11} & a_{13} \\ a_{31} & a_{33} \end{vmatrix} x^{13} + \begin{vmatrix} a_{12} & a_{13} \\ a_{32} & a_{33} \end{vmatrix} x^{23}$$

and

$$x'^{23} = \begin{vmatrix} a_{21} & a_{22} \\ a_{31} & a_{32} \end{vmatrix} x^{12} + \begin{vmatrix} a_{21} & a_{23} \\ a_{31} & a_{33} \end{vmatrix} x^{13} + \begin{vmatrix} a_{22} & a_{23} \\ a_{32} & a_{33} \end{vmatrix} x^{23}$$

Thus the second compound matrix $A^{\{1^2\}}$ of A is

$$A^{\{1^2\}} = \begin{bmatrix} \begin{vmatrix} a_{11} & a_{12} \\ a_{21} & a_{22} \end{vmatrix} & \begin{vmatrix} a_{11} & a_{13} \\ a_{21} & a_{23} \end{vmatrix} & \begin{vmatrix} a_{12} & a_{13} \\ a_{22} & a_{23} \end{vmatrix} \\ \begin{vmatrix} a_{11} & a_{12} \\ a_{31} & a_{32} \end{vmatrix} & \begin{vmatrix} a_{11} & a_{13} \\ a_{31} & a_{33} \end{vmatrix} & \begin{vmatrix} a_{12} & a_{13} \\ a_{32} & a_{33} \end{vmatrix} \\ \begin{vmatrix} a_{21} & a_{22} \\ a_{31} & a_{32} \end{vmatrix} & \begin{vmatrix} a_{21} & a_{23} \\ a_{31} & a_{33} \end{vmatrix} & \begin{vmatrix} a_{22} & a_{23} \\ a_{32} & a_{33} \end{vmatrix} \end{bmatrix} \tag{54}$$

and induces the transformation

$$\begin{bmatrix} x'^{12} \\ x'^{13} \\ x'^{23} \end{bmatrix} = A^{\{1^2\}} \begin{bmatrix} x^{12} \\ x^{13} \\ x^{23} \end{bmatrix}$$

Similarly the alternating functions of r vectors $X_1, X_2, \ldots, X_r$, each of which is transformed by the same matrix A, will be transformed by a matrix with $\binom{n}{r}$ rows and columns whose elements are the r-rowed minors of A. This matrix is then said to be the *rth compound matrix* of A and is designated as $A^{\{1^r\}}$.

The compound matrices $A^{\{1^r\}}$ clearly form a representation of $GL(n)$ since the compound matrix of a product is the product of two compound matrices, i.e.,

$$[AB]^{\{1^r\}} = A^{\{1^r\}}B^{\{1^r\}} \tag{55}$$

The spurs of the matrices $A^{\{1^r\}}$ thus give a group character.

In our discussion of the elementary symmetric functions a function $g(x)$ (see Eq. 25) was chosen so that the variables $\alpha_1, \alpha_2, \ldots, \alpha_n$ were the roots of the equation $g(x) = 0$. Using λ to denote the latent roots of the matrix A we have

$$g(\lambda) = |A - \lambda I| = \lambda^n - a_1\lambda^{n-1} + a_2\lambda^{n-2} \cdots + (-1)^n a_n = 0 \tag{56}$$

We have already identified the coefficients $a_1, a_2, \ldots, a_r, \ldots, a_n$ as elementary symmetric functions of the latent roots which are given by the sum of the principal r-rowed minors of A. In our example of the 3^2 order matrix A we have

$$\{1^2\} = a_2 = \begin{vmatrix} a_{11} & a_{12} \\ a_{21} & a_{22} \end{vmatrix} + \begin{vmatrix} a_{11} & a_{13} \\ a_{31} & a_{33} \end{vmatrix} + \begin{vmatrix} a_{22} & a_{23} \\ a_{32} & a_{33} \end{vmatrix}$$

which is just the spur of the compound matrix $A\{1^2\}$. Thus the S-function $\{1^2\}$ of the matrix A corresponds to a simple character of $GL(n)$. In general, we have the following:

The S-functions $\{1^r\}$ of the characteristic roots of the matrices A of GL(n) are simple characters of the group.

Since an S-function in n variables cannot have more than n parts we may also conclude that the S-function $\{1^n\}$ of A is a scalar and is simply equal to $|A|$.

4.3 INDUCED MATRICES

To find further characters of $GL(n)$ we must consider the properties of induced matrices. Consider a set of n variables $x_1, x_2, \ldots, x_n$ which are transformed by a matrix $A \equiv [a_{st}]$ such that

$$X' = AX$$

From these we may form $n(n + 1)/2$ powers and products of degree 2, e.g.,

$$x_1^2, x_2^2, \ldots, x_n^2, x_1x_2, x_1x_3, \ldots, x_{n-1}x_n$$

which are transformed into powers and products of degree 2 of the transformed variables, $x_1', x_2', \ldots, x_n'$. The elements of the transforming matrix on these $n(n + 1)/2$ terms are just polynomials of degree 2 in the elements of A. This matrix is said to be the *second induced matrix* of A and is designated as $A^{\{2\}}$.

Similarly the powers and products of degree r in $x_1, x_2, \ldots, x_n$ will be transformed by a matrix having $\binom{n + r - 1}{r}$ rows and columns of which the elements are polynomials of degree r in the elements of A. This matrix $A^{\{r\}}$ is then said to be the *rth induced matrix* of A.

Proceeding in a manner similar to our discussion of compound matrices we readily conclude that:

The S-functions $\{r\}$ of the characteristic roots of the matrices A of GL(n) are simple characters of the group.

4.4 INVARIANT MATRICES

The S-functions $\{1^r\}$ and $\{r\}$ corresponding to the spurs of the rth compound and the rth induced matrices of A have been shown to be simple characters $GL(n)$. In our discussion of the symmetric group we found it desirable to introduce the concept of the immanants of matrices to fill the gap between determinants and permanents and were then led to a complete determination of the group characters of S_n. To obtain the complete set of characters for $GL(n)$ requires the introduction of the concept of *invariant matrices*, first considered by Schur.[14]

We define an invariant matrix as follows. Let A be a square matrix of order n^2 and let $T(A)$ denote a matrix whose elements are polynomials in the elements of A. If for every pair of matrices, A, B, each of order n^2 we have

$$T(A)T(B) = T(AB) \tag{57}$$

then $T(A)$ is said to be an invariant matrix of A. Each invariant matrix of A will clearly give a matrix representation of $GL(n)$. The matrices $A^{\{1^r\}}$ and $A^{\{r\}}$ are particular cases of invariant matrices.

The direct product of two invariant matrices is necessarily an invariant matrix and every possible invariant matrix of degree r in the elements of A may be obtained by reduction of the direct product of r identical matrices A.

Schur[14] has demonstrated that:

If A is a matrix of order n^2 there are as many irreducible invariant matrices of A, of degree r, as there are partitions of r into not more than n parts and the spurs of these invariant matrices are the S-functions of weight r of the characteristic roots of A.

Details of the actual construction of invariant matrices corresponding to particular partitions have been given by Littlewood.[47–49] We note that we may deduce Littlewood and Richardson's[12] method of obtaining the characters of the symmetric group, given earlier, by simply considering A to be a permutation matrix and S_n as a subgroup of the full linear group.

4.5 DECOMPOSITION OF THE DIRECT PRODUCT FOR $GL(n)$

The decomposition of direct product representations into irreducible representations arises in many problems in atomic spectroscopy. The direct product of two invariant matrices $A^{\{\mu\}}$ and $A^{\{\lambda\}}$ of degrees m and n, respectively, may be decomposed into the direct sum of invariant matrices

$A^{\{\nu\}}$, i.e.,

$$A^{\{\mu\}} \times A^{\{\lambda\}} = \dot{\sum} \Gamma_{\mu\lambda\nu} A^{\{\nu\}} \tag{58}$$

where $\times$ denotes the direct product and $\dot{\sum}$ the direct sum. It is clear that the degree of the sum $\dot{\sum} \Gamma_{\mu\lambda\nu} A^{\{\nu\}}$ will be mn.

The spurs of the matrices $A^{\{\nu\}}$, $A^{\{\lambda\}}$, and $A^{\{\nu\}}$ are simply S-functions of the characteristic roots of the matrices A and hence the decomposition of the direct product of the group characters amounts to the ordinary multiplication of S-functions, i.e.,

$$\{\mu\} \cdot \{\lambda\} = \sum \Gamma_{\mu\lambda\nu} \{\nu\} \tag{59}$$

discussed earlier in terms of the rule for evaluating outer products of irreducible representations of the symmetric group. If the partitions (μ) and (λ) are of degrees m and n, then those of (ν) will be of degree $m + n$ and hence the decomposition of Eq. 59 may be checked by using Eq. 45 and regarding the decomposition as for the symmetric group.

4.6 DEGREE OF IRREDUCIBLE REPRESENTATIONS OF *GL*(*n*)

Equation 59 may also be checked by calculating the degrees of the representations of the group $GL(n)$ using the well-known result

$$D_{\{\lambda\}} = \prod_{h>j}^{n} (h - j + \lambda_j - \lambda_h)/(h - j) \tag{60a}$$

Equation 60 is tedious in its application and Robinson[11] has made use of the method of hook graphs to yield the alternative expression

$$D_{\{\lambda\}} = G_{(n)}^{\{\lambda\}}/H_{[\lambda]} \tag{60b}$$

where $H_{[\lambda]}$ is the hook of the Young diagram as defined in Eq. 8 and

$$G_{(n)}^{\{\lambda\}} = \prod_{i,j} (n + i - j) \tag{60c}$$

where n is the dimension of the group and i and j specify the position of each cell in terms of the ith column and the jth row, respectively.

As an example, consider the partition (421) associated with the {421} representation of $GL(5)$. We have from Eq. 8

$$H_{[421]} = 144$$

Labeling each cell of the Young tableau with the numbers (i, j) we obtain the diagram

1, 1	2, 1	3, 1	4, 1
1, 2	2, 2		
1, 3			

from which we deduce

$$\begin{aligned} G_{(5)}^{\{421\}} &= (5+1-1)(5+2-1)(5+3-1)(5+4-1)(5+1-2) \\ &\quad \times (5+2-2)(5+1-3) \\ &= 5 \cdot 6 \cdot 7 \cdot 8 \cdot 4 \cdot 5 \cdot 3 \\ &= 100800 \end{aligned}$$

to give the final result

$$D_{\{421\}} = 100800/144 = 700$$

5

The Restricted Groups

5.1 RESTRICTED GROUPS

Determining the irreducible representations of the full linear group in terms of invariant matrices $T(A)$, whose elements are homogeneous polynomials in the matrix elements of A, gives the complete set of irreducible representations which upon restriction of $GL(n)$ to any of its proper subgroups are completely reducible or analyzable. This does not exhaust the representations of $GL(n)$ but does cover completely all cases arising in the treatment of compact groups.

The full linear group $GL(n)$ contains many subgroups which may be found by simply restricting the class of matrices considered, or alternatively, by considering only transformations that hold certain functional forms invariant. For example, if we restrict our attention to matrices whose determinant is unity, we obtain the so-called unimodular linear group $SL(n)$ which is obviously a subgroup of $GL(n)$.

In all the problems we shall discuss, we need only consider the subgroup $U(n)$ made up of the infinite set of unitary matrices A of orde1 n^2 and its subgroups. A unitary transformation leaves invariant the norm of every complex vector upon which a unitary matrix operates; that is, if $[a_{st}]$ is a unitary matrix then under the transformation

$$[x_s'] = [a_{st}][x_t]$$

the form

$$[\bar{x}_t][x_s] = \sum x_i\bar{x}_i$$

is left invariant. We use a bar to denote the complex conjugate. A necessary and sufficient condition that the matrix $A \equiv [a_{st}]$ is unitary, is that

$$[\bar{a}_{ts}][a_{st}] = I = [\delta_{st}]$$

i.e., $\tilde{\bar{A}}A = I$, where the tilde ($\sim$) indicates the transposition of the rows and columns of the matrix. This condition implies that the matrix elements of a unitary matrix satisfy the relationship

$$\sum_r a_{rp}\bar{a}_{rq} = \delta_{pq}$$

and alternatively, since $[a_{st}][\bar{a}_{ts}] = I$,

$$\sum_r a_{pr}\bar{a}_{qr} = \delta_{pq}$$

These two conditions restrict the elements a_{st} of the matrix so that $|a_{st}^2| \leq 1$ and as a result the group is said to be compact.

The irreducible representations of the full linear group $GL(n)$ remain irreducible upon restriction to the group $U(n)$ and we may readily show the following:[13]

The complete set of irreducible representations of the group $U(n)$ is given by the set of independent irreducible invariant matrices of A and the simple characters of $U(n)$ are the S-functions of the characteristic roots of the unitary matrices A.

As a result we shall use the partitions $\{\lambda\}$ that are associated with the S-functions to label the irreducible representations of $U(n)$. It follows that the direct products and the degrees of the representations of $U(n)$ are evaluated identically with those of $GL(n)$.

5.2 THE ORTHOGONAL GROUP

The group of transformations that leaves the arbitrary nonsingular quadratic form

$$\sum g_{ij}x_ix_j \tag{61}$$

invariant is known as the *general orthogonal group* in n dimensions and is designated as $O(n)$. This group has as its elements the infinite set of orthogonal matrices A and is clearly a subgroup of $U(n)$.

Since, by the definition of an orthogonal matrix, $\tilde{A}A = 1$ we have

$$|A| = \pm 1 \tag{62}$$

Thus the set of orthogonal matrices that characterize $O(n)$ may be divided into two sets, those with determinant $+1$ and those with determinant -1.

The set of orthogonal matrices of determinant +1 clearly form a subgroup of $O(n)$ which is known as the *special orthogonal group* and designated as $SO(n)$ or $O^+(n)$. The special orthogonal group corresponds to the group of transformations on n variables which leaves the quadratic form

$$x_1^2 + x_2^2 + \cdots + x_n^2 \tag{63}$$

invariant and is not significantly different from the *rotation group* $R(n)$.

For the orthogonal group the invariant form given by Eq. 61 is clearly of type $\{2\}$ whereas for the special orthogonal group the determinant of the variables is also held invariant corresponding to a form of type $\{1^n\}$.

5.3 CHARACTERS OF $O(n)$

Consider the ternary quadratic form $\sum_{i,j}^3 g_{ij}x_ix_j$ where g_{ij} is a symmetric tensor of the type $\{2\}$. This form will be irreducible under the full linear group. Under the special orthogonal group, which holds the quadratic form $\sum_i^3 x_i^2$ invariant, the quadratic $\sum g_{ij}x_ix_j$ has the linear invariant

$$\tfrac{1}{3}(g_{11} + g_{22} + g_{33})(x_1^2 + x_2^2 + x_3^2)$$

The quadratic will be simple under the orthogonal group only if this invariant is removed; hence the form $\sum g_{ij}x_ix_j$ is reducible under the orthogonal group into two simple forms [2] and [0], [0] being the invariant. Thus we may relate the characters of the two groups by the equation

$$\{2\} = [2] + [0]$$

This means that the invariant matrix $A^{\{2\}}$ can be expressed as the direct sum of two simple matrices, i.e.,

$$A^{\{2\}} = A^{[2]} + A^{[0]}$$

Every representation of the full linear group is necessarily a representation of the orthogonal group being either simple or compound. Thus each S-function arising in $GL(n)$ will be a simple or compound character of $O(n)$. As a result it is possible to express every S-function arising in $GL(n)$ as a linear combination of characters of $O(n)$ and conversely every character of $O(n)$ can be expressed in terms of S-functions.

Littlewood[13,50] has shown that *if the number of variables on which the group is defined is $n = 2v$ or $n = 2v + 1$ then corresponding to every partition (λ) into not more than v parts there is a representation of the orthogonal group $O(n)$.*

The character of the group $O(n)$ corresponding to the partition (λ) will be designated by enclosing the partition in square brackets, e.g., $[\lambda]$. Littlewood has expressed these characters in terms of S-functions by developing generating functions for the concomitants of a quadratic. His results make use of Frobenius's notation for partitions[30] which we first sketch.

5.4 FROBENIUS'S NOTATION FOR PARTITIONS

We have seen that every partition may be represented by a Young tableau. To develop Frobenius's notation we represent each cell of the Young tableau by a dot. Then the diagonal of the dots in a graph beginning at the top left-hand corner is called the *leading diagonal*. The number of dots on the leading diagonal is known as the *rank* of the partition. If r is the rank of the partition and there are a_i dots to the *right* of the leading diagonal in the ith row and b_i dots below the leading diagonal in the ith column then the partition, in Frobenius notation is designated as

$$\begin{pmatrix} a_1, a_2, \ldots, a_r \\ b_1, b_2, \ldots, b_r \end{pmatrix}$$

For example, the partition (7 5 3 1^2) has the Frobenius graph

$$\begin{matrix} \circ & \cdot & \cdot & \cdot & \cdot & \cdot & \cdot \\ \cdot & \circ & \cdot & \cdot & \cdot & & \\ \cdot & \cdot & \circ & & & & \\ \cdot & & & & & & \\ \cdot & & & & & & \end{matrix}$$

and the partition is of rank 3 and in Frobenius notation is designated as

$$\begin{pmatrix} 6 & 3 & 0 \\ 4 & 1 & 0 \end{pmatrix}$$

5.5 CHARACTERS OF $O(n)$ (Continued)

We may now state Littlewood's result for the characters of $O(n)$ in terms of S-functions:

$$[\lambda] = \{\lambda\} + \sum (-1)^{p/2}\Gamma_{\gamma\eta\lambda}\{\eta\} \tag{64}$$

where the sum is over all S-functions of the set $\{\gamma\}$ *such that* $\{\lambda\}$ *appears in a product* $\{\gamma\}\{\eta\}$ *with coefficient* $\Gamma_{\gamma\eta\lambda}$, *and* p *is the weight of the partition* (γ). *The set of S-functions* $\{\gamma\}$ *comprises all S-functions that arise for the partitions which in Frobenius notation are of the form*

$$\begin{pmatrix} a+1 \\ a \end{pmatrix}, \quad \begin{pmatrix} a+1 & b+1 \\ a & b \end{pmatrix}, \quad \begin{pmatrix} a+1 & b+1 & c+1 \\ a & b & c \end{pmatrix}, \quad \ldots \tag{65}$$

i.e., the partitions

$$\{2\}, \{31\}, \{41^2\}, \{3^2\}, \{51^3\}, \{431\}, \ldots. \tag{66}$$

For example, consider the character [321] of $O(7)$. Corresponding to each Frobenius partition in Eq. 65 we have

$$\binom{1}{0}, \quad \{\gamma\} = \{2\} \quad \text{and} \quad \sum \Gamma_{\gamma\eta\lambda}\{\eta\} = \{31\} + \{2^2\} + \{21^2\}$$

$$\binom{2}{1}, \quad \{\gamma\} = \{31\} \quad \text{and} \quad \sum \Gamma_{\gamma\eta\lambda}\{\eta\} = \{2\} + \{1^2\}$$

leading to the result

$$[321] = \{321\} - \{2^2\} - \{31\} - \{21^2\} + \{2\} + \{1^2\}$$

This result may be checked by noting that the degree $D_{[\lambda]}$ of a representation of $O(n)$ for $n = 2\nu$ is given by[13]

$$D_{[\lambda]} = 2^{\nu} \prod_{r<s}^{\nu} \frac{[(\lambda_r - \lambda_s - r + s)(\lambda_r + \lambda_s + n - r - s)]}{(n-2)!\,(n-4)!\cdots(2)!} \tag{67}$$

if $\lambda_\nu \neq 0$ and one-half of this if $\lambda_\nu = 0$, while that for $n = 2\nu + 1$ is given by

$$D_{[\lambda]} = \prod_{r<s}^{\nu} (\lambda_r - \lambda_s - r + s)(\lambda_r + \lambda_s + n - r - s) \times \frac{\prod^{\nu} (2\lambda_r + n - 2r)}{(n-2)!\,(n-4)!\cdots(1)!} \tag{68}$$

The corresponding degrees $D_{\{\lambda\}}$ for $GL(n)$ may be found using Eq. 60.

5.6 BRANCHING RULE FOR $U(n) \to O(n)$

The branching rule for the decomposition of the irreducible representations of $U(n)$ into those of its subgroup $O(n)$ are given by the rule:[13,50]

$$\{\lambda\} = [\lambda] + \sum \Gamma_{\delta\eta\lambda}[\eta] \tag{69}$$

summed for S-functions of the set $\{\delta\}$ which involves all partitions into even parts only.
e.g.,

$$\{2\}, \{4\}, \{2^2\}, \{6\}, \{42\}, \{2^3\}, \{8\}, \ldots \tag{70}$$

For example, consider the S-function {321}; we have for each partition occurring in Eq. 70 the results

$$\{\delta\} = \{2\} \qquad \sum \Gamma_{\delta\eta\lambda}[\eta] = [31] + [21^2] + [2^2]$$

$$\{\delta\} = \{2^2\} \qquad \sum \Gamma_{\delta\eta\lambda}[\eta] = [2] + [1^2]$$

and thus

$$\{321\} = [321] + [31] + [21^2] + [2^2] + [2] + [1^2]$$

5.7 ASSOCIATED CHARACTERS OF $O(n)$

The character [0] of $O(n)$ has unity for every element of the group. The alternating tensor is not an absolute invariant but rather a relative invariant under the general orthogonal group and thus there is another character, designated as $[0]^*$, which has $+1$ for transformations of positive determinant and -1 for transformations of negative determinant. If $[\lambda]$ denotes any character of $O(n)$ then

$$[\lambda]^* = [\lambda][0]^* \tag{71}$$

is also a character. The *associated character*, $[\lambda]^*$, will differ from $[\lambda]$ unless the characters of all transformations of negative determinant are zero.

5.8 CHARACTERS OF THE ROTATION GROUP $R(n)$

If $n = 2\nu + 1$ the characters of the orthogonal group $O(n)$ comprise the expressions $[\lambda]$ and $[\lambda]^*$. In this case the simple characters of $O(n)$ are also simple characters of $R(n)$ (or $SO(n)$).

If $n = 2\nu$ and $\lambda_\nu = 0$ the simple characters of $O(n)$ are also simple characters of $R(n)$ but if $\lambda_\nu \neq 0$ the representation of $O(n)$ is self-associated and separates into two irreducible representations[50–52] of $R(n)$. Thus the corresponding character $[\lambda]$ may be expressed as the sum of two simple conjugate characters $[\lambda_1] + [\lambda_2]$. The two representations are interchanged by a transformation of negative determinant. The distinction between these two representations is indicated by making the separation of the character $[\lambda_1, \lambda_2, \ldots, \lambda_\nu]'$ of $O(n)$ under reduction to $R(n)$ according to the equation[52]

$$[\lambda_1, \lambda_2, \ldots, \lambda_\nu]' = [\lambda_1, \lambda_2, \ldots, \lambda_\nu] + [\lambda_1, \lambda_2, \ldots, -\lambda_\nu] \tag{72}$$

For example under the reduction $U(6) \to R(6)$ we have typically

$$\begin{aligned}\{32\} &\to [321]' + [31]' + [21^2]' + [2^2]' + [2]' + [1^2]' \\ &\to [321] + [32-1] + [31] + [211] + [21-1] + [2^2] \\ &\quad + [2] + [1^2]\end{aligned}$$

where we have used a dash to indicate the characters of $O(6)$.

A more extensive account of the properties of $R(n)$ is contained in Appendix I.

5.9 THE SYMPLECTIC GROUP

The group that leaves invariant the nonsingular linear complex

$$h_{ij}x^{ij} = (x_1y_2 - x_2y_1) + (x_3y_4 - x_4y_3) + \cdots + (x_{n-1}y_n - x_ny_{n-1}) \tag{73}$$

is known as the *symplectic group* in n variables and is designated as $Sp(n)$. The nonsingular linear complex is of type $\{1^2\}$ and the requirement of nonsingularity restricts the number of variables to being an even number, say $n = 2\nu$. The alternating tensor is a concomitant of h_{ij} and hence every transformation has determinant $+1$ and the difficulties which arise in distinguishing between orthogonal group and rotation group characters is absent in the symplectic group.

As with the orthogonal group, there is a character $\langle\lambda\rangle$ of the symplectic group $Sp(n)$ corresponding to every partition (λ) into not more than ν parts. The characters of the symplectic group are indicated by enclosing the partition (λ) in angular brackets, e.g., $\langle\lambda\rangle$.

Littlewood[13,49] has obtained the characters of the symplectic group in terms of S-functions using the formula

$$\langle\lambda\rangle = \{\lambda\} + \sum (-1)^{p/2}\, \Gamma_{\alpha\mu\lambda}\{\mu\} \tag{74}$$

where the sum is over the set of S-functions $\{\alpha\}$ corresponding all partitions which in the Frobenius notation are of the form

$$\begin{pmatrix} a \\ a+1 \end{pmatrix}, \quad \begin{pmatrix} a & b \\ a+1 & b+1 \end{pmatrix}, \quad \begin{pmatrix} a & b & c \\ a+1 & b+1 & c+1 \end{pmatrix}, \quad \ldots \tag{75}$$

e.g.,

$$\{1^2\}, \{21^2\}, \{31^3\}, \{2^3\}, \{41^4\}, \{32^21\}, \ldots \tag{76}$$

and ρ is the weight of the S-function. For example, consider the character $\langle 321\rangle$, corresponding to each partition in Eq. 76 we have

$$\{\alpha\} = \{1^2\} \qquad \sum \Gamma_{\alpha\mu\lambda}\{\mu\} = \{2^2\} + \{21^2\} + \{31\}$$

$$\{\alpha\} = \{21^2\} \qquad \sum \Gamma_{\alpha\mu\lambda}\{\mu\} = \{1\}$$

and hence

$$\langle 321\rangle = \{321\} - \{31\} - \{2^2\} - \{21^2\} + \{1\}$$

We may check the above result by calculating the degree $D_{\langle\lambda\rangle}$ of $\langle 321\rangle$ by the formula,[6]

$$D_{\langle\lambda\rangle} = \prod_{i=1}^{\nu} \frac{(\lambda_i + \nu - i + 1)}{\nu - i + 1} \times \prod_{k>i}^{\nu} \frac{(\lambda_i - \lambda_k + k - i)(\lambda_i + \lambda_k + 2\nu - i - k + 2)}{(k - i)(2\nu + 2 - i - k)} \tag{77}$$

and comparing the degrees of the characters $D_{\{\lambda\}}$ of $GL(2\nu)$ computed by Eq. 60.

We note that the characters of $Sp(2\nu)$ may be expressed in terms of the characters of $O(2\nu)$ and vice versa as has been shown by Littlewood[53] following a preliminary observation of Ibrahim.[54] In particular

$$\langle\lambda\rangle = [\lambda] + \sum \Gamma[\mu_{\xi\mu\lambda}] - \sum \Gamma_{\eta\mu\lambda}[\mu] \tag{78}$$

where $\{\xi\}$ involves all partitions into not more than two even parts and $\{\eta\}$ all partitions into exactly two odd parts.

Conversely we have

$$[\lambda] = \langle\lambda\rangle + \sum \Gamma_{\rho\mu\lambda}\langle\mu\rangle - \sum \Gamma_{\sigma\mu\lambda}\langle\mu\rangle \tag{79}$$

where $\{\rho\}$ involves all partitions of the form $(2^{2r}1^{2s})$ excluding the case $r = s = 0$ and $\{\sigma\}$ involves all partitions of the form $(2^{2r+1}1^{2s})$.

As examples of Eqs. 78 and 79 we have the results

$$\langle 21^2\rangle = [21^2] - [2] \quad \text{and} \quad [21^2] = \langle 21^2\rangle + \langle 2\rangle - \langle 0\rangle$$

5.10 BRANCHING RULE FOR $U(n) \to Sp(n)$

Following Littlewood[50] the S-functions that form the simple characters of $U(n)$ may be decomposed into characters of the $Sp(n)$ using the result

$$\{\lambda\} = \langle\lambda\rangle + \sum \Gamma_{\beta\mu\lambda}\langle\mu\rangle \tag{80}$$

where $\{\beta\}$ is summed over all partitions in which each part is repeated an even number of times.

$$\{1^2\},\ \{2^2\},\ \{1^4\},\ \{2^21^2\},\ \{3^2\}\ \cdots \tag{81}$$

For example, consider the character $\{321\}$ of $U(8)$, we have

$$\{\beta\} = \{1^2\} \qquad \sum \Gamma_{\beta\mu\lambda}\langle\mu\rangle = \langle 21^2\rangle + \langle 31\rangle + \langle 2^2\rangle$$

$$\{\beta\} = \{2^2\} \qquad \sum \Gamma_{\beta\mu\lambda}\langle\mu\rangle = \langle 2\rangle + \langle 1^2\rangle$$

and thus

$$\{321\} = \langle 321\rangle + \langle 31\rangle + \langle 2^2\rangle + \langle 21^2\rangle + \langle 2\rangle + \langle 1^2\rangle$$

5.11 MODIFICATION RULES AND NONSTANDARD SYMBOLS

Equations 64 and 74 allow the characters of the orthogonal and symplectic groups to be expressed in terms of S-functions while conversely, Eqs. 69 and 79 allow S-functions to be expressed in terms of characters of the orthogonal and symplectic groups. The character symbols that arise in the decompositions $U(n) \to O(n)$ and $U(n) \to Sp(n)$, where $n = 2\nu$ or $n = 2\nu + 1$, will frequently contain more than ν parts and as such do not correspond to the standard symbols of $O(n)$ or $Sp(n)$. In these cases the use of Eqs. 69 and 80 requires that either the S-functions having $>\nu$ parts be converted into S-functions having $\leq\nu$ parts or the S-functions be decomposed into characters of $O(n)$ or $Sp(n)$, as appropriate, and then modification rules used to convert the nonstandard symbols having $>\nu$ parts into standard symbols having $\leq\nu$ parts.

Murnaghan[55] has given a limited set of modification rules for the orthogonal groups without giving the general rule. These rules do not suffice for all practical cases and in a later publication Murnaghan[55a] has noted that Newell[55b,c] has found a complete rule for both the orthogonal and symplectic groups. Newell's results for the orthogonal group may be expressed in terms of two theorems, one for the case $n = 2\nu$ and the other for $n = 2\nu + 1$. We now state these theorems:

Theorem IA

For the orthogonal group $O_{2\nu}$ all $[\lambda]$ having more than ν parts vanish, except those of the form $[\lambda_1, \ldots, \lambda_\nu, (\rho)]$ where (ρ) is any partition which appears as

$$\begin{pmatrix} a+1 \\ a \end{pmatrix}, \quad \begin{pmatrix} a+1 & b+1 \\ a & b \end{pmatrix}, \quad \begin{pmatrix} a+1 & b+1 & c+1 \\ a & b & c \end{pmatrix}, \quad \ldots$$

in Frobenius' notation, i.e., the partitions:

$$(2), (31), (41^2), (3^2), (51^3), (431), \ldots$$

If r is the rank of (ρ) and ω its weight, then

$$[\lambda_1, \ldots, \lambda_\nu, (\rho)] = (-1)^{\omega/2}\varepsilon^r[\lambda_1, \ldots, \lambda_\nu]$$

where $\varepsilon = \mp 1$ is the determinant of the group element. Alternatively, we may say that the character is an associated character of $O_{2\nu}$ if r is odd.

Theorem IB

If $n = 2\nu + 1$, then all $[\lambda]$ having more than ν parts vanish except those of the form $[\lambda_1, \ldots, \lambda_\nu, (\eta)]$, where (η) is any self-conjugate partition, i.e., the partitions

$$(1), (21), (2^2), (31^2), (321), (41^3), \ldots$$

If r is the rank of (η) and ω the weight, then

$$[\lambda_1, \ldots, \lambda_\nu, (\eta)] = (-1)^{\omega/2-r}\varepsilon^\omega[\lambda_1, \ldots, \lambda_\nu]$$

One slight difficulty arises in the application of the above theorems, the partitions $[\lambda_1, \ldots, \lambda_\nu, (\rho)]$ and $[\lambda_1, \ldots, \lambda_\nu, (\eta)]$ need not be ordered in the standard descending order, i.e., they may be *disordered partitions*. We are at liberty to rearrange the partitions by one or more applications of the type

$$[\lambda_1, \ldots, \lambda_{j+1} - 1, \lambda_j + 1, \lambda_{j+2}, \ldots, \lambda_q] = -[\lambda_1, \ldots, \lambda_j, \lambda_{j+1}, \lambda_{j+2}, \ldots, \lambda_q]$$

For example, in applying Theorem IA to the partition $[\lambda_1, \ldots, \lambda_{\nu-1}, 1^2]$ we may write

$$[\lambda_1, \ldots, \lambda_{\nu-1}, 1^2] = -[\lambda_1, \ldots, \lambda_{\nu-1}, 02] = [\lambda_1, \ldots, \lambda_{\nu-1}]$$

or as a further example

$$\begin{aligned}[\lambda_1, \ldots, \lambda_{\nu-1}, 3^4] &= -[\lambda_1, \ldots, \lambda_{\nu-1}, 3324] \\ &= [\lambda_1, \ldots, \lambda_{\nu-1}, 3144] \\ &= -[\lambda_1, \ldots, \lambda_{\nu-1}, 0444] \\ &= -[\lambda_1, \ldots, \lambda_{\nu-1}]^*\end{aligned}$$

where we use an asterisk to indicate that the character is an associated character. Proceeding in this manner we may express any nonstandard symbol having $>\nu$ parts in terms of a standard symbol of $O(n)$ having $\leq\nu$ parts.

The corresponding modification rules for the symplectic group $Sp(2\nu)$ have been summarized by Newell[55c] in terms of the following theorem:

Theorem IC

All $\langle\lambda\rangle$ having more than ν parts vanish, except those which can be written in the form $\langle\lambda_1, \ldots, \lambda_\nu, (\alpha)\rangle$, which is equal to $(-1)^{\omega/2}\langle\lambda_1, \ldots, \lambda_\nu\rangle$. The partitions (α) are given in the Frobenius notation as

$$\begin{pmatrix} a \\ a+1 \end{pmatrix}, \quad \begin{pmatrix} a & b \\ a+1 & b+1 \end{pmatrix}, \quad \begin{pmatrix} a & b & c \\ a+1 & b+1 & c+1 \end{pmatrix}, \quad \ldots$$

e.g.,

$$(1^2), (21^2), (31^3), (2^3), (41^4), (32^21), \ldots$$

Rather than use the modification rules we may, if we wish, convert the S-functions having $>\nu$ parts into S-functions having $\leq\nu$ parts and then apply the branching rules for $U(n) \to O(n)$ or $Sp(n)$. We first note[7] that for S-functions defined on n variables we may make use of the equivalences

$$\{\lambda_1, \lambda_2, \ldots, \lambda_n\} \equiv \{\lambda_1 - \lambda_n, \lambda_2 - \lambda_n, \ldots, 0\}$$

and

$$\{\lambda_1, \lambda_2, \ldots, \lambda_n\} \equiv \{\lambda_1 - \lambda_n, \lambda_1 - \lambda_{n-1}, \ldots, \lambda_1 - \lambda_2, 0\}$$

While these equivalences simplify many practical cases they do not cover all cases. Littlewood[50] has developed a general method which we now give.

If $n = 2\nu$ express the S-function in the form:

$$\{r + \lambda_1, r + \lambda_2, \ldots, r + \lambda_\nu, r - \mu_\nu, r - \mu_{\nu-1}, \ldots, r - \mu_1\} \quad (82a)$$

or if $n = 2\nu + 1$ in the form:

$$\{r + \lambda_1, r + \lambda_2, \ldots, r + \lambda_\nu, r, r - \mu_\nu, r - \mu_{\nu-1}, \ldots, r - \mu_1\} \quad (82b)$$

If the change in sign for a transformation of negative determinant is ignored the values of these S-functions are independent of r and will be denoted by $\{\lambda; \mu\}$. Then

$$\{\lambda; \mu\} = \{\lambda\}\{\mu\} + \sum (-1)^p \, \Gamma_{\alpha\delta\lambda}\Gamma_{\beta\varepsilon\mu}\{\alpha\}\{\beta\} \quad (83)$$

where the sum is over all partitions (δ) of p where $(\varepsilon) = (\tilde{\delta})$.

As an example, consider the decomposition of $\{321^2\}$ under the reduction $U(5) \to R(5)$. First, following Murnaghan's method and using Eq. 69 we have

$$\{321^2\} = [321^2] + [31^2] + [2^21] + [21^3] + [21] + [1^3]$$

The characters of $R(5)$ are labeled by partitions into not more than two parts. Using Newell's rules we have the modifications

$$[321^2] \to 0; \; [31^2] \to [31]; \; [2^21] \to [2^2]; \; [21^2] \to [2]; \; [1^3] \to [1^2]$$

and thus under $U(5) \to R(5)$

$$\{321^2\} \to [31] + [2^2] + [21] + [2] + [1^2]$$

Alternatively, using Littlewood's method we have from Eq. 82b

$$\{321^2\} = \{1 + 2, 1 + 1, 1, 1 - 0, 1 - 1\} = \{21; 1\}$$

Using Eq. 83 gives

$$\begin{aligned}\{21; 1\} &= \{21\}\{1\} - \{2\}\{0\} - \{1^2\}\{0\} \\ &= \{31\} + \{2^2\} + \{21^2\} - \{2\} - \{1^2\}\end{aligned}$$

Furthermore,

$$\{21^2\} = \{1 + 1, 1 + 0, 1, 1 - 1, 1 - 1\} = \{1; 1^2\}$$

and

$$\{1; 1^2\} = \{1\}\{1^2\} - \{1\}\{0\} = \{21\} + \{1^3\} - \{1\} = \{21\} + \{1^2\} - \{1\}$$

Thus

$$\begin{aligned}\{321^2\} &= \{31\} + \{2^2\} + \{21\} - \{2\} - \{1\} \\ &= [31] + [2^2] + [21] + [2] + [1^2]\end{aligned}$$

in agreement with the result obtained from Newell's method.

5.12 BRANCHING RULES FOR THE GROUP G_2

The orthogonal group in seven dimension $O(7)$ contains the exceptional group G_2 as a subgroup.[1,15,56–59] We shall not dwell upon the properties of this group but content ourselves with stating the branching rules for $O(7) \to G_2 \to R_3$.

Judd[15] has shown that the irreducible representations of G_2 may be labeled by their highest weights (m_1, m_2). Weights for which m_2 is negative are avoided by regarding (u_1, u_2) as the highest weight where

$$(u_1, u_2) \equiv (m_1 + m_2, -m_2). \tag{84}$$

With the above notation established the simple characters $[\lambda_1\lambda_2\lambda_3]$ of $O(7)$ decompose into the simple characters (u_1u_2) of G_2 according to the rule

$$[\lambda_1\lambda_2\lambda_3] = \sum \{(i - k, j + k) + (j - k - 1, i - j)\} \tag{85}$$

where the sum runs over all integral values of i, j, and k satisfying

$$\lambda_1 \geq i \geq \lambda_2 \qquad \lambda_2 \geq j \geq \lambda_3 \qquad \lambda_3 \geq k \geq -\lambda_3$$

The equations

$$(u_1u_2) = -(u_2 - 1, u_1 + 1) \qquad (u_1, -1) = 0$$

and

$$(u_1, -2) = -(u_1 - 1, 0)$$

may be used to remove the characters whose arguments do not give admissible representations of G_2.

It would be interesting to consider the problem of expressing the characters of G_2 in terms of S-functions and to then obtain the branching rules and Kronecker product relations in terms of the theory of S-functions.

Shi Sheng-Ming[59] has given a detailed discussion of the decomposition of the characters of G_2 into those of R_3 and has given tables covering all weights up to (80). These decompositions cover most of those we shall require. Shi Sheng-Ming uses the weights (m_1, m_2) to label the representations of G_2 and these must be converted using Eq. 84 if the standard form is required.

5.13 PRODUCTS OF CHARACTERS FOR $O(n)$ AND $Sp(n)$

The decomposition of products of the simple characters of $O(n)$ and $Sp(n)$ plays a very important part in the application of group theory to atomic spectroscopy. Two methods are available.

In the first method Eqs. 64 and 74 are used to express the characters of $O(n)$ and $Sp(n)$, respectively, in terms of S-functions. The products of S-functions appearing in the product of the characters are then multiplied out and decomposed into the characters of the appropriate orthogonal or symplectic group using Eqs. 69 and 80.

For example, consider the product $[3]\cdot[1^2]$ for $O(7)$. Using Eqs. 64 and 69 we have

$$\begin{aligned}[3]\cdot[1^2] &= (\{3\}-\{1\})\{1^2\}\\ &= \{41\}+\{31^2\}-\{21\}-\{1^3\}\\ &= [41]+[31^2]+[3]+[21]\end{aligned}$$

or in the case of the product $\langle 3\rangle\langle 1^2\rangle$ for $Sp(6)$ we have from Eqs. 74 and 80

$$\begin{aligned}\langle 3\rangle\langle 1^2\rangle &= \{3\}(\{1^2\}-\{0\})\\ &= \{41\}+\{31^2\}-\{3\}\\ &= \langle 41\rangle+\langle 31^2\rangle+\langle 3\rangle+\langle 21\rangle\end{aligned}$$

The above method becomes very tedious when large partitions into several parts are being considered and does not indicate the remarkable similarity displayed by the two products $[3][1^2]$ and $\langle 3\rangle\langle 1^2\rangle$. Littlewood[45] has demonstrated a theorem that allows for the rapid evaluation of products of characters for both the orthogonal and the symplectic groups. We now state the theorem.

If

$$\sum \Gamma_{\xi\zeta\lambda}\,\Gamma_{\xi\nu\mu}\{\zeta\}\{\nu\} = \sum K_{\lambda\mu\rho}\{\rho\} \tag{86a}$$

the summation on the left being with respect to all possible S-functions including $\{\xi\}=\{0\}$, then for products of orthogonal group characters,

$$[\lambda][\mu] = \sum K_{\lambda\mu\rho}[\rho] \tag{86b}$$

and for products of symplectic group characters,

$$\langle\lambda\rangle\langle\mu\rangle = \sum K_{\lambda\mu\rho}\langle\rho\rangle \tag{86c}$$

Equations 86*b* and 86*c* demonstrate the equivalences of products of orthogonal and symplectic characters just noted. As an example of the application of Littlewood's theorem, consider again the product $[3]\cdot[1]^2$ of $O(7)$. Corresponding to $\{\xi\}=\{0\}$ we have $\{\zeta\}=\{3\}$ and $\{\nu\}=\{1^2\}$ and the product

$$\{3\}\{1^2\} = \{41\}+\{31^2\}$$

For $\{\xi\}=\{1\}$ we have $\{\zeta\}=\{2\}$ and $\{\nu\}=\{1\}$ and the product

$$\{2\}\{1\} = \{21\}+\{3\}$$

Using Eq. 86*b* we have the desired result

$$[3][1^2] = [41] + [31^2] + [3] + [21]$$

Products of the characters of the special group G_2 may be found by first expressing the characters (u_1u_2) in terms of the characters of $O(7)$ and then using Eqs. 86 to multiply the characters of $O(7)$ followed by use of Eq. 85 to obtain the result in terms of the characters of G_2. This somewhat round-about method suggests that the analysis of the characters of G_2 in terms of S-functions may be a profitable subject to pursue.

6

The Plethysm of S-Functions

We have so far discusssed two types of multiplication of S-functions corresponding to the formation of the inner and outer products of the characters of the symmetric group. A third type of multiplication of S-functions, known as the *plethysm of S-functions*, was discovered by Littlewood.[47] The mathematical operation of plethysm, which we shall shortly define, plays a fundamental part in physics whenever the theory of groups enters and can lead to remarkable simplification in the theory of complex spectra. Somewhat surprisingly, with evidently only two exceptions,[9,60] the powerful technique of plethysm has been unrecognized by physicists. We shall first outline some of the relevant theory of plethysm and then consider applications to specific problems in atomic spectroscopy.

6.1 BASIS FUNCTIONS AND THE FULL LINEAR GROUP $GL(n)$

The construction of basis functions $\varphi_\alpha^{(\Gamma)}$ that transform according to the αth row of an irreducible representation Γ of a group G is common in the application of group theory to problems in physics. If the irreducible representation Γ is of degree n the basis function $\varphi_\alpha^{(\Gamma)}$ may be used to construct a set of basis functions $\varphi_\rho^{\{1\}}$ that span the $\{1\}$ representation of $GL(n)$. For example, the basis functions $\varphi_\rho^{\{1\}}$ might have been constructed from the $(4l + 2)$ eignenfunctions of a single electron and span the $\{1\}$ representation of $GL(4l + 2)$.

The powers and products of degree r in the basis functions $\varphi_\rho^{\{1\}}$ will form the bases for the different irreducible representations $\{\lambda\}$ of $GL(n)$ contained in the decomposition of the rth Kronecker product

$$\{1\}^r = \sum f^\lambda\{\lambda\} \tag{87}$$

where the λ's are partitions of r and f^λ is the number of times a given irreducible representation $\{\lambda\}$ occurs in the decomposition. It follows from Eqs. 36 and 40 that f^λ may be identified with the degree of the irreducible representation $\{\lambda\}$ of the symmetric group of order r!

The irreducible representations $\{\lambda\}$ of $GL(n)$ will generally be reducible under restriction to the group G. In practice we are interested in constructing product functions of a particular symmetry type hence in picking out those product functions forming a basis for the representation of $GL(n)$ which turns up in the sum $\sum f^\lambda\{\lambda\}$ of Eq. 87. The decomposition of the representation $\{\nu\}$ of $GL(n)$, on restriction to G, can then be studied.

To be more specific, let us suppose that Γ corresponds to an irreducible representation $\{\mu\}$ of $GL(m)$ and that the functions $\varphi_\alpha^{\{\mu\}}$ form a basis for this representation. If the dimension of $\{\mu\}$ is $n(m \leq n$ for $\{\mu\} \neq \{0\})$, then this set of functions may be used to form a basis for the representation $\{1\}$ of $GL(n)$. We now construct powers and products of the functions $\varphi_\alpha^{\{\mu\}}$ of degree r, and from these product functions we choose a basis for the representation $\{\nu\}$ of $GL(n)$, where (ν) is a partition of r. In general, these product functions form a basis for a reducible representation of $GL(n)$. If we denote the set of functions forming a basis for the $\{\lambda\}$ representation of $GL(m)$ by $\varphi^{\{\lambda\}}$ and the set of powers and products of the functions $\varphi^{\{\mu\}}$ forming a basis for the $\{\nu\}$ representation of $GL(n)$ by $|\varphi^{\{\mu\}}|^{\{\nu\}}$, then clearly

$$|\varphi^{\{\mu\}}|^{\{\nu\}} = \sum \varphi^{\{\lambda\}} \tag{88}$$

The above result can be expressed equivalently in terms of invariant matrices. The functions $\varphi^{\{\mu\}}$ may be expressed as linear combinations of powers and products of functions $\varphi^{\{1\}}$ that form a basis for the $\{1\}$ representation of $GL(m)$. The matrix $A^{\{\mu\}}$ which transforms the $\varphi^{\{\mu\}}$'s will be an invariant matrix of the matrix A which transforms the functions $\varphi^{\{1\}}$. The operation which we perform on the functions $\varphi^{\{\lambda\}}$, to form the functions $|\varphi^{\{\mu\}}|^{\{\nu\}}$, is equivalent to forming the invariant matrix $|A^{\{\mu\}}|^{\{\nu\}}$ of the invariant matrix $A^{\{\lambda\}}$. An invariant matrix of an invariant matrix is, in general, reducible to the direct sum of other invariant matrices $A^{\{\lambda\}}$, each of which is the transformation matrix appropriate to the irreducible representation $\{\lambda\}$ with basis functions $\varphi^{\{\lambda\}}$. Thus Eq. 88 may be equivalently expressed in terms of invariant matrices by writing

$$|A^{\{\mu\}}|^{\{\nu\}} = \dot{\sum} A^{\{\lambda\}} \tag{89}$$

6.2 PLETHYSM OF S-FUNCTIONS

Equation 89 may be taken as defining a particular type of combination of S-functions. Taking the spurs of the invariant matrices, we have the definitive equation for the *plethysm of S-functions*:

$$\{\mu\} \otimes \{\nu\} = \sum \{\lambda\} \tag{90}$$

where the symbol ($\otimes$) is used to indicate the operation of plethysm and $\{\mu\} \otimes \{\nu\}$ is read as "$\{\mu\}$ plethys $\{\nu\}$."

Equation 90 can be interpreted as meaning that we take a set of functions of symmetry type $\{\mu\}$ and from these form symmetrized products of degree r and symmetry type $\{\nu\}$. From these we may form concomitants of type $\{\lambda\}$ which correspond to a particular selection of the tableaux occurring in the product $\{\mu\}^r$.

As an illustrative example, consider the case of two quadratics $a_{ij}x^ix^j$ and $b_{ij}x^ix^j$ which are both of symmetry type $\{2\}$. If we form linear products of these two forms we would expect to obtain new forms having the symmetries contained in the product

$$\{2\}\{2\} = \{4\} + \{31\} + \{2^2\}$$

Of the three forms linear in the two quadratics two will be symmetric,

$$a_{ij}b_{km}x^ix^jx^kx^m \qquad \text{and} \qquad a_{ij}b_{km}x^{ik}x^{jm}$$

in the ground forms and correspond to the symmetries $\{4\}$ and $\{2^2\}$, respectively, while one

$$a_{ij}b_{km}x^{ik}x^jx^m$$

will be antisymmetric and of type $\{31\}$. The two symmetric forms will arise in the plethysm

$$\{2\} \otimes \{2\} = \{4\} + \{2^2\}$$

and the antisymmetric form in the plethysm

$$\{2\} \otimes \{1^2\} = \{31\}$$

While we shall largely confine our attention to plethysm as applied to the full linear group and its subgroups it is worth noting that the operation of plethysm has a natural interpretation in terms of the symmetrized outer products of the irreducible representations of the symmetric group S_{mn} induced by the irreducible representations $\{\mu\} \times \{\mu\} \times \cdots$ (n factors) of the subgroup $S_m \times S_m \times \cdots$ (n factors).[61,62]

6.3 RULES OF PLETHYSM

The operation of plethysm is distributive on the *right* with respect to addition, subtraction, and multiplication. It is assumed that the operation $\otimes$ precedes ordinary multiplication from which we deduce the following rules:

$$\text{I. } A \otimes (BC) = (A \otimes B)(A \otimes C)$$
$$= A \otimes BA \otimes C \tag{91}$$

$$\text{II. } A \otimes (B \pm C) = A \otimes B \pm A \otimes C \tag{92}$$

$$\text{III. } (A \otimes B) \otimes C = A \otimes (B \otimes C) \tag{93}$$

The operation is not distributive with respect to addition or subtraction or multiplication on the left. Littlewood[13,27] has derived the additional rules that complete the definition of the algebra:

$$\text{IV. } (A + B) \otimes \{\lambda\} = \sum \Gamma_{\mu\nu\lambda}(A \otimes \{\mu\})(B \otimes \{\nu\}) \tag{94}$$

where $\Gamma_{\mu\nu\lambda}$ is the coefficient of $\{\lambda\}$ in $\{\mu\}\{\nu\}$.

$$\text{V. } (A - B) \otimes \{\lambda\} = \sum (-1)^r \Gamma_{\mu\nu\lambda}(A \otimes \{\mu\})(B \otimes \{\tilde{\nu}\}) \tag{95}$$

where $\{\tilde{\nu}\}$ is the partition of r conjugate to $\{\nu\}$.

$$\text{VI. } (AB) \otimes \{\lambda\} = \sum g_{\mu\nu\lambda}(A \otimes \{\mu\})(B \otimes \{\nu\}) \tag{96}$$

where $g_{\mu\nu\lambda}$ is the coefficient of $\{\lambda\}$ in the inner product $\{\mu\} \circ \{\nu\}$.

We note that the symbols on the *left* need not be S-functions but can be any suitable functions defined on the variables. For example, functions transforming according to any of the representations of subgroups of $GL(n)$.

6.4 THE EVALUATION OF PLETHYSMS OF S-FUNCTIONS

Zia-ud Din[64] and Ibrahim[65–67] have constructed tables for the expansion of $\{\mu\} \otimes \{\nu\}$ for all partitions up to a total degree of 18. Littlewood's articles also contain many examples of plethysms.[63] A considerable effort has been put into the development of methods for evaluating plethysms of S-functions.[68–84a] Here we shall be content to summarize some of the principal methods available.

Four theorems which may be used to obtain the separations of a given plethysm will now be quoted.

Theorem I[63]

If $\{\lambda\} \otimes \{\pi\} = \sum \{\nu\}$

then,

$$\sum_{\zeta,\nu} \Gamma_{1\zeta\nu}\{\zeta\} = \left[\sum_{\mu} \Gamma_{1\mu\pi}\{\lambda\} \otimes \{\mu\}\right]\left[\sum_{\gamma} \Gamma_{1\gamma\lambda}\{\gamma\}\right] \tag{97}$$

If (π) is a partition into only one part Eq. 97 simplifies to

$$\sum_{\zeta,\nu} \Gamma_{1\zeta\nu}\{\zeta\} = \{\lambda\} \otimes \{\pi - 1\}\left[\sum_{\gamma} \Gamma_{1\gamma\lambda}\{\gamma\}\right] \tag{98}$$

The above theorem, commonly known as "Littlewood's third method," gives a method for deducing the S-functions $\{\nu\}$ which appear in the required separation by providing a series $(\sum \Gamma_{1\zeta\nu}\{\zeta\})$ of S-functions obtained by removing a dot in all permissible ways from the graphs corresponding to each S-function appearing in the separation. The right-hand sides of Eqs. 97 and 98 are usually readily expandable; one is left only with the task of selecting a suitable set of $\{\nu\}$'s to match the left-hand side. A theorem due to Ibrahim[84] is particularly useful in defining the choice of the $\{\nu\}$'s.

Theorem II

The principal part of a product of two S-functions $\{\alpha\}$ and $\{\beta\}$ is defined as $\{\alpha_1 + \beta_1, \alpha_2 + \beta_2, \alpha_3 + \beta_3, \ldots\}$. The theorem then states the following:

The principal parts of the products of terms in the expansion $(\{\lambda\} \otimes \{\omega\}) \times (\{\mu\} \otimes \{\eta\})$ appear as terms in the expansion of

$$\{\lambda_1 + \mu_1, \lambda_2 + \mu_2, \lambda_3 + \mu_3, \ldots\} \otimes \{\nu\} \tag{99}$$

whenever $\chi^{(\omega)}\chi^{(\eta)} \supset \chi^{(\nu)}$ where $\chi^{(\omega)}$, $\chi^{(\eta)}$, and $\chi^{(\nu)}$ are the characters of the symmetric group of order $n!$ corresponding to the partitions (ω), (η), (ν) of n and (λ) and (μ) may be any other partitions.

Three special cases of this theorem have been given by Ibrahim.[68]

1. The principal parts in the product

$$(\{\lambda\} \otimes \{n\})(\{\mu\} \otimes \{1^n\}) \tag{100}$$

are terms in the expansion of $\{\lambda_1 + \mu_1, \lambda_2 + \mu_2, \lambda_3 + \mu_3, \cdots\} \otimes \{1^n\}$, where (n) is a partition into one part.

2. The principal parts in the product

$$(\{\lambda\} \otimes \{n\})(\{\mu\} \otimes \{n\}) \tag{101}$$

are terms in the expansion of $\{\lambda_1 + \mu_1, \lambda_2 + \mu_2, \lambda_3 + \mu_3, \ldots\} \otimes \{n\}$.

3. The principal parts in the product

$$(\{\lambda\} \otimes \{1^n\})(\{\mu\} \otimes \{1^n\}) \tag{102}$$

are terms in the expansion of $\{\lambda_1 + \mu_1, \lambda_2 + \mu_2, \lambda_3 + \mu_3, \ldots\} \otimes \{n\}$.

These results supply a list of *S*-functions that certainly appear in the reduction of the plethysm but they do not normally give a complete list or the frequency a given *S*-function occurs. They do, however, usually give a sufficient list to make the choice of $\{\nu\}$'s unambiguous.

Theorem III[68]

The principal parts in the expansion of the products

$$(\{\lambda\} \otimes \{\omega\})(\{\lambda\} \otimes \{\nu\}) \tag{103}$$

appear as terms in the separation of the plethysm,

$$\{\lambda\} \otimes \{\omega_1 + \nu_1, \omega_2 + \nu_2, \omega_3 + \nu_3, \ldots\}$$

Theorem IV[61,63]

If (λ) *is a partition of* r *and if*

$$\{\lambda\} \otimes \{\mu\} = \sum \{\nu\}$$

then if r *is even*

$$\{\tilde{\lambda}\} \otimes \{\mu\} = \sum \{\tilde{\nu}\} \tag{104}$$

and if r *is odd*

$$\{\tilde{\lambda}\} \otimes \{\tilde{\mu}\} = \sum \{\tilde{\nu}\} \tag{105}$$

Littlewood[63] has given two results that are of assistance in establishing further plethysms. If n is an integer, then

$$\{n\} \otimes \{2\} = \{2n\} + \{2n - 2, 2\} + \{2n - 4, 4\} + \cdots \tag{106}$$

to $(n + 1)/2$ or $(n + 2)/2$ terms, and

$$\{n\} \otimes \{1^2\} = \{2n - 1, 1\} + \{2n - 3, 3\} + \cdots \tag{107}$$

to $(n + 1)/2$ or $n/2$ terms. Using these two results together with Theorem IV we readily deduce that for *n odd*

$$\{1^n\} \otimes \{2\} = (2^1, 1^{2n-2}\} + \{2^3, 1^{2n-6}\} + \cdots \tag{108}$$

and

$$\{1^n\} \otimes \{1^2\} = \{1^{2n}\} + \{2^2, 1^{2n-4}\} + \{2^4, 1^{2n-8}\} + \cdots \tag{109}$$

both to $(n + 1)/2$ terms, while for *n even*

$$\{1^n\} \otimes \{2\} = \{1^{2n}\} + \{2^2, 1^{2n-4}\} + \{2^4, 1^{2n-8}\} + \cdots \tag{110}$$

to $(n + 2)/2$ terms, and

$$\{1^n\} \otimes \{1^2\} = \{2^1, 1^{2n-2}\} + \{2^3, 1^{2n-6}\} + \{2^5, 1^{2n-10}\} + \cdots \tag{111}$$

to $n/2$ terms.

6.5 THE DEGREE OF A PLETHYSM

The separation of the plethysm

$$\{\lambda\} \otimes \{\mu\} = \sum \{\nu\}$$

may be checked by calculating the degree $f^{(\{\lambda\}\otimes\{\mu\})}$ of the plethysm. Robinson[11,61] by considering the interpretation of the operation of plethysm in the symmetric group, has shown that

$$f^{(\{\lambda\}\otimes\{\mu\})} = \frac{(mn)!}{(m!)^n n!}(f^{\{\lambda\}})^n f^{\{\mu\}} \tag{112}$$

where m and n are the weights of the partitions (λ) and (μ), respectively, and $f^{\{\lambda\}}$ and $f^{\{\mu\}}$ are the degrees of the representations $\{\lambda\}$ and $\{\mu\}$ of the symmetric groups of order $m!$ and $n!$, respectively. The plethysm is then checked by noting that

$$f^{(\{\lambda\}\otimes\{\mu\})} = \sum f^{\{\nu\}} \tag{113}$$

6.6 EXAMPLES OF PLETHYSMS OF S-FUNCTIONS

We now consider a few examples to illustrate the application of the preceding theorems.

To evaluate the plethysm

$$\{31\} \otimes \{2\} = \sum \{\nu\}$$

we first apply Theorem I to give

$$\begin{aligned}\sum \Gamma_{1\zeta\nu}\{\zeta\} &= (\{31\} \otimes \{1\})(\sum \Gamma_{1\lambda(31)}\{\gamma\})\\ &= \{31\}(\{3\} + \{21\})\\ &= \{61\} + 2\{52\} + 2\{51^2\} + 2\{43\} + 3\{421\} + \{41^3\} + 2\{3^21\}\\ &\quad + \{32^2\} + \{321^2\}\end{aligned}$$

We now determine the set of S-functions $\{\nu\}$ that generate the list found in $\sum \Gamma_{1\zeta\nu}\{\zeta\}$.

Application of Theorem II, Eqs. 101 and 102, show that $\{31\} \otimes \{2\}$ must involve the S-functions associated with the principal parts of the products

$$(\{2\} \otimes \{2\})(\{1^2\} \otimes \{2\}) \quad \text{and} \quad (\{2\} \otimes \{1^2\})(\{1^2\} \otimes \{2\})$$

since

$$\{31\} \otimes \{2\} = \{2+1, 0+1\} \otimes \{2\}$$

Now, using Eqs. 101 and 102 we find

$$(\{2\} \otimes \{2\})(\{1^2\} \otimes \{2\}) = (\{4\} + \{2^2\})(\{1^4\} + \{2^2\}$$

having the principal parts $\{62\}$, $\{51^3\}$, $\{4^2\}$ and $\{3^21^2\}$. Using Eqs. 107 and 111 we find

$$(\{2\} \otimes \{1^2\})(\{1^2\} \otimes \{1^2\}) = \{21^2\}\{31\}$$

with the principal part $\{521\}$.

Thus $\{31\} \otimes \{2\}$ certainly contains the S-functions $\{62\}$, $\{51^3\}$, $\{4^2\}$, $\{3^21^2\}$, and $\{521\}$. We now determine the partitions $\{\zeta\}$ in $\sum \Gamma_{1\zeta\nu}\{\zeta\}$ for these S-functions by removing one cell in all permissible ways from the Young tableaux to give

$$\begin{aligned}
\{62\} &\rightarrow \{61\} + \{52\} \\
\{51^3\} &\rightarrow \{51^2\} + \{41^3\} \\
\{4^2\} &\rightarrow \{43\} \\
\{3^21^2\} &\rightarrow \{3^21\} + \{321^2\} \\
\{521\} &\rightarrow \{52\} + \{51^2\} + \{421\}
\end{aligned}$$

These account for all the S-functions in the list for $\sum \Gamma_{1\zeta\nu}\{\zeta\}$ except for

$$\{43\} + 2\{421\} + \{3^21\} + \{32^2\}$$

But it is clear that the following choice

$$\begin{aligned}
\{42^2\} &\rightarrow \{421\} + \{32^2\} \\
\{431\} &\rightarrow \{421\} + \{43\} + \{3^21\}
\end{aligned}$$

uniquely accounts for the remaining unassigned S-functions and hence

$$\{31\} \otimes \{2\} = \{62\} + \{521\} + \{51^3\} + \{4^2\} + \{431\} + \{42^2\} + \{3^21^2\}$$

This result may be verified by use of Eq. 113

$$f^{(\{31\}\otimes\{2\})} = 315 = 20 + 64 + 35 + 14 + 70 + 56 + 56$$

Using Eq. 104 we may readily deduce the further result that

$$\{21^2\} \otimes \{2\} = \{2^21^4\} + \{321^3\} + \{41^4\} + \{2^4\} + \{32^21\} + \{3^21^2\} + \{42^2\}$$

7

Plethysm and Restricted Groups

So far we have considered the plethysm of S-functions that characterize the full linear group $GL(n)$. The plethysm of a character of a restricted group can be obtained by expressing the character in terms of S-functions and then applying the sum and difference formulas of Eqs. 94 and 95.

As an example, consider the plethysm

$$[31] \otimes \{2\}$$

where [31] is a character of $O(5)$. Using Eq. 64 we have

$$[31] \otimes \{2\} = [(\{31\} + \{0\}) - (\{2\} + \{1^2\})] \otimes \{2\}$$

which upon applying Eq. 95 becomes

$$= (\{31\} + \{0\}) \otimes \{2\} + (\{2\} + \{1^2\}) \otimes \{1^2\} - (\{31\} + \{0\})(\{2\} + \{1^2\})$$

Equation 94 may now be used to yield

$$\begin{aligned} &\{31\} \otimes \{2\} + \{0\} + \{31\} + \{2\} \otimes \{1^2\} + \{1^2\} \otimes \{1^2\} + \{2\}\{1^2\} \\ &\quad - \{31\}\{2\} - \{31\}\{1^2\} - \{2\} - \{1^2\} \\ &\quad = \{62\} + \{521\} + \{51^3\} + \{4^2\} + \{431\} + \{42^2\} + \{3^21^2\} + 3\{31\} \\ &\quad + 2\{21^2\} + \{0\} - \{51\} - 2\{4^2\} - 2\{41^2\} - \{3^2\} - 2\{321\} \\ &\quad - \{31^3\} - \{2\} - \{1^2\} \end{aligned}$$

Decomposing these S-functions into the characters of $O(5)$ using Eq. 69

and converting the nonstandard symbols yields the final result as

$$\begin{aligned}[31] \otimes \{2\} = {} & [62] + [6] + [52] + [51] + [5] + [4^2] + [43] + 3[42] + [41] \\ & + 2[4] + 2[32] + [31] + [3] + 3[2^2] + [21] + 2[2] \\ & + [1] + [0]\end{aligned}$$

Plethysms of characters of the symplectic group may likewise be evaluated. For example, in the case of Sp_4 we have

$$\begin{aligned}\langle 2^2\rangle \otimes \{2\} &= (\{2^2\} - \{1^2\}) \otimes \{2\} \\ &= \{2^2\} \otimes \{2\} + \{1^2\} \otimes \{2\} - \{2^2\}\{1^2\} \\ &= \{4^2\} + \{42^2\} + \{3^21^2\} + \{2^4\} + \{21^2\} - \{3^2\} - \{321\} - \{2^21^2\} \\ &= \langle 4^2\rangle + \langle 4\rangle + \langle 2^2\rangle + \langle 0\rangle\end{aligned}$$

In general, the calculation of a plethysm for a restricted group is very tedious. However, Littlewood[45] has given two theorems for performing the plethysms $[\lambda] \otimes \{\mu\}$ and $\langle\lambda\rangle \otimes \{\mu\}$, where (μ) is a partition of 2, which rapidly gives the resolution of the Kronecker squares of characters of the orthogonal and symplectic groups into their symmetric and antisymmetric components.

Theorem I

If (μ) is a partition of 2, then

$$[\lambda] \otimes \{\mu\} = \sum \Gamma_{\lambda\mu\nu}[\nu] \tag{114}$$

where

$$\sum \Gamma_{\lambda\mu\nu}\{\nu\} = \sum (\Gamma_{\xi\eta\lambda}\{\eta\}) \otimes \{\mu\} + \sum \Gamma_{\xi\eta\lambda}\Gamma_{\xi\zeta\lambda}\{\eta\}\{\zeta\} \qquad (\eta) \neq (\zeta)$$

summed for all suitable S-functions $\{\xi\}$, $\{\eta\}$, $\{\zeta\}$, the last term not being repeated for the interchange of $\{\eta\}$ and $\{\zeta\}$.

Theorem II

If (μ) is a partition of 2, then

$$\langle\lambda\rangle \otimes \{\mu\} = \sum \Gamma_{\lambda\mu\nu}\langle\nu\rangle \tag{115}$$

where

$$\sum \Gamma_{\lambda\mu\nu}\{\nu\} = \sum (\Gamma_{\xi\eta\lambda}\{\eta\}) \otimes (\{\mu\} \circ \{\varepsilon\}) + \sum \Gamma_{\xi\eta\lambda}\Gamma_{\xi\zeta\lambda}\{\eta\}\{\zeta\}$$

in which $(\varepsilon) = (2)$ if (ξ) is of even weight, but $(\varepsilon) = (1^2)$ if (ξ) is of odd weight.

In applying these two theorems it is important to note that if the coefficient $\Gamma_{\xi\eta\lambda} = n$ in the first summation then $\Gamma_{\xi\eta\lambda}\{\eta\} = \{\eta\} + \{\eta\} + \cdots$ (n terms).

7.1 INTRANSITIVE GROUPS

An intransitive group[13,50] is the group of all transformations on two sets of variables each transformed independently by some basic group. The intransitive group is thus the direct product of two groups and its characters and representations are the products of two characters or representations of the basic groups. A character of an intransitive group can, therefore, be written as AB' where A and B' are characters of the basic groups.

Plethysms of the characters of the intransitive group can be obtained by using Eq. 96 for the product of functions on independent variables; i.e.,

$$(AB') \otimes \{\mu\} = \sum g_{\alpha\beta\mu}(A \otimes \{\alpha\})(B' \otimes \{\beta\}) \tag{116}$$

For example,

$$(\{1\}\{1'\}) \otimes \{3\} = \{3\}\{3\}' + \{21\}\{21\}' + \{1^3\}\{1^3\}'$$

or

$$\begin{aligned}(\{2\}\{2\}') \otimes \{2\} &= (\{2\} \otimes \{2\})(\{2\}' \otimes \{2\}) + (\{2\} \otimes \{1^2\})(\{2\}' \otimes \{1^2\})\\ &= (\{4\} + \{2^2\})(\{4\}' + \{2^2\}') + \{31\}\{31\}'\\ &= \{4\}\{4\}' + \{4\}\{2^2\}' + \{2^2\}\{4\}' + \{2^2\}\{2^2\}' + \{31\}\{31\}'\end{aligned}$$

As we shall see later, intransitive transformations are physically important when parts of a system, such as the spatial and spin coordinates of a particle, are uncoupled and we may consider the effects of independent group operations on each part.

7.2 SPIN REPRESENTATIONS

As well as the true representations of the orthogonal group which have a well-defined geometrical interpretation, there exist intrinsically double-valued representations known as *spin representations*.[51,52] The spin representations of the orthogonal groups are of considerable importance in the quantum-mechanical treatment of angular momentum. For example, in the three-dimensional rotation group the true representations correspond to the characters [0], [1], [2], [3], . . . and have 1, 3, 5, 7, . . . rowed representative matrices, respectively. The even-rowed representations correspond to the spin representations, with the two-rowed representation $[\frac{1}{2}]$ being the *basic spin representation*.

The direct product of a spin representation with itself will give a true representation and the direct product of a spin representation with a true representation will give another spin representation. Thus,

$$[\tfrac{1}{2}][\tfrac{3}{2}] = [1] + [2] \quad \text{and} \quad [\tfrac{3}{2}][1] = [\tfrac{1}{2}] + [\tfrac{3}{2}] + [\tfrac{5}{2}]$$

The spin representations of the orthogonal group are the transformation matrices which induce an orthogonal-type transformation among a set of anticommuting basis matrices with unit square.

Consider a set of $2\nu + 1$ anticommuting matrices $E_1, E_2, \ldots, E_{2\nu+1}$. Then the anticommuting property requires that

$$\{E_i, E_j\} = \pm 2\delta_{ij} I$$

For any orthogonal transformation

$$X_i' = \sum a_{ij} X_j$$

we can find a matrix T such that

$$T^{-1} E_i T = \sum a_{ij} E_j$$

The set of all matrices T defined in this way will form the basic spin representation of the group. The representation is double-valued since the transformation matrices T and $-T$ correspond to the same operation of the group, i.e.,

$$(-T)^{-1} E_i (-T) = T^{-1} E_i T$$

The matrices E_i forming the basis for the basic spin representation of the group in $n = 2\nu$ or $n = 2\nu + 1$ variables must have 2ν rows if the anticommutation relations are to be satisfied and hence the matrices T forming the basic spin representation must also have 2^ν rows. The spur of T is the basic spin character and will be denoted by $[(\frac{1}{2})^\nu]$.

A detailed account of the properties of spin representations is given in Appendix I.

7.3 THE TERNARY ORTHOGONAL GROUP

The ternary orthogonal group, including both the true and spin representations, is of particular importance in atomic spectroscopy. Starting with the obvious results

$$[\tfrac{1}{2}] \otimes \{2\} = [1]$$

$$[\tfrac{1}{2}] \otimes \{2\} \otimes \{2\} = [\tfrac{1}{2}] \otimes (\{4\} + \{2^2\}) = [\tfrac{1}{2}] \otimes \{4\} + [0]$$

$$= \{2\} = [2] + [0]$$

we have

$$[\tfrac{1}{2}] \otimes \{4\} = [2]$$

and generally

$$[\tfrac{1}{2}] \otimes \{n\} = [\tfrac{1}{2}n] \tag{117}$$

Thus there is an isomorphism between the ternary orthogonal group and the binary full linear group such that the representation $[n]$ of the ternary orthogonal group corresponds to the representation $\{2n\}$ of the binary group. Thus the basic spin representation $[\frac{1}{2}]$ of R_3 corresponds to the S-function $\{1\}$ of $GL(2)$.

The existence of this isomorphism renders trivial the calculation of products or plethysms of spin representations for the ternary orthogonal group.

For example, consider $[\frac{3}{2}] \otimes \{4\}$. The corresponding plethysm in the binary full linear group is $\{3\} \otimes \{4\}$ which separates as

$$\{3\} \otimes \{4\} = \{12\} + \{10 \cdot 2\} + \{93\} + \{84\} + \{82^2\} + \{741\} + \{6^2\} + \{642\} + \{4^3\}$$

Upon discarding the S-functions in more than two variables and remembering that the S-functions $\{\alpha_1, \alpha_2\}$ and $\{\alpha_1 - \alpha_2\}$ are equivalent we have

$$\{3\} \otimes \{4\} = \{12\} + \{8\} + \{6\} + \{4\} + \{0\}$$

and hence

$$[\tfrac{3}{2}] \otimes \{4\} = [6] + [4] + [3] + [2] + [0]$$

Similarly, we may readily show that

$$\begin{aligned}[\tfrac{3}{2}] \otimes \{1^4\} = \{3\} \otimes \{1^4\} &= \{91^3\} + \{831\} + \{741\} + \{731^2\} + \{6^2\} \\ &\quad + \{642\} + \{63^2\} + \{5^21^2\} + \{53^21\} + \{3^4\} \\ &= \{0\} = [0]\end{aligned}$$

7.4 THE QUARTERNARY ORTHOGONAL AND ROTATION GROUPS

The basic spinor for the full orthogonal group in four variables is of degree 4 and will be denoted as $[\frac{1}{2}, \frac{1}{2}]'$. The characters, $[\lambda_1, \lambda_2]'$, of the full orthogonal group in an even number of variables separate into two conjugate characters for the rotation group if $\lambda_2 \neq 0$. Thus

$$[\lambda_1, \lambda_2]' = [\lambda_1, \lambda_2] + [\lambda_1, -\lambda_2] \tag{118}$$

and

$$[\tfrac{1}{2}, \tfrac{1}{2}]' = [\tfrac{1}{2}, \tfrac{1}{2}] + [\tfrac{1}{2}, -\tfrac{1}{2}]$$

Littlewood[52] has shown that there is a 2:1 isomorphism between the quaternary rotation group and the double binary full linear group and in particular

$$[\tfrac{1}{2}, \tfrac{1}{2}] \otimes \{p\}[\tfrac{1}{2}, -\tfrac{1}{2}] \otimes \{q\} = [\tfrac{1}{2}(p+q), \tfrac{1}{2}(p-q)^{-}] \tag{119}$$

Now let $\{\lambda\}$ denote an S-function of the roots of the two-rowed spin matrix of character $[\frac{1}{2}, \frac{1}{2}]$ and $\{\mu\}'$ for the character $[\frac{1}{2}, -\frac{1}{2}]$. The characters of the double binary group will then correspond to the products $\{\lambda\}\{\mu\}'$.

We may now make the correspondences between the double binary group characters and those of the quaternary rotation group

$$[a, b] \rightarrow \{a + b\}\{a - b\}' \tag{120}$$

and

$$\{a\}\{b\}' \rightarrow [\tfrac{1}{2}(a + b), \tfrac{1}{2}(a - b)] \tag{121}$$

Equations (120) and (121) may then be used as the basis for computing the Kronecker products and plethysms of characters of the quaternary rotation group. Thus the Kronecker product $[21][2, -1]$ corresponds to

$$\begin{aligned} \{3\}\{1\}'\{1\}\{3\}' &= (\{4\} + \{31\})(\{31\}' + \{4\}') \\ &= (\{4\} + \{2\})(\{2\}' + \{4\}') \\ &= \{4\}\{2\}' + \{4\}\{4\}' + \{2\}\{4\}' + \{2\}\{2\}' \end{aligned}$$

hence

$$[21][2, -1] = [31] + [4] + [3, -1] + [2]$$

Noting Eqs. 120 and 121 we may readily deduce the general rule for Kronecker products for the quaternary rotation group as[84a]

$$[a, b][c, d] = \sum_{\alpha=0}^{t} \sum_{\beta=0}^{u} [a + c - \alpha - \beta, b + d - \alpha + \beta] \tag{122}$$

where t is the lesser of $(a + b)$ and $(c + d)$, and u is the lesser of $(a - b)$ and $(c - d)$.

In a similar manner the plethysm $[21] \otimes \{1^2\}$ is evaluated by writing

$$\begin{aligned} [21] \otimes \{1^2\} &\rightarrow \{3\}\{1\}' \otimes \{1^2\} \\ &= (\{3\} \otimes \{1^2\})\{2\}' + (\{3\} \otimes \{2\})\{1^2\}' \\ &= (\{51\} + \{3^2\})\{2\}' + (\{6\} + \{42\} + \{2^3\})\{1^2\}' \\ &= (\{4\} + \{0\})\{2\}' + (\{6\} + \{2\})\{1^2\}' \\ &= \{4\}\{2\}' + \{0\}\{2\}' + \{6\}\{0\}' + \{2\}\{0\}' \end{aligned}$$

where we have made use of Eqs. 106 and 107. Upon using Eq. 121 we obtain the result

$$[21] \otimes \{1^2\} = [31] + [1, -1] + [33] + [11]$$

Similarly,

$$[2, -1] \otimes \{1^2\} = [3, -1] + [1, -1] + [3, -3] + [11]$$

It is worth noting that if $[21]'$ is a character of the full orthogonal group in four variables then

$$\begin{aligned} [21]' \otimes \{1^2\} &= ([21] + [2, -1]) \otimes \{1^2\} \\ &= [21] \otimes \{1^2\} + [21][2, -1] + [2, -1] \otimes \{1^2\} \\ &= [4] + 2[31] + 2[3, -1] + [33] + [3, -3] \\ &\quad + 2[11] + 2[1, -1] \\ &= [4]' + 2[31]' + [33]' + 2[11]' \end{aligned}$$

7.5 THE ROTATION GROUP IN SIX DIMENSIONS

The rotation group in six dimensions $R(6)$ plays an important role in the group theoretical description of the electronic states of the $(d + s)^n$ configurations. Littlewood[52] has made use of the observation that this group is isomorphic with the four-dimensional full linear group, and in particular

$$[\tfrac{1}{2}\tfrac{1}{2}\tfrac{1}{2}] \otimes \{pqr\} = [\tfrac{1}{2}(p + q - r), \tfrac{1}{2}(p - q + r), \tfrac{1}{2}(p - q - r)] \quad (123)$$

Thus, noting Eq. 91, correspondences of the types[84a]

$$[abc] \to \{a + b, a - c, b - c\} \quad (124)$$

$$\{abc\} \to [\tfrac{1}{2}(a + b - c), \tfrac{1}{2}(a - b + c), \tfrac{1}{2}(a - b - c)] \quad (125)$$

may be used to evaluate Kronecker products and plethysms of the characters of $R(6)$. For example, we may use Eqs. 124 and 125 to readily establish the results

$$[111][111] = [222] + [211] + [200]$$
$$[11, -1][11, -1] = [22, -2] + [21, -1] + [200]$$
$$[111][11, -1] = [000] + [220] + [110]$$

and thus under the full orthogonal group $O(6)$

$$\begin{aligned}[111]'[111]' &= [111][111] + [11, -1][11, -1] + 2[111][11, -1] \\ &= [222]' + [211]' + 2[220] + 2[200] + 2[110] + 2[000].\end{aligned}$$

Equations 124 and 125 are of great value in making plethysms for $R(6)$. For example,

$$\begin{aligned}[211] \otimes \{2\} \to \{210\} \otimes \{2\} &= \{42\} + \{31^3\} + \{2^2\} + \{321\} \\ &= [311] + [111] + [200] + [210]\end{aligned}$$

7.6 BRANCHING RULES AND PLETHYSM

We have already outlined the methods for obtaining the character decompositions for $U(n) \to O(n)$, $U(n) \to Sp(n)$ and $O(7) \to G_2$. In applications of group theory to problems of atomic spectroscopy we are frequently interested in the decompositions of the characters of irreducible representations of a group in n-dimensions into those of a subgroup of lower dimensions or of a direct product of two groups on independent variables. For example, the character decompositions for $U_{4l+2} \to SU_2 \times R_{2l+1}$, $R_{8l+4} \to SU_2 \times Sp_{4l+2}$, $Sp_{4l+2} \to SU_2 \times R_{2l+1}$, $R_{2l+1} \to R_3$, $R_6 \to R_5$, and $G_2 \to R_3$ all have extensive applications in the theory of complex spectra.

Judd[15,85,86] has obtained a number of relevant decompositions from dimensional consideration and by making use of trivial decompositions as a basis for setting up chain calculations for establishing more complex decompositions. These methods are frequently tedious and not always unambiguous. Littlewood's algebra of plethysm gives a simple and complete solution to the general problem.

Consider a group G having a proper subgroup H. The character of the irreducible representation of G corresponding to the partition (1) may always be trivially decomposed into the characters of its subgroup H. For example, we may readily establish the following results: For,

$$U_{4l+2} \to SU_2 \times R_{2l+1} \qquad \{1\} \to [\tfrac{1}{2}]'[1] \tag{126}$$

$$R_{8l+4} \to SU_2 \times Sp_{4l+2} \qquad [1] \to [\tfrac{1}{2}]'\langle 1\rangle \tag{127}$$

$$Sp_{4l+2} \to SU_2 \times R_{2l+1} \qquad \langle 1\rangle \to [\tfrac{1}{2}]'[1] \tag{128}$$

$$R_{2l+1} \to R_3 \qquad [1] \to [l] \tag{129}$$

$$R_6 \to R_5 \qquad [1] \to [1] + [0] \tag{130}$$

$$G_2 \to R_3 \qquad (10) \to [3] \tag{131}$$

Let $\{\lambda\}$ represent the characters of G and $\{\rho\}$ those of its subgroup H. Then we have the important theorem:

Theorem

If under the restriction $G \to H$, the character $\{1\}$ decomposes as

$$\{1\} \to \{\alpha\} + \{\beta\} + \cdots + \{\omega\}$$

then the character $\{\lambda\}$ of G decomposes into the characters $\{\rho\}$ of H according to the characters of H contained in the plethysm

$$[\{\alpha\} + \{\beta\} + \cdots + \{\omega\}] \otimes \{\lambda\} \tag{132}$$

The plethysm of Eq. 132 may be evaluated in the usual manner by first expressing the characters of G and H in terms of S-functions, evaluating the plethysm to give S-functions pertaining to the characters of H. These S-functions are then expressed in terms of the characters of H to give the final result. The above theorem follows trivially from the discussion on pages 49–51.

7.7 EXAMPLES OF BRANCHING RULES

We now illustrate the application of the theorem to a few specific examples. Consider the decomposition of the character $\{1^n\}$ of U_{4l+2} into the characters of $SU_2 \times R_{2l+1}$. Noting Eqs. 126 and 132 we have

$$\begin{aligned}\{1^n\} &\to [\tfrac{1}{2}]'[1] \otimes \{1^n\} \\ &\to \{1\}'\{1\} \otimes \{1^n\}\end{aligned} \tag{133}$$

where we have made use of the isomorphism of SU_2 and $GL(2)$. The plethysm is now expanded using Eq. 96 to give

$$\{1^n\} \to \sum g_{\mu\nu(1^n)}(\{1\}' \otimes \{\mu\})(\{1\} \otimes \{\nu\}) \tag{133a}$$

where $g_{\mu\nu(1^N)}$ is the coefficient of $\{1^n\}$ in the inner product $\{\mu\} \circ \{\nu.\}$ But the inner product $\{\mu\} \circ \{\nu\}$ will only contain $\{1^n\}$ if $(\nu) = (\tilde{\mu})$ and then only once and hence

$$\{1^n\} \to \sum \{\tilde{\mu}\}'\{\mu\} \tag{134}$$

where the summation is over all partitions of n such that (1^n) occurs in the inner product $\{\mu\} \circ \{\tilde{\mu}\}$. Since for $GL(2)$ the S-functions $\{\tilde{\mu}\}'$ cannot have more than two parts only partitions $(\tilde{\mu})$ of n into not more than two parts need be considered. In the event of a partition $\{\tilde{\mu}\}$ having two parts $\{\tilde{\mu}\} \equiv \{\tilde{\mu}_1, \tilde{\mu}_2\}$ we have $\{\tilde{\mu}_1, \tilde{\mu}_2\} \equiv \{\tilde{\mu}_1 - \tilde{\mu}_2\}$. Recalling the isomorphism of SU_2 and $GL(2)$ we have the result

$$\{1^n\} \to \sum [\tfrac{1}{2}(\tilde{\mu}_1 - \tilde{\mu}_2)]'\{\mu\} \tag{135}$$

The formation of the decomposition is then completed. by reducing all the S-functions $\{\mu\}$ into not more than l parts and then expressing them in terms of the characters of R_{2l+1}; for example, for $U_{14} \to SU_2 \times R_7$ we have

$$\begin{aligned}\{1^3\} &\to [\tfrac{3}{2}]'\{111\} + [\tfrac{1}{2}]'\{210\} \\ &\to [\tfrac{3}{2}]'[111] + [\tfrac{1}{2}]'([210] + [000]) \\ &\to {}^4[111] + {}^2([210] + [000])\end{aligned} \tag{136}$$

where we have written the dimension $(2\lambda + 1)$ of the representation $[\lambda]'$ of SU_2 as a left superscript in accordance with the usual spectroscopic notation.

As a further example, consider the decomposition of the character [200] of R_7 into the characters of R_3. We have from Eqs. 129 and 132 that the relevant characters of R_3 occur in $[3] \otimes [200]$. Now,

$$[3] \to \{3\} - \{1\} \quad \text{and} \quad [200] \to \{200\} - \{0\}$$

$$\begin{aligned}\therefore \quad [3] \otimes [200] &= (\{3\} - \{1\}) \otimes (\{200\} - \{0\}) \\ &= (\{3\} - \{1\}) \otimes \{200\} - \{0\} \\ &= \{3\} \otimes \{2\} + \{1\} \otimes \{1^2\} - \{3\}\{1\} - \{0\} \\ &= \{6\} + \{42\} + \{1^2\} - \{4\} - \{31\} - \{0\} \\ &= [6] + [4] + [2]\end{aligned}$$

As a final example consider the decomposition of the character [111] of R_{28} into the characters of $SU_2 \times Sp_{14}$. From Eqs. 127 and 132 we have that the relevant characters of $SU_2 \times Sp_{14}$ occur in the plethysm

$$[\tfrac{1}{2}]'\langle 1\rangle \otimes [1^3]$$

Now

$$[\tfrac{1}{2}]'\langle 1\rangle \otimes [1^3] = \{1\}'\{1\} \otimes \{1^3\}$$

which from Eq. 133*a* is

$$= [\tfrac{3}{2}]'\{111\} + [\tfrac{1}{2}]'\{210\}$$

Decomposing $\{111\}$ and $\{210\}$ into the characters of Sp_{14} using Eqs. 80 and 81 gives

$$[1^3] \to {}^4(\langle 111\rangle + \langle 100\rangle) + {}^2(\langle 210\rangle + \langle 100\rangle)$$

for $R_{28} \to SU_2 \times Sp_{14}$.

7.8 BRANCHING RULES AND THE TERNARY ORTHOGONAL GROUP

The problem of decomposing a character of the unitary group in n-dimensions into those of the ternary orthogonal group $O(3)$ occurs frequently in theoretical spectroscopy. If the branching rules for $U(n) \to O(3)$ are known, the corresponding rules for $Sp(n) \to O(3)$ and $R(n) \to O(3)$ may be found by the method of differences. The determination of the branching rules $U(n) \to O(3)$ may be greatly simplified by first considering the branching rules for $GL(2)$ and then using the fact that the irreducible representations of $GL(2)$ remain irreducible upon restriction to $U(n)$ and that the group $O(3)$ is isomorphic to $GL(2)$.

The characters of $GL(2)$ correspond to S-functions having not more than two parts and since the S-functions having two parts may be reduced to S-functions having only one part by use of the equivalence $\{\mu_1, \mu_2\} \equiv \{\mu_1 - \mu_2\}$ we need only consider partitions into one part. Murnaghan[55a,79] has shown that this leads directly to the so-called Hermite's reciprocity principle that for $GL(2)$ we may write:

$$\{m\} \otimes \{k\} = \{k\} \otimes \{m\}$$

where m and k are positive integers. This result may then be used to establish the recursion formula:

$$\{m\} \otimes \{k\} = \{m-2\} \otimes \{k\} + \{m\} \otimes \{k-2\} + (\{m-1\} \otimes \{k-1\})(\{m+k-1\} - \{m+k-3\}) \quad (137)$$

Furthermore,

(i) if p is any positive integer $p \geq m + k - 1$, then

$$\{p\}(\{m + k - 1\} - \{m + k - 3\}) = \{p + m + k - 1\} + \{p - m - k + 1\}$$

(ii) if $p = m + k - 2$, then

$$\{p\}(\{m + k - 1\} - \{m + k - 3\}) = \{p + m + k - 1\}$$

(iii) if $p \leq m + k - 3$,

$$\{p\}(\{m + k - 1\} - \{m + k - 3\}) = \{p + m + k - 1\} - \{m + k - 3 - p\}$$

The above results may be readily used to decompose any character $\{k\}$ of $GL(n)$ into characters of $GL(2)$ once the decomposition of the unary character $\{1\}$ has been specified. The branching rules for the characters of $GL(n)$ corresponding to partitions into two parts may be determined by noting that

$$A \otimes (BC) = A \otimes BA \otimes C$$

and

$$\{p\} \otimes (\{m - 1\}\{1\}) = \{p\} \otimes (\{m\} + \{m - 1, 1\})$$

leading to the general result

$$\begin{aligned} \{p\} \otimes \{m - k, k\} &= \{p\} \otimes (\{m - k\}\{k\} - \{m - k + 1\}\{k - 1\}) \\ &= (\{p\} \otimes \{m - k\})(\{p\} \otimes \{k\}) \\ &\quad - (\{p\} \otimes \{m - k + 1\})(\{p\} \otimes \{k - 1\}) \end{aligned} \tag{138}$$

which involves plethysms for one-part partitions which may be evaluated using the recursion formula.

As an example, let us determine the decomposition of $\{22\}$ of $GL(5)$ into characters of $GL(2)$ if $\{1\} \to \{4\}$ under $GL(5) \to GL(2)$. The characters of $GL(2)$ obtained in the decomposition of $\{22\}$ of $GL(5)$ will be just the terms in

$$\begin{aligned} \{4\} \otimes \{22\} &= \{4\} \otimes (\{2\}\{2\} - \{3\}\{1\}) \\ &= (\{4\} \otimes \{2\})^2 - (\{4\} \otimes \{3\})\{4\} \\ &= (\{8\} + \{4\} + \{0\})^2 - (\{12\} + \{8\} + \{6\} + \{4\} + \{0\})\{4\} \\ &= 2\{0\} + 2\{4\} + \{6\} + 2\{8\} + \{12\} \end{aligned}$$

Recalling the isomorphism that exists between $O(3)$ and $GL(2)$, we have for $U(5) \to R(3)$

$$\{22\} \to 2[0] + 2[2] + [3] + 2[4] + [6]$$

The branching rules for partitions into more than two parts may be determined in a similar manner.

The decomposition of the character $\{1^m\}$ of $U(n)$ into characters of $O(3)$ plays a very important role in theoretical spectroscopy. In this case the decomposition may be found by noting that:[55a]

$$\{p\} \otimes \{1^m\} = \{p + 1 - m\} \otimes \{m\} \tag{139}$$

and the recursion relationship:

$$\{p\} \otimes \{1^m\} = (\{p - 1\} \otimes \{1^{m-1}\})\{p - m + 1\} - (\{p - 1\} \otimes \{1^m\})\{m - 2\}$$
$$m \leq p + 1$$

For example, to obtain the decomposition of $\{1^3\}$ of U_{11} into characters of $GL(2)$ when $\{1\} \to \{10\}$ under $GL(11) \to GL(2)$ we simply obtain these as the terms in the plethysm

$$\begin{aligned}\{10\} \otimes \{1^3\} &= \{8\} \otimes \{3\} \\ &= \{0\} + \{4\} + \{6\} + 2\{8\} + \{10\} + 2\{12\} + \{14\} \\ &\quad + \{16\} + \{18\} + \{20\} + \{24\}\end{aligned}$$

This result may be used to obtain the branching rules for $\{1^3\}$ of $U_{12} \to GL(2)$ where $\{1\} \to \{11\}$, by writing

$$\begin{aligned}\{11\} \otimes \{1^3\} &= (\{10\} \otimes \{1^2\})\{9\} - (\{10\} \otimes \{1^3\})\{1\} \\ &= \{3\} + \{5\} + \{7\} + 2\{9\} + 2\{11\} + \{13\} + 2\{15\} + \{17\} \\ &\quad + \{19\} + \{21\} + \{23\} + \{27\}\end{aligned}$$

The result of Eq. 139 gives an interesting connection between the angular momentum eigenvalues associated with the antisymmetric and symmetric states of a system of m equivalent particles. For example, the relation

$$\{10\} \otimes \{1^3\} = \{8\} \otimes \{3\}$$

implies that, for states of maximum multiplicity, there is a one-to-one correspondence between the orbital angular momentum eigenvalues associated with the antisymmetric orbital states of h^3 and those associated with the symmetric orbital states of g^3.

8

Classification of Atomic States of Equivalent Electron Configurations

8.1 CLASSIFICATION OF ATOMIC STATES

The conventional group-theoretical methods of classifying the atomic states of equivalent electron configurations have been reviewed by Judd.[15] A more elegant approach using second quantization methods has also been discussed by Judd.[87,88] In this chapter we wish to consider the classification in terms of Littlewood's algebra of plethysm. We shall first consider the problem of enumerating the Russell-Saunders SL terms of configurations of equivalent electrons and then discuss the cases of jj- and LL-coupling. While we shall restrict our discussion to problems in atomic spectroscopy there is no difficulty in extending the methods to problems in nuclear spectroscopy where both charge and spin functions arise.

The total wavefunction for a single electron can be written as the product of an orbital function and a spin function. If the orbital function transforms according to the representation $[l]$ of R_3 then the set of all possible wavefunctions for an electron with orbital quantum number l will span the $[\frac{1}{2}][l]$ representation of R_3 where $[\frac{1}{2}]$ is the spin representation spanned by the two spin functions.

The functions that describe the n-particle configuration will be simply the antisymmetric functions obtained by forming products of single particle functions defined on each of the n sets of coordinates. These functions will transform according to the representation

$$([\tfrac{1}{2}][l]) \otimes \{1^n\} \tag{140}$$

of R_3 which will generally be reducible. The analysis of the separation of this plethysm provides the classification of the atomic states.

8.2 *LS*-COUPLING

In this coupling scheme the orbital and spin spaces of the wavefunction are considered as independent spaces. As a result the single particle wavefunctions are said to transform according to the representation $[\tfrac{1}{2}]'[l]$ where the dash indicates that the transformation of the spin and orbital functions are to be considered independently. Correspondingly, the n-particle functions will transform according to the representation

$$([\tfrac{1}{2}])'[l]) \otimes \{1^n\} \tag{141}$$

In terms of the conventional group theoretical description the complete reduction of the above plethysm would correspond to the decomposition of the antisymmetric representation $\{1^n\}$ of U_{4l+2} into those of the direct product of two ternary orthogonal groups, one giving the spin classification and the other the orbital classification. For example, in the case of three equivalent d electrons we have

$$\begin{aligned}([\tfrac{1}{2}]'[2]) \otimes \{1^3\} &\to ([\tfrac{1}{2}]' \otimes \{3\})([2] \otimes \{1^3\}) + ([\tfrac{1}{2}]' \otimes \{21\})([2] \otimes \{21\}) \\ &\to {}^4([2] \otimes \{1^3\}) + {}^2([2] \otimes \{21\}) \\ &\to {}^4(PF) + {}^2(PD_2FGH)\end{aligned}$$

hence

$$\{1^3\} \to {}^4(PF) + {}^2(PD_2FGH)$$

While the above example gives the Russell-Saunders terms of d^3 immediately it fails to distinguish the two ${}^2D$ terms. To obtain additional classificatory symbols we must examine the reduction of the plethysm more closely.

8.3 THE SENIORITY CLASSIFICATION

The classification of the states of equivalent electron configurations may be extended by expressing the S-functions $\{1^n\}$ in terms of the characters of the symplectic group,

$$\{1^n\} = \sum_{\alpha=0} \langle 1^{n-2\alpha} \rangle \tag{142}$$

where the α's are positive integers such that $n \geq 2\alpha$. This amounts to making the decomposition $U_{4l+2} \rightarrow Sp_{4l+2}$. The plethysm of Eq. 141 may then be written as

$$([\tfrac{1}{2}]'[l]) \otimes \{1^n\} = \sum_{\alpha=0} ([\tfrac{1}{2}]'[l]) \otimes \langle 1^{n-2\alpha} \rangle$$
$$= \sum_{\alpha=0} ([\tfrac{1}{2}]'[l]) \otimes (\{1^{n-2\alpha}\} - \{1^{n-4\alpha}\}) \qquad (143)$$

The representations appearing in the separation of the terms corresponding to each value of α may be classified either with the symbol $\langle 1^{n-2\alpha} \rangle$ or equivalently with the *seniority number*[89]

$$v = n - 2\alpha \qquad (144)$$

In our example of d^3 we have

$$([\tfrac{1}{2}]'[2]) \otimes \{1^3\} \rightarrow ([\tfrac{1}{2}]'[2]) \otimes \langle 1^3 \rangle + ([\tfrac{1}{2}]'[2]) \otimes \langle 1 \rangle$$

and

$$([\tfrac{1}{2}]'[2]) \otimes \langle 1^3 \rangle \rightarrow {}^4(PF) + {}^2(PDFGH)$$
$$([\tfrac{1}{2}]'[2]) \otimes \langle 1 \rangle \rightarrow {}^2D$$

from which we conclude that all the terms of d^3 are of seniority $v = 3$ except for one ${}^2D$ which is of seniority $v = 1$.

8.4 ADDITIONAL CLASSIFICATORY SYMBOLS

The classification we have obtained so far amounts to simply determining the decomposition of the $\{1^n\}$ representation of U_{4l+2} under the reduction

$$U_{4l+2} \rightarrow Sp_{4l+2} \rightarrow SU_2 \times R_3$$

and we have used the algebra of plethysm to obtain the decomposition $Sp_{4l+2} \rightarrow SU_2 \times R_3$ directly rather than by the usual pedestrian methods. While this classification suffices for the d^n electron configurations it is not rich enough to provide an adequate classification of even the states of f^3. Clearly additional classificatory symbols must be introduced. This amounts to finding proper subgroups G such that $U_{4l+2} \supset G \supset R_3$. This problem has been examined exhaustively by Yan Zhi-Dan and his associates.[56–59] The only permissible chain of groups that leave S and L as good quantum numbers are

$$U_{4l+2} \begin{matrix} \nearrow Sp_{4l+2} \searrow \\ \searrow SU_2 \times SU_{2l+1} \nearrow \end{matrix} SU_2 \times (R_{2l+1} \rightarrow R_3) \qquad (145)$$

and in the particular case of $l = 3$ we have the additional subgroup G_2 such that

$$R_{2l+1} \rightarrow G_2 \rightarrow R_3 \qquad (146)$$

At each stage the branching rules are determined by the application of the theorem leading to Eq. 132 together with the trivial decomposition of Eqs. 126 to 131.

8.5 EXAMPLE OF f^3

Let us now consider the specific case of the classification of the terms of the configuration of three equivalent f electrons. We have to consider the decomposition of the $\{1^3\}$ representation of U_{14} under the reduction

$$U_{14} \begin{smallmatrix} \nearrow & Sp_{14} & \searrow \\ \searrow & SU_2 \times SU_7 & \nearrow \end{smallmatrix} SU_2 \times (R_7 \to G_2 \to R_3) \tag{147}$$

It follows from Eq. 80 that under the reduction $U_{14} \to Sp_{14}$

$$\{1^3\} \to \langle 1^3 \rangle + \langle 1 \rangle$$

The irreducible representation $\langle 1^3 \rangle$ and $\langle 1 \rangle$ of Sp_{14} may be decomposed into those of $SU_2 \times R_7$ by application of Eqs. 128 and 132 to give

$$\begin{aligned} \langle 1^3 \rangle &\to ([\tfrac{1}{2}]'[100]) \otimes \langle 1^3 \rangle \\ &\to (\{1\}'\{1\}) \otimes (\{1^3\} - \{1\}) \\ &\to \{3\}'\{1^3\} + \{21\}'\{21\} - \{1\}'\{1\} \\ &\to {}^4[111] + {}^2[210] \end{aligned}$$

where we have made use of Eq. 69 to reduce the S-functions to characters of R_7 and the isomorphism between the ternary orthogonal group and the binary full linear group. Similarly,

$$\langle 1 \rangle \to {}^2[100]$$

The irreducible representations of R_7 may then be decomposed into those of G_2 using Eq. 85 to give

$$\begin{aligned} [111] &\to (00) + (10) + (20) \\ [210] &\to (11) + (20) + (21) \\ [100] &\to (10) \end{aligned}$$

The decomposition of the irreducible representations of G_2 into those of R_3 are then made using Eqs. 131 and 132 to give

$$\begin{aligned} (00) &\to S \\ (10) &\to F \\ (11) &\to PH \\ (20) &\to DGI \\ (21) &\to DFGHKL \end{aligned}$$

An alternative classification may be made by noting that under $U_{14} \to SU_2 \times U_7$ the representation

$$\{1^3\} \to {}^4\{111\} + {}^2\{210\}$$

and under $U_7 \to R_7$

$$\{111\} \to [111]$$

and

$$\{210\} \to [210] + [100]$$

The classification under both schemes may be summarized in the following table where the seniority number v is used to label the representations of Sp_{14}:

U_{14}	$SU_2 \times U_7$	v	$SU_2 \times R_7$	$SU_2 \times G_2$	$SU_2 \times R_3$
$\{1^3\}$	${}^4\{111\}$	3	${}^4[111]$	${}^4(00)$	${}^4S$
				${}^4(10)$	${}^4F$
				${}^4(20)$	${}^4DGI$
	${}^2\{210\}$	3	${}^2[210]$	${}^2(11)$	${}^2PH$
				${}^2(20)$	${}^2DGI$
				${}^2(21)$	${}^2DFGHKL$
		1	${}^2[100]$	${}^2(10)$	${}^2F$

No difficulties arise in classifying the terms of other f^n configurations. We note that since under $G_2 \to R_3$

$$(31) \to PDF_2GH_2I_2K_2LMNO$$

and

$$(40) \to SDFG_2HI_2KL_2MNQ$$

there will be some terms that occur twice with the same symmetry symbols. The arbitrary separation of these doubly occurring terms is made by introducing an auxiliary label τ. If the irreducible representations of R_7 are symbolized by W and those of G_2 by U then the terms of f^n may be designated by the sequence

$$f^n \tau v WUSL \tag{148}$$

The complete list of terms for the f^n configurations have been given elsewhere[90] and those of d^n in a paper by Smith and Wybourne.[46]

Wybourne[91] has considered the classification of the Russell-Saunders terms of the g^n configurations using the chain of groups

$$U_{18} \to Sp_{18} \to SU_2 \times (R_9 \to R_3) \tag{149}$$

In this case the above chain of groups is completely inadequate in supplying an adequate set of classificatory symbols. Thus in the g^7 configuration there are 26 ${}^2K$ terms possessing the same symmetry symbols.

8.6 *jj*-COUPLING

In this coupling scheme the representation spaces of the orbital and spin functions are taken to be coupled and so the product $[\frac{1}{2}][l]$ in the plethysm $([\frac{1}{2}][l]) \otimes \{1^n\}$ giving the n-particle classification can be rewritten as the sum of two spin representations of the rotation group where

$$[\tfrac{1}{2}][l] = [l + \tfrac{1}{2}] + [l - \tfrac{1}{2}] \tag{150}$$

The plethysm giving the classification for the n-particle configuration can then be expanded using the formula for the sum of two functions defined on the same variables, Eq. 94, to give

$$\begin{aligned}([\tfrac{1}{2}][l]) \otimes \{1^n\} &= ([l + \tfrac{1}{2}] + [l - \tfrac{1}{2}]) \otimes \{1^n\} \\ &= \sum_{\alpha=0}^{n} ([l + \tfrac{1}{2}] \otimes \{1^\alpha\})([l - \tfrac{1}{2}] \otimes \{1^{n-\alpha}\})\end{aligned} \tag{151}$$

Equation 151 has a simple physical interpretation. The states of the n-particles configuration span the $\{1^n\}$ antisymmetric representation of U_{4l+2}. This space may be divided into two distinct spaces. In the first space (A space) the electrons all have a total angular momentum $j_+ = l + \frac{1}{2}$ while in the second space (B space) the electrons all have a total angular momentum $j_- = l - \frac{1}{2}$. The α electrons in the A space span the antisymmetric representation $\{1^\alpha\}$ of the unitary group $U^A_{2j_++1}$ while the $n - \alpha$ electrons in the B space span the antisymmetric representation $\{1^{n-\alpha}\}$ of the unitary group $U^B_{2j_-+1}$. Thus the plethysm of Eq. 151 amounts to making the reduction

$$U_{4l+2} \to U^A_{2j_++1} \times U^B_{2j_-+1} = U^A_{2l+2} \times U_{2l}{}^B \tag{152}$$

The eigenfunctions for the n-particle configuration are obtained by the conventional coupling of the angular momentum of the states in the two spaces A and B. We need only ensure that the Pauli exclusion principle is satisfied in the two spaces separately to be assured that the result eigenfunctions are totally antisymmetric. An additional classification of the states within the two spaces may be made by application of the reductions

$$U^A_{2l+2} \to Sp^A_{2l+2} \to R_3{}^A \tag{153a}$$

and

$$U_{2l}{}^B \to Sp_{2l}{}^B \to R_3{}^B \tag{153b}$$

The states of total angular momentum J are then found from the reduction

$$R_3{}^A \times R_3{}^B \to R_3{}^{AB} \tag{154}$$

The resultant states for the n-particle configuration may then be labeled by the sequence

$$(j_+)^\alpha \langle\sigma\rangle^A J^A (j_-)^{n-\alpha} \langle\sigma\rangle^B J^B;\ J^{AB} \tag{155}$$

Each choice of α ($\alpha \leq n$) could be said to define a particular *jj*-coupled configuration. We note that the states arising in j^{α} and $j^{2j+1-\alpha}$ are equivalent.

The relevant decompositions for Eqs. 153*a* and 153*b* may be obtained using Eq. 80 and 132 and noting that under $Sp_{2j+1} \rightarrow R_3$, $\langle 1 \rangle \rightarrow [j]$. The appropriate decompositions giving states of the $(\frac{5}{2})^n$ and $(\frac{7}{2})^n$ configurations are given in the table below:

U_6	Sp_6	R_3	U_8	Sp_8	R_3
$\{0\}$	$\langle 0 \rangle$	$[0]$	$\{0\}$	$\langle 0 \rangle$	$[0]$
$\{1\}$	$\langle 1 \rangle$	$[\frac{5}{2}]$	$\{1\}$	$\langle 1 \rangle$	$[\frac{3}{2}]$
$\{1^2\}$	$\langle 1^2 \rangle$	$[2], [4]$	$\{1^2\}$	$\langle 1^2 \rangle$	$[2], [4], [6]$
	$\langle 0 \rangle$	$[0]$		$\langle 0 \rangle$	$[0]$
$\{1^3\}$	$\langle 1^3 \rangle$	$[\frac{3}{2}][\frac{9}{2}]$	$\{1^3\}$	$\langle 1^3 \rangle$	$[\frac{3}{2}][\frac{5}{2}][\frac{9}{2}][\frac{11}{2}][\frac{15}{2}]$
	$\langle 1 \rangle$	$[\frac{5}{2}]$		$\langle 1 \rangle$	$[\frac{7}{2}]$
			$\{1^4\}$	$\langle 1^4 \rangle$	$[2], [4], [5], [8]$
				$\langle 1^2 \rangle$	$[2], [4], [6]$
				$\langle 0 \rangle$	$[0]$

We note that the J states of the $(\frac{5}{2})^n$ and $(\frac{7}{2})^n$ configurations are uniquely classified and hence the labels of Eq. 155 give an unambiguous classification of all the states of the $(\frac{5}{2})^{\alpha}(\frac{7}{2})^{n-\alpha}$ configurations that arise in f^n. It may also be shown[91] that in the configurations $(\frac{9}{2})^n$ no J states occur more than twice and hence if the g^n configuration is represented in terms of the *jj*-coupled configurations $(\frac{7}{2})^{\alpha}(\frac{9}{2})^{n-\alpha}$ no terms in Eq. 155 will be repeated with the same symmetry symbols more than twice.

Thus *jj*-coupling gives a much richer classification of the states arising from n equivalent electrons. We note as a consequent that the tables of coefficients of fractional parentage, which have been given in a book by de Shalit and Talmi[92] for $j \leq \frac{9}{2}$, are quite restricted in size. Computations for atoms having several electrons in the $5g^n$ (atomic number $Z \sim 126$) would clearly be simpler in the *jj*-coupling scheme than in the *LS*-coupling scheme though of course in the former the electrostatic interactions would no longer be in the block-diagonal form. However, as a compensation, the spin-orbit interaction would now be in block-diagonal form and of course *jj*-coupling forms the natural basis for the study of relativistic effects via Dirac's equations.

8.7 *LL*-COUPLING

A third type of coupling for equivalent electrons was suggested by Shudeman[93] and recently revived by Judd.[94] Its formal description is closely related to the case of *jj*-coupling just discussed. Two spaces are considered, one (the

A space) in which all the spins are "up" and the other (the B space) in which they are all "down." This means that the product $[\frac{1}{2}][l]$ is rewritten as $([l]^A + [l]^B)$ where $[l]^A$ corresponds to the orbital functions in "*spin up space*" and $[l]^B$ to the set in "*spin down space*." An operation of the spin space as a whole, in this coupling scheme, must be interpreted as an operation which exchanges functions between "*spin up*" and "*spin down*" spaces.

The wavefunctions will span the representation $([l]^A + [l]^B)$ of the imprimitive rotation group in two sets of variables. An imprimitive group being a group which contains all the operations of a group G on two sets of variables together with an operation which exchanges these two sets.[95]

The plethysm appropriate to the n-particle configuration may be expanded to yield

$$([l]^A + [l]^B) \otimes \{1^n\} = \sum_{\alpha=0}^{n} ([l]^A \otimes \{1^\alpha\})([l]^B \otimes \{1^{n-\alpha}\}) \tag{156}$$

The S-functions $\{1^n\}$, $\{1^\alpha\}$, and $\{1^{n-\alpha}\}$ may be taken as labeling antisymmetric representations of the unitary groups U_{4l+2}, U^A_{2l+1}, and U^B_{2l+1}, respectively, where

$$U_{4l+2} \rightarrow U^A_{2l+1} \times U^B_{2l+1} \tag{157}$$

Clearly no additional classification is possible by making the restriction

$$U^A_{2l+1} \times U^B_{2l+1} \rightarrow R^A_{2l+1} \times R^B_{2l+1}$$

since the antisymmetric representations of unitary groups remain irreducible upon restriction to the rotation group in the same number of dimensions. In fact a complete description of the terms of the f^n configurations may be obtained by simply considering the chain of groups

$$U_{4l+2} \rightarrow U^A_{2l+1} \times U^B_{2l+1} \rightarrow R_3{}^A \times R_3{}^B \rightarrow R_3 \tag{158}$$

and using as labels the sequence of quantum numbers

$$\{1^\alpha\}L^A\{1^{n-\alpha}\}L^B; LM_L \tag{159}$$

or equivalently

$$l^n L^A L^B; LM_L M_S \tag{160}$$

since $M_S = \frac{1}{2}n - \alpha$.

As in the case of *jj*-coupling we find that in *LL*-coupling no terms of g^n occur more than twice. Judd[94] has shown that in certain calculations the use of the *LL*-coupling scheme results in profound simplifications. The purity of a state in *LL*-coupling can be taken as an indication of the physical attainment of this scheme in actual spectra. Clearly the approximation of *LL*-coupling will only be realized when the operations of R_3 in the total spin space can be restricted to two operations, the identity and a 180° rotation about a line in the xy-plane. Thus a pure state in *LL*-coupling results from the restriction of the operations in the spin space to those of this simple finite group.

9

Classification of Atomic States for Mixed Configurations

9.1 MIXED CONFIGURATIONS

So far we have only considered the classification of the states of equivalent electron configurations of the generic type l^n. We now consider the more general case of configurations of the type $l_1^{n_1}l_2^{n_2} \cdots$ where the electrons occur in several distinct one electron orbitals $l_1, l_2 \cdots$. Plethysm provides a convenient method for classifying the states of these so called mixed configurations. In fact it would appear that the first application of plethysm in physics was in J. P. Elliott's treatment[9] of the $(d + s)^n$ configurations of nucleons. The particular cases of the mixed electron configurations of the type $(l_1 + l_2)^n$ and $(d + s)^n$ have been discussed in considerable detail by Feneuille[96–99] using conventional methods. We shall commence our discussion by considering some of the properties of configurations of the generic type $(l_1 + l_2 + l_3)^n$. Configurations of this type are of sufficient complexity to exhibit the principle features of the problems that arise in the group theoretical treatment of mixed configurations.[99a]

9.2 THE BASIC GROUP FOR $(l_1 + l_2 + l_3)^n$ CONFIGURATIONS

We understand $(l_1 + l_2 + l_3)^n$ to stand for the set of configurations that arise in the distribution of n electrons among the three orbitals l_1, l_2, and l_3

in all permissible ways and with no orbital l being occupied by more than $4l + 2$ electrons. The eigenfunctions of $(l_1 + l_2 + l_3)^n$ must span the antisymmetric representation $\{1^n\}$ of the unitary group $U_{4(l_1+l_2+l_3+\frac{3}{2})}$. The eigenfunctions of the n-particle system may then be classified according to their transformation properties under the various subgroups of the basic group $U_{4(l_1+l_2+l_3+\frac{3}{2})}$. The basic group contains a large number of subgroups and the particular set chosen for the classification must be chosen either on physical considerations or for computational simplicity.

Of the many chains of groups available the chain

$$U_{4(l_1+l_2+l_3+\frac{3}{2})} \rightarrow Sp_{4(l_1+l_2+l_3+\frac{3}{2})} \rightarrow SU_2 \times R_{2(l_1+l_2+l_3+\frac{3}{2})} \tag{161}$$

forms an obvious starting point. The relevant decompositions can be readily accomplished using Eqs. 80, 128, and 132. Clearly the chain of groups in Eq. 161 has the considerable advantage of introducing the symplectic symmetry, and hence the seniority number, in a natural way. As we shall see this advantage is sometimes offset by several serious disadvantages.

9.3 REDUCTION OF THE $R_{2(l_1+l_2+l_3+\frac{3}{2})}$ SYMMETRY

Having chosen the chain of groups given in Eq. 161 it remains to consider the possible reductions of the $R_{2(l_1+l_2+l_3+\frac{3}{2})}$ symmetry. The choice of reducing the symmetry depends on the particular choice of the order of coupling the states of the composite equivalent electron configurations $l_1^{n_1}$, $l_2^{n_2}$, $l_3^{n_3}$. One obvious choice would be

$$R_{2(l_1+l_2+l_3+\frac{3}{2})} \rightarrow R_{2l_1+1} \times R_{2(l_2+l_3+1)} \tag{162}$$

where we have clearly chosen to treat the configurations $l_2^{n_2}l_3^{n_3}$, collectively, and then coupled the resultant states to the l^{n_1} configuration. Alternative choices would be

$$R_{2(l_1+l_2+l_3+\frac{3}{2})} \rightarrow R_{2(l_1+l_2+1)} \times R_{2l_3+1} \tag{163}$$

and

$$R_{2(l_1+l_2+l_3+\frac{3}{2})} \rightarrow R_{2(l_1+l_3+1)} \times R_{2l_2+1} \tag{164}$$

Here we note that the different possible schemes arising in Eqs. 162–164 differ by their order of coupling and we could define transformation coefficients analogous to the Wigner $6j$-symbols.[100] We shall not, however, pursue this line of thought further.

Let us assume Eq. 162 as giving the desired symmetry reduction. The irreducible representations $[\lambda]$ of $R_{2(l_1+l_2+l_3+\frac{3}{2})}$ may be decomposed into those of the direct product $R_{2l_1+1} \times R_{2(l_2+l_3+1)}$ by noting that under the reduction of Eq. 162 we have

$$[1] \rightarrow [1][0]' + [0][1]' \tag{165}$$

where we use a dash to distinguish the symmetry symbols of $R_{2(l_2+l_3+1)}$ from those of R_{2l_1+1}. Thus $[\lambda]$ decomposes as

$$([1][0]' + [0][1]') \otimes [\lambda] \tag{166}$$

Clearly any subgroups of R_{2l_1+1} may be used to extend the classification of the $l_1^{n_1}$ states.

The group $R_{2(l_2+l_3+1)}$ may be further reduced by the following restrictions

$$R_{2(l_2+l_3+1)} \to R_{2l_2+1} \times R_{2l_3+1} \to R_3^{l_2} \times R_3^{l_3} \to R_3 \tag{167}$$

where the superscripts distinguish the three-dimensional rotation groups involved in the direct product $R_3 \times R_3$. Again, subgroups of R_{2l_2+1} and R_{2l_3+1} may be used to extend the classification if necessary.

9.4 EXAMPLE OF $(d + s + s')^n$

The configurations $(d + s + s')^n$ form one of the simplest examples of the group-theoretical description of a system involving the occupation of three inequivalent orbitals. Furthermore, spectra involving these configurations occur in the $3d$, $4d$, and $5d$ transition ions. The basic symmetry group will be U_{14} and noting Eqs. 161, 162, and 167 we have the chain of groups

$$U_{14} \to Sp_{14} \to SU_2 \times (R_7 \to R_6 \to R_5 \to R_3) \tag{168}$$

to consider. The necessary branching rules can be obtained using the methods of Section 7.7. In particular we have the decompositions

$Sp_{14} \to SU_2 \times R_7$

$\langle 0\rangle$ $^1[000]$

$\langle 1\rangle$ $^2[100]$

$\langle 1^2\rangle$ $^3[110] + {}^1[200]$

$\langle 1^3\rangle$ $^4[111] + {}^2[210]$

$\langle 1^4\rangle$ $^5[111] + {}^3[211] + {}^1[220]$

$\langle 1^5\rangle$ $^6[110] + {}^4[211] + {}^2[221]$

$\langle 1^6\rangle$ $^7[100] + {}^5[210] + {}^3[221] + {}^1[222]$

$\langle 1^7\rangle$ $^8[000] + {}^6[200] + {}^4[220] + {}^2[222]$

$R_7 \to R_6$

[000] [000]

[100] [100] + [000]

[110] [110] + [100]

$R_6 \to R_5$

[000] [00]

[100] [10] + [00]

[110] [11] + [10]

[111]	$[11 \pm 1] + [110]$	[111]	[11]
[200]	$[200] + [100] + [000]$	[200]	$[20] + [10] + [00]$
[210]	$[210] + [200] + [110] + [100]$	[210]	$[21] + [20] + [11] + [10]$
[211]	$[21 \pm 1] + [11 \pm 1] + [210] + [110]$	[211]	$[21] + [11]$
[220]	$[220] + [210] + [200]$	[220]	$[22] + [21] + [20]$
[221]	$[22 \pm 1] + [21 \pm 1] + [220] + [210]$	[221]	$[22] + [21]$
[222]	$[22 \pm 2] + [22 \pm 1] + [220]$	[222]	[22]

where we note that under $R_6 \rightarrow R_5$ the decompositions of $[\lambda_1\lambda_2\lambda_3]$ and $[\lambda_1\lambda_2\text{-}\lambda_3]$ are identical, a result which follows from Eqs. 123–125.

With the branching rules established, it is a simple matter to enumerate the states of $(d + s + s')^n$ for any value of n. We tabulate below the terms that arise in $(d + s + s')^3$ which occurs, for example in the $(5d + 6s + 7s)^3$ configurations of La I

U_{14}	Sp_{14}	$SU_2 \times R_7$	$SU_2 \times R_6$	$SU_2 \times R_5$	$SU_2 \times R_3$
$\{1^3\}$	$\langle 1^3 \rangle$	$^4[111]$	$^4[111]$	$^4[11]$	^4PF
			$^4[11-1]$	$^4[11]$	^4PF
			$^4[110]$	$^4[11]$	^4PF
				$^4[10]$	^4D
		$^2[210]$	$^2[210]$	$^2[21]$	^2PDFGH
				$^2[20]$	^2DG
				$^2[11]$	^2PF
				$^2[10]$	^2D
			$^2[200]$	$^2[20]$	^2DG
				$^2[10]$	^2D
				$^2[00]$	^2S
			$^2[110]$	$^2[11]$	^2PF
				$^2[10]$	^2D
			$^2[100]$	$^2[10]$	^2D
				$^2[00]$	^2S
	$\langle 1 \rangle$	$^2[100]$	$^2[100]$	$^2[10]$	^2D
				$^2[00]$	^2S
			$^2[000]$	$^2[00]$	^2S

9.5 EXAMPLE OF $(d + s)^n$

Whereas the classification of the $(d + s + s')^n$ configurations tends to be an academic exercise with doubtful application to real spectra the configurations $(d + s)^n$ undoubtedly form an important conglomerate of configurations in the first spectra of the iron group where the $3d^n$, $3d^{n-1}4s$, and $3d^{n-2}4s^2$ configurations are almost degenerate. In this case the appropriate chain of groups is

$$U_{12} \rightarrow Sp_{12} \rightarrow SU_2 \times (R_6 \rightarrow R_5 \rightarrow R_3) \tag{169}$$

The classification procedure is entirely analogous to the previous example and we list the terms for $(d + s)^3$ below. We note, of course, that the terms obtained above are simply a subset of those found for $(d + s + s')^n$.

U_{12}	Sp_{12}	$SU_2 \times R_6$	$SU_2 \times R_5$	$SU_2 \times R_3$
$\{1^3\}$	$\langle 1^3 \rangle$	$^4[111]$	$^4[11]$	^4PF
		$^4[11-1]$	$^4[11]$	^4PF
		$^2[210]$	$^2[21]$	^2PDFGH
			$^2[20]$	^2DG
			$^2[11]$	^2PF
			$^2[10]$	^2D
	$\langle 1 \rangle$	$^2[100]$	$^2[10]$	^2D
			$^2[00]$	^2S

9.6 PARITY AND MIXED CONFIGURATIONS

So far we have been careful to choose examples where the parity of all electron orbitals are the same. In this way we have guaranteed that all configurations in $(l_1 + l_2 + l_3)^n$ are of the same parity. Unfortunately, this is seldom the case. Two particularly important series of configurations $(d + p + s)^n$ and $(f + d + s)^n$ are found in the d-transition groups and in the f-transition groups, respectively. Clearly for any choice of n configurations of *even* and *odd* parities will occur and it would be unfortunate to treat these configurations as a single entity and to lose sight of the parity classification of the configurations which we know to be a good quantum number for all atomic interactions.

Thus the possibility of classifying the terms of $(d + p + s)^n$ under the chain of groups

$$U_{18} \rightarrow Sp_{18} \rightarrow SU_2 \times (R_9 \rightarrow R_5 \times R_4 \rightarrow R_3 \rightarrow R_3) \tag{170}$$

or of $(f + d + s)^n$ under the chain of groups

$$U_{26} \rightarrow Sp_{26} \rightarrow SU_2 \times (R_{13} \rightarrow R_7 \times R_6 \rightarrow G_2 \times R_5 \rightarrow R_3 \times R_3 \rightarrow R_3) \tag{171}$$

are of dubious value. A further objection to these schemes can be raised on physical grounds. The above classification scheme does not keep track of the number of electrons in the respective electron shells and while we may be interested in configurations of the types $f^n, f^{n-1}s, f^{n-1}d, f^{n-2}s^2, f^{n-2}s^2, f^{n-2}d^2$, and $f^{n-2}ds$ we are most unlikely to be interested in configurations of the type $f^{n-6}d^5s$. This suggests that alternative schemes should be investigated.

Since we may classify the states of composite configurations of the same parity collectively we might consider decomposing the unitary space $U_{4(l_1+l_2+l_3+\frac{3}{2})}$ into the direct product of two subspaces of opposite parity and keep track of the number of particles in the two subspaces. Thus if l_2 and l_3 have the same parity and opposite parity to l_1 we could make the division

$$U_{4(l_1+l_2+l_3+\frac{3}{2})} \to U_{4l_1+2} \times U_{4(l_2+l_3+1)} \tag{172}$$

This amounts to performing the plethysm

$$[\tfrac{1}{2}]'[[l_1] + ([l_2] + [l_3])] \otimes \{1^n\}$$

$$= \sum_{\alpha=0}^{n} [[\tfrac{1}{2}]'[l_1] \otimes \{1^\alpha\}][[\tfrac{1}{2}]'([l_2] + [l_3])] \otimes \{1^{n-\alpha}\} \tag{173}$$

where $\{1^\alpha\}$ labels the antisymmetric representations of U_{4l_1+2} and $\{1^{n-\alpha}\}$ those of $U_{4(l_2+l_3+1)}$. Each choice of the integer α amounts to specifying a particular mixed configuration $l_1^\alpha(l_2 + l_3)^{n-\alpha}$ whose terms are labeled by the symmetry symbols associated with the separation of the plethysm

$$[[\tfrac{1}{2}]'[l_1] \otimes \{1^\alpha\}][[\tfrac{1}{2}]'([l_2] + [l_3]) \otimes \{1^{n-\alpha}\}] \tag{174}$$

However, it is clear that the separation of the plethysm of Eq. 174 is simply equivalent to classifying the states of two configurations l_1^α and $(l_2 + l_3)^{n-\alpha}$ separately and then coupling the resultant states by conventional angular momentum theory. Schemes involving four inequivalent electron orbitals like $[(f + p) + (d + s)]^n$ are much richer in possibilities of classification but probably lacking in physical significance since in most cases the p and s orbitals will interact more strongly than the other orbitals and thus any semblance of a well-defined coupling scheme will be lost.

The $(d + s)^n$ configurations appear to be rich in their potential for a physically appropriate group-theoretical description and possibly constitute the only suitable candidate for an extensive treatment of an example of a mixed configuration by the methods we have outlined. The $(g + d + s)^n$ configurations is another obvious system but it would not be encountered in spectra until atoms of atomic number $Z \sim 118$ are reached, when, for the first time, g orbitals would be expected to occur in the ground state configurations. In the case of $(d + s)^n$ the chain of groups $U_{12} \to SU_2 \times [SU_6 \to SU_3 \to R_3]$ could form a useful alternative to Eq. 169.

Some advantage might be gained in the $(f + d + s)^n$ configurations that arise in the rare earths and actinides by treating the $(d + s)^{n-\alpha}$ configurations as a single entity and then coupling the resultant states onto the f^α core using the plethysm of Eq. 174 as the basis for the classification. For example, the configurations $fds^2 + fd^2s + fd^3$ which occur in Ce I could be treated in this way by using the representation $\{1\} \times \{1^3\}$ of $U_{14} \times U_{12}$ and then using the results obtained for $(d + s)^3$ to obtain the decomposition of the $\{1^3\}$ representation of U_{12}.

10

Symmetry Properties of Interactions in Equivalent Electron Configurations

10.1 TRANSFORMATION PROPERTIES OF SINGLE-PARTICLE OPERATORS

Having made use of symmetry principles to classify the states of a configuration we have then to consider the symmetry properties of actual atomic interactions acting on the antisymmetrized states. In what follows we shall attempt to extract the maximum information concerning the properties of the matrix elements of various atomic interactions from purely symmetry considerations. The fundamental starting point is Racah's observation[101] that all atomic interactions can be expressed in terms of tensor operators. Since that time the theory of tensor operators has been the subject of a number of treatises[15,102–109] and as a result we shall only sketch the necessary details, following the notation of Judd.[15] We shall be interested in the symmetry properties of tensor operators rather than the enumeration of the matrix elements of the operators.

We define a double tensor $\mathbf{T}^{(\kappa k)}$ as having $(2\kappa + 1)(2k + 1)$ components $T^{(\kappa k)}_{\pi q}$ and acting as a tensor of rank κ in spin space and as a tensor of rank k in orbital space and satisfying the commutation relations,

$$\begin{aligned}
[L_z, T^{(\kappa k)}_{\pi q}] &= qT^{(\kappa k)}_{\pi q} \\
[L_\pm, T^{(\kappa k)}_{\pi q}] &= [k(k+1) - q(q \pm 1)]^{\frac{1}{2}} T^{(\kappa k)}_{\pi q\pm 1} \\
[S_z, T^{(\kappa k)}_{\pi q}] &= \pi T^{(\kappa k)}_{\pi q} \\
[S_\pm, T^{(\kappa k)}_{\pi q}] &= [\kappa(\kappa+1) - \pi(\pi \pm 1)]^{\frac{1}{2}} T^{(\kappa k)}_{\pi\pm 1 q}
\end{aligned} \tag{175}$$

It is convenient to define a one-particle double tensor $\mathbf{w}^{(\kappa k)}$ whose reduced matrix element is defined by the relation[15]

$$\langle sl \| w^{(\kappa k)} \| s'l' \rangle = \{[\kappa, k]\}^{\frac{1}{2}} \delta(s, s')(l, l') \tag{176}$$

where we make use of the convention of representing quantities of the form $(2a + 1)(2b + 1) \cdots$ as $[a, b, \cdots]$. Many electron operators $\mathbf{W}^{(\kappa k)}$ may be defined in terms of the single-particle operators $\mathbf{w}^{(\kappa k)}$ by writing

$$\mathbf{W}_{\pi q}^{(\kappa k)} = \sum_i (\mathbf{w}_{\pi q}^{(\kappa k)})_i \tag{177}$$

where the summation is over the coordinates of all the electrons. We note that the operators $\mathbf{w}^{(\kappa k)}$ have been defined in Eq. 176 so as to have nonvanishing matrix elements when operating within a configuration of equivalent electrons of the generic type l^n.

The $(4l + 2)^2$ components of the double tensor operators $\mathbf{W}^{(\kappa k)}$ form the generators of the group U_{4l+2}. Since the n-electron states are necessarily antisymmetric the matrix elements of the $W_{\pi q}^{(\kappa k)}$ will vanish unless they transform according to one or more of the representations $\{\lambda\}$ of U_{4l+2} contained in the product

$$\{1^n\}\{1^n\}^\dagger \equiv \{1^n\}\{1^{4l+2-n}\} \tag{178}$$

where $\{1^n\}^\dagger$ is the representation of U_{4l+2} adjoint to that of $\{1^n\}$. Ordinary S-function multiplication leads to the result

$$\{1^n\}\{1^n\}^\dagger \to \sum_{a=0}^{n} \{2^a 1^{4l+2-n-2a}\} \tag{179}$$

or in the simple case of $n = 1$

$$\{1\}\{1^{4l+2-n-1}\} \to \{21^{4l}0\} + \{0\} \tag{180}$$

The single particle operator $w_{00}^{(00)}$ is a complete scalar and transforms under U_{4l+2} as the $\{0\}$ representation while the remaining $(4l + 2)^2 - 1$ components $w_{\pi q}^{(\kappa k)}$ transform together as the $\{21^{4l}0\}$ representation. It follows from Eq. 177 that the $W_{\pi q}^{(\kappa k)}$ operators will transform in an equivalent fashion.[99a]

The transformation properties of the tensor operators $\mathbf{W}^{(\kappa k)}$ may be further investigated by reducing the symmetry of the group U_{4l+2} in the same way as used to classify the states of l^2. Thus under the reduction $U_{4l+2} \to Sp_{4l+2}$ we have

$$\{21^{4l}0\} \to \langle 2 \rangle + \langle 1^2 \rangle \quad \text{and} \quad \{0\} \to \langle 0 \rangle \tag{181}$$

while under the reduction $Sp_{4l+2} \to SU_2 \times R_{2l+1}$ we have

$$\langle 1^2 \rangle \to {}^1[2] + {}^3[1^2]; \langle 0 \rangle \to {}^1[0]; \langle 2 \rangle \to {}^3[2] + {}^1[1^2] + {}^3[0] \tag{182}$$

Furthermore, under the reduction $R_{2l+1} \rightarrow R_3$ we have

$$[2] \rightarrow 2, 4, \ldots, 2l; \; [0] \rightarrow 0; \; [1^2] \rightarrow 1, 3, \ldots, 2l-1 \tag{183}$$

where we have omitted the brackets [ ] in giving the representations of R_3. Thus the transformation properties of the double tensors $\mathbf{W}^{(\kappa k)}$ may be described by the following table:

U_{4l+2}	Sp_{4l+2}	$SU_2 \times R_{2l+1}$	$SU_2 \times R_3$
$\{21^{4l}0\}$	$\langle 2 \rangle$	$^3[2]$	$\mathbf{W}^{(12)}, \mathbf{W}^{(14)}, \ldots, \mathbf{W}^{(1,2l)}$
		$^3[0]$	$\mathbf{W}^{(10)}$
		$^1[1^2]$	$\mathbf{W}^{(01)}, \mathbf{W}^{(03)}, \ldots, \mathbf{W}^{(0,2l-1)}$
	$\langle 1^2 \rangle$	$^3[1^2]$	$\mathbf{W}^{(11)}, \mathbf{W}^{(13)}, \ldots, \mathbf{W}^{(1,2l-1)}$
		$^1[2]$	$\mathbf{W}^{(02)}, \mathbf{W}^{(04)}, \ldots, \mathbf{W}^{(0,2l)}$
$\{0\}$	$\langle 0 \rangle$	$^1[0]$	$\mathbf{W}^{(00)}$

In the particular case of f electrons where the group G_2 is used to classify the states of the f^n configurations a further classification of the $\mathbf{W}^{(\kappa k)}$operators is possible. Proceeding as above we readily find that the four sets of tensors

$$\mathbf{W}^{(\kappa 2)}, \mathbf{W}^{(\kappa 4)}, \mathbf{W}^{(\kappa 6)}$$
$$\mathbf{W}^{(\kappa 1)}, \mathbf{W}^{(\kappa 5)}$$
$$\mathbf{W}^{(\kappa 3)}$$
$$\mathbf{W}^{(00)}$$

for a given component π form the components of tensors transforming as the (20), (11), (10), and (00) representations of G_2, respectively.

10.2 QUASI-SPIN CLASSIFICATION OF OPERATORS

Until quite recently it was thought that there were no groups higher than U_{4l+2} that could be usefully employed in the symmetry description of states and operators in atomic spectroscopy. Nevertheless numerous relationships were found to exist between the matrix elements of operators acting in configurations containing a different number of equivalent particles. Furthermore, these relationships always seemed to be intimately connected with seniority number v and the number of particles. While explanations of these relationships did exist their demonstration was lacking in generality and transparency.

Since these relationships depend on the number of particles we would suspect that their explanation must be connected with the properties of a higher group that effects transformations among the 2^{4l+2} states of the l-shell. Clearly the 2^{4l+2} states of the l-shell will span the $\{1\}$ representation of the unitary group in 2^{4l+2} dimensions but this group is obviously too all embracing and we must search for a suitable subgroup which in turn contains U_{4l+2}, and more specifically Sp_{4l+2}, as a subgroup.

Judd[85,87,88] made a successful search for this subgroup by considering the commutation properties of the annihilation and creation operators and forming a set that was closed under commutation. These operators were then identified as the generators of the rotation group in $8l + 5$ dimensions R_{8l+5}. We shall leave the reader to refer to Judd's article[88] for explicit details. Judd then showed that the states that belong to the l-shell form a basis for the basic spin representation $[\frac{1}{2}\,\frac{1}{2}\cdots\frac{1}{2}]$ of R_{8l+5}. Upon restriction of R_{8l+5} to its subgroup R_{8l+4} we have the obvious branching rule

$$[\tfrac{1}{2}\,\tfrac{1}{2}\cdots\tfrac{1}{2}] \rightarrow [\tfrac{1}{2}\,\tfrac{1}{2}\cdots\tfrac{1}{2}] + [\tfrac{1}{2}\,\tfrac{1}{2}\cdots-\tfrac{1}{2}] \tag{184}$$

Judd then showed that the states of l^n with n *even* formed a basis for the $[\frac{1}{2}\,\frac{1}{2}\cdots\frac{1}{2}]$ representation of R_{8l+4} while those with n *odd* formed a basis for the $[\frac{1}{2}\,\frac{1}{2}\cdots-\frac{1}{2}]$ representation.

Labeling the states of l^n by the representations of Eq. 184 gives little new information. However, Judd noticed that it was possible to select from the operators that generate the group R_{8l+4} a subset of operators that generated the subgroup $SU_2 \times Sp_{4l+2}$ thereby establishing the connection with symplectic symmetry and the higher group R_{8l+5}. Thus we now have at our disposal the chain of groups

$$R_{8l+5} \rightarrow R_{8l+4} \rightarrow SU_2 \times (Sp_{4l+2} \rightarrow SU_2 \times R_{2l+1} \rightarrow SU_2 \times R_3 \rightarrow R_3) \tag{185}$$

and of course in the case of f-electrons the exceptional group G_2.

The group SU_2 which arises in the restriction $Sp_{4l+2} \rightarrow SU_2 \times R_{2l+1}$ is well known in the labeling of the spin eigenfunctions of multielectron systems. In this case the eigenfunctions are labeled by the spin quantum numbers S and M_s where the spin angular momentum operators S_+, S_-, and S_z satisfy the well-known commutation properties

$$[S_+, S_-] = 2S_z \quad \text{and} \quad [S_z, S_\pm] = \pm S_\pm \tag{186}$$

The group SU_2 which arises in the restriction $R_{8l+4} \rightarrow SU_2 \times Sp_{4l+2}$ may likewise be used to label the eigenfunctions of the states of l^n. Judd,[88] following the earlier work of Flowers and Szpikowski,[110] and Lawson and Macfarlane,[111] constructed from coupled products of annihilation and creation operators the components, $Q_\pm$ and Q_z, of a quantity $\mathbf{Q}$ known as the *quasi-spin*. The components of $\mathbf{Q}$ satisfy commutation relations identical to those of the spin angular momentum given in Eq. 186. The eigenvalues Q and M_Q of $\mathbf{Q}$ and Q_z are available as quantum numbers for labeling the states of the l^n configuration. In fact for a given number n of equivalent electrons l the quantum numbers Q and M_Q are simply given by

$$Q = (2l + 1 - v)/2 \quad \text{and} \quad M_Q = -(2l + 1 - n)/2 \tag{187}$$

where v is the seniority number. Again the reader is referred to Judd's works for extensive details. The concept of quasi-spin may be readily extended to mixed configurations as noted, for example, by Feneuille.[96–99]

The concept of quasi-spin is of application not only in describing the states of multielectron configurations but more importantly in describing the properties of operators acting on states classified in the quasi-spin scheme. Since the states of a configuration l^n transform according to either the representation $[\frac{1}{2}\,\frac{1}{2}\cdots\frac{1}{2}]$ (for *n even*) of R_{8l+4} or the representation $[\frac{1}{2}\,\frac{1}{2}\cdots -\frac{1}{2}]$ (for *n odd*) operators acting on these states must transform according to the irreducible representations of R_{8l+4} arising out of the decomposition of the Kronecker squares of these representations; i.e., they must transform as the right-hand side of

$$[\tfrac{1}{2}\,\tfrac{1}{2}\cdots\pm\tfrac{1}{2}]\times[\tfrac{1}{2}\,\tfrac{1}{2}\cdots\pm\tfrac{1}{2}]=[0\cdots 0]+[110\cdots 0]+[11110\cdots 0] + [11\cdots 1\pm 1] \quad (188)$$

where the upper sign or the lower sign is taken throughout. The above result is a trivial extension of our earlier discussion of spin representations. The irreducible representations appearing in Eq. 188 may be rapidly decomposed into those of $SU_2\times Sp_{4l+2}$ by use of Eqs. 127 and 132 to give the typical reduction (for $l\geq 2$),

$$\begin{array}{ll}
R_{8l+4} \rightarrow & SU_2\times Sp_{4l+2} \\
[0\cdots 0] & {}^1\langle 0\rangle \quad (189a) \\
[110\cdots 0] & {}^1\langle 2\rangle + {}^3(\langle 0\rangle + \langle 1^2\rangle) \quad (189b) \\
[11110\cdots 0] & {}^1(\langle 0\rangle + \langle 1^2\rangle + \langle 2^2\rangle) + {}^3(\langle 1^2\rangle + \langle 2\rangle + \langle 21^2\rangle) \\
 & + {}^5(\langle 0\rangle + \langle 1^2\rangle + \langle 1^4\rangle) \quad (189c)
\end{array}$$

the quasi-spin multiplicity $(2K+1)$ being given as a left superscript to the representations of Sp_{4l+2}.

Judd[85] has shown that in the quasi-spin formalism an N-particle operator may be expressed as sums over $2N$-fold products of triple tensor operators $\mathbf{a}^{(qsl)}$ and hence in discussing the properties of N-particle operators with respect to R_{8l+4} symmetry we may restrict our attention to representations of R_{8l+4} of the type $[11\cdots 10\cdots 0]$ having not more than $2N$ symbols 1.

10.3 QUASI-SPIN PROPERTIES OF THE MATRIX ELEMENTS OF SINGLE-PARTICLE OPERATORS

We may now use the results of the preceding section to discuss the quasi-spin properties of the matrix elements of single-particle operators of the form

$\mathbf{W}^{(\kappa k)}$. It is clear from Eq. 189*b* that the operators $\mathbf{W}^{(\kappa k)}$ with $\kappa + k$ *odd*, which have symplectic symmetry $\langle 2 \rangle$, transform like tensors of quasi-spin rank $K = 0$ while the operators $\mathbf{W}^{(\kappa k)}$ with $\kappa + k$ *even* which have symplectic symmetry $\langle 1^2 \rangle$, transform like tensors of quasi-spin rank $K = 1$.

If an operator behaves as a tensor of quasi-spin rank K then the dependence of the matrix elements of the operator acting between a bra and ket of quasi-spin classification $\langle QM_Q|$ and $|Q'M_Q\rangle$ upon the number of equivalent electrons n must be wholly contained in the expression

$$(-1)^{Q-M_Q}\begin{pmatrix} Q & K & Q' \\ -M_Q & 0 & M_Q \end{pmatrix} \tag{190}$$

a result that follows directly from the well-known Wigner-Eckart theorem.[15]

The values of Q, K, and Q' must satisfy the normal triangular condition for nonvanishing $3j$-symbols, i.e.,

$$Q + Q' \geq K \geq |Q - Q'| \tag{191}$$

From this condition we immediately conclude that the matrix elements of operators transforming as quasi-spin tensors of rank $K = 0$ are diagonal with respect to the symplectic symmetry, or equivalently the seniority number v, and are independent of n. Likewise the matrix elements of operators transforming like tensors of quasi-spin rank $K = 1$ will be null unless the seniority selection rule

$$\Delta v = 0, \pm 2 \tag{192}$$

is satisfied.

For the particular case of the half-filled shell $n = 2l + 1$, hence from Eq. 187 $M_Q = 0$. The $3j$-symbol of Eq. 190 will, in this case, vanish unless $Q + K + Q'$ is even from which we may immediately conclude that the matrix elements of operators of quasi-spin $K = 1$ will vanish when $n = 2l + 1$ except for the case $\Delta v = \pm 2$.

Since the n-dependence of the matrix elements of tensor operators of quasi-spin rank K is wholly contained in Eq. 190 we may readily relate the matrix elements of the operators $\mathbf{W}^{(\kappa k)}$, with $\kappa + k$ even, in one configuration to those of another to give in particular the results

$$\frac{\langle l_v{}^n\Theta\|\, W^{(\kappa k)}\,\|l_v{}^n\Theta'\rangle}{\langle l^v\Theta\|\, W^{(\kappa k)}\,\|l^v\Theta'\rangle} = \frac{(-1)^{\frac{1}{2}(n-v)}\begin{pmatrix} \frac{1}{2}(2l+1-v) & 1 & \frac{1}{2}(2l+1-v) \\ \frac{1}{2}(2l+1-n) & 0 & -\frac{1}{2}(2l+1-n) \end{pmatrix}}{\begin{pmatrix} \frac{1}{2}(2l+1-v) & 1 & \frac{1}{2}(2l+1-v) \\ \frac{1}{2}(2l+1-v) & 0 & -\frac{1}{2}(2l+1-v) \end{pmatrix}}$$

$$= \frac{2l+1-n}{2l+1-v} \tag{193}$$

and

$$\frac{\langle l_v{}^n\Theta \| W^{(\kappa k)} \| l_{v-2}^n\Theta'\rangle}{\langle l_v{}^v\Theta \| W^{(\kappa k)} \| l_{v-2}^v\Theta'\rangle} = (-1)^{\frac{1}{2}(n-v)} \frac{\begin{pmatrix} \frac{1}{2}(2l+1-v) & 1 & \frac{1}{2}(2l-1-v) \\ \frac{1}{2}(2l+1-n) & 0 & -\frac{1}{2}(2l+1-n) \end{pmatrix}}{\begin{pmatrix} \frac{1}{2}(2l+1-v) & 1 & \frac{1}{2}(2l-1-v) \\ \frac{1}{2}(2l+1-v) & 0 & -\frac{1}{2}(2l+1-v) \end{pmatrix}}$$

$$= \tfrac{1}{2}[(n+2-v)(4l+4-n-v)/(2l+2-v)]^{\frac{1}{2}} \quad (194)$$

These few examples should suffice to show the advantages that can be obtained when operators correspond to a unique quasi-spin rank K. In cases where an operator does not correspond to a unique quasi-spin rank it may be well worthwhile to decompose it into a set of operators having well-defined quasi-spin ranks.

10.4 SELECTION RULES AND THE MATRIX ELEMENTS OF SINGLE-PARTICLE OPERATORS

We have shown in the preceding section how the introduction of the quasi-spin concept allows us to exploit the properties of symplectic symmetry and to prove without actual numerical calculation the null properties of large numbers of matrix elements. We now consider the application of the general Wigner-Eckart theorem to obtain further selection rules and relationships between sets of matrix elements.

It is always possible for us to decompose an operator **W** into operators **O** which have well-defined transformation properties under some group G which is also used in describing the transformation properties of the manifold of states in which **O** acts. Let us suppose the operator $\mathbf{O}(\gamma_2\Gamma_2\rho_2)$ transforms as the ρ_2 component of the Γ_2 representation of G where γ_2 serves to distinguish cases where the Γ_2 representation occurs more than once in the description of the operators. Similarly, let us associate the bra $\langle\psi|$ with $\langle\gamma_1\Gamma_1\rho_1|$ and the ket $|\psi'\rangle$ with $|\gamma_3\Gamma_3\rho_3\rangle$, then the Wigner-Eckart theorem states that a matrix element $\langle\gamma_1\Gamma_1\rho_1| \mathbf{O}(\gamma_2\Gamma_2\rho_2) |\gamma_3\Gamma_3\rho_3\rangle$ may be factored into sums of products of two factors, one A_α which is independent of ρ_1, ρ_2, and ρ_3 and the other a recoupling coefficient $\langle\Gamma_2\Gamma_3\alpha\Gamma_1\rho_1 | \Gamma_2\rho_2, \Gamma_3\rho_3\rangle$ to give the general result[112–115]

$$\langle\gamma_1\Gamma_1\rho_1| \mathbf{O}(\gamma_2\Gamma_2\rho_2) |\gamma_3\Gamma_3\rho_3\rangle = \sum_\alpha A_\alpha\langle\Gamma_2\Gamma_3\alpha\Gamma_1\rho_1 | \Gamma_2\rho_2, \Gamma_3\rho_3\rangle \quad (195)$$

The number of terms in the summation over α will be given by the number of times, $c(\Gamma_1\Gamma_2\Gamma_3)$ the identity representation Γ_0 occurs in the triple Kronecker product $\Gamma_1 \times \Gamma_2 \times \Gamma_3$ or equivalently the number of times Γ_2 occurs in the

Kronecker product $\Gamma_1 \times \Gamma_3$. If $c(\Gamma_1\Gamma_2\Gamma_3) = 0$ then we are assured that the matrix elements are null for all choices of γ's and ρ's in Eq. 195. The Kronecker products for the groups $Sp(n)$ and $R(n)$ may be readily determined using the methods of Sections 5.13 and 7.3–7.6.

As an example, consider the tensor operator $\mathbf{W}^{(12)}$, which occurs in discussions of magnetic hyperfine structure,[15] and which we have seen to transform as the [200] and (20) representations of the groups R_7 and G_2 in the case of f-electrons. The numbers $c([\lambda][\lambda'][200])$ and $c(UU'(20))$ may be presented in tabular form as below.

The numbers $c([\lambda][\lambda'][200])$

$[\lambda]$ \ $[\lambda']$	[000]	[100]	[110]	[200]	[111]	[210]	[211]	[220]	[221]	[222]
[000]	—	—	—	1	—	—	—	—	—	—
[100]	—	1	—	—	—	1	—	—	—	—
[110]	—	—	1	1	—	—	1	—	—	—
[200]	1	—	1	1	—	—	—	1	—	—
[111]	—	—	—	—	1	1	1	—	—	—
[210]	—	1	—	—	1	2	—	—	1	—
[211]	—	—	1	—	1	—	2	1	1	—
[220]	—	—	—	1	—	—	1	1	—	1
[221]	—	—	—	—	—	1	1	—	2	1
[222]	—	—	—	—	—	—	—	1	1	1

The numbers $c(UU'(20))$

U \ U'	(00)	(10)	(11)	(20)	(21)	(30)	(22)	(31)	(40)
(00)	—	—	—	1	—	—	—	—	—
(10)	—	1	1	1	1	1	—	—	—
(11)	—	1	1	1	1	1	—	1	—
(20)	1	1	1	2	2	1	1	1	1
(21)	—	1	1	2	2	2	1	2	1
(30)	—	1	1	1	2	2	1	2	1
(22)	—	—	—	1	1	1	1	1	1
(31)	—	—	1	1	2	2	1	3	2
(40)	—	—	—	1	1	1	1	2	2

These tables were first obtained by Judd.[115] Further tables may be found in the works of Racah,[1] McLellan,[116] Judd,[15,85] and Nutter.[117]

Inspection of the table for $c([\lambda][\lambda'][200])$ shows no examples of selection rules for null matrix elements that are not already covered by the normal selection rules on v, S, and L. However, the table for $c(UU'(20))$ contains several cases of additional selection rules. For example, $c((11)(22)(20)) = 0$ implies that the entire set of matrix elements

$$\langle f^4(211)(11)^3L\| W^{(12)} \|f^4(220)(22)^1L'\rangle = 0$$

for $L = PH$ and $L' = SDGHILN$ as may be verified by an inspection of the tables of Jucys, et al.[118] Other examples of selection rules for the tensor operators $\mathbf{W}^{(ok)}$ (k even) have been given by Judd[115] and by McLellan[116] for the tensor operator $\mathbf{W}^{(11)}$ which occurs in the spin-orbit interaction.

Investigations of the triple products $\Gamma_1 \times \Gamma_2 \times \Gamma_3$ do not necessarily exhaust the possibility for detecting null matrix elements without prior calculation. Judd and Wadzinski[119] have shown that additional selection rules are sometimes possible even when $c(\Gamma_1\Gamma_2\Gamma_3) \geq 1$. Their result makes use of the resolution of the Kronecker squares of irreducible representations into their symmetric and antisymmetric parts. Smith and Wybourne[120] have given a detailed account of this problem in terms of plethysm. Consider the matrix element of Eq. 195 again and suppose the ρ's label representations of a subgroup g of the group G associated with the Γ representations. We have then the result

$$\langle \gamma_1\Gamma_1\rho_1|\ \mathbf{O}(\gamma_2\Gamma_2\rho_2)\ |\gamma_2\Gamma_2\rho_2\rangle = 0$$

unless *either*

$$\Gamma_2 \otimes \{2\} \supset \Gamma_1 \quad \text{and} \quad \rho_2 \otimes \{2\} \supset \rho_1 \tag{196a}$$

or

$$\Gamma_2 \otimes \{1^2\} \supset \Gamma_1 \quad \text{and} \quad \rho_2 \otimes \{1^2\} \supset \rho_1 \tag{196b}$$

This is equivalent to stating that functions that transform as Γ_1 under G must be of the same symmetry classification as those spanning the ρ_1 representation of the subgroup g.

The matrix elements

$$\langle f^6[211](30)^3L\| W^{(15)} \|f^6[110](11)^3H\rangle \tag{197}$$

where $L = PFGHIKM$, form a particularly interesting example of the above selection rule, The tensor operator $\mathbf{W}^{(15)}$ transforms like the states $|[110](11)^3H\rangle$ and we readily find that

$$[110] \otimes \{1^2\} = [110] + [211] \quad \text{and} \quad (11) \otimes \{1^2\} = (11) + (30) \tag{198}$$

indicating that the representations [211] and (30) occurring in the bra of Eq. 197 occur in the antisymmetric parts of the Kronecker squares of the representations [110] and (11) and thus Eqs. 196*a*,*b* are satisfied, as in fact is the seniority rule. However, since the operator $\mathbf{W}^{(15)}$ transforms like a ^3H state under $SU_2 \times R_3$ we must also examine the Kronecker square $\{^3H\}^2$. The antisymmetric components will occur in the separation of the plethysm

$$([\tfrac{1}{2}]'[5]) \otimes \{1^2\} = {}^1(SDGILN) + {}^3(PFHKM) \tag{199}$$

these being just the allowed terms in the configuration h^2. From this result we can conclude, without any numerical calculation, that the matrix elements

$$\langle f^6[211](30)^3L\| \, W^{(15)} \, \|f^6[110](11)^3H\rangle = 0$$

for $L = G$ and I. In an exactly similar manner we may demonstrate that the matrix element

$$\langle f^6[220](22)^1H\| \, W^{(15)} \, \|f^6[110](11)^3H\rangle = 0$$

10.5 THE WIGNER-ECKART THEOREM AND ISOSCALAR FACTORS

We are still far from exhausting the possibilities of the Wigner-Eckart theorem. Let us now enlarge our description of the Wigner-Eckart theorem as given in Eq. 195 and suppose that while the Γ's label irreducible representations belonging to a group G the ρ's label irreducible representations of a subgroup g of G. We now introduce quantities ω to label the components of the ρ's. The dependence of the matrix elements of an operator $\mathbf{O}$ upon the quantities ω is found from Eq. 195 to be

$$\langle \gamma_1\Gamma_1\tau_1\rho_1\omega_1| \, \mathbf{O}(\gamma_2\Gamma_2\tau_2\rho_2\omega_2) \, |\gamma_3\Gamma_3\tau_3\rho_3\omega_3\rangle = \sum_\alpha A_\alpha\langle\alpha\rho_1\omega_1 \,|\, \rho_2\omega_2 + \rho_3\omega_3\rangle \tag{200}$$

where we use the labels τ_1, τ_2, and τ_3 to make the classification complete. A_α is independent of the components ω of the irreducible representations ρ, and may be written as a reduced matrix element

$$A_\alpha = \langle\gamma_1\Gamma_1\tau_1\rho_1\| \, \mathbf{O}_\alpha(\gamma_2\Gamma_2\tau_2\rho_2) \, \|\gamma_3\Gamma_3\tau_3\rho_3\rangle \tag{201}$$

We may now apply the Wigner-Eckart theorem to factor out the dependence of A_α upon the irreducible representations ρ which are components of the irreducible representations Γ to give

$$A_\alpha = \sum_\beta A_{\alpha\beta}\langle\beta\Gamma_1\tau_1\rho_1 \,|\, \Gamma_2\tau_2\rho_2 + \Gamma_3\tau_3\rho_3\rangle \tag{202}$$

where

$$A_{\alpha\beta} = \langle\gamma_1\Gamma_1\tau_1\| \, \mathbf{O}_{\alpha\beta}(\gamma_2\Gamma_2\tau_2) \, \|\gamma_3\Gamma_3\tau_3\rangle \tag{203}$$

Thus Eq. 200 becomes

$$\langle\gamma_1\Gamma_1\tau_1\rho_1\omega_1| \, \mathbf{O}(\gamma_2\Gamma_2\tau_2\rho_2\omega_2) \, |\gamma_3\Gamma_3\tau_3\rho_3\omega_3\rangle = \sum_{\alpha,\beta} A_{\alpha\beta}\langle\alpha\rho_1\omega_1 \,|\, \rho_2\omega_2 + \rho_3\omega_3\rangle\langle\beta\Gamma_1\tau_1\rho_1 \,|\, \Gamma_2\tau_2\rho_2 + \Gamma_3\tau_3\rho_3\rangle \tag{204}$$

The coefficients appearing on the right are known as *isoscalar factors.*

The Wigner-Eckart theorem may be applied repeatedly throughout a chain of connected groups each time giving rise to a product of reduced matrix elements times an isoscalar factor. We now illustrate its application in the case of the f^n configurations where the states are classified by the chain of groups

$$Sp_{14} \to SU_2 \times (R_7 \to G_2 \to R_3) \to R_2 \times R_2$$

Consider the typical matrix element of the operator $\mathbf{W}^{(\kappa k)}$

$$\langle f^n \langle\sigma_1\rangle[\lambda_1]U_1\tau_1S_1L_1M_{S_1}M_{L_1}| \, W^{(\kappa k)}_{\pi q}(\langle\sigma_2\rangle[\lambda_2]U_2\tau_2\kappa k\pi q) \times |f^n\langle\sigma_3\rangle[\lambda_3]U_3\tau_3S_3L_3M_{S_3}M_{L_3}\rangle$$

applying the Wigner-Eckart theorem twice to factor out the dependence of the matrix element on M_S and M_L and remembering that under the group R_3 the Kronecker products $D^{(L)} \times D^{(L')}$ and $D^{(S)} \times D^{(S')}$ contain no repeated representations we find we may express the matrix element as

$$\begin{aligned}= \langle f^n\langle\sigma_1\rangle[\lambda_1]U_1\tau_1S_1L_1\| \, W^{(\kappa k)}(\langle\sigma_2\rangle[\lambda_2]U_2\tau_2\kappa k) \, \|f^n\langle\sigma_3\rangle[\lambda_3]U_3\tau_3S_3L_3\rangle \\ \times \langle S_1M_{S_1} \mid \kappa\pi + S_3M_{S_3}\rangle\langle L_1M_{L_3} \mid kq + L_3M_{L_3}\rangle(-1)^{\kappa+k+S_1-S_3+L_1-L_3} \\ \times \{[S_1, L_1]\}^{-\frac{1}{2}} \qquad (205)\end{aligned}$$

where the last two factors are just the usual Clebsch-Gordan coefficients and we have introduced a phase factor, and the quantity $\{[S_1, L_1]\}^{-\frac{1}{2}}$, to ensure that our reduced matrix elements of $\mathbf{W}^{(\kappa k)}$ have their customary definition.[15] The reduced matrix element of $\mathbf{W}^{(\kappa k)}$ may be factorized by repeated application of the Wigner-Eckart theorem to give

$$\begin{aligned}\langle f^n\langle\sigma_1\rangle[\lambda_1]U_1\tau_1S_1L_1\| \, W^{(\kappa k)}(\langle\sigma_2\rangle[\lambda_2]U_2\tau_2\kappa k) \, \|f^n\langle\sigma_3\rangle[\lambda_3]U_3\tau_3S_3L_3\rangle \\ = \sum_{\alpha,\beta,\gamma} A_{\alpha\beta\gamma}\langle\alpha U\tau_1L_1 \mid U_2\tau_2k + U_3\tau_3L_3\rangle\langle\beta[\lambda_1]U_1\tau_1 \mid [\lambda_2]U_2\tau_2 + [\lambda_3]U_3\tau_3\rangle \\ \times \langle\gamma\langle\sigma_1\rangle[\lambda_1]S_1 \mid \langle\sigma_2\rangle[\lambda_2]\kappa + \langle\sigma_3\rangle[\lambda_3]S_3\rangle(-1)^{\kappa+k+S_1-S_3+L_1-L_3}\{[S_1, L_1]\}^{-\frac{1}{2}} \\ (206)\end{aligned}$$

where the number of terms in the summation over α is $c(U_1U_2U_3)$, over β is $c([\lambda_1][\lambda_2][\lambda_3])$ and over γ is $c(\langle\sigma_1\rangle\langle\sigma_2\rangle\langle\sigma_3\rangle)$.

Equation 206 is particularly informative as it displays the detailed structure of the matrix element in terms of independent isoscalar factors. Clearly, if any one isoscalar factor vanishes the entire matrix element vanishes. Thus instead of considering selection rules for individual matrix elements we may consider the selection rules as they apply to individual isoscalar factors. Thus the vanishing of the matrix elements

$$\langle f^6[211](30)^3GI\| \, W^{(15)} \, \|f^6[110](11)^3H\rangle$$

implies that the isoscalar factors

$$\langle (30)GI \,|\, (11)H + (11)H \rangle = 0$$

Since the same isoscalar factors arise in the matrix elements

$$\langle f^5[221](30)^2GI \| W^{(15)} \| f^5[211](11)^4H \rangle$$

and

$$\langle f^5[221](30)^2GI \| W^{(05)} \| f^5[221](11)^2H \rangle$$

they must also be null.

Equations like Eq. 206 may be made the basis for establishing many relationships between different sets of matrix elements. For example, the tensor operators $\mathbf{W}^{(ok)}$ with k even transform as the [200] and (20) representations of R_7 and G_2 and since $c((20)(31)(20)) = 1$ we may write the matrix elements

$$\langle f^6[221](31)^3L \| W^{(ok)} \| f^6[221](20)^3L' \rangle = A\langle (31)L \,|\, (20)k + (20)L' \rangle$$

where A is independent of L, L', and k. Similarly,

$$\langle f^6[221](31)^3L \| W^{(ok)} \| f^6[211](20)^3L' \rangle = A'\langle (31)L \,|\, (20)k + (20)L' \rangle$$

and

$$\langle f^5[221](31)^2L \| W^{(ok)} \| f^5[221](20)^2L' \rangle = A''\langle (31)L \,|\, (20)k + (20)L' \rangle$$

While the isoscalar factors do vanish for $k = L'$ and L even they do not in general vanish. Since the isoscalar factors are the same in all three sets of matrix elements we would expect the three sets of matrix elements to be simply proportional to one another.

Calculation of just two matrix elements shows in fact that

$$\langle f^6[221](31)^3L \| W^{(ok)} \| f^6[221](20)^3L' \rangle = \frac{\sqrt{2}}{2}\langle f^6[221](31)^3L \| W^{(ok)} \| f^6[211](20)^3L' \rangle$$

for $L = PDF_2GH_2I_2K_2LMNO$, $k = 2, 4, 6$, and $L' = DGI$.

Somewhat surprisingly we find

$$\langle f^5[221](31)^2P \| W^{(02)} \| f^5[221](20)^2D \rangle = 0$$

and since the corresponding matrix element in f^6 is nonzero we conclude

$$\langle f^5[221](31)^2L \| W^{(ok)} \| f^5[221](20)^2L' \rangle = 0 \tag{207}$$

which of course implies that $A'' = 0$.

Judd[87,121] has shown that the matrix elements of $\mathbf{W}^{(ok)}$ and $\mathbf{W}^{(1k)}$ (for k even) are related in different configurations l^{n_a} and l^{n_b} by the equation

$$\frac{\langle l^{n_a}v_1[\lambda]\tau S_1L\| W^{(1k)} \| l^{n_a}v_1[\lambda]\tau' S_1L'\rangle}{\langle l^{n_b}v_2[\lambda]\tau S_2L\| W^{(0k)} \| l^{n_b}v_2[\lambda]\tau' S_2L'\rangle} = -\left[\frac{(2l+1-v_2)(2l+2-v_2)(2l+3-v_2)}{(2l+1-n_b)^2}\right]^{\frac{1}{2}} \tag{208}$$

Using this result, together with Eq. 207 we conclude that the matrix element

$$\langle f^6[221](31)^3L\| W^{(1k)} \| f^6[221](20)^3L'\rangle = 0$$

for k even. These somewhat surprising null matrix elements have been explained by Judd in a private communication by expanding the bra and the ket as a linear combination of LL coupled states.

The result of Eq. 206 can also be used to establish relationships between sets of matrix elements even when $c(U_1U_2U_3) > 1$ or $c([\lambda_1][\lambda_2][\lambda_3]) > 1$. Several examples of these relationships have been given by Judd,[15,115] the relationship

$$\begin{aligned}\langle f^6[210](20)^5L\| W^{(02)} \| [210](20)^5L'\rangle &= \tfrac{2}{3}(\langle f^3[111](20)^4L\| W^{(02)} \| f^3[111](20)^4L'\rangle \\ &\quad + \langle f^3[210](20)^2L\| W^{(0k)} \| f^3[210](20)^2L'\rangle)\end{aligned}$$

being a typical example. In this case $c((20)(20)(20)) = 2$.

Nielson and Koster[122] have given tables for the reduced matrix elements of the tensor operators

$$\mathbf{U}^{(k)} = \sqrt{2/(2k+1)}\,\mathbf{W}^{(0k)} \quad \text{and} \quad \mathbf{V}^{(11)} = \mathbf{W}^{(11)}/6^{\frac{1}{2}} \tag{209}$$

for all configurations of the types p^n, d^n, and f^n while Jucys et al.[118] have given tables for $\mathbf{V}^{(1k)}$ for up to four equivalent f-electrons and thus the methods just outlined are of doubtful computational value. They are, however, useful in as far as they shed further light on the symmetry structure of matrix elements and as such can be expected to ultimately result in a more profound formulation of the theory of complex spectra. The compiled tables of matrix elements with the use of Eq. 206, can be used as a source of unnormalized isoscalar factors which in turn may be used in the construction of symmetry adapted operators.

10.6 THE COULOMB INTERACTION

The tensor operators $\mathbf{W}^{(\kappa k)}$ form the basic building blocks for describing the symmetry properties of atomic interactions. Any atomic interaction may be expressed as a linear combination of suitable sets of products of the components of the $\mathbf{W}^{(\kappa k)}$. From these sets we may construct, by forming suitable linear combinations, new operators that not only describe the real interactions but also have well-defined transformation properties under the symmetry operations of the groups used to classify the atomic states.

The Coulomb interaction in l^n-type configurations forms an interesting example of the application of symmetry principles to the description of atomic interactions. We proceed by first expressing the Coulomb repulsion in terms of the tensor operators $\mathbf{W}^{(\kappa k)}$ to give for the matrix elements[15]

$$\langle l^n\psi| \sum_{i>j=1}^{n} \frac{e^2}{r_{ij}} |l^n\psi'\rangle = \sum_k a_k \langle l^n\psi| \sum_{i>j=1}^{n} (\mathbf{w}_i^{(0k)} \cdot \mathbf{w}_i^{(0k)}) |l^n\psi'\rangle \tag{210}$$

where

$$a_k = \frac{2}{2k+1} (l\| C^{(k)} \|l)^2 F^k \tag{211}$$

and

$$(l\| C^{(k)} \|l') = (-1)^l \{[l, l']\}^{\frac{1}{2}} \begin{pmatrix} l & k & l' \\ 0 & 0 & 0 \end{pmatrix} \tag{212}$$

The quantities F^k are the well-known Slater radial integrals.[123] The triangular conditions on the $3j$-symbol together with the necessary condition that $l + k + l'$ be *even* restricts the values of the summation index k in Eq. 210 to even values in the range $2l \geq k \geq 0$. Thus the Coulomb repulsion energies of the terms of a configuration l^n may be expressed in terms of $l + 1$ Slater radial F^k integrals. The numerical results may be more conveniently expressed in terms of the modified integrals F_k where

$$F_k = F^k/D_k$$

where the denominators D_k are chosen according to the prescription of Condon and Shortley.[123]

The numerical calculations for two-electron configurations are straightforward[15] and for d^2 and f^2 the term energies are found to be

d^2:
$$^1S = F_0 + 14F_2 + 126F_4$$
$$^1D = F_0 - 3F_2 + 36F_4$$
$$^1G = F_0 + 4F_2 + F_4$$
$$^3P = F_0 + 7F_2 - 84F_4$$
$$^3F = F_0 - 8F_2 - 9F_4$$

f^2:
$$^1S = F_0 + 60F_2 + 198F_4 + 1716F_6$$
$$^1D = F_0 + 19F_2 - 99F_4 + 715F_6$$
$$^1G = F_0 - 30F_2 + 97F_4 + 78F_6$$
$$^1I = F_0 + 25F_2 + 9F_4 + F_6$$
$$^3P = F_0 + 45F_2 + 33F_4 - 1287F_6$$
$$^3F = F_0 - 10F_2 - 33F_4 - 286F_6$$
$$^3H = F_0 - 25F_2 - 51F_4 - 13F_6$$

The corresponding term energies for g^2 have been given by Shortley and

Fried.[124] We now wish to examine the application of symmetry principles to the problem of constructing a set of operators which have well-defined symmetry properties and whose matrix elements yield the result of Eq. 210.

The Coulomb interaction, as with other atomic interactions, must be totally *symmetric* with respect to pairs of electron coordinates since it acts between pairs of totally antisymmetric atomic states. As a result the two-particle part of the Coulomb interaction must transform like a tensor of symmetry type $\{2\}$. Since the Coulomb interaction is a scalar it must transform under R_3 like an S-state. Furthermore, the Coulomb interaction operators are built out of scalar products of the $(l + 1)$ basic operators

$$\mathbf{W}^{(00)} \quad \text{and} \quad \mathbf{W}^{(02)}, \ldots, \mathbf{W}^{(0,2l)}$$

which collectively transform as the [0] and [20] representations of R_{2l+1}.

From the above argument we may conclude that the Coulomb interaction must be expressible in terms of a set of $(l + 1)$ operators $(e_0, e_1, \ldots, e_l)$ that transform according to representations contained in the separation of the plethysm

$$\{2\} \otimes \{2\} = \{4\} + \{22\} \tag{213}$$

It follows from Eq. 69 that under the restriction $U_{2l+1} \to R_{2l+1}$

$$\{4\} \to [4] + [2] + [0] \quad \text{and} \quad \{22\} \to [22] + [2] + [0] \tag{214}$$

The representation [2] of R_{2l+1} cannot yield an S-state under restriction to R_3 since among the states of l^2 there can only be one S-state and it is associated with the [0] representation of R_{2l+1}. Thus only scalar operators transforming as $[0]^2$, [22], [4] under R_{2l+1}, and giving rise to one, or more, S-states under restriction to R_3, are available for the symmetrization of the Coulomb interaction.

In the particular case of the configurations d^n, f^n, and g^n the operators $e_0, e_1, \ldots, e_l$ must transform according to the symmetry classifications:

$$\begin{array}{llllll}
d^n: & [00]S\, e_0 & f^n: & [000](00)S\, e_0 & g^n: & [0000]S\, e_0 \\
 & [00]S\, e_1 & & [000](00)S\, e_1 & & [0000]S\, e_1 \\
 & [22]S\, e_2 & & [400](40)S\, e_2 & & [4000]S\, e_2 \\
 & & & [220](22)S\, e_3 & & [2200]S\, e_3 \\
 & & & & & [2200]S\, e_4
\end{array} \tag{215}$$

where the representations of G_2 have been included for the case of f-electrons.

Equation 215 does not display the symplectic symmetry that is to be associated with the various operators. The operators $\mathbf{W}^{(ok)}$ with k even form only a subset of the set of operators with $(\kappa + k)$ even which transform as $\langle 1^2 \rangle$ under Sp_{4l+2} and hence we cannot expect all the operators in Eq. 215 to have well-defined symplectic symmetry. Two-particle operators constructed from the operators $\mathbf{W}^{(\kappa k)}$ with $\kappa + k \neq 0$ and even must be associated with the representations of Sp_{4l+2} arising in the plethysm

$$\begin{aligned}\langle 1^2 \rangle \otimes \{2\} &= \{1^4\} + \{22\} - \{1^2\} \\ &= \langle 1^4 \rangle + \langle 2^2 \rangle + \langle 1^2 \rangle + \langle 0 \rangle \end{aligned} \tag{216}$$

These representations of Sp_{4l+2} may be decomposed into those of $SU_2 \times R_{2l+1}$ following the methods of Section 7.7 to give

$$\begin{array}{ll} Sp_{4l+2} \rightarrow & SU_2 \times R_{2l+1} \\ \langle 0 \rangle & {}^1[0] \\ \langle 1^2 \rangle & {}^1[2] + {}^3[11] \\ \langle 1^4 \rangle & {}^1[22] + {}^3[211] + {}^5[1^4] \\ \langle 2^2 \rangle & {}^1([4] + [22] + [2] + [0]) + {}^3([211] + [31] + [11] + [2]) \\ & \qquad + {}^5([22] + [2] + [0]) \end{array}$$

It is at once apparent that operators transforming as the [4] representation of R_{2l+1} have well-defined symplectic symmetry, transforming as $\langle 2^2 \rangle$ under Sp_{4l+2}. The operators transforming as [0] and [22] under R_{2l+1} do not have well-defined symplectic symmetry since ${}^1[0]$ occurs in both $\langle 0 \rangle$ and $\langle 2^2 \rangle$ while ${}^1[22]$ occurs in $\langle 2^2 \rangle$ and $\langle 1^4 \rangle$. To obtain the advantages of well-defined symplectic symmetry it is necessary to decompose the operators transforming as ${}^1[22]$ into a part transforming as $\langle 2^2 \rangle$ and a second part transforming as $\langle 1^4 \rangle$. Explicit constructions are to be found in the articles by Racah[1] and Judd.[15,85,87,88]

We may extend our discussion of the symmetry properties of the Coulomb interaction operators further by considering their transformation properties with respect to the group R_{8l+4} to determine the quasi-spin ranks K that are to be associated with the operators $e_0, e_1, \ldots, e_l$. Since our attention is limited to two-particle operators we may restrict our consideration to the decomposition of the representations [0], [11], and [1111] of R_{8l+4} under restriction to $SU_2 \times Sp_{4l+2}$. The relevant decompositions have already appeared in Eqs. 189*a–c*. Inspection of these results show that $\langle 2^2 \rangle$ occurs only with quasi-spin $K = 0$ and hence the operators transforming as [4] under R_{2l+1} have well-defined quasi-spin and symplectic symmetry. We may immediately conclude that the matrix elements of e_2 for f^n and g^n are diagonal in the seniority scheme and independent of n for fixed seniority.

The operators transforming as [22] under R_{2l+1} have parts transforming with quasi-spin ranks of $K = 0$ and 2. The explicit construction of the operators having well-defined quasi-spin involves first constructing an operator $-\Omega$ of quasi-spin rank $K = 0$, and transforming as $\langle 2^2 \rangle$ and [22] under Sp_{4l+2} and R_{2l+1} and then expressing the operator e_l as

$$e_l = (e_l + \Omega) - \Omega \tag{217}$$

The operator $e_l + \Omega$ will then be of quasi-spin rank $K = 2$ and transform as $\langle 1^4 \rangle$ and [22] under Sp_{4l+2} and R_{2l+1}. Again, explicit constructions have been given by Racah[1] and Judd.[15]

Having made an explicit construction of the operators $e_0, e_1, \ldots e_l$ the matrix elements of the Coulomb repulsion, given in Eq. 210, will be replaced by the matrix elements of the linear combination,

$$e_0 E_0 + e_1 E_1 + \cdots + e_l E_l \tag{218}$$

where the coefficients $E_0, E_1, \ldots, E_l$ are particular linear combinations of the Slater radial integrals F_k. We now consider the determination of the coefficients $E_0, E_1, \ldots, E_l$, avoiding the explicit construction of the corresponding operators $e_0, e_1, \ldots, e_l$.

We first consider the selection rules appropriate to the symmetry description of the different operators. The states of maximum multiplicity of l^n transform as the $[1^n]$ representation of R_{2l+1} and since the Kronecker product

$$[1^n] \times [1^n] \not\ni [4]$$

the matrix elements of the operator transforming as [4] will be null for the states of maximum multiplicity. Since the Kronecker product $[1^n] \times [1^n]$ for $4l \geq n \geq 2$ will generally contain [00] and [22] we conclude that the matrix elements of e_0 and the operators transforming as the [22] representation of R_{2l+1} will normally be nonzero for states of maximum multiplicity. The matrix elements of e_0 may be conveniently taken to be independent of all the quantum numbers defining the states of a given l^n and completely diagonal. The choice of E_0 is guided by physical considerations, we choose it so that for l^2 the diagonal matrix elements of e_0 are unity and E_0 corresponds to the energy of the bary-center of the terms of l^2. Noting the formulas for d^2 and f^2 term energies we take for

$$d^n: \quad E_0 = F_0 - 7F_2/2 - 63F_4/2 \tag{219}$$

and for

$$f^n: \quad E_0 = F_0 - 10F_2 - 33F_4 - 286F_6 \tag{220}$$

The matrix elements of e_0 for a configuration l^n will then be just $(n/2)(n-1)$.

The matrix elements of e_1 must be independent of all quantum numbers other than S and v and hence we may take their matrix elements to be, to within an additive constant, proportional to the matrix elements of $\mathbf{Q}^2 - \mathbf{S}^2$ where the sign and the proportionality constant are at our disposal. The eigenvalues of $\mathbf{Q}^2 - \mathbf{S}^2$ will be just $Q(Q+1) - S(S+1)$, where Q is as defined in Eq. 187. By insisting that the matrix elements of e_1 vanish for states of maximum multiplicity we find we may take

$$e_1 = \mathbf{Q}^2 - \mathbf{S}^2 + \frac{2n(l+2) - (2l+1)(2l+3)}{4} \tag{221}$$

The matrix elements of e_1 are completely diagonal and will clearly, for given l^n, be just

$$= \frac{v}{4}(v+2) + \frac{(n-v)(2l+3)}{2} - S(S+1) \tag{222}$$

Since the matrix elements of all the operators, barring those transforming as [0], under R_{2l+1} necessarily vanish for the ^1S term of the configuration l^2 we may conclude that the Coulomb repulsion energy for ^1S is just

$$^1S = E_0 + (2l+3)E_1 \tag{223}$$

Comparison of this result with the expressions for the energy of the ^1S term in d^2 and f^2 in terms of the Slater radial integrals permits the identifications

$$d^n\colon\quad E_1 = \frac{5(F_2 + 9F_4)}{2} \qquad f^n\colon\quad E_1 = \frac{70F_2 + 231F_4 + 2002F_6}{2} \tag{224}$$

to be made.

The coefficient E_2 in d^n may be found by first noting that owing to Eq. 202 the matrix elements of e_2 for the ^1DG states of d^2 that transform as [20] under R_5 may be written as

$$\langle d^2[20]^1LM_L|\, e_2\, |d^2[20]^1LM_L\rangle = A\langle [20]L \,|\, [22]S + [20]L\rangle \tag{225}$$

where the L-dependence of the matrix elements is entirely contained in the isoscalar factor. In d^4 we have, from Eq. 206

$$\langle d^4[20]^1L\|\, W^{(oL)}\, \|d^4[22]^1S\rangle = B\sqrt{[L]}\langle [20]L \,|\, [22]S + [20]L\rangle \tag{226}$$

Comparing Eqs. 225 and 226 we immediately obtain

$$\frac{\langle d^2[20]^1DM_L|\, e_2\, |d^2[20]^1DM_L\rangle}{\langle d^2[20]^1GM_L|\, e_2\, |d^2[20]^1GM_L\rangle} = \frac{9}{5}\frac{\langle d^4[20]^1D\|\, W^{(04)}\, \|d^4[22]^1S\rangle}{\langle d^4[20]^1G\|\, W^{(04)}\, \|d^4[22]^1S\rangle}$$

$$= -\frac{9}{5} \tag{227}$$

where the reduced matrix elements of $\mathbf{W}^{(02)}$ and $\mathbf{W}^{(04)}$ are obtained from Nielson and Koster's tabulation of the analogous $\mathbf{U}^{(k)}$ operators.[122] Equation 227 fixes the ratio of the matrix elements. Making use of this ratio and the formulas for the term energies of d^2 we readily obtain the result

$$d^n\colon\quad E_2 = \frac{F_2 - 5F_4}{2} \tag{228}$$

The term energies of d^2 then become

$$
\begin{aligned}
{}^1S &= E_0 + 7E_1 & {}^3P &= E_0 + 21E_2 \\
{}^1D &= E_0 + 2E_1 - 9E_2 & {}^3F &= E_0 - 9E_2 \\
{}^1G &= E_0 + 2E_1 + 5E_2 & &
\end{aligned}
$$

In an exactly similar manner we may express the term energies of f^2 as

$$
\begin{aligned}
{}^1S &= E_0 + 9E_1 & {}^3P &= E_0 + 33E_3 \\
{}^1D &= E_0 + 2E_1 + 286E_2 - 11E_3 & {}^3F &= E_0 \\
{}^1G &= E_0 + 2E_1 - 260E_2 - 4E_3 & {}^3H &= E_0 - 9E_3 \\
{}^1I &= E_0 + 2E_1 + 70E_2 + 7E_3 & &
\end{aligned}
$$

where we find for f^n

$$
E_2 = \frac{F_2 - 3F_4 + 7F_6}{9} \quad \text{and} \quad E_3 = \frac{5F_2 + 6F_4 - 91F_6}{3} \tag{229}
$$

We note that since the Coulomb repulsion energies of the terms of maximum multiplicity depend only on the matrix elements of e_0, which is constant for all terms, and the operators transforming as [22] under R_{2l+1} the *relative* separations of these terms will depend solely on a number of parameters equal to the number of S-states contained in the representation [22] of R_{2l+1}. That is, one in the case of d^n and f^n configurations and two in the case of g^n configurations.[91,94,124a] Such a result is certainly not self-evident from the formulas when written in terms of the Slater radial integrals and is demonstration of how a careful application of symmetry principles can sometimes yield quite unexpected results of striking simplicity. Since the complete electrostatic energy matrices for p^n, d^n, and f^n configurations have been tabulated by Nielson and Koster[122] we shall not pursue the explicit calculation of the matrix elements of the Coulomb interaction further.

10.7 SYMMETRIZATION OF THE ORBIT-ORBIT INTERACTION

The orbit-orbit interaction arises in the relativistic extension of quantum mechanics and corresponds to a correction due to the retardation of the electromagnetic field produced by an electron.[125] The contribution to the Hamiltonian may be written as

$$
H_{oo} = \frac{-e^2}{2mc}\left[\frac{\mathbf{p}_i \cdot \mathbf{p}_j}{r_{ij}} + \frac{\mathbf{r}_{ij} \cdot (\mathbf{r}_{ij} \cdot \mathbf{p}_i)\mathbf{p}_j}{r_{ij}^{\;3}}\right] \tag{230}
$$

For a configuration l^n the orbit-orbit interaction may be expressed in tensorial form to give [126–129]

$$H_{oo} = -8 \sum_k \frac{(2k+1)}{(k+2)} l(l+1)(2l+1)\langle l \| C^{(k)} \| l \rangle^2 \begin{Bmatrix} k & k+1 & 1 \\ l & l & l \end{Bmatrix}^2 M^k \times \sum_{i>j} (\mathbf{w}_i^{(0k+1)} \cdot \mathbf{w}_j^{(0k+1)}) \quad (231)$$

where the Marvin integrals M^k are defined as [130]

$$M^k = \frac{e^2\hbar^2}{8m^2c^2} \langle nl^2 | \frac{r_<^k}{r_>^{k+3}} | nl^2 \rangle \quad (232)$$

General results for mixed configurations have been given by Armstrong and Feneuille.[131]

Let us now consider the application of symmetry principles to the orbit-orbit interaction and try to represent it in terms of operators having well-defined symmetry properties. Since H_{oo} is a two particle scalar operator and the operators $\mathbf{W}^{(0k+1)}$, with *k even*, transform as $[1^2]$ under R_{2l+1}, the symmetrized operators must transform under R_{2l+1} as the representations

$$[11] \otimes \{2\} = \{1^4\} + \{22\} = [22] + [2] + [0] + [1^4] \quad (233)$$

Of these only [22], $[1^4]$, and [0] can give rise to S-states when the restriction $R_{2l+1} \to R_3$ is made. Thus for d^n and f^n configurations

$$\begin{array}{llll} d^n: & [00]S\, e_3 & f^n: & [000](00)S\, e_4 \\ & [22]S\, e_4 & & [111](00)S\, e_5 \\ & & & [220](22)S\, e^6 \end{array} \quad (234)$$

By noting that under Sp_{4l+2}

$$\langle 2 \rangle \otimes \{2\} = \langle 4 \rangle + \langle 22 \rangle + \langle 1^2 \rangle + \langle 0 \rangle$$

we conclude that [00] has symplectic symmetry corresponding to $\langle 0 \rangle$ and $\langle 2^2 \rangle$, while [22] and [111] are associated with $\langle 2^2 \rangle$ only.

Let us now determine the linear combinations,

$$\sum_k a_k (\mathbf{w}_i^{(0k+1)} \cdot \mathbf{w}_j^{(0k+1)})$$

that transform as [00] and [22] under R_5; i.e., consider the case of d^n configurations. It is clear from Eq. 200 that the linear combination

$$\sum_k \{\mathbf{w}_i^{(0k+1)} \mathbf{w}_j^{(0k+1)}\}^{(00)0}{}_0 \langle [11](k+1) + [11](k+1) \,|\, [\lambda]S0 \rangle$$
$$= \sum_k (-1)^{k+1} (2k+3)^{-\frac{1}{2}} (\mathbf{w}_i^{(0k+1)} \cdot \mathbf{w}_j^{(0k+1)}) \times \langle [11](k+1) + [11](k+1) \,|\, [\lambda]S0 \rangle \quad (235)$$

will have the same transformation properties as the states $|[\lambda]S0\rangle$. Thus we have only to determine the coupling coefficients

$$\langle[11](k+1)+[11](k+1) \mid [\lambda]S0\rangle$$

to obtain an appropriate linear combination satisfying Eq. 234. For the case of $[\lambda] \equiv [00]$ the quantity

$$\frac{(-1)^{k+1}}{(2k+3)^{\frac{1}{2}}}\langle[11](k+1)+[11](k+1) \mid [00]S0\rangle$$

is clearly independent of k and hence we may take

$$e_3 = 4\sum_{i>j}[(\mathbf{w}_i^{(01)}\cdot\mathbf{w}_j^{(01)})+(\mathbf{w}_i^{(03)}\cdot\mathbf{w}_j^{(03)})] \tag{236}$$

since we have the normalization of the matrix elements at our disposal.

The quantities $(-1)^{k+1}\langle[11](k+1)+[11](k+1) \mid [22]S0\rangle/(2k+3)^{\frac{1}{2}}$ may be found by noting that in d^4

$$\begin{aligned}\langle d^4[11]^3(k+1)\| W^{(1k+1)} \| d^4[22]^1S\rangle \\ = A(2k+3)^{\frac{1}{2}}\langle[22]S+[11](k+1) \mid [11](k+1)\rangle\end{aligned} \tag{237}$$

where A is independent of k. Racah[1] has shown that the isoscalar factors obey the following reciprocity relation

$$\begin{aligned}\langle[\lambda]\tau L+[\lambda']\tau'L' \mid [\lambda'']\tau''L''\rangle = (-1)^{L''-L'-L+x}\left\{\frac{[L]}{[L']}\frac{g([\lambda''])}{g([\lambda])}\right\}^{\frac{1}{2}} \\ \times\langle[\lambda'']\tau''L''+[\lambda']\tau'L' \mid [\lambda]\tau L\rangle\end{aligned} \tag{238}$$

where x is independent of the L's and $g([\lambda''])$ and $g([\lambda])$ are the dimensions of the representations $[\lambda'']$ and $[\lambda]$ of R_{2l+1}. Using this result in Eq. 237 yields

$$\langle d^4[11]^3(k+1)\| W^{(1k+1)} \| d^4[22]^1S\rangle = B\langle[11]k+[11]k \,\|\, [22]S\rangle \tag{239}$$

where B is a proportionality constant independent of k. The reduced matrix elements of $\mathbf{W}^{(11)}$ and $\mathbf{W}^{(13)}$ may be readily evaluated using the known formulas[15] and the tables of Nielson and Koster[122] to give

$$\langle d^4[11]^3P\| W^{(11)} \| d^4[22]^1S\rangle = -7/(35)^{\frac{1}{2}}$$

and

$$\langle d^4[11]^3F\| W^{(13)} \| d^4[22]^3F\rangle = 3/(15)^{\frac{1}{2}}$$

Using these results in Eq. 239 and then in Eq. 235 shows immediately that we may take

$$e_4 = 4\sum_{i>j}[7(\mathbf{w}_i^{(01)}\cdot\mathbf{w}_j^{(01)})-3(\mathbf{w}_i^{(03)}\cdot\mathbf{w}_j^{(03)})] \tag{240}$$

to be a tensor transforming as [22] under R_5.

The scalar products involving *odd* rank tensors may be readily expressed in terms of the Casimir operators[15] of R_5 and R_3 by noting that for d^n

$$\sum_{i>j}(\mathbf{w}_i^{(01)} \cdot \mathbf{w}_j^{(01)}) = \frac{L(L+1)}{40} - \frac{3n}{20} \tag{241}$$

and

$$\sum_{i>j}[(\mathbf{w}_i^{(01)} \cdot \mathbf{w}_j^{(01)}) + (\mathbf{w}_i^{(03)} \cdot \mathbf{w}_j^{(03)})] = \tfrac{3}{4}G(R_5) - \frac{n}{2} \tag{242}$$

where the eigenvalues of the Casimir operator $G(R_5)$ are given by[15]

$$6G(R_5)u = \left[\frac{v}{2}(12 - v) - 2S(S+1)\right]u \tag{243}$$

Thus we may write the eigenvalues of e_3 and e_4 as

$$\begin{aligned} e_3u &= [3G(R_5) - 2n]u \\ e_4u &= [L(L+1) - 9G(R_5)]u \end{aligned} \tag{244}$$

It remains then only to determine the linear combinations of the M^k integrals to be associated with e_3 and e_4. By specific calculation in d^2, or by use of the results of Wybourne[132] (*N.B.* a correction to this paper is mentioned in refs. 129 and 133), we find

$$E_3 = -2(7M_0 - 6M_2) \quad \text{and} \quad E_4 = -42(M_0 + 2M_2) \tag{245}$$

where we have put

$$M^0 = 7M_0 \quad \text{and} \quad M^2 = 49M_2$$

to simplify the numerical values. Thus we have

$$H_{oo} = e_3E_3 + e_4E_4 \tag{246}$$

10.8 SPIN-DEPENDENT TWO-PARTICLE INTERACTIONS

The orbit-orbit interaction is but one type of interaction that arises in the relativistic extension of quantum mechanics. A description of the various terms that arise in the relativisitic treatment, together with their expansions in terms of tensor operators has been given by Armstrong[134,135] and Armstrong and Feneuille.[131] We shall limit our attention to the symmetry description of the tensor operator expressions for the spin-spin and spin-other-orbit interactions.

For a configuration of equivalent electrons l^n the contribution to the Hamiltonian due to the spin-spin interaction may be written as [136,137]

$$H_{ss} = -2\sum_k [(k+1)(k+2)(2k+3)]^{\frac{1}{2}}(l\| C^{(k)} \|l)(l\| C^{(k+2)} \|l)M^k \times \sum_{i\neq j}\{\mathbf{w}_i^{(1k)}\mathbf{w}_j^{(1k+2)}\}^{(22)0} \tag{247}$$

where M^k is again the Marvin integral given in Eq. 232. Since k is *even* the operators $\mathbf{W}^{(1k)}$ and $\mathbf{W}^{(1k+2)}$ transform as $\langle 2\rangle$ and $[2]$ under the groups Sp_{4l+2} and R_{2l+1}, respectively. The two particle operators transform like a ^5D term under $SU_2 \times R_3$. The symplectic symmetry properties of the two-particle operators are found by separating the plethysm

$$\langle 2\rangle \otimes \{2\} = \langle 0\rangle + \langle 1^2\rangle + \langle 2^2\rangle + \langle 4\rangle$$

Operators transforming as $\langle 4\rangle$ will necessarily have null matrix elements in l^n while $\langle 0\rangle$ and $\langle 1^2\rangle$ cannot yield a quintet state. Thus H_{ss} corresponds to an operator having pure symplectic symmetry $\langle 2^2\rangle$ and quasi-spin $K = 0$. Thus we conclude that the matrix elements of H_{ss} are diagonal in the seniority scheme and for given seniority number v are independent of n.

The restriction $Sp_{4l+2} \rightarrow SU_2 \times R_{2l+1}$ for $\langle 2^2\rangle$ leads to just two symmetry types capable of yielding, upon making further restriction to $SU_2 \times R_3$, ^5D terms. These are

$$\langle 2^2\rangle[22]^5D \quad \text{and} \quad \langle 2^2\rangle[2]^5D \tag{248}$$

The representation $^5[2]$ of $SU_2 \times R_{2l+1}$ can never yield more than one ^5D term. For d^n configurations $^5[22]$ yields only one ^5D term upon restriction to $SU_2 \times R_3$ and hence the spin-spin operator may be decomposed into two parts each of well-defined symmetry under Sp_{10} and $SU_2 \times R_5$. The structure of these two operators has been discussed by Judd.[136]

For f^n configurations $^5[22]$ yields three ^5D terms upon making the restriction $SU_2 \times R_7 \rightarrow SU_2 \times R_3$. These three ^5D terms may be distinguished by introducing the group G_2 to yield the symmetry classification of the four operators as

$$\langle 2^2\rangle[200](20)^5D; \qquad \langle 2^2\rangle[220](20)^5D; \qquad \langle 2^2\rangle[220](21)^5D;$$
$$\langle 2^2\rangle[220](22)^5D \tag{249}$$

The formation of operators having these symmetry properties has been discussed by Judd, et al.[119,137]

The spin-other-orbit interaction for a configuration l^n is rather more complicated. The contribution to the Hamiltonian may be written as[136,137]

$$\begin{aligned} H_{soo} = \sum_k [(k+1)(2l+k+2)(2l-k)]^{\frac{1}{2}} \sum_{i \neq j} [\{\mathbf{w}_i^{(0k+1)}\mathbf{w}_j^{(1k)}\}^{(11)0} \\ \times \{M^{k-1}(l\| C^{(k+1)} \|l)^2 + 2M^k(l\| C^{(k)} \|l)^2 \\ + \{\mathbf{w}_i^{(0k)}\mathbf{w}_j^{(1k+1)}\}^{(11)0}\{M^k(l\| C^{(k)} \|l)^2 \\ + 2M^{k-1}(l\| C^{(k+1)} \|l)^2\}] \end{aligned} \tag{250}$$

where k may take on both *even* and *odd* values. The two particle operators transform like $^3P_0$ states while the basic one-particle operators consist of *odd* rank operators having $\langle 1^2\rangle$ symmetry and the scalar operator $\mathbf{w}^{(00)}$ having $\langle 0\rangle$ symmetry.

For future usage we shall use a plus sign to stand for *even* parity operators and a minus sign for *odd* parity operators. The two-particle operators will be thus designated as $++$ if they are composed of bilinear products of even parity operators and $--$ if they are composed of bilinear products of odd parity operators. Operators of the type $+-$ will be non-Hermitian and will thus be excluded from our discussion since we restrict our attention to real observables.

The symmetry description of the two-particle $--$ operators determined by examining the plethysm

$$\langle 2\rangle \otimes \{2\} = \langle 0\rangle + \langle 1^2\rangle + \langle 2^2\rangle + \langle 4\rangle$$

for representations that contain $^3P_0$ states upon further decomposition. Clearly we may limit our attention to the symmetry types $\langle 1^2\rangle$ and $\langle 2^2\rangle$. Consideration of the restriction of $Sp_{4l+2} \rightarrow SU_2 \times R_{2l+1}$ allows us to limit our attention to the symmetry types

$$\langle 1^2\rangle^3[11]; \qquad \langle 2^2\rangle^3([11] + [211] + [31])$$

Specialization to the case of d^n configurations gives us the symmetry types for $--$ operators as

$$(--)\langle 1^2\rangle[11]^3P; \qquad \langle 2^2\rangle[11]^3P; \qquad \langle 2^2\rangle[21]^3P; \qquad \langle 2^2\rangle[31]^3P \qquad (251)$$

The symmetry description of the $++$ two-particle operators is determined from an examination of the plethysm

$$\langle 1^2\rangle \otimes \{2\} = \langle 0\rangle + \langle 1^2\rangle + \langle 2^2\rangle + \langle 1^4\rangle$$

and

$$\langle 0\rangle\langle 1^2\rangle = \langle 1^2\rangle$$

Specialization to the d^n configurations gives us the symmetry types

$$(++)\langle 1^2\rangle[11]^3P; \qquad \langle 1^4\rangle[21]^3P; \qquad \langle 2^2\rangle[11]^3P; \qquad \langle 2^2\rangle[21]^3P; \qquad \langle 2^2\rangle[31]^3P \qquad (252)$$

The symmetry type $\langle 1^2\rangle[11]^3P$ arises once in $\langle 0\rangle\langle 1^2\rangle$ and once in $\langle 1^2\rangle \otimes \{2\}$.

Comparison of Eqs. 251 and 252 might lead us to conclude that 10 distinct operators must be constructed to describe the spin-other-orbit interaction in terms of operators having well-defined symmetries. The situation is not quite this complicated. First we note that the operators having $\langle 2^2\rangle$ symmetry under Sp_{4l+2} have quasi-spin $K = 0$ while the operator having $\langle 1^4\rangle$ symmetry has quasi-spin $K = 2$. The symmetry $\langle 1^2\rangle$ is associated with quasi-spin $K = 0$, 1, and 2. The states of d^2 transform under R_5 as the [0], [2], and [1²] representations and we have the various Kronecker products

$$\begin{aligned}
&[0][0] = [0]; \quad [1^2][1^2] = 2[0] + [1] + 2[2] + 2[2^2] + [1^2] + [21] \\
&[2][0] = [2]; \quad [1^2][2] = [2] + [1^2] + [21] + [31] \qquad (253) \\
&[1^2][0] = [1^2]; \quad [2][2] = [0] + [1^2] + [2] + [22] + [31] + [4]
\end{aligned}$$

Inspection of the entries in Eq. 253 might lead us to conclude that four operators transforming as $\langle 1^2\rangle[11]^3P$ are required. However, operators transforming like $^3P_0$ states must vanish when evaluated between the singlet states of d^2, which transform as [0] and [2] under R_5 and hence there can only be three independent operators transforming as $\langle 1^2\rangle[11]^3P$. Clearly it would be advantageous to construct from the four operators transforming as $\langle 1^2\rangle[11]^3P$ three linear combinations that would yield three operators of well-defined quasi-spin.

Inspection of Eq. 253 shows further that only one operator transforming as [31] under R_5 is required and hence the operators $(++)\langle 2^2\rangle[31]^3P$ and $(--)\langle 2^2\rangle[31]^3P$ cannot be independent and may be combined into a single operator transforming as $\langle 2^2\rangle[31]^3P$ with quasi-spin $K = 0$. Proceeding in this manner we find seven independent operators are required to express the spin-other-orbit interaction in terms of operators having well-defined quasi-spin, and transformation properties under Sp_{10} and R_5. These seven operators may be designated as

$$^1\langle 2^2\rangle[11]^3P; \quad ^1\langle 2^2\rangle[21]^3P; \quad ^1\langle 2^2\rangle[31]^3P; \quad ^5\langle 1^4\rangle[21]^3P;$$
$$^1\langle 1^2\rangle[11]^3P; \quad ^3\langle 1^2\rangle[11]^3P; \quad ^5\langle 1^2\rangle[11]^3P \quad (254)$$

where the quasi-spin multiplicity $(2K + 1)$ has been indicated as a left superscript. Judd[136] has given specific details concerning the computation of the matrix elements of operators having the above symmetry properties.

10.9 GENERALIZED TWO-PARTICLE OPERATORS

The preceding examples of the symmetry description of two-particle operators may be regarded as particular cases of operators formed from linear combinations of two-particle operators of the generic type

$$\sum_{i \neq j} \{\mathbf{w}_i^{(\kappa_1 k_1)} \mathbf{w}_j^{(\kappa_2 k_2)}\}_Q^{(\kappa_{12} k_{12})K} \quad (255)$$

However, the number of symmetry types of two-particle operators is limited by the symmetry classification of the states between which the operators act and it is of some interest to determine the possible symmetry types available and the number of independent two-particle operators that can be associated with a given symmetry type.

We shall limit our discussion to scalar operators and hence only consider two-particle operators with $\kappa_{12} = k_{12}$ and $K = Q = 0$. Since the maximum multiplicity in l^2 is 3 we are limited to operators transforming like $^1S_0$, $^3P_0$, and $^5D_0$ states. The limitation of our attention to scalar operators amounts to neglecting consideration of interactions that destroy the total angular momentum J as a good quantum number, e.g., hyperfine interactions and crystal field effects.

Since the operators $\mathbf{W}^{(\kappa k)}$ with $\kappa + k$ *even* have symplectic symmetry $\langle 0 \rangle$ and $\langle 1^2 \rangle$ while those with $\kappa + k$ *odd* have $\langle 2 \rangle$ symmetry we may obtain a complete description of the possible symmetry classifications available for the description of the particle operators by examining the plethysm

$$(\langle 0 \rangle + \langle 1^2 \rangle + \langle 2 \rangle) \otimes \{2\}$$
$$= \langle 0 \rangle \otimes \{2\} + \langle 1^2 \rangle \otimes \{2\} + \langle 2 \rangle \otimes \{2\} + \langle 0 \rangle\langle 1^2 \rangle + \langle 0 \rangle\langle 2 \rangle + \langle 1^2 \rangle\langle 2 \rangle \quad (256)$$

We may divide the terms appearing on the right-hand side into three groups

$$(++)\ \langle 0 \rangle \otimes \{2\}; \qquad \langle 1^2 \rangle \otimes \{2\}; \qquad \langle 0 \rangle\langle 1^2 \rangle \quad (257a)$$
$$(--)\ \langle 2 \rangle \otimes \{2\} \quad (257b)$$
$$(+-)\ \langle 0 \rangle\langle 2 \rangle; \qquad \langle 1^2 \rangle\langle 2 \rangle \quad (257c)$$

The operators whose symmetry description is contained in Eq. 257*c* will necessarily be non-Hermitian and may be excluded from further consideration. Explicit evaluation of the remaining plethysms shows that only the symplectic symmetries $\langle 0 \rangle$, $\langle 1^2 \rangle$, $\langle 1^4 \rangle$, and $\langle 2^2 \rangle$ are available for the symmetry description of two-particle operators. Of these, only $\langle 2^2 \rangle$ and $\langle 1^4 \rangle$ can be associated with operators transforming like ${}^5D_0$ states and $\langle 1^2 \rangle$, $\langle 1^4 \rangle$, and $\langle 2^2 \rangle$ with operators transforming like ${}^3P_0$ states.

The symmetry classification of generalized two-particle operators may be further extended by considering the restrictions $Sp_{4l+2} \to SU_2 \times R_{2l+1} \to SU_2 \times R_3$ and in the case of f^n configurations the group G_2 is available. It is instructive to first consider the case of generalized two-particle operators acting in the p^n configurations. The complete classification under $Sp_6 \to SU_2 \times R_3$ yields the following list of symmetry types available for the symmetry description of all possible two-particle operators acting in p^n.

Type	Plethysm	Sp_6	$SU_2 \times R_3$
$(++)$	$\langle 0 \rangle \otimes \{2\}$	$\langle 0 \rangle$	${}^1S$
	$\langle 1^2 \rangle \otimes \{2\}$	$\langle 0 \rangle$	${}^1S$
		$\langle 2^2 \rangle$	${}^1S$
$(--)$	$\langle 2 \rangle \otimes \{2\}$	$\langle 2 \rangle$	${}^1S$
		$\langle 2^2 \rangle$	${}^1S$
$(++)$	$\langle 0 \rangle\langle 1^2 \rangle$	$\langle 1^2 \rangle$	${}^3P$
	$\langle 1^2 \rangle \otimes \{2\}$	$\langle 1^2 \rangle$	${}^3P$
		$\langle 2^2 \rangle$	${}^3P$
$(--)$	$\langle 2 \rangle \otimes \{2\}$	$\langle 1^2 \rangle$	${}^3P$
		$\langle 2^2 \rangle$	${}^3P$
$(++)$	$\langle 1^2 \rangle \otimes \{2\}$	$\langle 2^2 \rangle$	${}^5D$
$(--)$	$\langle 2 \rangle \otimes \{2\}$	$\langle 2^2 \rangle$	${}^5D$

The above list contains 12 entries. However, since only seven independent matrix elements are required to describe the energies of the states of p^2 it is clear that these cannot be 12 independent operators.

We first note that p^2 contains just three terms ^1S, ^1D, and ^3P and hence there cannot be more than three independent operators transforming like a ^1S state. Similarly, there cannot be more than three independent operators transforming like ^3P states and one like a ^5D state. This means, for example, that the matrix elements of an operator of symmetry type $(++)\langle 2^2\rangle^5D$ must be proportional to those of an operator of symmetry type $(--)\langle 2^2\rangle^5D$ and hence it must be possible to combine the two operators into a single operator of symmetry type $\langle 2^2\rangle^5D$.

There is no difficulty in extending these methods of classifying generalized two-particle operators to the d^n and f^n configurations. For d^n we readily obtain the following list:

Type	Plethysm	Sp_{10}	R_5	$SU_2 \times R_3$
$(++)$	$\langle 0\rangle \otimes \{2\}$	$\langle 0\rangle$	[00]	^1S
	$\langle 1^2\rangle \otimes \{2\}$	$\langle 0\rangle$	[00]	^1S
		$\langle 2^2\rangle$	[00]	^1S
			[22]	^1S
		$\langle 1^4\rangle$	[00]	^1S
$(--)$	$\langle 2\rangle \otimes \{2\}$	$\langle 0\rangle$	[00]	^1S
		$\langle 2^2\rangle$	[00]	^1S
			[22]	^1S
$(++)$	$\langle 0\rangle\ \langle 1^2\rangle$	$\langle 1^2\rangle$	[11]	^3P
	$\langle 1^2\rangle \otimes \{2\}$	$\langle 1^2\rangle$	[11]	^3P
		$\langle 2^2\rangle$	[11]	^3P
			[21]	^3P
			[31]	^3P
		$\langle 1^4\rangle$	[21]	^3P
$(--)$	$\langle 2\rangle \times \{2\}$	$\langle 1^2\rangle$	[11]	^3P
		$\langle 2^2\rangle$	[11]	^3P
			[21]	^3P
			[31]	^3P
$(++)$	$\langle 1^2\rangle \times \{2\}$	$\langle 2^2\rangle$	[20]	^5D
			[22]	^5D
		$\langle 1^4\rangle$	[10]	^5D
$(--)$	$\langle 2\rangle \times \{2\}$	$\langle 2^2\rangle$	[20]	^5D
			[22]	^5D

This list of 23 symmetry types clearly represents the transformation properties of a number of operators that must be simply related. There are 14 significant matrix elements in d^2 and a careful analysis of the above list indicates that there are certainly not more than 14 independent operators. Whether this number could be further reduced could only be determined by explicit construction of the operators, a tedious but straightforward task.

10.10 EFFECTIVE TWO-PARTICLE OPERATORS

It is common practice in atomic spectroscopy to allow for the effects of configuration interaction on the energy level structure of equivalent electron configurations of the type l^n by treating the Slater radial integrals associated with the Coulomb interaction as parameters to be adjusted to the experimental data and introducing effective N-particle operators with further associated parameters.[138–140] If we limit our attention to effective two-particle operators we have, in addition to $(l + 1)$ symmetrized Coulomb operators made up of scalar products of *even* rank orbital tensor operators, l symmetrized scalar operators which are constructed from scalar products of *odd* rank orbital tensor operators. These l symmetrized scalar operators transform in precisely the same manner as the symmetrized forms of the orbit-orbit interaction and part of the contact spin-spin interaction. Thus we may conclude that the parameters associated with the l symmetrized effective two-particle operators absorb not only the effects of perturbations produced by Coulomb interaction with other configurations, but also a number of intra-l^n interactions that are normally excluded in making specific calculations. Clearly if the radial integrals are treated as parameters and the l symmetrized effective two-particle operators are included we are not permitted to add to our energy matrices the orbit-orbit or part of the contact spin-spin interactions as they are already accommodated by the parameters.

It is apparent from the preceding section that it is possible to construct scalar two-particle operators which involve bilinear products of the tensor operators $\mathbf{w}^{(\kappa k)}$ with $\kappa = 0$ and 1. Effective operators of this type, among others, arise when the effects of configuration interaction by the spin-dependent interactions are taken into account. Again we must conclude that part of the effects of these interactions will be accommodated by the parameters associated with the usual effective two-particle operators.

Generally, the use of effective operators tends to accommodate more effects than the original reasons that inspired their introduction. This constitutes both the strength and the weakness of the method of effective operators. While the method is well-adapted to correlating a considerable mass of energy level data in terms of a few parameters the physical significance and composition of the parameters is obscured.

Future work will undoubtedly be directed more toward *ab initio* calculations using atomic wave functions to calculate all radial integrals. Klapisch's approach[141] of expressing the central field potential in terms of a few parameters and then calculating explicitly both the radial and angular parts using an iterative method to modify the central field parameters would seem to be capable of considerable development and still retaining all the advantages brought about by the separation of the radial and angular variables.

It is quite apparent that once the effects of spin-spin, spin-other-orbit, and configuration interaction are taken into account by regarding the radial parts as parameters there will be as many parameters as there are independent matrix elements in l^2 and thus no significance could be associated with the results of adjusting the parameters to experimental data since the procedure would be entirely equivalent to treating the complete set of two-particle matrix elements as parameters. Any results obtained would only indicate the extent to which N-particle operators with $N > 2$ are necessary in the description of atomic spectra.

10.11 EFFECTIVE THREE-PARTICLE OPERATORS

Rajnak and Wybourne[138] have demonstrated that the second-order effects produced by single-particle excitations may be represented in terms of effective three-particle operators. Extensive analyses of these operators have been made by Feneuille[142] for the case of d^n configurations and by Judd[85] for f^n configurations. Smith and Wybourne[46] have given a general account of the application of symmetry principles to the classification of N-particle operators.

The N-particle orbital operators may be constructed from N-fold products of single particle tensor operators $\boldsymbol{v}^{(k)}$ whose components $v_q^{(k)}$ are defined by the equation

$$\langle lm| \, v_q^{(k)} \, |lm'\rangle = (-1)^{l-m}(2k+1)^{\frac{1}{2}}\begin{pmatrix} l & k & l \\ -m & q & m \end{pmatrix} \tag{258}$$

If $\mathscr{A}$ represents a set of operators chosen from the set of $v_q^{(k)}$'s defined above then it is clear that the representation of R_3 spanned by the completely symmetric linear combinations of products of degree N in the single particle operators will be given by the plethysm

$$\mathscr{A} \otimes \{N\} \tag{259}$$

In particular, if the full set of operators transforming according to the reducible representation $[l][l]$ of R_3 is used the complete set of all possible N-particle operators will be given by

$$[l][l] \otimes \{N\} \tag{260}$$

Every operator in the set which is constructed so that it involves $\mathbf{v}^{(0)}$ α times will be equivalent to an $(N - \alpha)$ particle operator. It is convenient to exclude these reducible operators and to consider only those operators which are not equivalent to operators involving a smaller number of particles. If the operator $\mathbf{v}^{(0)}$ is excluded then the plethysm

$$\left(\sum_{k=1}^{2l+1}[k]\right) \otimes \{N\} \tag{261}$$

will serve to classify the irreducible N-particle operators.

A prescription for the formation and classification of irreducible N-particle operators can be obtained by noting that

$$[l][l] = [l] \otimes (\{2\} + \{11\}) \tag{262}$$

$$= [l] \otimes ([0] + [2] + [11]) \tag{263}$$

where the symmetry symbol $\{2\}$ produces the representations $[k]$ where k is *even* and $\{11\}$ produces those for k *odd*. The operator $\mathbf{v}^{(0)}$ may be excluded in the classification of irreducible N-particle operators by simply excluding the symmetry symbol [0] in the plethysm. When this is done the plethysm appropriate to the classification of irreducible N-particle operators becomes

$$[l] \otimes ([2] + [11]) \otimes \{N\} \tag{264}$$

If the plethysm

$$([2] + [11]) \otimes \{N\} \tag{265}$$

is performed first a set of symmetry symbols $[\mu]$ are obtained which can be used to classify the operators that transform according to the representations appearing in the plethysm

$$[l] \otimes [\mu] \tag{266}$$

in R_3. In general some of these symbols will be nonstandard and will have to be modified as discussed in Section 5.11.

In a perturbation theory that is restricted to Coulomb interactions only those operators that transform as scalars need to be considered. As a result only those plethysms $[l] \otimes [\mu]$ which yield a [0] representation in R_3 are selected from the plethysm of Eq. 264. For each one of these there will be an irreducible N-particle scalar operator that can be classified with the symmetry symbol $[\mu]$. In some cases it is possible to obtain a finer classification by expanding the symbols $[\mu]$ of R_{2l+1} into the characters of a subgroup.

If the plethysm of Eq. 264 is expanded as follows

$$[l] \otimes ([2] + [11]) \otimes \{N\} = [l] \otimes \left[\sum_{\alpha=0}^{N}([2] \otimes \{\alpha\})([11] \otimes \{N - \alpha\})\right] \tag{267}$$

it becomes clear that the operators transforming according to the representations given by the plethysm

$$[l] \otimes [([2] \otimes \{\alpha\})([11] \otimes \{N - \alpha\})] \tag{268}$$

will involve α single-particle operators with k *even* and $(N - \alpha)$ with k *odd* and the operator would be said to be of type $(++ \cdots + - - \cdots -)$ where there are $\alpha(+)$'s and $(N - \alpha)(-)$'s.

Irreducible N-particle operators involving products of an *odd* number of tensor operators of *odd* rank (i.e., $(N - \alpha)$ is *odd*) will be non-Hermitian and as such would not be expected to appear in a perturbation expansion and will be excluded from further discussion.

The symmetry symbols occurring in the plethysm of Eq. 268 may be readily interpreted in terms of conventional group theory. The N-particle operators span a $4l(l + 1)$ – dimensional space and in fact belong to the symmetric $\{N\}$ representation of the unitary group $U_{4l(l+1)}$. The symmetry symbols $\{\alpha\}$ and $\{N - \alpha\}$ of Eq. 268 label the symmetric representations $\{\alpha\}$ of $U^A_{l(2l+3)}$ and $\{N - \alpha\}$ of $U^B_{l(2l+1)}$, respectively. Since the direct product group $U^A_{l(2l+3)} \times U^B_{l(2l+1)}$ is a subgroup of $U_{4l(l+1)}$ the plethysms $[2] \otimes \{\alpha\}$ and $[11] \otimes \{N - \alpha\}$ can be interpreted as particular decompositions of the representation $\{\alpha\}$ of $U^A_{l(2l+3)}$ and $\{N - \alpha\}$ of $U^B_{l(2l+1)}$ into those of R_{2l+1}. Schematically, the resolution of the plethysm may be interpreted as selecting certain of the representations derived from the group decomposition

$$U_{4l(l+1)} \rightarrow U^{A}_{l(2l+3)} \times U^{B}_{l(2l+1)} \rightarrow R^{A}_{2l+1} \times R^{B}_{2l+1} \begin{matrix} \nearrow & R_{2l+1} & \searrow \\ \searrow & R_3{}^A \times R_3{}^B & \nearrow \end{matrix} R_3 \tag{269}$$

of the symmetric representation $\{N\}$ of $U_{4l(l+1)}$.

When $\alpha = N$ in Eq. 268 the N-particle operators, which will be constructed entirely from N-tuple products of *even* rank single-particle operators, may be labeled by the representations $[\lambda]$ of R^A_{2l+1} contained in the decomposition of the representation $\{N\}$ of $U^A_{l(2l+3)}$ according to the plethysm $[2] \times \{N\}$. In other cases the operators will be labeled by the sequence

$$([\lambda][\lambda'])[\mu]0 \tag{270}$$

where $[\lambda]$ is the representation of R^A_{2l+1} contained in the decomposition of $[\alpha]$ of $U^A_{l(2l+3)}$ according to the plethysm $[2] \otimes \{\alpha\}$, $[\lambda']$ that of $R^B_{l(2l+1)}$ contained in the plethysm $[11] \otimes \{N - \alpha\}$, $[\mu]$ the representation of R_{2l+1} contained in the reduction of the product representation $[\lambda][\lambda']$ and 0 indicates the identity representation $[0]$ of R_3 obtained in performing the plethysm $[l] \otimes [\mu]$.

The typical plethysms $[2] \otimes \{n\}$ and $[11] \otimes \{n\}$ occurring in Eq. 268 may be readily evaluated by noting that

$$\{2\} \otimes \{n\} = \sum \{\mu\} \tag{271}$$

and

$$[11] \otimes \{n\} = \{11\} \otimes \{n\} = \sum \{\tilde{\mu}\} \tag{272}$$

where the summation in Eq. 271 is over all partitions (μ) of $2n$ into even parts while in Eq. 272 the summation is over the partitions $(\tilde{\mu})$ conjugate to those arising in Eq. 271. The plethysm $[2] \otimes \{n\}$ is then obtained as

$$[2] \otimes \{n\} = \{2\} \otimes (\{n\} - \{n-1\}) \tag{273}$$

In the table below we give the classification obtained for Hermitian scalar N-particle operators ($N \leq 3$) for d^n configurations.

N	Plethysm	Label	Classification	Type
0	$[2] \times ([0] \otimes [0])$	e_0	$[0]0$	
2	$[2] \otimes ([2] \otimes \{2\})$	e_1	$[0]0$	$(++)$
		e_2	$[22]0$	
	$[2] \otimes ([11] \otimes \{2\})$	e_3	$[0]0$	$(--)$
		e_4	$[22]0$	
3	$[2] \otimes ([2] \otimes \{3\})$	t_1	$[0]0$	$(+++)$
		t_2	$[22]0$	
		t_3	$[42]0$	
		t_4	$[6]0$	
	$[2] \otimes \{2\}([11] \otimes \{2\})$	t_5	$([2][2])[0]0$	$(+--)$
		t_6	$([2][2])[22]0$	
		t_7	$([2][2])[22]0$	
		t_8	$([2][1])[3]0$	
		t_9	$([2][22])[42]0$	

The symmetry types labeled e_0–e_2 are exactly the same as those found for the Coulomb interaction in Section 10.6, Eq. 215 while the symmetry types labeled e_3 and e_4 are the same as those occurring in Eq. 234. The symmetry types labeled t_1–t_4 describe the transformation properties of three-particle operators constructed from linear combinations of triple products of single particle tensor operators $\mathbf{v}^{(k)}$ where k is *even* while the symmetry types t_5–t_9 described the transformation properties of three-particle operators constructed from linear combinations of two tensor operators $\mathbf{v}^{(k)}$ with k *odd* and one with k *even*.

The operator t_4 transforms as $[6]$ under R_5 and since $c([\lambda][\lambda'][6] = 0)$ for all choices of $[\lambda]$ and $[\lambda']$ occurring in d^n its matrix elements will all be null. In second-order perturbation theory the only three-particle effective operators that arise are of type $(+++)$. The four operators e_0, e_1, e_3, and t_1 all transform as $[0]0$ and thus their matrix elements are completely diagonal in $[\lambda]$

and independent of L. Since there are just three different representations $[\lambda]$ in d^3 only three independent operators can exist. Thus if we choose to associate parameters with e_0, e_1, e_3, and t_1 rather than the explicit forms of their radial parts we may dispense with t_1 as its effect may be absorbed in the parameters associated with e_0, e_1, and e_3. Thus, to second-order, just two three-particle operators t_2 and t_3 need to be considered. Feneuille[142] has considered the explicit construction of operators transforming as t_2 and t_3 while Smith and Wybourne[46] have examined the properties of operators transforming as $(+--)$. In this case only the additional operators t_6 and t_8 are required as t_5, t_7, and t_9 may be absorbed in the parameters associated with the other two- and three-particle operators.

Clearly the classification of spin-dependent three particle operators constructed from triple products of the double tensors $\mathbf{W}^{(\kappa k)}$ could readily be made. However, it is doubtful that such a calculation would be worthwhile as in the parametrized form of the effective operators the number of parameters would be excessive for all but the most complex configurations where a detailed knowledge is lacking.

One possibility that might be considered would be to treat the independent matrix elements of l^2 as parameters along the lines of the discussion of Section 10.10 and add to these the three-particle scalar operators as further parameters. This would have the effect of giving a complete description of all possible two-particle interactions together with a complete description of the second-order Coulomb configuration interaction. Such an approach would undoubtedly give a very good description of the energy levels including term separations and multiplet separations, for systems having three or more equivalent d or f electrons. This description could be useful for the prediction of unknown levels and their associated properties though of course much of the information concerning the details of the relevant interactions would be lost in the parameters.

11

Symmetry Properties of Interactions in Mixed Configurations

11.1 TRANSFORMATION PROPERTIES OF SINGLE-PARTICLE OPERATORS

There are no substantial difficulties in extending the methods of the preceding chapter to mixed configurations of the types discussed in Chapter 9. Here we shall limit our attention to configurations of the type $(l_1 + l_2)^n$. The basic single-particle operators $\mathbf{w}^{(\kappa k)}(l_1, l_2)$ are defined by the relation

$$\langle sl\| w^{(\kappa k)}(l_1, l_2) \|sl'\rangle = \delta(l, l_1)\, \delta(l', l_2)\{[\kappa, k]\}^{\frac{1}{2}} \tag{274}$$

The complete set of one-particle operators will comprise the sets $\mathbf{w}^{(\kappa k)}(l_1, l_1)$ $(\kappa = 0,\ 1;\ 0 \leq k \leq 2l_1)$, $\mathbf{w}^{(\kappa k)}(l_2, l_2)$ $(\kappa = 0,\ 1;\ 0 \leq k \leq 2l_2)$, $\mathbf{w}^{(\kappa k)}(l_1, l_2)$ $(\kappa = 0,\ 1,\ |l_1 - l_2| \leq k \leq l_1 + l_2)$, and $\mathbf{w}^{(\kappa k)}(l_2, l_1)$ $(\kappa = 0,\ 1;\ |l_1 - l_2| \leq k \leq l_1 + l_2)$.

The operators $\mathbf{w}^{(\kappa k)}(l_1, l_2)$ and $\mathbf{w}^{(\kappa k)}(l_2, l_1)$ are not necessarily Hermitian and it is desirable to replace them by the linear combinations[97,99a]

$$\mathbf{w}^{\pm(\kappa k)}(l_1, l_2) = [\mathbf{w}^{(\kappa k)}(l_1, l_2) \pm (-1)^{\kappa+k+l_1-l_2}\mathbf{w}^{(\kappa k)}(l_2, l_1)]/2^{\frac{1}{2}} \tag{275}$$

The states of $(l_1 + l_2)^n$ will be supposed to be classified by the chain of groups

$$U_{4(l_2+l_2+1)} \rightarrow Sp_{4(l_2+l_2+1)} \rightarrow SU_2(R_{2(l_2+l_2+1)} \rightarrow R_{2l_2+1} \times R_{2l_2+1} \rightarrow R_3) \tag{276}$$

The antisymmetric states of the $(l_1 + l_2)^2$ configuration may be found by

Table 11-1 Transformation Properties of One-Particle Operators in $(l_1 + l_2)^n$ Configurations

$U_{4(l_1+l_2+1)}$	$Sp_{4(l_1+l_2+1)}$	$SU_2 \times R_{2(l_1+l_2+1)}$	$R_{2l_1+1} \times R_{2l_2+1}$	$SU_2 \times R_3$
$\{21^{4(l_1+l_2)+2}\}$	$\langle 1^2 \rangle$	$^3[1^3]$	$[1^2] \times [0]$	$\mathbf{w}^{(1k)}(l_1, l_1) \quad 1 \le k \le 2l_1 - 1 \quad (k \text{ odd})$
			$[1] \times [1]$	$\overset{+}{\mathbf{w}}{}^{(1k)}(l_1, l_2) \quad \lvert l_1 - l_2 \rvert \le k \le l_1 + l_2 \quad (k \text{ even})$
			$[0] \times [1^2]$	$\mathbf{w}^{(1k)}(l_2, l_2) \quad 1 \le k \le 2l_2 - 1 \quad (k \text{ odd})$
		$^1[2]$	$[2] \times [0]$	$\mathbf{w}^{(0k)}(l_1, l_1) \quad 2 \le k \le 2l_1 \quad (k \text{ even})$
			$[1] \times [1]$	$\overset{+}{\mathbf{w}}{}^{(0k)}(l_1, l_2) \quad \lvert l_1 - l_2 \rvert \le k \le l_1 + l_2 \quad (k \text{ even})$
			$[0] \times [2]$	$\mathbf{w}^{(0k)}(l_2, l_2) \quad 2 \le k \le 2l_2 \quad (k \text{ even})$
			$[0] \times [0]$	$[(2l_2 + 1)^{\frac{1}{2}}\mathbf{w}^{(00)}(l_1, l_1)$ $- (2l_1 + 1)^{\frac{1}{2}}\mathbf{w}^{(00)}(l_2, l_2)][(2l_1 + 1) + (2l_2 + 1)]^{-\frac{1}{2}}$
	$\langle 2 \rangle$	$^1[1^2]$	$[1^2] \times [0]$	$\mathbf{w}^{(0k)}(l_1, l_1) \quad 1 \le k \le 2l_1 - 1 \quad (k \text{ odd})$
			$[1] \times [1]$	$\overset{-}{\mathbf{w}}{}^{(0k)}(l_1, l_2) \quad \lvert l_1 - l_2 \rvert \le k \le l_1 + l_2 \quad (k \text{ odd})$
			$[0] \times [1^2]$	$\mathbf{w}^{(0k)}(l_2, l_2) \quad 1 \le k \le 2l_2 - 1 \quad (k \text{ odd})$
		$^3[2]$	$[2] \times [0]$	$\mathbf{w}^{(1k)}(l_1, l_1) \quad 2 \le k \le 2l_1 \quad (k \text{ even})$
			$[1] \times [1]$	$\overset{-}{\mathbf{w}}{}^{(1k)}(l_1, l_2) \quad \lvert l_1 - l_2 \rvert \le k \le l_1 + l_2 \quad (k \text{ odd})$
			$[0] \times [2]$	$\mathbf{w}^{(1k)}(l_2, l_2) \quad 2 \le k \le 2l_2 \quad (k \text{ even})$
			$[0] \times [0]$	$[(2l_2 + 1)^{\frac{1}{2}}\mathbf{w}^{(00)}(l_1, l_1) - (2l_1 + 1)^{\frac{1}{2}}\mathbf{w}^{(00)}(l_2, l_2)][(2l_1 + 1) + (2l_2 + 1)]^{-\frac{1}{2}}$
		$^3[0]$	$[0] \times [0]$	$[(2l_1 + 1)^{\frac{1}{2}}\mathbf{w}^{(00)}(l_1, l_1)$ $+ (2l_2 + 1)^{\frac{1}{2}}\mathbf{w}^{(00)}(l_2, l_2)][(2l_1 + 1) + (2l_2 + 1)]^{-\frac{1}{2}}$
$\{0\}$	$\langle 0 \rangle$	$^1[0]$	$[0] \times [0]$	$[(2l_1 + 1)^{\frac{1}{2}}\mathbf{w}^{(00)}(l_1, l_1) + (2l_2 + 1)^{\frac{1}{2}}\mathbf{w}^{(00)}(l_2, l_2)][(2l_1 + 1) + (2l_2 + 1)]^{-\frac{1}{2}}$

Table 11-2 Transformation Properties of One-Particle Operators in $(d + s)^n$ Configurations

U_{12}	Sp_{12}	$SU_2 \times R_6$	R_5	$SU_2 \times R_3$
$\{21^{10}\}$	$\langle 1^2 \rangle$	${}^3[1^2]$	[1]	$\overset{+}{\mathbf{w}}^{(12)}(ds)$
			$[1^2]$	$\mathbf{w}^{(11)}(dd), \mathbf{w}^{(13)}(dd)$
		${}^1[2]$	[0]	$\{\mathbf{w}^{(00)}(dd) - 5^{\frac{1}{2}}\mathbf{w}^{(00)}(ss)\}/6^{\frac{1}{2}}$
			[1]	$\overset{+}{\mathbf{w}}^{(02)}(ds)$
			[2]	$\mathbf{w}^{(02)}(dd), \mathbf{w}^{(04)}(dd)$
	$\langle 2 \rangle$	${}^1[1^2]$	[1]	$\overset{-}{\mathbf{w}}^{(02)}(ds)$
			$[1^2]$	$\mathbf{w}^{(01)}(dd), \mathbf{w}^{(03)}(dd)$
		${}^3[2]$	[0]	$\{\mathbf{w}^{(10)}(dd) - 5^{\frac{1}{2}}\mathbf{w}^{(10)}(ss)\}/6^{\frac{1}{2}}$
			[1]	$\overset{-}{\mathbf{w}}^{(12)}(ds)$
			[2]	$\mathbf{w}^{(12)}(dd), \mathbf{w}^{(14)}(dd)$
		${}^3[0]$	[0]	$\{5^{\frac{1}{2}}\mathbf{w}^{(10)}(dd) + \mathbf{w}^{(10)}(ss)\}/6^{\frac{1}{2}}$
$\{0\}$	$\langle 0 \rangle$	${}^1[0]$	[0]	$\{5^{\frac{1}{2}}\mathbf{w}^{(00)}(dd) + \mathbf{w}^{(00)}(ss)\}/6^{\frac{1}{2}}$

following the methods of Chapter 9. The single-particle operators may then be classified by identifying their spin and orbital ranks with the corresponding representations of $SU_2 \times R_3$ found from Eq. 276.

11.2 COULOMB INTERACTION IN $(d + s)^n$ CONFIGURATIONS

The Coulomb interaction in $(l_1 + l_2)^n$ configurations may be expressed in terms of scalar products of the tensor operators given in Table 11-1 as transforming as ${}^1[2]$ and ${}^1[0]$ under $SU_2 \times R_{2(l_1+l_2+1)}$. Just like in the case of equivalent electron configurations discussed earlier we may form a set of symmetrized scalar two-particle operators associated with the different symmetry types that arise in the separation of the basic plethysm

$$\begin{aligned}([2] + [0]) \otimes \{2\} &= [2] \otimes \{2\} + [0] \otimes \{2\} + [2][0] \\ &= [4] + [2^2] + 2[2] + 2[0] \qquad (277)\end{aligned}$$

Each of the representations of $R_{2(l_1+l_2+1)}$ appearing in the right-hand side of Eq. 277 may be decomposed into representations of $R_{2l_1+1} \times R_{2l_2+1}$ using the methods of Section 7.6 and noting that under $R_{2(l_1+l_2+1)} \to R_{2l_1+1} \times R_{2l_2+1}$ we take $[1] \to [1][0]' + [0][1]'$, where the representations of R_{2l_2+1} are distinguished from those of R_{2l_1+1} by a dash. Proceeding in this manner we obtain

$$[0] \to [0][0]'$$
$$[2] \to [2][0]' + [0][2]' + [1][1]' + [0][0]'$$
$$[2^2] \to [2^2][0]' + [0][2^2]' + [21][1]' + [1][21]' + [1^2][1^2]' + [2][2]' + [1][1]' + [0][2]' + [2][0]'$$
$$[4] \to [4][0]' + [0][4]' + [3][1]' + [1][3]' + [2][2]' + [1][1]' + [0][0]' + [2][0]' + [0][2]'$$

Each representation of $R_{2l_1+1} \times R_{2l_2+1}$ may be decomposed into those of $R_3 \times R_3$ by noting that under $R_{2l_1+1} \to R_3$ we have $[1] \to [l_1]$ and under $R_{2l_2+1} \to R_3$ we have $[1]' \to [l_2]'$. Finally, each representation of $R_3 \times R_3$ may be decomposed into those of R_3 and the symmetry types associated with the S-states that occur in the final decomposition identified.

The problem of determining the symmetry types for the configuration $(d + s)^n$ becomes particularly simple since we need only consider the decompositions under $R_6 \to R_5 \to R_3$ to give the complete list of symmetry types as

$[000] \otimes \{2\}$	$[000][00]S$	e_0
$[200][000]$	$[200][00]S$	e_2
$[200] \otimes \{2\}$	$[000][00]S$	e_1
	$[200][00]S$	e_3
	$[220][22]S$	e_4
	$[400][00]S$	e_5
	$[400][30]S$	e_6

The symmetry properties of the operators $e_0, e_1, \ldots, e_6$ may further be studied by noting that the representation $^1[000]$ of $SU_2 \times R_6$ is associated with the symplectic symmetry $\langle 0 \rangle$ under Sp_{12} while $^1[200]$ occurs as one of the terms in the decomposition of $\langle 1^2 \rangle$ under $Sp_{12} \to SU_2 \times R_6$. Since

$$\langle 0 \rangle \otimes \{2\} = \langle 0 \rangle$$
$$\langle 1^2 \rangle \langle 0 \rangle = \langle 1^2 \rangle$$
$$\langle 1^2 \rangle \otimes \{2\} = \langle 1^4 \rangle + \langle 2^2 \rangle + \langle 1^2 \rangle + \langle 0 \rangle$$

we conclude that the representations $^1[000]$, $^1[200]$, $^1[220]$, and $^1[400]$ that arise in the classification of the Coulomb interaction must be terms in the decomposition of the symplectic symmetries $\langle 0 \rangle$, $\langle 1^2 \rangle$, $\langle 2^2 \rangle$, and $\langle 1^4 \rangle$ of

$Sp_{12} \rightarrow SU_2 \times R_6$. Making the respective decompositions we find

$$\langle 0 \rangle \rightarrow {}^1[000]$$
$$\langle 1^2 \rangle \rightarrow {}^1[200] + {}^3[110]$$
$$\langle 1^4 \rangle \rightarrow {}^1[220] + {}^3([211] + [21-1]) + {}^5[110]$$
$$\langle 2^2 \rangle \rightarrow {}^1([400] + [220] + [200] + [000]) + {}^3([211] + [21-1] + [310] + [110] + [200]) + {}^5([220] + [200] + [000])$$

Thus the operators e_5 and e_6 which have [400] symmetry under R_6 must be associated with the pure symplectic symmetry $\langle 2^2 \rangle$ of Sp_{12}. The operators e_2 may be associated with the symplectic symmetry $\langle 2^2 \rangle$ and $\langle 1^4 \rangle$.

If the representations [0], $[1^2]$, and $[1^4]$ of R_{24} are decomposed into those of $SU_2 \times Sp_{12}$ it is found that the symplectic symmetry $\langle 2^2 \rangle$ is associated with the quasi-spin $K = 0$, $\langle 1^4 \rangle$ with $K = 2$, and $\langle 1^2 \rangle$ with $K = 0$, 1, and 2. Thus the matrix elements of e_5 and e_6 will be diagonal in the seniority scheme, the operators e_2 and e_3 may each be represented as a linear combination of operators having quasi-spin ranks $K = 0$, 1, and 2 while e_4 will be representable as the sum of two operators one of quasi-spin $K = 0$ and the other with $K = 2$.

Details of the explicit construction of the operators $e_0, e_1, \ldots, e_6$ have been given by Feneuille.[97] Judd and Armstrong[143] have discussed further the properties of e_4 from the point of view of LL-coupling.

11.3 EFFECTIVE TWO-PARTICLE OPERATORS IN MIXED CONFIGURATIONS

The introduction of effective scalar two-particle operators constructed from the tensor operators $\mathbf{w}^{(0k)}(ll)$ where k is odd leads to a considerable improvement in the parametrized description of the energy levels of l^n type configurations. In $(l_1 + l_2)^n$-type configurations the effective two-particle operators are constructed from the tensor-operators $\mathbf{w}^{(0k)}(l_1, l_1)$, $\mathbf{w}^{-(0k)}(l_1, l_2)$, and $\mathbf{w}^{(0k)}(l_2, l_2)$ which together transform as ${}^1[1^2]$ under $SU_2 \times R_{2(l_1+l_2+1)}$. The symmetry types available for the description of the effective scalar two-particle oprators are found from the separation of the plethysm

$$[1^2] \otimes \{2\} = [1^4] + [2^2] + [2] + [0] \tag{278}$$

Again, each of the representations of $R_{2(l_1+l_2+1)}$ appearing on the right-hand side may be decomposed into representations of $R_{2l_2+1} \times R_{2l_2+1}$ noting that

$$[1^4] \rightarrow [1^4][0]' + [0][1^4]' + [1^3][1]' + [1][1^3]' + [1^2][1^2]'$$

Each representation of $R_{2l_1+1} \times R_{2l_2+1}$ may then be decomposed into representations of $R_3 \times R_3$ and then to R_3. The symmetry types associated with each S-state obtained in the final decomposition may then be identified.

In the particular case of the $(d + s)^n$ configurations, consideration of the decompositions under $R_6 \to R_5 \to R_3$ leads to the symmetry classification

$$
\begin{array}{lll}
[1^2] \otimes \{2\} & [000][00]S & e_7 \\
 & [200][00]S & e_8 \\
 & [220][22]S & e_9
\end{array}
$$

If the above three operators are added to the seven symmetrized Coulomb operators of the preceding section we find there are ten independent scalar two-particle operators at our disposal for describing the term structure of the $(d + s)^n$ configurations. This number is equivalent to the number of independent matrix elements that leave S and L as good quantum numbers in $(d + s)^2$.

There is no difficulty in extending the methods of Section 10.9 to the classification of generalized two-particle operators[144] in $(d + s)^n$. Likewise the symmetry classification of the various spin-dependent two-particle interactions follows, apart from minor modifications, the methods of Section 10.8.

11.4 SCALAR THREE-PARTICLE OPERATORS IN $(d + s)^n$ CONFIGURATIONS

As a final example of the application of symmetry principles to mixed configurations we consider the determination of the symmetry description of the scalar three-particle operators that may be constructed from the one-particle operators transforming as ${}^1[2]$ under $SU_2 \times R_6$. We neglect the operators transforming as ${}^1[0]$ under $SU_2 \times R_6$ as it will only yield three-particle operators that are reducible to two- and one-particle operators. The symmetry types involved will be associated with just those terms appearing in the separation of the plethysm

$$
\begin{aligned}
[200] \otimes \{3\} = [000] + 2[200] + [220] + [222] + [22-2] + [310] \\
+ [400] + [420] + [600] \quad (279)
\end{aligned}
$$

Each of the characters of R_6 appearing on the right-hand side of Eq. 279 may be decomposed into characters of R_5 and then into those of R_3. The operators having symmetry-type [600] will all have null matrix elements and may be neglected. When this is done the symmetry types that yield an S-state in the final decomposition are found to be:

[000][00]S	t'_1
[200][00]S	t'_2
[200][00]S	t'_3
[220][22]S	t'_4
[222][22]S	t'_5
[22 − 2][22]S	t'_6
[310][30]S	t'_7
[400][00]S	t'_8
[400][30]S	t'_9
[420][22]S	t'_{10}
[420][30]S	t'_{11}
[420][42]S	t'_{12}

Under the restriction $R_6 \to R_5$ we have

$$[200] \to [00] + [10] + [20]$$

The presence of the [00] representation of R_5 in the above decomposition suggests that the list of symmetry types will contain some symmetries associated with two- and one-particle operators. We may determine the symmetry types available for the description of three-particle operators which are not reducible to two- or one-particle operators by considering the plethysm

$$([200] - [000]) \otimes \{3\} = [000] + [200] + [222] + [22-2] + [310] + [420] + [600] \quad (280)$$

Retaining only those symmetry types which yield an S-state restriction to R_3, we obtain the 11 symmetry types.

[000][00]S	t_1
[200][00]S	t_2
[222][22]S	t_3
[22 − 2][22]S	t_4
[310][30]S	t_5
[420][22]S	t_6
[420][30]S	t_7
[420][42]S	t_8
[600][00]S	t_9
[600][30]S	t_{10}
[600][60]S	t_{11}

The operators having symmetry type [600] will all yield null matrix elements in $(d + s)^n$ and may be discarded. Only three independent operators transforming as [000][00] are required in $(d + s)^3$ and hence the effects of t_1 may be absorbed by the parameters associated with e_0, e_1, and e_7. For similar reasons the effects of the operator t_2 may be absorbed by the parameters associated with e_2, e_3, and e_8.

The three operators t_3, t_4, and t_6 all transform as [22] under R_5 and since only two independent operators of this symmetry type, in addition to the operators e_4 and e_9 are required, we may combine them into two appropriate linear combinations. The explicit construction of the linear combinations, with minor differences of notation and definition, has been made by Feneuille.[99] In all we find 10 effective two-particle operators and five effective three-particle operators are required to represent the effects of configuration interaction on the energy levels of the $(d + s)^n$ configurations.

References

1. G. Racah, "Theory of Complex Spectra IV," *Phys. Rev.*, **76,** 1352 (1949).
2. G. Racah, *Group Theory and Spectroscopy*, Ergeb. der exakten Naturwiss, Vol. 37, Springer, Berlin, 1965.
3. S. Lie and G. Scheffers, *Vorlesungen über continuierliche Gruppen*, Teubner Verlagsgesellschaft, Leipzig, 1938.
4. E. Cartan, *Sur la Structure des Groupes de Transformation finis et continuo*, Thesis, Nony, Paris, 1894.
5. H. Weyl, *Gruppentheorie und Quantermechanike*, S. Hirzel Verlag, Leipzig, 1931; translated by H. P. Robertson, *The Theory of Groups and Quantum Mechanics*, reprinted by Dover Publications, New York.
6. H. Weyl, *The Classical Groups*, Princeton University Press, Princeton, N.J., 1946.
7. H. A. Jahn, "Theoretical Studies in Nuclear Structure I.," *Proc. Roy. Soc. (London)* **A201,** 516 (1950).
8. B. H. Flowers, "Studies in *jj*-coupling I," *Proc. Roy. Soc. (London)*, **A212,** 248 (1952).
9. J. P. Elliott, "Collective Motion in the Nuclear Shell Model," *Proc. Roy. Soc.* (London), **A245,** 128 (1958).
10. D. E. Rutherford, *Substitutional Analysis*, Edinburgh University Press, Edinburgh, 1948.
11. G. de B. Robinson, *Representation Theory of the Symmetric Group*, Edinburgh University Press, Edinburgh, 1961.
12. D. E. Littlewood and A. R. Richardson, "Group Characters and Algebra," *Phil. Trans. Roy. Soc.* (London), **A233,** 99 (1934).
13. D. E. Littlewood, *The Theory of Group Characters*, 2nd ed., Oxford University Press, Oxford, 1958.
14. I. Schur, *Über eine Klasse von Matrizen die sich einer gegebenen Matrix zuordnen lassen*, Inaugural-Dissertation, Berlin, 1901.
15. B. R. Judd, *Operator Techniques in Atomic Spectroscopy*, McGraw-Hill, New York, 1962.
16. M. Hamermesh, *Group Theory*, Addison-Wesley, Reading, Mass., 1962.
17. H. Boerner, *Representations of Groups*, North-Holland Publishing Co., Amsterdam, 1963.
18. J. S. Frame, G. de B. Robinson, and R. M. Thrall, "The Hook Graphs of the Symmetric Group," *Can. J. Math.*, **6,** 316 (1954).
19. Details of the relevant derivations may be found in references 5, 11, 13, 16, and 17.
20. D. E. Littlewood and A. R. Richardson, "Immanants of Some Special Matrices," *Quart. J. Math.* (Oxford), **5,** 269 (1934).
21. T. Muir, *A Treatise on the Theory of Determinants*, Dover Publications Inc., New York, 1960.

22. A. C. Aitken, *Determinants and Matrices*, Oliver and Boyd, London, 1946.
23. M. Zia-ud-Din, "The Characters of the Symmetric Group of order 11!," *Proc. Lond., Math. Soc.*, **39,** (2) 200 (1935).
24. M. Zia-ud-Din, "The Characters of the Symmetric Groups of Degree 12 and 13," *Proc. London Math. Soc.*, **42,** (2) 340 (1937).
25. K. Kondo, "Table of Characters of the Symmetric Group of Degree 14," *Proc. Phys. Math. Soc., Japan*, **22,** (3) 585 (1940).
26. R. L. Bivins, N. Metropolis, P. R. Stein, and M. B. Wells, "Characters of the Symmetric Groups of Degree 15 and 16," *Math. Tables Other Aids Computation*, **8,** 212 (1954).
27. D. E. Littlewood, *A University Algebra*, William Heinemann, London, 1950
28. D. E. Littlewood, *The Skeleton Key of Mathematics*, Hutchison, London, 1949.
29. P. A. MacMahon, *Combinatory Analysis*, Vol. I and II, Cambridge University Press, Cambridge, 1915.
30. G. Frobenius, "Über die Charaktere der Symmetrischen Gruppe," *Sitz Ber. Preuss. Akad.*, Berlin, p. 516 (1900).
31. F. D. Murnaghan, "On the Representations of the Symmetric Group," *Am. J. Math.*, **59,** 437 (1937).
32. F. D. Murnaghan, "The Characters of the Symmetric Group," *Am. J. Math.*, **59,** 739 (1937).
33. G. de B. Robinson, "On the Representations of the Symmetric Group," *Am. J. Math.*, **60,** 745 (1938); **69,** 286 (1947); **70,** 277 (1948).
34. A. Gamba, "Sui Caratteri delle Rappresentaziona del Gruppo Simmetrico," *Atti Accad. Naz. Lincei, Rend., Cl. Sci. Fis. Natur.* VIII, Ser. **12,** 167 (1952).
35. F. D. Murnaghan, "The Characters of the Symmetric Group," *An. Acad. Brasil. Ci.*, **23,** 1 (1951).
36. D. E. Littlewood, "Modular Representations of the Symmetric Group," *Proc. Roy. Soc.* (London), **209,** 333 (1951).
37. F. D. Murnaghan, "The Analysis of the Kronecker Product of Irreducible Representations of the Symmetric Group," *Am. J. Math.*, **60,** 761 (1938).
38. F. D. Murnaghan, "The Analysis of the Kronecker Product of Irreducible Representations of the Symmetric Group," *Proc. Nat. Acad. Sci. USA*, **41,** 515 (1955).
39. A. Gamba and L. A. Radicati, "Sopra un Teorema per le Reduzione di talune Rappresentazione del Gruppo Simmetrico," *Atti. Accad. Naz. Lincei Rend.*, **14,** 632 (1953).
40. F. D. Murnaghan, "On the Kronecker Products of Irreducible Representations of the Symmetric Group," *Proc. Natl. Acad. Sci. USA.*, **42,** 93 (1956).
41. G. de B. Robinson and O. E. Taulbee, "The Reductions of the Inner Product of two Irreducible Representations of S_n," *Proc. Natl. Acad. Sci.*, **40,** 723 (1954).
42. D. E. Littlewood, "The Kronecker Product of Symmetric Group Representations," *J. London Math. Soc.*, **31,** 89 (1956).
43. D. E. Littlewood, "Plethysm and the Inner Product of S-functions," *J. London Math. Soc.*, **32,** 18 (1957).
44. D. E. Littlewood, "The Inner Plethysm of S-functions," *Can. J. Math.*, **10,** 1 (1958).
45. D.E.Littlewood, "Products and Plethysms of Characters with Orthogonal, Symplectic and Symmetric Groups," *Can. J. Math.*, **10,** 17 (1958).
46. P. R. Smith and B. G. Wybourne, "Plethysm and the Theory of Complex Spectra," *J. Math. Phys.*, **9,** 1040 (1968).
47. D. E. Littlewood, "Polynomial Concomitants and Invariant Matrices," *J. London Math. Soc.*, **11,** 49 (1936).
48. D. E. Littlewood, "On Induced and Compound Matrices," *J. London Math. Soc.*, **11,** 370 (1936).

49. D. E. Littlewood, "The Construction of Invariant Matrices," *Proc. London Math. Soc.*, **43,** (2), 226 (1937).
50. D. E. Littlewood, "On Invariant Theory under Restricted Groups," *Trans. Roy. Soc.* (London), **A239,** 387 (1943).
51. D. E. Littlewood, "On the Concomitants of Spin Tensors," *Proc. London Math. Soc.*, **49,** (2), 307 (1947).
52. D. E. Littlewood, "Invariant Theory under Orthogonal Groups," *Proc. London Math. Soc.*, **50,** 349 (1948).
53. D. E. Littlewood, "On Orthogonal and Symplectic Group Characters," *J. Lond. Math. Soc.* **30,** 121 (1955).
54. E. M. Ibrahim, "On a Theorem of Murnaghan," *Proc. Nat. Acad. Sci.* **40,** 306 (1954).
55. F. D. Murnaghan, *The Theory of Group Representations*, Johns Hopkins Press, Baltimore, 1938.
55a. F. D. Murnaghan, *The Unitary and Rotation Groups*, Spartan Books, Washington, D.C. 1962.
55b. M. J. Newell, "On the Representations of the Orthogonal and Symplectic Groups," *Proc. Roy. Irish Acad.*, **54A,** 143 (1951).
55c. M. J. Newell, "Modification Rules for Orthogonal and Symplectic Groups," *Proc. Roy. Irish Acad.*, **54A,** 153 (1951).
56. Zhang Qing-Yu, "A Group Theory Problem in Quantum Mechanics I," *Chinese Math.*, **5,** 424 (1964).
57. Yan Zhi-Da, "A Problem of Lie Groups II," *Chinese Math.*, **3,** 130 (1963).
58. Kuang Zhi-Quan, Some Discussion about "A Problem of Lie Groups II," *Chinese Math.*, **6,** 263 (1965).
59. Shi Sheng-Ming, "On the Induced Representations of a Semisimple Lie Algebra in its Three-Dimensional Principal Subalgebra, and a Calculation for the Case of G_2," *Chinese Math.*, **6,** 610 (1965).
60. M. Kretzschmar, "Gruppentheoretische Unterschungen zum Schalenmodell," *Z. Physik*, **158,** 284 (1960).
61. G. de B. Robinson, "On the Disjoint Product of Irreducible Representations of the Symmetric Group," *Can. J. Math.* **1,** 166 (1949).
62. G. de B. Robinson, "Induced Representations and Invariants," *Can. J. Math.* **2,** 334 (1950).
63. D. E. Littlewood, "Invariant Theory, Tensors and Group Characters," *Trans. Roy. Soc.* (London), **A239,** 305 (1943).
64. M. Zia-ud Din, "Invariant Matrices and *S*-Functions," *Proc. Edinburgh Math. Soc.*, **5,** 43 (1936).
65. E. M. Ibrahim, "Tables for the Plethysm of *S*-function of Degrees 10 and 12," *Proc. Math. Phys. Soc. Egypt*, **5,** 85 (1954).
66. E. M. Ibrahim, "*S*-Functional Plethysms of Degrees 14 and 15," *Proc. Math. Phys. Soc. Egypt*, **10,** 137 (1959).
67. E. M. Ibrahim, "Tables for the Plethysm of *S*-Functions," *Royal Soc.* (London), Depository of Unpublished Tables.
68. E. M. Ibrahim, "The Plethysm of *S*-Functions," *Quart. J. Math.* Oxford, **3,** (2), 50 (1952).
69. H. O. Foulkes, "The New Multiplication of *S*-Functions," *J. London Math. Soc.*, **26,** 132 (1951).
70. H. O. Foulkes, "Differential Operators Associated with *S*-Functions," *J. Lond. Math. Soc.*, **24,** 136 (1949).
71. H. O. Foulkes, "Plethysm of *S*-Functions," *Trans. Roy. Soc.* (London), **A246,** 555 (1954).

72. H. O. Foulkes, "Concomitants of the Quintic and Sextic up to Degree Four in the Coefficients of the Ground Form," *J. London Math. Soc.*, **25,** 205 (1950).
73. M. J. Newell, "A Theorem on the Plethysm of S-Functions," *Quart. J. Math.* Oxford, **2,** 161 (1951).
74. J. A. Todd, "A Note on the Algebra of S-Functions," *Proc. Cambridge Phil. Soc.*, **45,** 328 (1949).
75. D. G. Duncan, "Note on a Formula by Todd," *J. London Math. Soc.*, **27,** 235 (1952).
76. D. G. Duncan, "On D. E. Littlewood's Algebra of S-Functions," *Can. J. Math.*, **4,** 504 (1952).
77. D. G. Duncan, "Note on the Algebra of S-Functions," *Can. J. Math.*, **6,** 509 (1954).
78. F. D. Murnaghan, "On the Analysis of Representations of the Linear Group," *Proc. Natl. Acad. Sci.* USA, **37,** 51 (1951).
79. F. D. Murnaghan, "A Generalization of Hermite's Law of Reciprocity," *Proc. Natl. Acad. Sci.* USA, **37,** 439 (1951).
80. E. M. Ibrahim, "Note on a Paper by Murnaghan," *Proc. Am. Math. Soc.*, **7,** 1000 (1956).
81. F. D. Murnaghan, "On the Analyses of $\{m\} \otimes \{1^k\}$ and $\{m\} \otimes \{k\}$," *Proc. Natl. Acad. Sci.* USA, **40,** 721 (1954).
82. R. H. Makar and S. A. Missiha, "The Coefficient of the S-Function $\{nm - k - r, k, r\}$, $k \leq m$, in the Analysis of $\{m\} \otimes \{\nu\}$, where $\{\nu\}$ is any Partition of n and $m = 5$ or 6," *Proc. K. Accad. Wet.*, **61,** 77 (1958).
83. R. H. Makar, "On the Analysis of the Representations of the Linear Group of Dimension 2," *Proc. K. Accad. Wet.*, **61,** 475 (1958).
84. E. M. Ibrahim, "On D. E. Littlewood's Algebra of S-Functions," *Am. Math. Soc. Proc.*, **7,** 199 (1956).
84a. B. G. Wybourne and P. H. Butler, "The Configurations $(d + s)^N$ and the Group R_6," *J. Phys.*, **30,** 181 (1968).
85. B. R. Judd, "Three-Particle Operators for Equivalent Electrons," *Phys. Rev.*, **141,** 4 (1966).
86. B. R. Judd, "Zeeman Effect as a Prototype for Intra-Atomic Interactions," *Physica*, **33,** 174 (1967).
87. B. R. Judd, *Second Quantization and Atomic Spectroscopy*, The Johns Hopkins Press, Baltimore, Md., 1966.
88. B. R. Judd, "Group Theory in Atomic Spectroscopy," in *Group Theory and Its Applications*, E. M. Loebl, Ed., Academic Press, New York, 1968.
89. G. Racah, "Theory of Complex Spectra, III," *Phys. Rev.*, **63,** 367 (1943).
90. B. G. Wybourne, *Spectroscopic Properties of the Rare Earths*, Wiley, New York, 1965.
91. B. G. Wybourne, "Group Theoretical Classification of the Atomic States of g^n Configurations," *J. Chem. Phys.*, **45,** 1100 (1966).
92. A. de Shalit and I. Talmi, *Nuclear Shell Theory*, Academic Press, New York, 1963.
93. C. L. B. Shudeman, "Equivalent Electrons and their Spectroscopic Terms," *J. Franklin Inst.*, **224,** 501 (1937).
94. B. R. Judd, "Atomic Shell Theory Recast," *Phys. Rev.*, **162,** 28 (1967).
95. D. E. Littlewood, "The Characters and Representations of Imprimitive Groups," *Proc. London Math. Soc.*, **6,** (3), 251 (1956).
96. S. Feneuille, "Application de la Theorie des Groupes de Lie aux Configurations Melangees," *J. Phys.*, **28,** 61 (1967).
97. S. Feneuille, "Symetrie des Operateirs de L'Interaction Coulombienne pour les Configurations $(d + s)^N$," *J. Phys.*, **28,** 315 (1967).
98. S. Feneuille, "Interaction de Configurations Lointaines pour les Configurations $(l + l')^N$," *J. Phys.*, **28,** 497 (1967).

99. S. Feneuille, "Operateurs a Trois Particules pour les Configurations $(d + s)^N$," *J. Phys.*, **28**, 701 (1967).
99a. P. H. Butler and B. G. Wybourne, "Generalized Racah Tensors and the Structure of Mixed Configurations," *J. Math. Phys.*, in press.
100. M. Rotenberg, R. Bivins, N. Metropolis, and J. K. Wooten, *The* $3 - j$ *and* $6 - j$ *Symbols*, The Technology Press, Massachusetts Institute of Technology, Boston, Mass., 1959. (We take the opportunity of noting that the entry on page 73 for $(6\ 9/2\ 9/2 \mid 3\ -7/2\ 1/2)$ should read $\underline{1}000$, $\underline{11}$, and not 1000, $\underline{11}$).
101. G. Racah, "Theory of Complex Spectra II," *Phys. Rev.*, **62**, 438 (1942).
102. M. E. Rose, *Elementary Theory of Angular Momentum*, Wiley, New York, 1957.
103. U. Fano and G. Racah, *Irreducible Tensorial Sets*, Academic Press, New York, 1959.
104. A. R. Edmonds, *Angular Momentum in Quantum Mechanics*, 2nd ed., Princeton University Press, Princeton, N.J., 1960.
105. F. R. Innes and C. W. Ufford, "Microwave Zeeman Effect and the Theory of Complex Spectra," *Phys. Rev.*, **111**, 194 (1958).
106. W. H. Heintz and R. L. Gibbs, "Angular Coefficients of Atomic Matrix Elements Involving Interelectronic Coordinates," *J. Mat. Phys.*, **8**, 1817 (1967).
107. D. M. Brink and G. R. Satchler, *Angular Momentum*, 2nd ed., Oxford University Press, London, 1968.
108. A. Jucys, J. Levinsonas, and V. Vanagas, *Mathematical Apparatus of the Theory of Angular Momentum*, Vilna, 1960. A translation by Sen and Sen has been published in the Israel Program for Scientific Translations, Jerusalem, 1962.
109. L. C. Biedenharn and H. Van Dam, *Quantum Theory of Angular Momentum*, Academic Press, New York, 1965.
110. B. H. Flowers and S. Szpikowski, "Quasi-spin in *LS* Coupling," *Proc. Phys. Soc.* (London), **84**, 673 (1964).
111. R. D. Lawson and M. H. MacFarlane, "The Quasi-Spin Formalism and the Dependence of Nuclear Matrix elements on Particle Number," *Nuc. Phys.*, **66**, 80 (1965).
112. C. Eckart, "The Application of Group Theory to the Quantum Dynamics of Monatomic Systems," *Rev. Mod. Phys.*, **2**, 305, (1930).
113. E. P. Wigner, *Group Theory*, Academic Press, New York, 1959.
114. A. P. Stone, "Tensor Operators under Semi-Simple Groups," *Proc. Cambridge Phil. Soc.*, **57**, 460 (1961).
115. B. R. Judd, "The Matrix Elements of Tensor Operators for the Electronic Configurations f^n," *Proc. Phys. Soc.* (London), **74**, 330 (1959).
116. A. G McLellan, "Selection Rules for Spin-Orbit Matrix Elements for the Configuration f^n," *Proc. Phys. Soc.* (London), **76**, 419 (1960).
117. P. B. Nutter, "The Reduction of Product Representations in the Continuous Groups R_7 and G_2," Raytheon Technical Memorandum T-544 (1964) (unpublished).
118. R. I. Karaziya, Ya. I. Vizbaraite, Z. B. Rudzikas, and A. P. Jucys, "Talitsy dlya rascheta matrichnykh zlementov operatorov atomnykh velichin," *Acad. Nauk Lit. SSR, Inst. Fiz. Mat. Moscow* (1967).
119. B. R. Judd and H. T. Wadzinski, "A Class of Null Spectroscopic Coefficients," *J. Math. Phys.*, **8**, 2125 (1967).
120. P. R. Smith and B. G. Wybourne, "Selection Rules and the Decomposition of the Kronecker Square of Irreducible Representations," *J. Math. Phys.*, **8**, 2434 (1967).
121. B. R. Judd, "Double Tensor Operators for Configurations of Equivalent Electrons," *J. Math. Phys.*, **3**, 557 (1962).

122. C. W. Nielson and G. F. Koster, *Spectroscopic Coefficients for the p^n, d^n, and f^n Configurations*, The M.I.T. Press, Massachusetts Institute of Technology, Cambridge, Mass., 1963.
123. E. U. Condon and G. H. Shortley, *The Theory of Atomic Spectra*, Cambridge University Press, New York, 1935.
124. G. H. Shortley and B. Fried, "Extension of the Theory of Complex Spectra," *Phys. Rev.*, **54,** 739 (1938).
124a. B. G. Wybourne, "Hermite's Reciprocity Law and the Angular-Momentum States of Equivalent Particle Configurations," *J. Math. Phys.*, **10,** 467 (1969).
125. H. A. Bethe and E. E. Salpeter, *Quantum Mechanics of One- and Two-Electron Systems*, Springer, Berlin, 1957.
126. S. Yanagawa, "Orbit-Orbit Interactions in Atomic l^n Configurations," *J. Phys. Soc.* Japan, **10,** 1029 (1955).
127. R. Dagis, Z. Rudzikas, J. Vizbaraite, and A. Jucys, "Orbita-orbita Saveikos Efektas Ekvivalemtimiu Electronu Atveju," *Liet. Fiz. Rinkinys*, **3,** 159 (1963).
128. R. Dagis, Z. Rudzikas, R. Katiljus, and A. Jucys, "Saveikos Orbita-orbita Operatoriaus Dvielektroniu Matriciniu Elementu Skaiciavimo Klausimu," *Liet. Fiz. Rinkinys*, **3,** 365 (1963).
129. Z. Rudzikas, J. Vizbaraite, and A. Jucys, "Saveikos Orbita-orbita Atominiuose Spektruose Tolesnis Nagrinejimas," *Liet. Fiz. Rinkinys*, **5,** 315 (1965).
130. H. H. Marvin, "Mutual Magnetic Interactions of Electrons," *Phys. Rev.*, **71,** 102 (1947).
131. L. Armstrong and S. Feneuille, "Magnetic Interactions in Mixed Configurations," *Phys. Rev.*, **173,** 58 (1968).
132. B. G. Wybourne, "Orbit-Orbit Interactions and the "Linear Theory" of Configuration Interaction," *J. Chem. Phys.*, **40,** 1457 (1964).
133. S. Fraga and G. Malli, "Orbit-Orbit Interaction in Many-Electron Atoms," *J. Chem. Phys.*, **46,** 4754 (1967).
134. L. Armstrong, "Relativistic Effects in Atomic Structure I, *J. Math. Phys.*, **7,** 1891 (1966).
135. L. Armstrong, "Relativistic Effects in Atomic Fine Structure II," *J. Math. Phys.*, **9,** 1083 (1968).
136. B. R. Judd, "Zeeman Effect as a Prototype for Intra-Atomic Interactions," *Physica*, **33,** 174 (1967).
137. B. R. Judd, H. M. Crosswhite, and H. Crosswhite, "Intra-Atomic Magnetic Interactions for *f* Electrons," *Phys. Rev.*, **169,** 130 (1968).
138. K. Rajnak and B. G. Wybourne, "Configuration Interaction Effects in l^N-Configurations," *Phys. Rev.*, **132,** 280 (1963).
139. G. Racah and J. Stein, "Effective Electrostatic Interactions in l^N Configurations," *Phys. Rev.*, **156,** 58 (1967).
140. B. R. Judd, "Effective Operators for Configurations of Equivalent Electrons," NATO Summer Institute on Correlations in Atoms and Molecules, Frascati (1967).
141. M. Klapisch, "Une nouvelle methode pour le calcul des fonctions radiales et la classifications des spectres atomiques," *Compt. Rend. Acad. Sci.*, **265,** 914 (1967).
142. S. Feneuille, "Operators a trois particules pour des electrons et equivalents," *Compt. Rend. Acad. Sci.*, **262,** 23 (1966).
143. B. R. Judd and L. Armstrong, "Matrix Factorizations for the Coulomb Interaction between Electrons in Atoms," *Proc. Roy. Soc.* (London), **A309,** 185 (1969).
144. B. G. Wybourne, "Symmetry Classification of Two-Particle Operators in Atomic Spectroscopy," *J. Phys.*, **39,** 39 (1969).

145. L. Armstrong and B. R. Judd, "Quasiparticles in Atomic Shell Theory," *Proc. Roy. Soc.* (London), in press.
146. M. J. Cunningham and B. G. Wybourne, "Quasi-particle Formalism and Atomic Shell Theory," *J. Math. Phys.*, in press.
147. S. Feneuille, "Traitement des Configurations $(d + s)^N$ dans le Formalisme des Quasi-Particules," *J. Phys.*, in press.
148. P. H. Butler and B. G. Wybourne, "Applications of S-Functional Analysis to Continuous Groups in Physics," *J. Phys.*, **30**, 795 (1969).
149. P. H. Butler and B. G. Wybourne, "Reduction of the Kronecker Products for Rotation Groups," *J. Phys.*, **30**, 655 (1969).

Appendix I Spin and Difference Characters for the Rotation Groups

The spin representations of the rotation groups have been of little significance in the development of theories of complex spectra.[88] However, recent work on the application of the quasiparticle formalism to atomic shell theory[145–147] has shown that the study of the properties of spin representations of the odd and even dimensional rotation groups ($R_{2\nu+1}$ and $R_{2\nu}$) is likely to assume some significance in future applications of theoretical spectroscopy. In simple applications (where $\nu \leq 3$) the methods of Chapter Seven based on groups isomorphic to the rotation groups suffice. The general case is somewhat more complex and has recently been elaborated upon by Butler and Wybourne.[148,149]

KRONECKER PRODUCTS OF SPIN REPRESENTATIONS OF $O_{2\nu}$ AND $O_{2\nu+1}$

Associated with $O_{2\nu}$ or $O_{2\nu+1}$ there is a basic spin representation $\Delta' \equiv [(\frac{1}{2})^\nu]'$ of degree 2^ν. It is not difficult to show that[13]

$$\Delta'^2 = \sum_0^{2\nu} \{1^r\} = \{1^\nu\} + 2\sum_0^{\nu-1}\{1^r\} \tag{1}$$

for $O_{2\nu}$ and

$$\Delta'^2 = \sum_0^{\nu} \{1^{2r}\} = \sum_0^{\nu}\{1^r\} \tag{2}$$

for $O_{2\nu+1}$, where no attempt has been made to distinguish the ordinary and associated characters of the orthogonal group. The spin characters for the other spin representations of the full orthogonal group may all be expressed as products of the basic spin representation Δ' with S-functions by making the expansion

$$[\lambda_1 + \tfrac{1}{2}, \lambda_2 + \tfrac{1}{2}, \ldots, \lambda_\nu + \tfrac{1}{2}]' = \Delta' \sum (-1)^{\frac{1}{2}(p+r)}\Gamma_{\varepsilon\eta\lambda}\{\eta\} \tag{3}$$

where the summation is over all S-functions $\{\varepsilon\}$ involving self-conjugate partitions of weight p and rank r. Thus in the case of O_8 we have

$$[\tfrac{5}{2}\,\tfrac{3}{2}\,\tfrac{3}{2}\,\tfrac{1}{2}]' = [(\tfrac{1}{2})^4]'(\{21^2\} - \{1^3\} - \{21\} + \{1\})$$

The product of a basic spin character with an S-function may be expressed as a sum of spin characters by writing

$$\Delta'\{\lambda\} = \sum \Gamma_{\zeta\eta\lambda}[\eta_1 + \tfrac{1}{2}, \eta_2 + \tfrac{1}{2}, \ldots, \eta_\nu + \tfrac{1}{2}]' \tag{4}$$

where the summation is over all S-functions $\{\zeta\}$. Thus in the case of O_8 we have

$$[(\tfrac{1}{2})^4]'\{21^2\} = [\tfrac{5}{2}\,\tfrac{3}{2}\,\tfrac{3}{2}\,\tfrac{1}{2}]' + [\tfrac{5}{2}\,\tfrac{3}{2}\,\tfrac{1}{2}\,\tfrac{1}{2}]' + 2[\tfrac{3}{2}\,\tfrac{3}{2}\,\tfrac{1}{2}\,\tfrac{1}{2}]' + [\tfrac{3}{2}\,\tfrac{3}{2}\,\tfrac{3}{2}\,\tfrac{1}{2}]'$$
$$+ [\tfrac{5}{2}\,\tfrac{1}{2}\,\tfrac{1}{2}\,\tfrac{1}{2}]' + 2[\tfrac{3}{2}\,\tfrac{1}{2}\,\tfrac{1}{2}\,\tfrac{1}{2}]'$$

Equations 3 and 4, used in conjunction with Eqs. 1 and 2 provide a systematic method for the reduction of the Kronecker product of any pair of spin representations or of a spin representation with a true representation.

DIFFERENCE CHARACTERS AND TRUE REPRESENTATIONS OF $R_{2\nu}$

The representations $[\lambda_1, \lambda_2, \ldots, \lambda_\nu]'$ of $O_{2\nu}$ remain irreducible upon restriction to the group $R_{2\nu}$ except where $\lambda_\nu \neq 0$ when the representation of $O_{2\nu}$ separates into two irreducible representations of $R_{2\nu}$, i.e.,

$$[\lambda_1, \lambda_2, \ldots, \lambda_\nu]' = [\lambda_1, \lambda_2, \ldots, \lambda_\nu] + [\lambda_1, \lambda_2, \ldots, -\lambda_\nu] \tag{5}$$

Let us define the difference between the two conjugate characters as[13,55]

$$[\lambda_1, \lambda_2, \ldots, \lambda_\nu]'' = [\lambda_1, \lambda_2, \ldots, \lambda_\nu] - [\lambda_1, \lambda_2, \ldots, -\lambda_\nu] \tag{6}$$

Then we have

$$[\lambda_1, \lambda_2, \ldots, \lambda_\nu] = \tfrac{1}{2}([\lambda_1, \lambda_2, \ldots, \lambda_\nu]' + [\lambda_1, \lambda_2, \ldots, \lambda_\nu]'') \tag{7a}$$

and

$$[\lambda_1, \lambda_2, \ldots, -\lambda_\nu] = \tfrac{1}{2}([\lambda_1, \lambda_2, \ldots, \lambda_\nu]' - [\lambda_1, \lambda_2, \ldots, \lambda_\nu]'') \tag{7b}$$

From first principles

$$[1^\nu]'' = \prod_{r=1}^{\nu} (2i \sin \varphi_r) \tag{8}$$

The product of the difference character $[1^\nu]''$ times an S-function $\{\lambda\}$ will yield a difference character, simple or compound, of $R_{2\nu}$. In general, we find that the difference characters associated with the true representations of $R_{2\nu}$ are expressible as

$$[\lambda_1 + 1, \lambda_2 + 1, \ldots, \lambda_\nu + 1]'' = [1^\nu]''(\textstyle\sum(-1)^{p/2}\Gamma_{\alpha\eta\lambda}\{\eta\}) \tag{9}$$

where the summation is over the S-functions of the set $\{\alpha\}$ given in Eq. 72 of the text.

The product of the difference character $[1^\nu]''$ with an S-function $\{\lambda\}$ may be expanded as a series of difference characters of $R_{2\nu}$ by writing[13]

$$[1^\nu]''\{\lambda\} = \sum \Gamma_{\beta\eta\lambda}[\eta_1 + 1, \eta_2 + 1, \ldots, \eta_\nu + 1]'' \tag{10}$$

where the summation is over all S-functions of the set $\{\beta\}$ which involves all partitions into an even number of parts of any given magnitude.

Equations 7a, 7b, 9, and 10 together contain sufficient information to make the reduction of the Kronecker product of any pair of true representations of $R_{2\nu}$ once the Kronecker square of $[1^\nu]''$ has been expressed in terms of characters of $R_{2\nu}$. We first note that

$$\begin{aligned}([1^\nu]'')^2 &= [1^\nu]'[1^\nu]' - 4[1^\nu][1^{\nu-1} - 1] \\ &= [1^\nu][1^\nu] + [1^{\nu-1} - 1][1^{\nu-1} - 1] - 2[1^\nu][1^{\nu-1} - 1]\end{aligned} \tag{11}$$

Butler and Wybourne[149] have shown that

$$[1^\nu][1^\nu] = \sum \Gamma_{\beta\mu\lambda}\{\mu\} \tag{12a}$$

and

$$[1^\nu][1^{\nu-1} - 1] = \sum \Gamma_{\beta\mu\lambda}\{\mu\} \tag{12b}$$

where the summation is over all the S-functions of the set $\{\beta\}$ and the partition (λ) defines the principal part of the product (i.e., $[2^\nu]$ for Eq. 12a and $[2^{\nu-1}]$ for Eq. 12b). The terms in $[1^{\nu-1} - 1][1^{\nu-1} - 1]$ are the same as those of Eq. 12a except for every partition with $\lambda_\nu \neq 0$ we replace λ_ν by $-\lambda_\nu$.

DIFFERENCE CHARACTERS AND SPIN REPRESENTATIONS OF $R_{2\nu}$

Equations 5–7b are equally valid for the spin representations of $R_{2\nu}$. The basic difference character $\Delta'' = [(\frac{1}{2})^v]''$ becomes

$$\Delta'' = \prod_{r=1}^{\nu} (2i \sin \varphi_r) \tag{13}$$

The difference characters associated with the other spin representations of $R_{2\nu}$ may be expressed as

$$[\lambda_1 + \tfrac{1}{2}, \lambda_2 + \tfrac{1}{2}, \ldots, \lambda_\nu + \tfrac{1}{2}]'' = \Delta''(\sum (-1)^{\frac{1}{2}(p-r)}\Gamma_{\varepsilon\eta\lambda}\{\eta\}) \tag{14}$$

where the summation is again over all S-functions $\{\varepsilon\}$ involving self-conjugate partitions of weight p and rank r.

It is not difficult to establish that the Kronecker square of the difference character is expressible in terms of S-functions as

$$(\Delta'')^2 = \{1^\nu\} + 2\sum_{r=0}^{\nu-1}(-1)^{\nu+r}\{1^r\} \tag{15}$$

and that

$$\Delta''\Delta' = [1^{\nu}]'' \tag{16}$$

These two results, together with the use of Eqs. 5–7*b*, 10, and the results of Eqs. 62 and 67 of the text give a systematic method for the reduction of the Kronecker products of any pair of spin representations of $R_{2\nu}$.

KRONECKER PRODUCTS OF TRUE AND SPIN REPRESENTATIONS OF $R_{2\nu}$

There is no difficulty in extending the preceding results to calculate the Kronecker product of a true representation of $R_{2\nu}$ with a spin representations if the results

$$\begin{aligned}[1^{\nu}]''\,\Delta' &= [1^{\nu}]'\,\Delta'' \\ &= \Delta''\left(\{1^{\nu}\} + 2\sum_{r=0}^{\nu-1}\{1^{r}\}\right)\end{aligned} \tag{17}$$

and

$$[1^{\nu}]''\,\Delta'' = \Delta'\left(\{1^{\nu}\} + 2\sum_{r=0}^{\nu-1}(-1)^{\nu+r}\{1^{r}\}\right) \tag{18}$$

are noted together with the relation

$$\Delta''\{\lambda\} = \sum(-1)^{p}\Gamma_{\zeta\eta\lambda}[\eta_1 + \tfrac{1}{2}, \eta_2 + \tfrac{1}{2}, \ldots, \eta_{\nu} + \tfrac{1}{2}]'' \tag{19}$$

where the summation is over the set of all *S*-functions $\{\zeta\}$.

RESOLUTION OF THE KRONECKER SQUARE OF SPIN REPRESENTATIONS FOR $O_{2\nu}$ AND $O_{2\nu+1}$

The symmetric terms in the Kronecker square of a spin representation of the full orthogonal group are found to be[149]

$$\begin{aligned}[\lambda_1 + \tfrac{1}{2}, \lambda_2 + \tfrac{1}{2}, \ldots, \lambda_{\nu} + \tfrac{1}{2}]' \otimes \{2\} &= [\Delta' \textstyle\sum(-1)^{\frac{1}{2}(p+r)}\Gamma_{\varepsilon\eta\lambda}\{\eta\}] \otimes \{2\} \\ &= (\Delta' \otimes \{2\})([\textstyle\sum(-1)^{\frac{1}{2}(p+r)}\Gamma_{\varepsilon\eta\lambda}\{\eta\}] \otimes \{2\}) \\ &\quad + (\Delta' \otimes \{1^2\})([\textstyle\sum(-1)^{\frac{1}{2}(p+r)}\gamma_{\varepsilon\eta\lambda}\{\eta\}] \otimes \{2^2\})\end{aligned} \tag{20a}$$

and for the antisymmetric terms

$$\begin{aligned}&[\lambda_1 + \tfrac{1}{2}, \lambda_2 + \tfrac{1}{2}, \ldots, \lambda_{\nu} + \tfrac{1}{2}]' \otimes \{1^2\} \\ &\quad = (\Delta' \otimes \{2\})([\textstyle\sum(-1)^{\frac{1}{2}(p+r)}\Gamma_{\varepsilon\eta\lambda}\{\eta\}] \otimes \{1^2\}) \\ &\quad\quad + (\Delta' \otimes \{1^2\})([\textstyle\sum(-1)^{\frac{1}{2}(p+r)}\Gamma_{\varepsilon\eta\lambda}\{\eta\}] \otimes \{2\})\end{aligned} \tag{20b}$$

The plethysms involving the *S*-functions $\{\eta\}$ may be evaluated using the methods of Chapter Six. The plethysms for the basic spin representations follow from first principles[13,149] to give for $O_{2\nu}$

$$\Delta' \otimes \{2\} = [1^{\nu}]' + [1^{\nu-1}]' + [1^{\nu-3}]' + 2[1^{\nu-4}]' + [1^{\nu-5}]' + [1^{\nu-7}]' + 2[1^{\nu-8}]' + \cdots \quad (21a)$$

and

$$\Delta' \otimes \{1^2\} = [1^{\nu-1}]' + 2[1^{\nu-2}]' + [1^{\nu-3}]' + [1^{\nu-5}]' + 2[1^{\nu-6}]' + \cdots \quad (22b)$$

and for $O_{2\nu+1}$

$$\Delta' \otimes \{2\} = [1^{\nu}]' + [1^{\nu-3}]' + [1^{\nu-4}]' + [1^{\nu-7}]' + [1^{\nu-8}]' + \cdots \quad (22a)$$

and

$$\Delta' \otimes \{1^2\} = [1^{\nu-1}]' + [1^{\nu-2}]' + [1^{\nu-5}]' + [1^{\nu-6}]' + 1^{\nu-9}]' + \cdots \quad (22b)$$

These two sets of results, combined with those of Eqs. 20*a* and 20*b* allow any Kronecker square of the spin representations of $O_{2\nu}$ or $O_{2\nu+1}$ to be separated into their symmetric and antisymmetric terms. Since the representations of $O_{2\nu+1}$ are irreducible upon restriction to $R_{2\nu+1}$ there is no added difficulty in resolving the Kronecker squares of the spin representations of $R_{2\nu+1}$.

The resolution of the Kronecker squares of the true and spin representations of $R_{2\nu}$ is a considerably more tedious task though the complete solution to this problem is known and the reader is referred to the literature.[149]

BRANCHING RULES FOR SPIN REPRESENTATIONS UNDER $R_n \to R_3$

The branching rules for the decomposition of the spin characters of R_n into characters of R_3 are especially important in the application of the quasi-particle formalism to problems in atomic shell theory.[145,146]

The representations of R_3 depend on a single parametric angle θ and the character associated with the representation $D^{[J]}$ is

$$[J] = \sum_{\rho=0}^{2J} e^{i(J-\rho)\theta} \quad (23)$$

where J may be integral or half-integral and ρ runs through positive integers. The representations of $R_{2\nu}$ and $R_{2\nu+1}$ each involve ν parametric angles θ_ε and in particular the character [1] of the vector representation $\Gamma^{[1]}$ of $R_{2\nu}$ is of the form

$$[1] = \sum_{\varepsilon=1}^{\nu} e^{\pm i\theta\varepsilon} \quad (24)$$

while for $R_{2\nu+1}$ the corresponding character is

$$[1] = \sum_{\varepsilon=1}^{\nu} e^{\pm i\theta\varepsilon} + 1 \tag{25}$$

Under the restriction $R_n \to R_3$ the ν parametric angles become related and it is then possible to express the results of Eqs. 24 and 25 in terms of a single parametric angle θ. If under $R_n \to R_3$ we have

$$\begin{aligned} [1] &\to \sum g_J[J] \\ &= \sum g_J \sum_{\rho=0}^{2J} e^{J(i-\rho)\theta} \end{aligned} \tag{26}$$

comparison with Eqs. 24 or 25 allows the relationships between the ν parametric angles to be fixed immediately; for example, if under $R_{2l+1} \to R_3$ we have $[1] \to [l]$, comparison of Eq. 25 with Eq. 23 shows that the required relationship is

$$\theta_\varepsilon = \varepsilon\theta \tag{27}$$

Thus, if under $R_9 \to R_3$ we have $[1] \to [4]$, we must take

$$\theta_1 = \theta \quad \theta_2 = 2\theta \quad \theta_3 = 3\theta \quad \theta_4 = 4\theta \tag{28}$$

If, however, under $R_9 \to R_3$ we have $[1] \to [0] + [1] + [2]$, comparison of Eq. 25 with Eq. 23 leads to the choice

$$\theta_1 = 2\theta \quad \theta_2 = \theta_3 = \theta \quad \theta_4 = 0 \tag{29}$$

Table I-1 Decomposition of the Basic Spin Representation under $R_{2v+1} \to R_3$ when $[1] \to [l]$

l	$\Delta \to$
2	$[\frac{3}{2}]$
3	$[0] + [3]$
4	$[2] + [5]$
5	$[\frac{5}{2}] + [\frac{9}{2}] + [\frac{15}{2}]$
6	$[\frac{3}{2}] + [\frac{9}{2}] + [\frac{11}{2}] + [\frac{15}{2}] + [\frac{21}{2}]$
7	$[2] + [4] + [5] + [7] + [8] + [9] + [11] + [14]$
8	$[0] + [3] + [4] + [5] + [6] + [7] + [8] + [9] + [10] + [11] + [12] + [13] + [15] + [18]$
9	$[\frac{3}{2}] + [\frac{5}{2}] + 2[\frac{9}{2}] + [\frac{11}{2}] + [\frac{13}{2}] + 2[\frac{15}{2}] + 2[\frac{17}{2}] + [\frac{19}{2}] + 2[\frac{21}{2}] + [\frac{23}{2}] + [\frac{25}{2}] + 2[\frac{27}{2}] + [\frac{29}{2}] + [\frac{31}{2}] + [\frac{33}{2}] + [\frac{35}{2}] + [\frac{39}{2}] + [\frac{45}{2}]$
10	$[\frac{1}{2}] + [\frac{5}{2}] + 2[\frac{7}{2}] + [\frac{9}{2}] + 2[\frac{11}{2}] + 2[\frac{13}{2}] + 2[\frac{15}{2}] + 2[\frac{17}{2}] + 3[\frac{19}{2}] + 2[\frac{21}{2}] + 2[\frac{23}{2}] + 3[\frac{25}{2}] + 2[\frac{27}{2}] + 2[\frac{29}{2}] + 2[\frac{31}{2}] + [\frac{33}{2}] + 2[\frac{35}{2}] + 2[\frac{37}{2}] + [\frac{39}{2}] + [\frac{41}{2}] + [\frac{43}{2}] + [\frac{45}{2}] + [\frac{49}{2}] + [\frac{55}{2}]$

Table I-2 Decompositions of Basic Spin Representations under $R_{2\nu} \rightarrow R_3$

ν	$[1] \rightarrow$	$\Delta \rightarrow$
3	$[0] + [2]$	$[\frac{3}{2}]$
4	$[1] + [2]$	$[1] + [2]$
	$[0] + [3]$	$[0] + [3]$
5	$[1] + [3]$	$[\frac{1}{2}] + [\frac{5}{2}] + [\frac{7}{2}]$
	$[0] + [4]$	$[2] + [5]$
6	$[1] + [4]$	$[\frac{3}{2}] + [\frac{5}{2}] + [\frac{9}{2}] + [\frac{11}{2}]$
	$[2] + [3]$	$2[\frac{3}{2}] + [\frac{5}{2}] + [\frac{7}{2}] + [\frac{9}{2}]$
7	$[2] + [4]$	$[\frac{1}{2}] + [\frac{3}{2}] + [\frac{5}{2}] + 2[\frac{7}{2}] + [\frac{9}{2}] + [\frac{11}{2}] + [\frac{13}{2}]$
	$[1] + [5]$	$[2] + [3] + [4] + [5] + [7] + [8]$
8	$[0] + [1] + [2] + [3]$	$4[1] + 6[2] + 4[3] + 4[4] + 2[5]$
	$[2] + [5]$	$[1] + [2] + 2[3] + 2[4] + [5] + 2[6] + [7] + [8] + [9]$
	$[1] + [6]$	$[1] + [2] + [4] + 2[5] + [6] + [7] + [8] + [10] + [11]$
	$[3] + [4]$	$[1] + 3[2] + 2[3] + 2[4] + 3[5] + [6] + [7] + [8]$

The character associated with the basic spin character Δ of $R_{2\nu+1}$ may be expressed in terms of parametric angles θ_ε to give

$$\Delta = \sum \exp\left[\frac{i}{2}(\pm\theta_1 \pm \theta_2 \pm \cdots \pm \theta_\nu)\right] \tag{30}$$

where the summation is over all possible combinations of plus and minus signs. In the case of $R_{2\nu}$ we have

$$\Delta_1 = \sum \exp\left[\frac{i}{2}(\pm\theta_1 \pm \theta_2 \pm \cdots \pm \theta_\nu)\right] \tag{31}$$

where the summation is over all combinations of signs involving an *even* number of minus signs. The conjugate spin character Δ_2 is of the same form except that the summation is over all possible combinations of signs involving an *odd* number of minus signs.

If the decomposition of the character of the vector representation $\Gamma^{[1]}$ of $R_{2\nu}$ or $R_{2\nu+1}$ under restriction to R_3 is defined, the relationships among the ν parametric angles is fixed and may be used in Eqs. 30 or 31, as the case may be. Comparison with Eq. 23 leads immediately to the decomposition rule for the spin character. Typical results are given in Tables I-1 and I-2.

Appendix II Tables of Group Properties

P. H. Butler

Physics Department, University of Canterbury, Christchurch, New Zealand

Tables of the properties of continuous groups, especially of the dimensions, branching rules, and Kronecker products of irreducible representations have appeared scattered throughout the literature. These tables have, almost without exception, been calculated by hand by a remarkable diversity of methods. Littlewood's great achievement has been the development of a systematic method for calculating the properties of groups, a method that is fully amenable to computation on modern high-speed computers. In the following pages we present a collection of tables of group properties which have been machine computed and checked. Our hope is that these tables will be of value to physicists, and dare we suggest some chemists, in a wide range of applications.

PRODUCTION OF THE TABLES

The tables were all calculated on the University of Canterbury's IBM 360/44 computer and printed "on-line" with an IBM electric typewriter using a modified typing element to produce the appropriate group symbols. The final copy was then photographically reproduced from the original "on-line" sheets. The computer was programmed to dimensionally check all calculations other than those of the dimensions themselves. The machine would print an error signal if the dimensional checks were not satisfied.

ORDERING THE PARTITIONS

The tables are arranged in one of two orders. Most tables are arranged in dictionary order, but tables of properties of S-functions are ordered in the sequence of partitions of increasing integers with the individual partitions of given integer n in reverse dictionary order. Thus when characters are left out because of simple equivalences, partitions into the larger number of parts will be omitted.

TABLES OF DIMENSIONS OF IRREDUCIBLE REPRESENTATIONS

The tables of the dimensions of the groups $S_n(0 \leq n \leq 14)$, $U_n(3 \leq n \leq 18)$, $R_n(5 \leq n \leq 15$ and *odd*), O_n $(4 \leq n \leq 8$ and *even*) and $Sp_n(4 \leq n \leq 18$ and *even*) were computed by using the formulas given in the text and appear as Tables A-1 to A-34.

The table of dimensions of the irreducible representations of S_n has been shortened by putting the conjugate of each partition adjacent to the original partition. The tables involving the dimensions and branching rules for the groups U_n have all been shortened by listing only partitions that cannot be further reduced by application of the equivalence relations

$$\{\lambda_1, \lambda_2, \ldots, \lambda_n\} \equiv \{\lambda_1 - \lambda_n, \lambda_2 - \lambda_n, \ldots, 0\}^\dagger$$

and

$$\{\lambda_1, \lambda_2, \ldots, \lambda_n\} \equiv \{\lambda_1 - \lambda_n, \lambda_1 - \lambda_{n-1}, \ldots, \lambda_1 - \lambda_2, 0\}^\dagger$$

The reader will recall that for $R_{2\nu}$ the dimensions of the representations $[\lambda_1, \lambda_2, \ldots, \pm\lambda_\nu]$ with $\lambda_\nu \neq 0$ are equal and just half of the corresponding representation $[\lambda_1, \lambda_2, \ldots, \lambda_\nu]'$ of $0_{2\nu}$.

PROPERTIES OF S-FUNCTIONS

Littlewood and Richardson's result for the diagrammatic calculation of the outer multiplication of S-functions forms the key to most of Littlewood's subsequent results. The rules for performing the outer multiplication have been incorporated into a computer program which is capable of extremely rapid calculation of any product. An extensive tabulation of the outer multiplication of S-functions is given in Table B-1.

Results for the inner multiplication of S-functions appear in Table B-2. This tabulation has been shortened by noting that if

$$\{\lambda\} \circ \{\mu\} = \sum g_{\lambda\mu\nu}\{\nu\}$$

the coefficients $g_{\lambda\mu\nu}$ satisfy the symmetry relationship

$$g_{\lambda\mu\nu} = g_{\mu\lambda\nu} = g_{\lambda\nu\mu}$$

and the conjugation relation

$$g_{\lambda\mu\nu} = g_{\lambda\tilde{\mu}\tilde{\nu}}$$

The calculation of the inner multiplication were facilitated by first expanding the S-function appearing on the left of the inner multiplication sign $\circ$ into products of the symmetric functions h_r defined by Eq. 29. This operation is readily performed by noting that

$$\{\lambda\} = |h_{\lambda_s - s + t}|$$

the s and t being subscripts for the rows and columns of the determinant, respectively. A tabulation of these expansions appears in Table B-3. The quantities $(abc \cdots z)$ are all to be interpreted as $h_a h_b h_c \cdots h_z$; e.g., $(321) \equiv \{3\}\{2\}\{1\} \equiv h_3 h_2 h_1$.

The formulas in Eq. 64 and 74 are of fundamental importance in calculating the properties of orthogonal and symplectic group characters and allow us to express orthogonal and symplectic group characters in terms of S-functions. A tabulation of the expansion of orthogonal group characters in terms of S-functions is given in Table B-4 and the corresponding tabulation for symplectic group characters given in Table B-5.

The branching rules involving the reduction of the groups U_n, Sp_n, and R_n to the three-dimensional rotation group were calculated using the properties of plethysms of S-functions within the group GL_2. These plethysms, which appear in Table B-6, were evaluated by noting that[47] if

$$\{n\} \otimes \{p\} = \sum K_r\{r\}$$

where n and p are integers then K_r is the coefficient of ρ^{r+1} in the expansion of

$$\frac{(1 - \rho^n)(1 - \rho^{n+1}) \cdots (1 - \rho^{n+p-1})}{(1 - \rho^2)(1 - \rho^3) \cdots (1 - \rho^p)}$$

This polynomial division is well adapted to machine calculation and was used in preference to the recursive method discussed in Section 7.8. The corresponding plethysms $[n] \otimes \{p\}$ for R_3 can readily be found from Table B-6 by first replacing $[n] \otimes \{p\}$ by $\{2n\} \otimes \{p\}$ and then replacing each S-function $\{r\}$ by $[r/2]$. Alternatively we note that if

$$[n] \otimes \{p\} = \sum K_r[r]$$

then K_r is the coefficient of ρ^{-r} in the expansion of

$$\rho^{-np}(1 - \rho) \prod_{i=1}^{p} \frac{(1 - \rho^{2n+i})}{(1 - \rho^i)}$$

Littlewood's results frequently require the evaluation of the sum of S-functions $\{\nu\}$ which when multiplied by a particular S-function $\{\mu\}$ give a particular S-function $\{\lambda\}$, the coefficient of $\{\nu\}$ being the coefficient of $\{\lambda\}$ in the outer product. Hence we may define the division of S-functions $\{\lambda\}/\{\mu\}$ to be:

$$\{\lambda\}/\{\mu\} = \sum_{\nu} \Gamma_{\mu\nu\lambda}\{\nu\}$$

where $\Gamma_{\mu\nu\lambda}$ is the same as the coefficient in the outer product

$$\{\mu\}\{\nu\} = \sum_{\lambda} \Gamma_{\mu\nu\lambda}\{\lambda\}$$

An extensive tabulation of S-function division is given in Table B-7.

BRANCHING RULES

The branching rules for the reductions $U_n \to R_n$, O_n or Sp_n and for $R_{2\nu+1} \to R_3$ and $Sp_{2\nu} \to R_3$ appear in Tables C-1 to C-24. The branching rules involving reduction of the group U_n were obtained by use of Eqs. 69 and 86 with partitions having greater than ν parts (where $n = 2\nu$ or $2\nu + 1$) being reduced to ν or fewer parts by use of Littlewood's general method.

The branching rules for $R_n \to R_3$, and $Sp_n \to R_3$ were all calculated by first expressing the character of R_n or Sp_n in terms of S-functions which were then expanded as products of the symmetric functions h_r. In each case it is assumed that under $R_{2\nu+1} \to R_3$ we have $[1] \to [\nu]$ and under $Sp_{2\nu} \to R_3$ we have $[1] \to [\nu - \frac{1}{2}]$. The branching rules are finally obtained by making use of the fact that

$$[a] \otimes (h_{r_1}h_{r_2} \cdots h_{r_p}) = ([a] \otimes h_{r_1})([a] \otimes h_{r_2}) \cdots ([a] \otimes h_{r_p})$$

evaluating the individual plethysms and then reducing the product.

KRONECKER PRODUCTS

The Kronecker products for selected representations of the groups $R_{2\nu+1}$ and $Sp_{2\nu}$ are given in Tables D-1 to D-15 as well as those for the special cases of R_4, R_6, and O_8.

THE GROUP G_2

Tabulations relating to the exceptional group G_2 have been gathered together in Table E-1 to E-5. The Kronecker products for G_2 and the branching rules for the reduction $G_2 \to R_3$ were calculated by first expressing the relevant characters of G_2 in terms of those of R_7, performing the necessary operations on the R_7 characters and then converting back to G_2 characters in the case of the Kronecker products or directly to R_3 in the case of $G_2 \to R_3$.

The conversion of $G_2 \to R_7$ was made by noting that under the reduction $R_7 \to G_2$ we certainly find $[u_1, u_2, 0] \supset (u_1, u_2)$ with additional terms of G_2 of lower weight which may in turn be further reduced to characters of R_7; for example, since under $R_7 \to G_2$ we have

$$[310] \to (21) + (30) + (31), \quad [300] \to (30)$$
$$[210] \to (11) + (20) + (21), \quad [200] \to (20)$$
$$[110] \to (11) + (10) \quad \text{and} \quad [100] \to (10)$$

we readily find

$$(31) = [310] - [300] - [210] + [200] + [110] - [100]$$

CONCLUDING REMARKS

In presenting these tables, we are well aware of their shortcomings in regard to the omission of tables that some workers in the field may require. In particular, we have made no attempt to tabulate properties of spin representations, general plethysms, or branching rules for product groups. The program used to produce these tables is in a state of continuous development and requests for extensions of these tables or for other tables should be directed to the authors.

List of Tables

Dimensions of Irreducible Representations

Table Number	Dimensions of the Irreducible Representations of	Table Number	Dimensions of the Irreducible Representations of
A-1	S_n	A-18	O_4
A-2	U_3	A-19	R_5
A-3	U_4	A-20	O_6
A-4	U_5	A-21	R_7
A-5	U_6	A-22	O_8
A-6	U_7	A-23	R_9
A-7	U_8	A-24	R_{11}
A-8	U_9	A-25	R_{13}
A-9	U_{10}	A-26	R_{15}
A-10	U_{11}	A-27	Sp_4
A-11	U_{12}	A-28	Sp_6
A-12	U_{13}	A-29	Sp_8
A-13	U_{14}	A-30	Sp_{10}
A-14	U_{15}	A-31	Sp_{12}
A-15	U_{16}	A-32	Sp_{14}
A-16	U_{17}	A-33	Sp_{16}
A-17	U_{18}	A-34	Sp_{18}

Properties of *S*-Functions

Table
Number

B-1 Outer Multiplication of *S*-Functions
B-2 Inner Multiplication of *S*-Functions
B-3 Expansion of *S*-Functions into Products of the Symmetric Functions h_r
B-4 Expansion of Orthogonal Group Characters into *S*-Functions
B-5 Expansion of Symplectic Group Characters into *S*-Functions
B-6 Plethysms within the Group GL_2
B-7 *S*-Function Division

Branching Rules

Table Number	Branching Rules for the Reduction	Table Number	Branching Rules for the Reduction
C-1	$U_3 \rightarrow R_3$	C-17	$R_7 \rightarrow R_3$
C-2	$U_4 \rightarrow O_4$	C-18	$R_9 \rightarrow R_3$
C-3	$U_5 \rightarrow R_5$	C-19	$R_{11} \rightarrow R_3$
C-4	$U_6 \rightarrow O_6$	C-20	$R_{13} \rightarrow R_3$
C-5	$U_7 \rightarrow R_7$	C-21	$R_{15} \rightarrow R_3$
C-6	$U_8 \rightarrow O_8$	C-22	$R_{17} \rightarrow R_3$
C-7	$U_9 \rightarrow R_9$	C-23	$R_{19} \rightarrow R_3$
C-8	$U_{11} \rightarrow R_{11}$	C-24	$R_{21} \rightarrow R_3$
C-9	$U_{13} \rightarrow R_{13}$	C-25	$Sp_4 \rightarrow R_3$
C-10	$U_4 \rightarrow Sp_4$	C-26	$Sp_6 \rightarrow R_3$
C-11	$U_6 \rightarrow Sp_6$	C-27	$Sp_8 \rightarrow R_3$
C-12	$U_8 \rightarrow Sp_8$	C-28	$Sp_{10} \rightarrow R_3$
C-13	$U_{10} \rightarrow Sp_{10}$	C-29	$Sp_{12} \rightarrow R_3$
C-14	$U_{12} \rightarrow Sp_{12}$	C-30	$Sp_{14} \rightarrow R_3$
C-15	$U_{14} \rightarrow Sp_{14}$	C-31	$Sp_{16} \rightarrow R_3$
C-16	$R_5 \rightarrow R_3$	C-32	$Sp_{18} \rightarrow R_3$

Kronecker Products

Table Number	Reduction of the Kronecker Products for	Table Number	Reduction of the Kronecker Products for
D-1	R_4	D-9	Sp_6
D-2	R_5	D-10	Sp_8
D-3	R_6	D-11	Sp_{10}
D-4	R_7	D-12	Sp_{12}
D-5	O_8	D-13	Sp_{14}
D-6	R_9	D-14	Sp_{16}
D-7	R_{11}	D-15	Sp_{18}
D-8	Sp_4		

Properties of the Exceptional Group G-2

Table Number	
E-1	Dimensions of the Irreducible Representations of G_2
E-2	Branching Rules for the Reduction $R_7 \rightarrow G_2$
E-3	Branching Rules for the Reduction $G_2 \rightarrow R_3$
E-4	Reduction of the Kronecker Products for G_2
E-5	Characters of G_2 expressed in terms of R_7 Characters

A-1 Dimensions of the Irreducible Representations of S_n.

{0}	{0}	1
{1}	{1}	1
{2}	{11}	1
{3}	{111}	1
{21}	{21}	2
{4}	{1111}	1
{31}	{211}	3
{22}	{22}	2
{5}	{11111}	1
{41}	{2111}	4
{32}	{221}	5
{311}	{311}	6
{6}	{111111}	1
{51}	{21111}	5
{42}	{2211}	9
{411}	{3111}	10
{33}	{222}	5
{321}	{321}	16
{7}	{1111111}	1
{61}	{211111}	6
{52}	{22111}	14
{511}	{31111}	15
{43}	{2221}	14
{421}	{3211}	35
{4111}	{4111}	20
{331}	{322}	21
{8}	{11111111}	1
{71}	{2111111}	7
{62}	{221111}	20
{611}	{311111}	21
{53}	{22211}	28
{521}	{32111}	64
{5111}	{41111}	35
{44}	{2222}	14
{431}	{3221}	70
{422}	{3311}	56
{4211}	{4211}	90
{332}	{332}	42
{9}	{111111111}	1
{81}	{21111111}	8
{72}	{2211111}	27
{711}	{3111111}	28
{63}	{222111}	48
{621}	{321111}	105
{6111}	{411111}	56
{54}	{22221}	42
{531}	{32211}	162
{522}	{33111}	120
{5211}	{42111}	189
{51111}	{51111}	70
{441}	{3222}	84
{432}	{3321}	168
{4311}	{4221}	216
{333}	{333}	42
{10}	{1111111111}	1
{91}	{211111111}	9
{82}	{22111111}	35
{811}	{31111111}	36
{73}	{2221111}	75
{721}	{3211111}	160
{7111}	{4111111}	84
{64}	{222211}	90
{631}	{322111}	315
{622}	{331111}	225
{6211}	{421111}	350
{61111}	{511111}	126
{55}	{22222}	42
{541}	{32221}	288
{532}	{33211}	450
{5311}	{42211}	567
{5221}	{43111}	525
{52111}	{52111}	448
{442}	{3322}	252
{4411}	{4222}	300
{433}	{3331}	210
{4321}	{4321}	768
{11}	{11111111111}	1
{101}	{2111111111}	10
{92}	{221111111}	44
{911}	{311111111}	45
{83}	{22211111}	110
{821}	{32111111}	231
{8111}	{41111111}	120
{74}	{2222111}	165
{731}	{3221111}	550
{722}	{3311111}	385
{7211}	{4211111}	594
{71111}	{5111111}	210
{65}	{222221}	132
{641}	{322211}	693
{632}	{332111}	990
{6311}	{422111}	1232
{6221}	{431111}	1100
{62111}	{521111}	924
{611111}	{611111}	252
{551}	{32222}	330
{542}	{33221}	990
{5411}	{42221}	1155
{533}	{33311}	660
{5321}	{43211}	2310
{53111}	{52211}	1540
{5222}	{44111}	825
{443}	{3332}	462
{4421}	{4322}	1320
{4331}	{4331}	1188
{12}	{111111111111}	1
{111}	{21111111111}	11
{102}	{2211111111}	54
{1011}	{3111111111}	55
{93}	{222111111}	154
{921}	{321111111}	320
{9111}	{411111111}	165
{84}	{22221111}	275
{831}	{32211111}	891
{822}	{33111111}	616
{8211}	{42111111}	945
{81111}	{51111111}	330
{75}	{2222211}	297
{741}	{3222111}	1408
{732}	{3321111}	1925
{7311}	{4221111}	2376
{7221}	{4311111}	2079
{72111}	{5211111}	1728
{711111}	{6111111}	462
{66}	{222222}	132
{651}	{322221}	1155
{642}	{332211}	2673
{6411}	{422211}	3080
{633}	{333111}	1650
{6321}	{432111}	5632
{63111}	{522111}	3696
{6222}	{441111}	1925
{62211}	{531111}	3564
{621111}	{621111}	2100
{552}	{33222}	1320
{5511}	{42222}	1485
{543}	{33321}	2112
{5421}	{43221}	5775
{54111}	{52221}	3520
{5331}	{43311}	4158
{5322}	{44211}	4455
{53211}	{53211}	7700
{444}	{3333}	462
{4431}	{4332}	2970
{4422}	{4422}	2640
{13}	{1111111111111}	1
{121}	{211111111111}	12
{112}	{22111111111}	65
{1111}	{31111111111}	66
{103}	{2221111111}	208
{1021}	{3211111111}	429
{10111}	{4111111111}	220
{94}	{222211111}	429
{931}	{322111111}	1365
{922}	{331111111}	936
{9211}	{421111111}	1430
{91111}	{511111111}	495
{85}	{22222111}	572
{841}	{32221111}	2574
{832}	{33211111}	3432
{8311}	{42211111}	4212
{8221}	{43111111}	3640
{82111}	{52111111}	3003
{811111}	{61111111}	792
{76}	{2222221}	429
{751}	{3222211}	2860
{742}	{3322111}	6006
{7411}	{4222111}	6864
{733}	{3331111}	3575
{7321}	{4321111}	12012
{73111}	{5221111}	7800
{7222}	{4411111}	4004
{72211}	{5311111}	7371
{721111}	{6211111}	4290
{7111111}	{7111111}	924
{661}	{322222}	1287
{652}	{332221}	5148
{6511}	{422221}	5720
{643}	{333211}	6435
{6421}	{432211}	17160
{64111}	{522211}	10296
{6331}	{433111}	11440
{6322}	{442111}	12012
{63211}	{532111}	20592
{631111}	{622111}	9360
{62221}	{541111}	9009
{553}	{33322}	3432
{5521}	{43222}	8580
{55111}	{52222}	5005
{544}	{33331}	2574
{5431}	{43321}	15015
{5422}	{44221}	12870
{54211}	{53221}	21450
{5332}	{44311}	11583
{53311}	{53311}	16016
{4441}	{4333}	3432
{4432}	{4432}	8580

A-1 Dimensions of the Irreducible Representations of S_n (contd).

Partition	Conjugate	Dimension
{14}	{11111111111111}	1
{131}	{2111111111111}	13
{122}	{221111111111}	77
{1211}	{311111111111}	78
{113}	{22211111111}	273
{1121}	{32111111111}	560
{11111}	{41111111111}	286
{104}	{2222111111}	637
{1031}	{3221111111}	2002
{1022}	{3311111111}	1365
{10211}	{4211111111}	2079
{101111}	{5111111111}	715
{95}	{222221111}	1001
{941}	{322211111}	4368
{932}	{332111111}	5733
{9311}	{422111111}	7007
{9221}	{431111111}	6006
{92111}	{521111111}	4928
{911111}	{611111111}	1287
{86}	{22222211}	1001
{851}	{32222111}	6006
{842}	{33221111}	12012
{8411}	{42221111}	13650
{833}	{33311111}	7007
{8321}	{43211111}	23296
{83111}	{52211111}	15015
{8222}	{44111111}	7644
{82211}	{53111111}	14014
{821111}	{62111111}	8085
{8111111}	{71111111}	1716
{77}	{2222222}	429
{761}	{3222221}	4576
{752}	{3322211}	14014
{7511}	{4222211}	15444
{743}	{3332111}	16016
{7421}	{4322111}	42042
{74111}	{5222111}	24960
{7331}	{4331111}	27027
{7322}	{4421111}	28028
{73211}	{5321111}	47775
{731111}	{6221111}	21450
{72221}	{5411111}	20384
{722111}	{6311111}	21021
{7211111}	{7211111}	9504
{662}	{332222}	6435
{6611}	{422222}	7007
{653}	{333221}	15015
{6521}	{432221}	36608
{65111}	{522221}	21021
{644}	{333311}	9009
{6431}	{433211}	50050
{6422}	{442211}	42042
{64211}	{532211}	69498
{641111}	{622211}	28665
{6332}	{443111}	35035
{63311}	{533111}	48048
{63221}	{542111}	63063
{632111}	{632111}	59904
{62222}	{551111}	14014
{554}	{33332}	6006
{5531}	{43322}	27027
{5522}	{44222}	21450
{55211}	{53222}	35035
{5441}	{43331}	21021
{5432}	{44321}	48048
{54311}	{53321}	64064
{54221}	{54221}	68640
{5333}	{44411}	15015
{4442}	{4433}	12012

A-2 Dimensions of the Irreducible Representations of U_3.

{000}	1
{100}	3
{200}	6
{210}	8
{300}	10
{310}	15
{400}	15
{410}	24
{420}	27
{500}	21
{510}	35
{520}	42
{600}	28
{610}	48
{620}	60
{630}	64
{700}	36
{710}	63
{720}	81
{730}	90
{800}	45
{810}	80
{820}	105
{830}	120
{840}	125
{900}	55
{910}	99
{920}	132
{930}	154
{940}	165
{1000}	66
{1010}	120
{1020}	162
{1030}	192
{1040}	210
{1050}	216
{1100}	78
{1110}	143
{1120}	195
{1130}	234
{1140}	260
{1150}	273
{1200}	91
{1210}	168
{1220}	231
{1230}	280
{1240}	315
{1250}	336
{1260}	343
{1300}	105
{1310}	195
{1320}	270
{1330}	330
{1340}	375
{1350}	405
{1360}	420
{1400}	120
{1410}	224
{1420}	312
{1430}	384
{1440}	440
{1450}	480
{1460}	504
{1470}	512
{1500}	136
{1510}	255
{1520}	357
{1530}	442
{1540}	510
{1550}	561
{1560}	595
{1570}	612

A-3 Dimensions of the Irreducible Representations of U_4.

{0000}	1
{1000}	4
{1100}	6
{2000}	10
{2100}	20
{2110}	15
{2200}	20
{3000}	20
{3100}	45
{3110}	36
{3200}	60
{3210}	64
{3300}	50
{4000}	35
{4100}	84
{4110}	70
{4200}	126
{4210}	140
{4220}	84
{4300}	140
{4310}	175
{4400}	105
{5000}	56
{5100}	140
{5110}	120
{5200}	224
{5210}	256
{5220}	160
{5300}	280
{5310}	360
{5320}	300
{5400}	280
{5410}	384
{5500}	196
{6000}	84
{6100}	216
{6110}	189
{6200}	360
{6210}	420
{6220}	270
{6300}	480
{6310}	630
{6320}	540
{6330}	300
{6400}	540
{6410}	756
{6420}	729
{6500}	504
{6510}	735
{6600}	336
{7000}	120
{7100}	315
{7110}	280
{7200}	540
{7210}	640
{7220}	420
{7300}	750
{7310}	1000
{7320}	875

{7330}	500
{7400}	900
{7410}	1280
{7420}	1260
{7430}	960
{7500}	945
{7510}	1400
{7520}	1470
{7600}	840
{7610}	1280
{7700}	540
{8000}	165
{8100}	440
{8110}	396
{8200}	770
{8210}	924
{8220}	616
{8300}	1100
{8310}	1485
{8320}	1320
{8330}	770
{8400}	1375
{8410}	1980
{8420}	1980
{8430}	1540
{8440}	825
{8500}	1540
{8510}	2310
{8520}	2464
{8530}	2156
{8600}	1540
{8610}	2376
{8620}	2640
{8700}	1320
{8710}	2079
{8800}	825
{9000}	220
{9100}	594
{9110}	540
{9200}	1056
{9210}	1280
{9220}	864
{9300}	1540
{9310}	2100
{9320}	1890
{9330}	1120
{9400}	1980
{9410}	2880
{9420}	2916
{9430}	2304
{9440}	1260
{9500}	2310
{9510}	3500
{9520}	3780
{9530}	3360
{9540}	2450
{9600}	2464
{9610}	3840
{9620}	4320
{9630}	4096
{9700}	2376
{9710}	3780
{9720}	4374
{9800}	1980

{9810}	3200
{9900}	1210
{10000}	286
{10100}	780
{10110}	715
{10200}	1404
{10210}	1716
{10220}	1170
{10300}	2080
{10310}	2860
{10320}	2600
{10330}	1560
{10400}	2730
{10410}	4004
{10420}	4095
{10430}	3276
{10440}	1820
{10500}	3276
{10510}	5005
{10520}	5460
{10530}	4914
{10540}	3640
{10550}	1911
{10600}	3640
{10610}	5720
{10620}	6500
{10630}	6240
{10640}	5200
{10700}	3744
{10710}	6006
{10720}	7020
{10730}	7020
{10800}	3510
{10810}	5720
{10820}	6825
{10900}	2860
{10910}	4719
{101000}	1716
{11000}	364
{11100}	1001
{11110}	924
{11200}	1820
{11210}	2240
{11220}	1540
{11300}	2730
{11310}	3780
{11320}	3465
{11330}	2100
{11400}	3640
{11410}	5376
{11420}	5544
{11430}	4480
{11440}	2520
{11500}	4459
{11510}	6860
{11520}	7546
{11530}	6860
{11540}	5145
{11550}	2744
{11600}	5096
{11610}	8064
{11620}	9240
{11630}	8960
{11640}	7560

{11650}	5376
{11700}	5460
{11710}	8820
{11720}	10395
{11730}	10500
{11740}	9450
{11800}	5460
{11810}	8960
{11820}	10780
{11830}	11200
{11900}	5005
{11910}	8316
{11920}	10164
{111000}	4004
{111010}	6720
{111100}	2366
{12000}	455
{12100}	1260
{12110}	1170
{12200}	2310
{12210}	2860
{12220}	1980
{12300}	3500
{12310}	4875
{12320}	4500
{12330}	2750
{12400}	4725
{12410}	7020
{12420}	7290
{12430}	5940
{12440}	3375
{12500}	5880
{12510}	9100
{12520}	10080
{12530}	9240
{12540}	7000
{12550}	3780
{12600}	6860
{12610}	10920
{12620}	12600
{12630}	12320
{12640}	10500
{12650}	7560
{12660}	3920
{12700}	7560
{12710}	12285
{12720}	14580
{12730}	14850
{12740}	13500
{12750}	10935
{12800}	7875
{12810}	13000
{12820}	15750
{12830}	16500
{12840}	15625
{12900}	7700
{12910}	12870
{12920}	15840
{12930}	16940
{121000}	6930
{121010}	11700
{121020}	14580
{121100}	5460
{121110}	9295
{121200}	3185

Dimensions of the Irreducible Representations of U_5.

{00000}	1
{10000}	5
{11000}	10
{20000}	15
{21000}	40
{21100}	45
{21110}	24
{22000}	50
{22100}	75
{30000}	35
{31000}	105
{31100}	126
{31110}	70
{32000}	175
{32100}	280
{32110}	175
{33000}	175
{33100}	315
{40000}	70
{41000}	224
{41100}	280
{41110}	160
{42000}	420
{42100}	700
{42110}	450
{42200}	560
{42210}	480
{42220}	200
{43000}	560
{43100}	1050
{43110}	720
{43200}	1120
{43210}	1024
{44000}	490
{44100}	980
{44200}	1176
{50000}	126
{51000}	420
{51100}	540
{51110}	315
{52000}	840
{52100}	1440
{52110}	945
{52200}	1200
{52210}	1050
{52220}	450
{53000}	1260
{53100}	2430
{53110}	1701
{53200}	2700
{53210}	2520
{53220}	1215
{54000}	1470
{54100}	3024
{54110}	2205
{54200}	3780
{54210}	3675
{55000}	1176
{55100}	2520
{55200}	3360
{60000}	210
{61000}	720
{61100}	945
{61110}	560
{62000}	1500
{62100}	2625
{62110}	1750
{62200}	2250
{62210}	2000
{62220}	875
{63000}	2400
{63100}	4725
{63110}	3360
{63200}	5400
{63210}	5120
{63220}	2520
{63300}	3750
{63310}	4000
{63320}	2625
{63330}	1000
{64000}	3150
{64100}	6615
{64110}	4900
{64200}	8505
{64210}	8400
{64220}	4410
{64300}	7875
{64310}	8750
{64320}	6125
{65000}	3360
{65100}	7350
{65110}	5600
{65200}	10080
{65210}	10240
{65300}	10500
{65310}	12000
{66000}	2520
{66100}	5670
{66200}	8100
{66300}	9000
{70000}	330
{71000}	1155
{71100}	1540
{71110}	924
{72000}	2475
{72100}	4400
{72110}	2970
{72200}	3850
{72210}	3465
{72220}	1540
{73000}	4125
{73100}	8250
{73110}	5940
{73200}	9625
{73210}	9240
{73220}	4620
{73300}	6875
{73310}	7425
{73320}	4950
{73330}	1925
{74000}	5775
{74100}	12320
{74110}	9240
{74200}	16170
{74210}	16170
{74220}	8624
{74300}	15400
{74310}	17325
{74320}	12320
{74330}	5390
{75000}	6930
{75100}	15400
{75110}	11880
{75200}	21560
{75210}	22176
{75220}	12320
{75300}	23100
{75310}	26730
{75320}	19800
{76000}	6930
{76100}	15840
{76110}	12474
{76200}	23100
{76210}	24255
{76300}	26400
{76310}	31185
{77000}	4950
{77100}	11550
{77200}	17325
{77300}	20625
{80000}	495
{81000}	1760
{81100}	2376
{81110}	1440
{82000}	3850
{82100}	6930
{82110}	4725
{82200}	6160
{82210}	5600
{82220}	2520
{83000}	6600
{83100}	13365
{83110}	9720
{83200}	15840
{83210}	15360
{83220}	7776
{83300}	11550
{83310}	12600
{83320}	8505
{83330}	3360
{84000}	9625
{84100}	20790
{84110}	15750
{84200}	27720
{84210}	28000
{84220}	15120
{84300}	26950
{84310}	30625
{84320}	22050
{84330}	9800
{84400}	17325
{84410}	21000
{84420}	17010
{84430}	10080
{84440}	3675
{85000}	12320
{85100}	27720
{85110}	21600
{85200}	39424
{85210}	40960
{85220}	23040
{85300}	43120
{85310}	50400
{85320}	37800
{85330}	17920
{85400}	36960
{85410}	46080
{85420}	38880
{85430}	24576
{86000}	13860
{86100}	32076
{86110}	25515
{86200}	47520
{86210}	50400
{86220}	29160
{86300}	55440
{86310}	66150
{86320}	51030
{86400}	53460
{86410}	68040
{86420}	59049
{87000}	13200
{87100}	31185
{87110}	25200
{87200}	47520
{87210}	51200
{87300}	57750
{87310}	70000
{87400}	59400
{87410}	76800
{88000}	9075
{88100}	21780
{88200}	33880
{88300}	42350
{88400}	45375
{90000}	715
{91000}	2574
{91100}	3510
{91110}	2145
{92000}	5720
{92100}	10400
{92110}	7150
{92200}	9360
{92210}	8580
{92220}	3900
{93000}	10010
{93100}	20475
{93110}	15015
{93200}	24570
{93210}	24024
{93220}	12285
{93300}	18200
{93310}	20020
{93320}	13650
{93330}	5460
{94000}	15015
{94100}	32760
{94110}	25025
{94200}	44226
{94210}	45045
{94220}	24570
{94300}	43680
{94310}	50050
{94320}	36400
{94330}	16380
{94400}	28665
{94410}	35035
{94420}	28665
{94430}	17199
{94440}	6370
{95000}	20020
{95100}	45500
{95110}	35750
{95200}	65520
{95210}	68640
{95220}	39000
{95300}	72800
{95310}	85800
{95320}	65000
{95330}	31200
{95400}	63700

A-4 Dimensions of the Irreducible Representations of U_5 (contd).

{95410}	80080	{96310}	120120	{97210}	102960	{98200}	90090
{95420}	68250	{96320}	93600	{97220}	61425	{98210}	99099
{95430}	43680	{96330}	46800	{97300}	117000	{98300}	114400
{95440}	18200	{96400}	98280	{97310}	143000	{98310}	141570
{96000}	24024	{96410}	126126	{97320}	113750	{98400}	125125
{96100}	56160	{96420}	110565	{97400}	122850	{98410}	165165
{96110}	45045	{96430}	73710	{97410}	160160	{99000}	15730
{96200}	84240	{97000}	25740	{97420}	143325	{99100}	38610
{96210}	90090	{97100}	61425	{98000}	23595	{99200}	61776
{96220}	52650	{97110}	50050	{98100}	57200	{99300}	80080
{96300}	99840	{97200}	94770	{98110}	47190	{99400}	90090

A-5 Dimensions of the Irreducible Representations of U_6.

{000000}	1	{433000}	4410	{544100}	30870	{642220}	18480
		{433100}	6804	{544110}	19600	{643000}	69300
{100000}	6	{433110}	3969	{550000}	5292	{643100}	115500
{110000}	15	{440000}	1764	{551000}	15120	{643110}	71280
{111000}	20	{441000}	4704	{551100}	17010	{643200}	107800
		{441100}	5040	{552000}	25200	{643210}	88704
{200000}	21	{442000}	7056	{552100}	37800	{643220}	36960
{210000}	70	{442100}	10080	{552200}	27000	{643300}	57750
{211000}	105	{442200}	6720	{553000}	30240	{643310}	53460
{211100}	84	{443000}	7056	{553100}	51030	{643320}	29700
{211110}	35	{443100}	11340	{553200}	48600	{644000}	48510
{220000}	105	{444000}	4116	{554000}	26460	{644100}	86240
{221000}	210			{554100}	47628	{644110}	55440
{221100}	189	{500000}	252	{555000}	14112	{644200}	90552
{222000}	175	{510000}	1050			{644210}	77616
		{511000}	1800	{600000}	462	{644220}	34496
{300000}	56	{511100}	1575	{610000}	1980	{650000}	16632
{310000}	210	{511110}	700	{611000}	3465	{651000}	48510
{311000}	336	{520000}	2520	{611100}	3080	{651100}	55440
{311100}	280	{521000}	5760	{611110}	1386	{651110}	29106
{311110}	120	{521100}	5670	{620000}	4950	{652000}	83160
{320000}	420	{521110}	2688	{621000}	11550	{652100}	126720
{321000}	896	{522000}	6000	{621100}	11550	{652110}	74844
{321100}	840	{522100}	7875	{621110}	5544	{652200}	92400
{321110}	384	{522110}	4200	{622000}	12375	{652210}	72765
{322000}	840	{522200}	4500	{622100}	16500	{653000}	103950
{322100}	1050	{522210}	3200	{622110}	8910	{653100}	178200
{322110}	540	{522220}	1050	{622200}	9625	{653110}	112266
{330000}	490	{530000}	4410	{622210}	6930	{653200}	173250
{331000}	1176	{531000}	11340	{622220}	2310	{653210}	145530
{331100}	1176	{531100}	11907	{630000}	9240	{653300}	99000
{332000}	1470	{531110}	5880	{631000}	24255	{653310}	93555
{332100}	1960	{532000}	15750	{631100}	25872	{654000}	97020
{333000}	980	{532100}	22050	{631110}	12936	{654100}	177408
		{532110}	12250	{632000}	34650	{654110}	116424
{400000}	126	{532200}	14175	{632100}	49280	{654200}	194040
{410000}	504	{532210}	10500	{632110}	27720	{654210}	169785
{411000}	840	{532220}	3675	{632200}	32340	{655000}	58212
{411100}	720	{533000}	12600	{632210}	24255	{655100}	110880
{411110}	315	{533100}	19845	{632220}	8624	{655110}	74844
{420000}	1134	{533110}	11760	{633000}	28875	{660000}	13860
{421000}	2520	{533200}	17010	{633100}	46200	{661000}	41580
{421100}	2430	{533210}	13440	{633110}	27720	{661100}	48510
{421110}	1134	{533220}	5292	{633200}	40425	{662000}	74250
{422000}	2520	{540000}	5880	{633210}	32340	{662100}	115500
{422100}	3240	{541000}	16128	{633220}	12936	{662200}	86625
{422110}	1701	{541100}	17640	{633300}	19250	{663000}	99000
{422200}	1800	{541110}	8960	{633310}	17325	{663100}	173250
{422210}	1260	{542000}	25200	{633320}	9240	{663200}	173250
{422220}	405	{542100}	36750	{633330}	2695	{663300}	103125
{430000}	1764	{542110}	21000	{640000}	13860	{664000}	103950
{431000}	4410	{542200}	25200	{641000}	38808	{664100}	194040
{431100}	4536	{542210}	19200	{641100}	43120	{664200}	218295
{431110}	2205	{543000}	26880	{641110}	22176	{665000}	83160
{432000}	5880	{543100}	44100	{642000}	62370	{665100}	161700
{432100}	8064	{543110}	26880	{642100}	92400	{666000}	41580
{432110}	4410	{543200}	40320	{642110}	53460		
{432200}	5040	{543210}	32768	{642200}	64680		
{432210}	3675	{544000}	17640	{642210}	49896		

Dimensions of the Irreducible Representations of U_7.

Representation	Dimension
{0000000}	1
{1000000}	7
{1100000}	21
{1110000}	35
{2000000}	28
{2100000}	112
{2110000}	210
{2111000}	224
{2111100}	140
{2111110}	48
{2200000}	196
{2210000}	490
{2211000}	588
{2211100}	392
{2220000}	490
{2221000}	784
{3000000}	84
{3100000}	378
{3110000}	756
{3111000}	840
{3111100}	540
{3111110}	189
{3200000}	882
{3210000}	2352
{3211000}	2940
{3211100}	2016
{3211110}	735
{3220000}	2646
{3221000}	4410
{3221100}	3402
{3221110}	1323
{3300000}	1176
{3310000}	3528
{3311000}	4704
{3311100}	3360
{3320000}	5292
{3321000}	9408
{3321100}	7560
{3330000}	4116
{3331000}	8232
{4000000}	210
{4100000}	1008
{4110000}	2100
{4111000}	2400
{4111100}	1575
{4111110}	560
{4200000}	2646
{4210000}	7350
{4211000}	9450
{4211100}	6615
{4211110}	2450
{4220000}	8820
{4221000}	15120
{4221100}	11907
{4221110}	4704
{4222000}	10500
{4222100}	11025
{4222110}	4900
{4222200}	4725
{4222210}	2800
{4222220}	735
{4300000}	4704
{4310000}	14700
{4311000}	20160
{4311100}	14700
{4311110}	5600

Representation	Dimension
{4320000}	23520
{4321000}	43008
{4321100}	35280
{4321110}	14336
{4322000}	33600
{4322100}	36750
{4322110}	16800
{4322200}	16800
{4322210}	10240
{4330000}	20580
{4331000}	42336
{4331100}	37044
{4331110}	15680
{4332000}	44100
{4332100}	51450
{4332110}	24500
{4400000}	5292
{4410000}	17640
{4411000}	25200
{4411100}	18900
{4420000}	31752
{4421000}	60480
{4421100}	51030
{4422000}	50400
{4422100}	56700
{4422200}	27000
{4430000}	37044
{4431000}	79380
{4431100}	71442
{4432000}	88200
{4432100}	105840
{4440000}	24696
{4441000}	56448
{4442000}	70560
{5000000}	462
{5100000}	2310
{5110000}	4950
{5111000}	5775
{5111100}	3850
{5111110}	1386
{5200000}	6468
{5210000}	18480
{5211000}	24255
{5211100}	17248
{5211110}	6468
{5220000}	23100
{5221000}	40425
{5221100}	32340
{5221110}	12936
{5222000}	28875
{5222100}	30800
{5222110}	13860
{5222200}	13475
{5222210}	8085
{5222220}	2156
{5300000}	12936
{5310000}	41580
{5311000}	58212
{5311100}	43120
{5311110}	16632
{5320000}	69300
{5321000}	129360
{5321100}	107800
{5321110}	44352
{5322000}	103950
{5322100}	115500
{5322110}	53460
{5322200}	53900
{5322210}	33264

Representation	Dimension
{5322220}	9240
{5330000}	64680
{5331000}	135828
{5331100}	120736
{5331110}	51744
{5332000}	145530
{5332100}	172480
{5332110}	83160
{5332200}	90552
{5332210}	58212
{5332220}	17248
{5400000}	19404
{5410000}	66528
{5411000}	97020
{5411100}	73920
{5411110}	29106
{5420000}	124740
{5421000}	242550
{5421100}	207900
{5421110}	87318
{5422000}	207900
{5422100}	237600
{5422110}	112266
{5422200}	115500
{5422210}	72765
{5430000}	155232
{5431000}	339570
{5431100}	310464
{5431110}	135828
{5432000}	388080
{5432100}	473088
{5432110}	232848
{5432200}	258720
{5432210}	169785
{5440000}	116424
{5441000}	271656
{5441100}	258720
{5441110}	116424
{5442000}	349272
{5442100}	443520
{5442110}	224532
{5500000}	19404
{5510000}	69300
{5511000}	103950
{5511100}	80850
{5520000}	138600
{5521000}	277200
{5521100}	242550
{5522000}	247500
{5522100}	288750
{5522200}	144375
{5530000}	194040
{5531000}	436590
{5531100}	407484
{5532000}	519750
{5532100}	646800
{5532200}	363825
{5540000}	194040
{5541000}	465696
{5541100}	452760
{5542000}	623700
{5542100}	808500
{5550000}	116424
{5551000}	291060
{5552000}	415800
{6000000}	924
{6100000}	4752
{6110000}	10395
{6111000}	12320

Representation	Dimension
{6111100}	8316
{6111110}	3024
{6200000}	13860
{6210000}	40425
{6211000}	53900
{6211100}	38808
{6211110}	14700
{6220000}	51975
{6221000}	92400
{6221100}	74844
{6221110}	30240
{6222000}	67375
{6222100}	72765
{6222110}	33075
{6222200}	32340
{6222210}	19600
{6222220}	5292
{6300000}	29568
{6310000}	97020
{6311000}	137984
{6311100}	103488
{6311110}	40320
{6320000}	166320
{6321000}	315392
{6321100}	266112
{6321110}	110592
{6322000}	258720
{6322100}	291060
{6322110}	136080
{6322200}	137984
{6322210}	86016
{6322220}	24192
{6330000}	161700
{6331000}	344960
{6331100}	310464
{6331110}	134400
{6332000}	377300
{6332100}	452760
{6332110}	220500
{6332200}	241472
{6332210}	156800
{6332220}	47040
{6333000}	215600
{6333100}	291060
{6333110}	151200
{6333200}	206976
{6333210}	143360
{6333220}	48384
{6333300}	75460
{6333310}	58800
{6333320}	26460
{6333330}	6272
{6400000}	49896
{6410000}	174636
{6411000}	258720
{6411100}	199584
{6411110}	79380
{6420000}	336798
{6421000}	665280
{6421100}	577368
{6421110}	244944
{6422000}	582120
{6422100}	673596
{6422110}	321489
{6422200}	332640
{6422210}	211680
{6422220}	61236
{6430000}	436590
{6431000}	970200
{6431100}	898128

A-6 Dimensions of the Irreducible Representations of U_7 (contd).

{6431110}	396900
{6432000}	1131900
{6432100}	1397088
{6432110}	694575
{6432200}	776160
{6432210}	514500
{6432220}	158760
{6433000}	727650
{6433100}	1010394
{6433110}	535815
{6433200}	748440
{6433210}	529200
{6433220}	183708
{6433300}	291060
{6433310}	231525
{6433320}	107163
{6440000}	349272
{6441000}	827904
{6441100}	798336
{6441110}	362880
{6442000}	1086624
{6442100}	1397088
{6442110}	714420
{6442200}	827904
{6442210}	564480
{6442220}	181440
{6443000}	931392
{6443100}	1347192
{6443110}	734832
{6443200}	1064448
{6443210}	774144
{6443220}	279936
{6500000}	66528
{6510000}	242550
{6511000}	369600
{6511100}	291060
{6511110}	117600
{6520000}	498960
{6521000}	1013760
{6521100}	898128
{6521110}	387072
{6522000}	924000
{6522100}	1091475
{6522110}	529200
{6522200}	554400
{6522210}	358400
{6530000}	727650
{6531000}	1663200
{6531100}	1571724
{6531110}	705600
{6532000}	2021250
{6532100}	2546775
{6532110}	1286250
{6532200}	1455300
{6532210}	980000
{6533000}	1386000
{6533100}	1964655
{6533110}	1058400
{6533200}	1496880
{6533210}	1075200
{6533300}	606375
{6533310}	490000
{6540000}	776160
{6541000}	1892352
{6541100}	1862784
{6541110}	860160
{6542000}	2587200
{6542100}	3395700
{6542110}	1764000
{6542200}	2069760
{6542210}	1433600
{6543000}	2365440
{6543100}	3492720
{6543110}	1935360
{6543200}	2838528
{6543210}	2097152
{6550000}	523908
{6551000}	1330560
{6551100}	1347192
{6551110}	635040
{6552000}	1940400
{6552100}	2619540
{6552110}	1389150
{6553000}	1995840
{6553100}	3031182
{6553110}	1714608
{6600000}	60984
{6610000}	228690
{6611000}	355740
{6611100}	284592
{6620000}	490050
{6621000}	1016400
{6621100}	914760
{6622000}	952875
{6622100}	1143450
{6622200}	592900
{6630000}	762300
{6631000}	1778700
{6631100}	1707552
{6632000}	2223375
{6632100}	2845920
{6632200}	1660120
{6633000}	1588125
{6633100}	2286900
{6633200}	1778700
{6633300}	741125
{6640000}	914760
{6641000}	2276736
{6641100}	2276736
{6642000}	3201660
{6642100}	4268880
{6642200}	2656192
{6643000}	3049200
{6643100}	4573800
{6643200}	3794560
{6650000}	823284
{6651000}	2134440
{6651100}	2195424
{6652000}	3201660
{6652100}	4390848
{6653000}	3430350
{6653100}	5292540
{6660000}	457380
{6661000}	1219680
{6662000}	1905750
{6663000}	2178000

A-7 Dimensions of the Irreducible Representations of U_8.

{00000000}	1
{10000000}	8
{11000000}	28
{11100000}	56
{11110000}	70
{20000000}	36
{21000000}	168
{21100000}	378
{21110000}	504
{21111000}	420
{21111100}	216
{21111110}	63
{22000000}	336
{22100000}	1008
{22110000}	1512
{22111000}	1344
{22111100}	720
{22200000}	1176
{22210000}	2352
{22211000}	2352
{22220000}	1764
{30000000}	120
{31000000}	630
{31100000}	1512
{31110000}	2100
{31111000}	1800
{31111100}	945
{31111110}	280
{32000000}	1680
{32100000}	5376
{32110000}	8400
{32111000}	7680
{32111100}	4200
{32111110}	1280
{32200000}	7056
{32210000}	14700
{32211000}	15120
{32211100}	8820
{32211110}	2800
{32220000}	11760
{32221000}	16128
{32221100}	10584
{32221110}	3584
{33000000}	2520
{33100000}	9072
{33110000}	15120
{33111000}	14400
{33111100}	8100
{33200000}	15876
{33210000}	35280
{33211000}	37800
{33211100}	22680
{33220000}	31752
{33221000}	45360
{33221100}	30618
{33300000}	14112
{33310000}	35280
{33311000}	40320
{33320000}	42336
{33321000}	64512
{33330000}	24696
{40000000}	330
{41000000}	1848
{41100000}	4620
{41110000}	6600
{41111000}	5775
{41111100}	3080
{41111110}	924
{42000000}	5544
{42100000}	18480
{42110000}	29700
{42111000}	27720
{42111100}	15400
{42111110}	4752
{42200000}	25872
{42210000}	55440
{42211000}	58212
{42211100}	34496
{42211110}	11088
{42220000}	46200
{42221000}	64680
{42221100}	43120
{42221110}	14784
{42222000}	34650
{42222100}	30800
{42222110}	11880
{42222200}	10780
{42222210}	5544
{42222220}	1232
{43000000}	11088
{43100000}	41580
{43110000}	71280
{43111000}	69300
{43111100}	39600
{43111110}	12474
{43200000}	77616
{43210000}	177408
{43211000}	194040
{43211100}	118272
{43211110}	38808
{43220000}	166320
{43221000}	242550
{43221100}	166320
{43221110}	58212
{43222000}	138600
{43222100}	126720
{43222110}	49896
{43222200}	46200
{43222210}	24255
{43300000}	77616
{43310000}	199584
{43311000}	232848
{43311100}	147840
{43311110}	49896
{43320000}	249480
{43321000}	388080
{43321100}	277200
{43321110}	99792
{43322000}	249480
{43322100}	237600
{43322110}	96228
{43330000}	155232
{43331000}	271656
{43331100}	206976
{43331110}	77616
{44000000}	13860
{44100000}	55440
{44110000}	99000
{44111000}	99000
{44111100}	57750
{44200000}	116424
{44210000}	277200
{44211000}	311850
{44211100}	194040
{44220000}	277200
{44221000}	415800
{44221100}	291060
{44222000}	247500
{44222100}	231000
{44222200}	86625
{44300000}	155232
{44310000}	415800
{44311000}	498960
{44311100}	323400
{44320000}	554400
{44321000}	887040
{44321100}	646800
{44322000}	594000
{44322100}	577500
{44330000}	388080
{44331000}	698544
{44331100}	543312
{44400000}	116424
{44410000}	332640
{44411000}	415800
{44420000}	498960
{44421000}	831600
{44422000}	594000
{44430000}	465696
{44431000}	873180
{44440000}	232848

 Dimensions of the Irreducible Representations of U_9.

Representation	Dimension
{000000000}	1
{100000000}	9
{110000000}	36
{111000000}	84
{111100000}	126
{200000000}	45
{210000000}	240
{211000000}	630
{211100000}	1008
{211110000}	1050
{211111000}	720
{211111100}	315
{211111110}	80
{220000000}	540
{221000000}	1890
{221100000}	3402
{221110000}	3780
{221111000}	2700
{221111100}	1215
{222000000}	2520
{222100000}	6048
{222110000}	7560
{222111000}	5760
{222200000}	5292
{222210000}	8820
{300000000}	165
{310000000}	990
{311000000}	2772
{311100000}	4620
{311110000}	4950
{311111000}	3465
{311111100}	1540
{311111110}	396
{320000000}	2970
{321000000}	11088
{321100000}	20790
{321110000}	23760
{321111000}	17325
{321111100}	7920
{321111110}	2079
{322000000}	16632
{322100000}	41580
{322110000}	53460
{322111000}	41580
{322111100}	19800
{322111110}	5346
{322200000}	38808
{322210000}	66528
{322211000}	58212
{322211100}	29568
{322211110}	8316
{330000000}	4950
{331000000}	20790
{331100000}	41580
{331110000}	49500
{331111000}	37125
{331111100}	17325
{332000000}	41580
{332100000}	110880
{332110000}	148500
{332111000}	118800
{332111100}	57750
{332200000}	116424
{332210000}	207900
{332211000}	187110
{332211100}	97020
{333000000}	41580

Representation	Dimension
{333100000}	124740
{333110000}	178200
{333111000}	148500
{333200000}	174636
{333210000}	332640
{333211000}	311850
{333300000}	116424
{333310000}	249480
{400000000}	495
{410000000}	3168
{411000000}	9240
{411100000}	15840
{411110000}	17325
{411111000}	12320
{411111100}	5544
{411111110}	1440
{420000000}	10692
{421000000}	41580
{421100000}	80190
{421110000}	93555
{421111000}	69300
{421111100}	32076
{421111110}	8505
{422000000}	66528
{422100000}	171072
{422110000}	224532
{422111000}	177408
{422111100}	85536
{422111110}	23328
{422200000}	166320
{422210000}	291060
{422211000}	258720
{422211100}	133056
{422211110}	37800
{422220000}	187110
{422221000}	221760
{422221100}	128304
{422221110}	38880
{422222000}	97020
{422222100}	74844
{422222110}	25515
{422222200}	22176
{422222210}	10080
{422222220}	1944
{430000000}	23760
{431000000}	103950
{431100000}	213840
{431110000}	259875
{431111000}	198000
{431111100}	93555
{431111110}	25200
{432000000}	221760
{432100000}	608256
{432110000}	831600
{432111000}	675840
{432111100}	332640
{432111110}	92160
{432200000}	665280
{432210000}	1212750
{432211000}	1108800
{432211100}	582120
{432211110}	168000
{432220000}	831600
{432221000}	1013760
{432221100}	598752
{432221110}	184320
{432222000}	462000
{432222100}	363825
{432222110}	126000

Representation	Dimension
{432222200}	110880
{432222210}	51200
{433000000}	249480
{433100000}	769824
{433110000}	1122660
{433111000}	950400
{433111100}	481140
{433111110}	136080
{433200000}	1122660
{433210000}	2182950
{433211000}	2079000
{433211100}	1122660
{433211110}	330750
{433220000}	1683990
{433221000}	2138400
{433221100}	1299078
{433221110}	408240
{433222000}	1039500
{433222100}	841995
{433222110}	297675
{433300000}	798336
{433310000}	1746360
{433311000}	1774080
{433311100}	997920
{433311110}	302400
{433320000}	1796256
{433321000}	2433024
{433321100}	1539648
{433321110}	497664
{440000000}	32670
{441000000}	152460
{441100000}	326700
{441110000}	408375
{441111000}	317625
{441111100}	152460
{442000000}	365904
{442100000}	1045440
{442110000}	1470150
{442111000}	1219680
{442111100}	609840
{442200000}	1219680
{442210000}	2286900
{442211000}	2134440
{442211100}	1138368
{442220000}	1633500
{442221000}	2032800
{442221100}	1219680
{442222000}	952875
{442222100}	762300
{442222200}	237160
{443000000}	548856
{443100000}	1764180
{443110000}	2646270
{443111000}	2286900
{443111100}	1176120
{443200000}	2744280
{443210000}	5488560
{443211000}	5336100
{443211100}	2927232
{443220000}	4410450
{443221000}	5717250
{443221100}	3528360
{443222000}	2858625
{443222100}	2352240
{443300000}	2195424
{443310000}	4939704
{443311000}	5122656
{443311100}	2927232
{443320000}	5292540
{443321000}	7318080

A-8 Dimensions of the Irreducible Representations of U_9 (contd).

{443321100}	4704480	{444211000}	5717250	{444320000}	7840800
{444000000}	457380	{444220000}	4900500	{444321000}	11151360
{444100000}	1568160	{444221000}	6534000	{444400000}	1646568
{444110000}	2450250	{444222000}	3403125	{444410000}	4116420
{444111000}	2178000	{444300000}	2927232	{444420000}	5292540
{444200000}	2744280	{444310000}	6860700		
{444210000}	5717250	{444311000}	7318080		

A-9 Dimensions of the Irreducible Representations of U_{10}.

{0000000000}	1	{2111000000}	1848	{2211111000}	4950
		{2111100000}	2310	{2211111100}	1925
{1000000000}	10	{2111110000}	1980	{2220000000}	4950
{1100000000}	45	{2111111000}	1155	{2221000000}	13860
{1110000000}	120	{2111111100}	440	{2221100000}	20790
{1111000000}	210	{2111111110}	99	{2221110000}	19800
{1111100000}	252	{2200000000}	825	{2221111000}	12375
		{2210000000}	3300	{2222000000}	13860
{2000000000}	55	{2211000000}	6930	{2222100000}	27720
{2100000000}	330	{2211100000}	9240	{2222110000}	29700
{2110000000}	990	{2211110000}	8250	{2222200000}	19404

A-10 Dimensions of the Irreducible Representations of U_{11}.

{00000000000}	1	{21111100000}	4752	{22200000000}	9075
		{21111110000}	3465	{22210000000}	29040
{10000000000}	11	{21111111000}	1760	{22211000000}	50820
{11000000000}	55	{21111111100}	594	{22211100000}	58080
{11100000000}	165	{21111111110}	120	{22211110000}	45375
{11110000000}	330	{22000000000}	1210	{22211111000}	24200
{11111000000}	462	{22100000000}	5445	{22220000000}	32670
		{22110000000}	13068	{22221000000}	76230
{20000000000}	66	{22111000000}	20328	{22221100000}	98010
{21000000000}	440	{22111100000}	21780	{22221110000}	81675
{21100000000}	1485	{22111110000}	16335	{22222000000}	60984
{21110000000}	3168	{22111111000}	8470	{22222100000}	104544
{21111000000}	4620	{22111111100}	2904		

A-11 Dimensions of the Irreducible Representations of U_{12}.

{000000000000}	1	{211111100000}	9009	{222100000000}	56628
		{211111110000}	5720	{222110000000}	113256
{100000000000}	12	{211111111000}	2574	{222111000000}	151008
{110000000000}	66	{211111111100}	780	{222111100000}	141570
{111000000000}	220	{211111111110}	143	{222111110000}	94380
{111100000000}	495	{220000000000}	1716	{222111111000}	44044
{111110000000}	792	{221000000000}	8580	{222200000000}	70785
{111111000000}	924	{221100000000}	23166	{222210000000}	188760
		{221110000000}	41184	{222211000000}	283140
{200000000000}	78	{221111000000}	51480	{222211100000}	283140
{210000000000}	572	{221111100000}	46332	{222211110000}	196625
{211000000000}	2145	{221111110000}	30030	{222220000000}	169884
{211100000000}	5148	{221111111000}	13728	{222221000000}	339768
{211110000000}	8580	{221111111100}	4212	{222221100000}	382239
{211111000000}	10296	{222000000000}	15730	{222222000000}	226512

A-12 Dimensions of the Irreducible Representations of U_{13}.

{0000000000000}	1	{2111111110000}	9009	{2221110000000}	356928
		{2111111111000}	3640	{2221111000000}	390390
{1000000000000}	13	{2111111111100}	1001	{2221111100000}	312312
{1100000000000}	78	{2111111111110}	168	{2221111110000}	182182
{1110000000000}	286	{2200000000000}	2366	{2221111111000}	75712
{1111000000000}	715	{2210000000000}	13013	{2222000000000}	143143
{1111100000000}	1287	{2211000000000}	39039	{2222100000000}	429429
{1111110000000}	1716	{2211100000000}	78078	{2222110000000}	736164
		{2211110000000}	111540	{2222111000000}	858858
{2000000000000}	91	{2211111000000}	117117	{2222111100000}	715715
{2100000000000}	728	{2211111100000}	91091	{2222111110000}	429429
{2110000000000}	3003	{2211111110000}	52052	{2222200000000}	429429
{2111000000000}	8008	{2211111111000}	21294	{2222210000000}	981552
{2111100000000}	15015	{2211111111100}	5915	{2222211000000}	1288287
{2111110000000}	20592	{2220000000000}	26026	{2222211100000}	1145144
{2111111000000}	21021	{2221000000000}	104104	{2222220000000}	736164
{2111111100000}	16016	{2221100000000}	234234	{2222221000000}	1288287

A-13 Dimensions of the Irreducible Representations of U_{14}.

{00000000000000}	1	{21111111111000}	5005	{22211111100000}	637637
		{21111111111100}	1260	{22211111110000}	331240
{10000000000000}	14	{21111111111110}	195	{22211111111000}	124215
{11000000000000}	91	{22000000000000}	3185	{22220000000000}	273273
{11100000000000}	364	{22100000000000}	19110	{22221000000000}	910910
{11110000000000}	1001	{22110000000000}	63063	{22221100000000}	1756755
{11111000000000}	2002	{22111000000000}	140140	{22221110000000}	2342340
{11111100000000}	3003	{22111100000000}	225225	{22221111000000}	2277275
{11111110000000}	3432	{22111110000000}	270270	{22221111100000}	1639638
		{22111111000000}	245245	{22221111110000}	869505
{20000000000000}	105	{22111111100000}	168168	{22222000000000}	1002001
{21000000000000}	910	{22111111110000}	85995	{22222100000000}	2576574
{21100000000000}	4095	{22111111111000}	31850	{22222110000000}	3864861
{21110000000000}	12012	{22111111111100}	8085	{22222111000000}	4008004
{21111000000000}	25025	{22200000000000}	41405	{22222111100000}	3006003
{21111100000000}	38610	{22210000000000}	182182	{22222200000000}	2147145
{21111110000000}	45045	{22211000000000}	455455	{22222210000000}	4294290
{21111111000000}	40040	{22211100000000}	780780	{22222211000000}	5010005
{21111111100000}	27027	{22211110000000}	975975	{22222220000000}	2760615
{21111111110000}	13650	{22211111000000}	910910		

A-14 Dimensions of the Irreducible Representations of U_{15}.

{000000000000000}	1	{211111111111100}	1560	{222111111110000}	573300
		{211111111111110}	224	{222111111111000}	196000
{100000000000000}	15	{220000000000000}	4200	{222200000000000}	496860
{110000000000000}	105	{221000000000000}	27300	{222210000000000}	1821820
{111000000000000}	455	{221100000000000}	98280	{222211000000000}	3903900
{111100000000000}	1365	{221110000000000}	240240	{222211100000000}	5855850
{111110000000000}	3003	{221111000000000}	429000	{222211110000000}	6506500
{111111000000000}	5005	{221111100000000}	579150	{222211111000000}	5465460
{111111100000000}	6435	{221111110000000}	600600	{222211111100000}	3478020
		{221111111000000}	480480	{222211111110000}	1656200
{200000000000000}	120	{221111111100000}	294840	{222220000000000}	2186184
{210000000000000}	1120	{221111111110000}	136500	{222221000000000}	6246240
{211000000000000}	5460	{221111111111000}	46200	{222221100000000}	10540530
{211100000000000}	17472	{221111111111100}	10800	{222221110000000}	12492480
{211110000000000}	40040	{222000000000000}	63700	{222221111000000}	10930920
{211111000000000}	68640	{222100000000000}	305760	{222221111100000}	7154784
{211111100000000}	90090	{222110000000000}	840840	{222222000000000}	5725720
{211111110000000}	91520	{222111000000000}	1601600	{222222100000000}	12882870
{211111111000000}	72072	{222111100000000}	2252250	{222222110000000}	17177160
{211111111100000}	43680	{222111110000000}	2402400	{222222111000000}	16032016
{211111111110000}	20020	{222111111000000}	1961960	{222222200000000}	9202050
{211111111111000}	6720	{222111111100000}	1223040	{222222210000000}	16359200

A-15 Dimensions of the Irreducible Representations of U_{16}.

{0000000000000000}	1	{2111111111111110}	255	{2222000000000000}	866320
		{2200000000000000}	5440	{2222100000000000}	3465280
{1000000000000000}	16	{2210000000000000}	38080	{2222110000000000}	8168160
{1100000000000000}	120	{2211000000000000}	148512	{2222111000000000}	13613600
{1110000000000000}	560	{2211100000000000}	396032	{2222111100000000}	17017000
{1111000000000000}	1820	{2211110000000000}	777920	{2222111110000000}	16336320
{1111100000000000}	4368	{2211111000000000}	1166880	{2222111111000000}	12128480
{1111110000000000}	8008	{2211111100000000}	1361360	{2222111111100000}	6930560
{1111111000000000}	11440	{2211111110000000}	1244672	{2222111111110000}	2998800
{1111111100000000}	12870	{2211111111000000}	891072	{2222200000000000}	4504864
		{2211111111100000}	495040	{2222210000000000}	14158144
{2000000000000000}	136	{2211111111110000}	209440	{2222211000000000}	26546520
{2100000000000000}	1360	{2211111111111000}	65280	{2222211100000000}	35395360
{2110000000000000}	7140	{2211111111111100}	14144	{2222211110000000}	35395360
{2111000000000000}	24752	{2220000000000000}	95200	{2222211111000000}	27029184
{2111100000000000}	61880	{2221000000000000}	495040	{2222211111100000}	15767024
{2111110000000000}	116688	{2221100000000000}	1485120	{2222220000000000}	14158144
{2111111000000000}	170170	{2221110000000000}	3111680	{2222221000000000}	35395360
{2111111100000000}	194480	{2221111000000000}	4862000	{2222221100000000}	53093040
{2111111110000000}	175032	{2221111100000000}	5834400	{2222221110000000}	56632576
{2111111111000000}	123760	{2221111110000000}	5445440	{2222221111000000}	45048640
{2111111111100000}	68068	{2221111111000000}	3960320	{2222222000000000}	27810640
{2111111111110000}	28560	{2221111111100000}	2227680	{2222222100000000}	55621280
{2111111111111000}	8840	{2221111111110000}	952000	{2222222110000000}	66745536
{2111111111111100}	1904	{2221111111111000}	299200	{2222222200000000}	34763300

A-16 Dimensions of the Irreducible Representations of U_{17}.

{00000000000000000}	1	{22100000000000000}	52020	{22221110000000000}	29755440
		{22110000000000000}	218484	{22221111000000000}	41327000
{10000000000000000}	17	{22111000000000000}	631176	{22221111100000000}	44633160
{11000000000000000}	136	{22111100000000000}	1352520	{22221111110000000}	37870560
{11100000000000000}	680	{22111110000000000}	2231658	{22221111111000000}	25247040
{11110000000000000}	2380	{22111111000000000}	2892890	{22221111111100000}	13109040
{11111000000000000}	6188	{22111111100000000}	2975544	{22221111111110000}	5202000
{11111100000000000}	12376	{22111111110000000}	2434536	{22222000000000000}	8836464
{11111110000000000}	19448	{22111111111000000}	1577940	{22222100000000000}	30296448
{11111111000000000}	24310	{22111111111100000}	801108	{22222110000000000}	62486424
		{22111111111110000}	312120	{22222111000000000}	92572480
{20000000000000000}	153	{22111111111111000}	90168	{22222111100000000}	104144040
{21000000000000000}	1632	{22111111111111100}	18207	{22222111110000000}	90889344
{21100000000000000}	9180	{22200000000000000}	138720	{22222111111000000}	61855248
{21110000000000000}	34272	{22210000000000000}	776832	{22222111111100000}	32626944
{21111000000000000}	92820	{22211000000000000}	2524704	{22222200000000000}	32821152
{21111100000000000}	190944	{22211100000000000}	5770752	{22222210000000000}	90258168
{21111110000000000}	306306	{22211110000000000}	9918480	{22222211000000000}	150430280
{21111111000000000}	388960	{22211111000000000}	13224640	{22222211100000000}	180516336
{21111111100000000}	393822	{22211111100000000}	13885872	{22222211110000000}	164105760
{21111111110000000}	318240	{22211111110000000}	11541504	{22222211111000000}	114874032
{21111111111000000}	204204	{22211111111000000}	7574112	{22222220000000000}	77364144
{21111111111100000}	102816	{22211111111100000}	3884160	{22222221000000000}	171920320
{21111111111110000}	39780	{22211111111110000}	1525920	{22222221100000000}	232092432
{21111111111111000}	11424	{22211111111111000}	443904	{22222221110000000}	225059328
{21111111111111100}	2295	{22220000000000000}	1456560	{22222222000000000}	118195220
{21111111111111110}	288	{22221000000000000}	6311760	{22222222100000000}	212751396
{22000000000000000}	6936	{22221100000000000}	16230240		

A-17 Dimensions of the Irreducible Representations of U_{18}.

Representation	Dimension	Representation	Dimension	Representation	Dimension
{000000000000000000}	1	{221100000000000000}	313956	{222211111000000000}	113070672
		{221110000000000000}	976752	{222211111100000000}	107931096
{100000000000000000}	18	{221111000000000000}	2267460	{222211111110000000}	82233216
{110000000000000000}	153	{221111100000000000}	4081428	{222211111111000000}	49814352
{111000000000000000}	816	{221111110000000000}	5819814	{222211111111100000}	23721120
{111100000000000000}	3060	{221111111000000000}	6651216	{222211111111110000}	8697744
{111110000000000000}	8568	{221111111100000000}	6122142	{222220000000000000}	16604784
{111111000000000000}	18564	{221111111110000000}	4534920	{222221000000000000}	61674912
{111111100000000000}	31824	{221111111111000000}	2686068	{222221100000000000}	138768552
{111111110000000000}	43758	{221111111111100000}	1255824	{222221110000000000}	226141344
{111111111000000000}	48620	{221111111111110000}	453492	{222221111000000000}	282676680
		{221111111111111000}	122094	{222221111100000000}	277537104
{200000000000000000}	171	{221111111111111100}	23085	{222221111110000000}	215862192
{210000000000000000}	1938	{222000000000000000}	197676	{222221111111000000}	132838272
{211000000000000000}	11628	{222100000000000000}	1186056	{222221111111100000}	64047024
{211100000000000000}	46512	{222110000000000000}	4151196	{222222000000000000}	71954064
{211110000000000000}	135660	{222111000000000000}	10279152	{222222100000000000}	215862192
{211111000000000000}	302328	{222111100000000000}	19273410	{222222110000000000}	395747352
{211111100000000000}	529074	{222111110000000000}	28267668	{222222111000000000}	527663136
{211111110000000000}	739024	{222111111000000000}	32978946	{222222111100000000}	539655480
{211111111000000000}	831402	{222111111100000000}	30837456	{222222111110000000}	431724384
{211111111100000000}	755820	{222111111110000000}	23128092	{222222111111000000}	271211472
{211111111110000000}	554268	{222111111111000000}	13837320	{222222200000000000}	200443464
{211111111111000000}	325584	{222111111111100000}	6523308	{222222210000000000}	489972912
{211111111111100000}	151164	{222111111111110000}	2372112	{222222211000000000}	734959368
{211111111111110000}	54264	{222111111111111000}	642447	{222222211100000000}	801773856
{211111111111111000}	14535	{222200000000000000}	2372112	{222222211110000000}	668144880
{211111111111111100}	2736	{222210000000000000}	11069856	{222222220000000000}	367479684
{211111111111111110}	323	{222211000000000000}	30837456	{222222221000000000}	734959368
{220000000000000000}	8721	{222211100000000000}	61674912	{222222221100000000}	901995588
{221000000000000000}	69768	{222211110000000000}	94225560	{222222222000000000}	449141836

A-18 Dimensions of the Irreducible Representations of O_4.

Rep.	Dim.	Rep.	Dim.	Rep.	Dim.	Rep.	Dim.	Rep.	Dim.
[00]	1	[64]	66	[98]	72	[123]	320	[145]	400
		[65]	48	[99]	38	[124]	306	[146]	378
[10]	4	[66]	26			[125]	288	[147]	352
[11]	6			[100]	121	[126]	266	[148]	322
		[70]	64	[101]	240	[127]	240	[149]	288
[20]	9	[71]	126	[102]	234	[128]	210	[1410]	250
[21]	16	[72]	120	[103]	224	[129]	176	[1411]	208
[22]	10	[73]	110	[104]	210	[1210]	138	[1412]	162
		[74]	96	[105]	192	[1211]	96	[1413]	112
[30]	16	[75]	78	[106]	170	[1212]	50	[1414]	58
[31]	30	[76]	56	[107]	144				
[32]	24	[77]	30	[108]	114	[130]	196	[150]	256
[33]	14			[109]	80	[131]	390	[151]	510
		[80]	81	[1010]	42	[132]	384	[152]	504
[40]	25	[81]	160			[133]	374	[153]	494
[41]	48	[82]	154	[110]	144	[134]	360	[154]	480
[42]	42	[83]	144	[111]	286	[135]	342	[155]	462
[43]	32	[84]	130	[112]	280	[136]	320	[156]	440
[44]	18	[85]	112	[113]	270	[137]	294	[157]	414
		[86]	90	[114]	256	[138]	264	[158]	384
[50]	36	[87]	64	[115]	238	[139]	230	[159]	350
[51]	70	[88]	34	[116]	216	[1310]	192	[1510]	312
[52]	64			[117]	190	[1311]	150	[1511]	270
[53]	54	[90]	100	[118]	160	[1312]	104	[1512]	224
[54]	40	[91]	198	[119]	126	[1313]	54	[1513]	174
[55]	22	[92]	192	[1110]	88			[1514]	120
		[93]	182	[1111]	46	[140]	225	[1515]	62
[60]	49	[94]	168			[141]	448		
[61]	96	[95]	150	[120]	169	[142]	442		
[62]	90	[96]	128	[121]	336	[143]	432		
[63]	80	[97]	102	[122]	330	[144]	418		

A-19 Dimensions of the Irreducible Representations of R_5.

Representation	Dimension
[00]	1
[10]	5
[11]	10
[20]	14
[21]	35
[22]	35
[30]	30
[31]	81
[32]	105
[33]	84
[40]	55
[41]	154
[42]	220
[43]	231
[44]	165
[50]	91
[51]	260
[52]	390
[53]	455
[54]	429
[55]	286
[60]	140
[61]	405
[62]	625
[63]	770
[64]	810
[65]	715
[66]	455
[70]	204
[71]	595
[72]	935
[73]	1190
[74]	1326
[75]	1309
[76]	1105
[77]	680
[80]	285
[81]	836
[82]	1330
[83]	1729
[84]	1995
[85]	2090
[86]	1976
[87]	1615
[88]	969
[90]	385
[91]	1134
[92]	1820
[93]	2401
[94]	2835
[95]	3080
[96]	3094
[97]	2835
[98]	2261
[99]	1330
[100]	506
[101]	1495
[102]	2415
[103]	3220
[104]	3864
[105]	4301
[106]	4485
[107]	4370
[108]	3910
[109]	3059
[1010]	1771
[110]	650
[111]	1925
[112]	3125
[113]	4200
[114]	5100
[115]	5775
[116]	6175
[117]	6250
[118]	5950
[119]	5225
[1110]	4025
[1111]	2300
[120]	819
[121]	2430
[122]	3960
[123]	5355
[124]	6561
[125]	7524
[126]	8190
[127]	8505
[128]	8415
[129]	7866
[1210]	6804
[1211]	5175
[1212]	2925

A-20 Dimensions of the Irreducible Representations of O_6.

Representation	Dimension
[000]	1
[100]	6
[110]	15
[111]	20
[200]	20
[210]	64
[211]	90
[220]	84
[221]	140
[222]	70
[300]	50
[310]	175
[311]	252
[320]	300
[321]	512
[322]	280
[330]	300
[331]	540
[332]	378
[333]	168
[400]	105
[410]	384
[411]	560
[420]	729
[421]	1260
[422]	720
[430]	960
[431]	1750
[432]	1280
[433]	630
[440]	825
[441]	1540
[442]	1232
[443]	792
[444]	330
[500]	196
[510]	735
[511]	1080
[520]	1470
[521]	2560
[522]	1500
[530]	2156
[531]	3960
[532]	2970
[533]	1540
[540]	2450
[541]	4608
[542]	3780
[543]	2560
[544]	1188
[550]	1911
[551]	3640
[552]	3120
[553]	2340
[554]	1430
[555]	572
[600]	336
[610]	1280
[611]	1890
[620]	2640
[621]	4620
[622]	2750
[630]	4096
[631]	7560
[632]	5760
[633]	3080
[640]	5200
[641]	9828
[642]	8190
[643]	5720
[644]	2808
[650]	5376
[651]	10290
[652]	8960
[653]	6930
[654]	4480
[655]	2002
[660]	3920
[661]	7560
[662]	6750
[663]	5500
[664]	3960
[665]	2340
[666]	910
[700]	540
[710]	2079
[711]	3080
[720]	4374
[721]	7680
[722]	4620
[730]	7020
[731]	13000
[732]	10010
[733]	5460
[740]	9450
[741]	17920
[742]	15092
[743]	10752
[744]	5460
[750]	10935
[751]	21000
[752]	18480
[753]	14580
[754]	9750
[755]	4620
[760]	10584
[761]	20480
[762]	18480
[763]	15360
[764]	11440
[765]	7168
[766]	3120
[770]	7344
[771]	14280
[772]	13090
[773]	11220
[774]	8840
[775]	6188
[776]	3570
[777]	1360
[800]	825
[810]	3200
[811]	4752
[820]	6825
[821]	12012
[822]	7280
[830]	11200
[831]	20790
[832]	16128
[833]	8918
[840]	15625
[841]	29700
[842]	25200
[843]	18200
[844]	9450
[850]	19200
[851]	36960
[852]	32768
[853]	26208
[854]	17920
[855]	8800
[860]	20825
[861]	40392
[862]	36720
[863]	30940
[864]	23562
[865]	15300
[866]	7072
[870]	19200
[871]	37422
[872]	34560
[873]	30030
[874]	24192
[875]	17550
[876]	10752
[877]	4590
[880]	12825
[881]	25080
[882]	23408
[883]	20748
[884]	17290
[885]	13300
[886]	9120
[887]	5168
[888]	1938

A-21 Dimensions of the Irreducible Representations of R_7.

Representation	Dimension
[000]	1
[100]	7
[110]	21
[111]	35
[200]	27
[210]	105
[211]	189
[220]	168
[221]	378
[222]	294
[300]	77
[310]	330
[311]	616
[320]	693
[321]	1617
[322]	1386
[330]	825
[331]	2079
[332]	2310
[333]	1386
[400]	182
[410]	819
[411]	1560
[420]	1911
[421]	4550
[422]	4095
[430]	3003
[431]	7722
[432]	9009
[433]	6006
[440]	3003
[441]	8008
[442]	10296
[443]	9009
[444]	4719
[500]	378
[510]	1750
[511]	3375
[520]	4312
[521]	10395
[522]	9625
[530]	7560
[531]	19683
[532]	23625
[533]	16632
[540]	10010
[541]	27027
[542]	35750
[543]	33033
[544]	19305
[550]	8918
[551]	24570
[552]	34125
[553]	35035
[554]	27027
[555]	13013
[600]	714
[610]	3366
[611]	6545
[620]	8568
[621]	20825
[622]	19635
[630]	15912
[631]	41769
[632]	51051
[633]	37128
[640]	23562
[641]	64141
[642]	86394
[643]	82467
[644]	51051
[650]	27846
[651]	77350
[652]	109395
[653]	116025
[654]	94809
[655]	51051
[660]	22848
[661]	64260
[662]	93500
[663]	104720
[664]	95472
[665]	68068
[666]	30940
[700]	1254
[710]	5985
[711]	11704
[720]	15561
[721]	38038
[722]	36309
[730]	29925
[731]	79002
[732]	97755
[733]	72618
[740]	47025
[741]	128744
[742]	175560
[743]	171171
[744]	109725
[750]	62244
[751]	173888
[752]	248976
[753]	269724
[754]	228228
[755]	130416
[760]	67830
[761]	191862
[762]	282625
[763]	323323
[764]	305235
[765]	230945
[766]	117572
[770]	52326
[771]	149226
[772]	223839
[773]	264537
[774]	264537
[775]	223839
[776]	151164
[777]	65892
[800]	2079
[810]	10010
[811]	19656
[820]	26411
[821]	64827
[822]	62426
[830]	51975
[831]	137781
[832]	171990
[833]	129654
[840]	84700
[841]	232848
[842]	320320
[843]	316932
[844]	207900
[850]	119119
[851]	334152
[852]	482664
[853]	530621
[854]	459459
[855]	272272
[860]	145530
[861]	413343
[862]	614250
[863]	713097
[864]	688905
[865]	540540
[866]	292383
[870]	149226
[871]	427329
[872]	646646
[873]	775523
[874]	793611
[875]	696388
[876]	499681
[877]	244188
[880]	109725
[881]	316008
[882]	484120
[883]	593047
[884]	628425
[885]	585200
[886]	470288
[887]	305235
[888]	128877

 Dimensions of the Irreducible Representations of O_8.

Representation	Dimension
[0000]	1
[1000]	8
[1100]	28
[1110]	56
[1111]	70
[2000]	35
[2100]	160
[2110]	350
[2111]	448
[2200]	300
[2210]	840
[2211]	1134
[2220]	840
[2221]	1344
[2222]	588
[3000]	112
[3100]	567
[3110]	1296
[3111]	1680
[3200]	1400
[3210]	4096
[3211]	5600
[3220]	4536
[3221]	7350
[3222]	3360
[3300]	1925
[3310]	6160
[3311]	8624
[3320]	8910
[3321]	14784
[3322]	7392
[3330]	6600
[3331]	11550
[3332]	7392
[3333]	2772
[4000]	294
[4100]	1568
[4110]	3675
[4111]	4800
[4200]	4312
[4210]	12936
[4211]	17820
[4220]	15092
[4221]	24640
[4222]	11550
[4300]	7840
[4310]	25725
[4311]	36288
[4320]	39200
[4321]	65536

Representation	Dimension
[4322]	33600
[4330]	32340
[4331]	57024
[4332]	37422
[4333]	14784
[4400]	8918
[4410]	30576
[4411]	43680
[4420]	51597
[4421]	87360
[4422]	46800
[4430]	56056
[4431]	100100
[4432]	68640
[4433]	30030
[4440]	35035
[4441]	64064
[4442]	48048
[4443]	27456
[4444]	9438
[5000]	672
[5100]	3696
[5110]	8800
[5111]	11550
[5200]	10752
[5210]	32768
[5211]	45360
[5220]	39424
[5221]	64680
[5222]	30800
[5300]	21840
[5310]	72800
[5311]	103194
[5320]	114400
[5321]	192192
[5322]	100100
[5330]	99840
[5331]	176904
[5332]	117936
[5333]	48048
[5400]	32928
[5410]	114688
[5411]	164640
[5420]	199584
[5421]	339570
[5422]	184800
[5430]	229376
[5431]	411600
[5432]	286720
[5433]	129360
[5440]	160160
[5441]	294294
[5442]	224224

Representation	Dimension
[5443]	132132
[5444]	48048
[5500]	32928
[5510]	117600
[5511]	170100
[5520]	215600
[5521]	369600
[5522]	206250
[5530]	275968
[5531]	498960
[5532]	356400
[5533]	169400
[5540]	254800
[5541]	471744
[5542]	368550
[5543]	228800
[5544]	92664
[5550]	142688
[5551]	267540
[5552]	218400
[5553]	150150
[5554]	80080
[5555]	26026
[6000]	1386
[6100]	7776
[6110]	18711
[6111]	24640
[6200]	23400
[6210]	72072
[6211]	100100
[6220]	88452
[6221]	145600
[6222]	70070
[6300]	50400
[6310]	169785
[6311]	241472
[6320]	272160
[6321]	458752
[6322]	241472
[6330]	245700
[6331]	436800
[6332]	294294
[6333]	122304
[6400]	85050
[6410]	299376
[6411]	431200
[6420]	531441
[6421]	907200
[6422]	498960
[6430]	631800
[6431]	1137500
[6432]	800800
[6433]	368550

Representation	Dimension
[6440]	467775
[6441]	862400
[6442]	664048
[6443]	399168
[6444]	150150
[6500]	112896
[6510]	407484
[6511]	591360
[6520]	762048
[6521]	1310720
[6522]	739200
[6530]	1009008
[6531]	1830400
[6532]	1321320
[6533]	640640
[6540]	987840
[6541]	1835008
[6542]	1448832
[6543]	917504
[6544]	384384
[6550]	619164
[6551]	1164800
[6552]	960960
[6553]	673920
[6554]	371800
[6555]	128128
[6600]	102816
[6610]	376992
[6611]	549780
[6620]	727056
[6621]	1256640
[6622]	719950
[6630]	1018368
[6631]	1856400
[6632]	1361360
[6633]	680680
[6640]	1113840
[6641]	2079168
[6642]	1667666
[6643]	1089088
[6644]	482664
[6650]	925344
[6651]	1749300
[6652]	1466080
[6653]	1060290
[6654]	618800
[6655]	238238
[6660]	479808
[6661]	913920
[6662]	785400
[6663]	598400
[6664]	388960
[6665]	198016
[6666]	61880

A-23 Dimensions of the Irreducible Representations of R_9.

Representation	Dimension
[0000]	1
[1000]	9
[1100]	36
[1110]	84
[1111]	126
[2000]	44
[2100]	231
[2110]	594
[2111]	924
[2200]	495
[2210]	1650
[2211]	2772
[2220]	1980
[2221]	4158
[2222]	2772
[3000]	156
[3100]	910
[3110]	2457
[3111]	3900
[3200]	2574
[3210]	9009
[3211]	15444
[3220]	12012
[3221]	25740
[3222]	18018
[3300]	4004
[3310]	15444
[3311]	27456
[3320]	27027
[3321]	60060
[3322]	46332
[3330]	23595
[3331]	56628
[3332]	56628
[3333]	28314
[4000]	450
[4100]	2772
[4110]	7700
[4111]	12375
[4200]	8748
[4210]	31500
[4211]	54675
[4220]	44352
[4221]	96228
[4222]	69300
[4300]	18018
[4310]	71500
[4311]	128700
[4320]	132132
[4321]	297297
[4322]	235950
[4330]	128700
[4331]	312741
[4332]	321750
[4333]	169884
[4400]	22932
[4410]	95550
[4411]	175500
[4420]	196196
[4421]	450450
[4422]	375375
[4430]	252252
[4431]	625482
[4432]	675675
[4433]	396396
[4440]	182182
[4441]	468468
[4442]	557700
[4443]	429429
[4444]	184041
[5000]	1122
[5100]	7140
[5110]	20196
[5111]	32725
[5200]	23868
[5210]	87516
[5211]	153153
[5220]	127296
[5221]	278460
[5222]	204204
[5300]	54978
[5310]	222156
[5311]	403172
[5320]	424116
[5321]	962115
[5322]	777546
[5330]	437580
[5331]	1072071
[5332]	1123122
[5333]	612612
[5400]	92820
[5410]	393822
[5411]	729300
[5420]	835380
[5421]	1933750
[5422]	1640925
[5430]	1137708
[5431]	2844270
[5432]	3128697
[5433]	1896180
[5440]	918918
[5441]	2382380
[5442]	2888028
[5443]	2297295
[5444]	1042899
[5500]	102816
[5510]	448800
[5511]	841500
[5520]	1005312
[5521]	2356200
[5522]	2057000
[5530]	1527552
[5531]	3866616
[5532]	4375800
[5533]	2800512
[5540]	1633632
[5541]	4288284
[5542]	5348200
[5543]	4492488
[5544]	2275416
[5550]	1039584
[5551]	2784600
[5552]	3646500
[5553]	3403400
[5554]	2275416
[5555]	884884
[6000]	2508
[6100]	16302
[6110]	46683
[6111]	76076
[6200]	56430
[6210]	209475
[6211]	368676
[6220]	311220
[6221]	684684
[6222]	508326
[6300]	137940
[6310]	564300
[6311]	1029952
[6320]	1100385
[6321]	2510508
[6322]	2054052
[6330]	1175625
[6331]	2896740
[6332]	3072300
[6333]	1711710
[6400]	260832
[6410]	1120392
[6411]	2086656
[6420]	2427516
[6421]	5651360
[6422]	4855032
[6430]	3423420
[6431]	8607456
[6432]	9585576
[6433]	5933928
[6440]	2934360
[6441]	7651072
[6442]	9389952
[6443]	7629336
[6444]	3586440
[6500]	383724
[6510]	1695750
[6511]	3197700
[6520]	3879876
[6521]	9145422
[6522]	8083075
[6530]	6104700
[6531]	15540822
[6532]	17805375
[6533]	11639628
[6540]	6928350
[6541]	18290844
[6542]	23094500
[6543]	19815081
[6544]	10392525
[6550]	4938024
[6551]	13302432
[6552]	17635800
[6553]	16812796
[6554]	11639628
[6555]	4803656
[6600]	383724
[6610]	1726758
[6611]	3282972
[6620]	4081428
[6621]	9699690
[6622]	8729721
[6630]	6802380
[6631]	17459442
[6632]	20369349
[6633]	13755924
[6640]	8633790
[6641]	22980804
[6642]	29546748
[6643]	26189163
[6644]	14549535
[6650]	8162856
[6651]	22170720
[6652]	29930472
[6653]	29476980
[6654]	21616452
[6655]	9976824
[6660]	4744224
[6661]	13046616
[6662]	18120300
[6663]	18845112
[6664]	15418728
[6665]	9422556
[6666]	3426384

A-24 Dimensions of the Irreducible Representations of R_{11}.

Representation	Dimension
[00000]	1
[10000]	11
[11000]	55
[11100]	165
[11110]	330
[11111]	462
[20000]	65
[21000]	429
[21100]	1430
[21110]	3003
[21111]	4290
[22000]	1144
[22100]	5005
[22110]	11583
[22111]	17160
[22200]	7865
[22210]	23595
[22211]	37752
[22220]	23595
[22221]	47190
[22222]	28314
[30000]	275
[31000]	2025
[31100]	7128
[31110]	15400
[31111]	22275
[32000]	7150
[32100]	33033
[32110]	78650
[32111]	117975
[32200]	57915
[32210]	178750
[32211]	289575
[32220]	188760
[32221]	382239
[32222]	235950
[33000]	13650
[33100]	70070
[33110]	175175
[33111]	268125
[33200]	162162
[33210]	525525
[33211]	868725
[33220]	616616
[33221]	1274130
[33222]	825825
[33300]	182182
[33310]	650650
[33311]	1115400
[33320]	1002001
[33321]	2147145
[33322]	1533675
[33330]	715715
[33331]	1656369
[33332]	1533675
[33333]	674817
[40000]	935
[41000]	7293
[41100]	26520
[41110]	58344
[41111]	85085
[42000]	28798
[42100]	137445
[42110]	333234
[42111]	503965
[42200]	255255
[42210]	802230
[42211]	1310309
[42220]	875160
[42221]	1786785
[42222]	1123122
[43000]	72930
[43100]	386750
[43110]	984555
[43111]	1519375
[43200]	948090
[43210]	3128697
[43211]	5214495
[43220]	3792360
[43221]	7900750
[43222]	5214495
[43300]	1191190
[43310]	4332042
[43311]	7487480
[43320]	6891885
[43321]	14889875
[43322]	10830105
[43330]	5214495
[43331]	12167155
[43332]	11471889
[43333]	5214495
[44000]	112200
[44100]	628320
[44110]	1645600
[44111]	2571250
[44200]	1718496
[44210]	5834400
[44211]	9845550
[44220]	7468032
[44221]	15752880
[44222]	10696400
[44300]	2858856
[44310]	10696400
[44311]	18718700
[44320]	17969952
[44321]	39309270
[44322]	29415100
[44330]	15169440
[44331]	35837802
[44332]	34763300
[44333]	16686384
[44400]	2598960
[44410]	10210200
[44411]	18232500
[44420]	19059040
[44421]	42542500
[44422]	33426250
[44430]	21237216
[44431]	51196860
[44432]	52144950
[44433]	27810640
[44440]	12388376
[44441]	30970940
[44442]	34763300
[44443]	24334310
[44444]	9038458

Dimensions of the Irreducible Representations of R_{13}.

Representation	Dimension
[000000]	1
[100000]	13
[110000]	78
[111000]	286
[111100]	715
[111110]	1287
[111111]	1716
[200000]	90
[210000]	715
[211000]	2925
[211100]	7722
[211110]	14300
[211111]	19305
[220000]	2275
[221000]	12285
[221100]	36036
[221110]	70070
[221111]	96525
[222000]	23660
[222100]	91091
[222110]	195195
[222111]	278850
[222200]	117117
[222210]	325325
[222211]	501930
[222220]	286286
[222221]	552123
[222222]	306735
[300000]	442
[310000]	3927
[311000]	17017
[311100]	46410
[311110]	87516
[311111]	119119
[320000]	16575
[321000]	94809
[321100]	287300
[321110]	568854
[321111]	790075
[322000]	204204
[322100]	812175
[322110]	1772199
[322111]	2552550
[322200]	1106105
[322210]	3128697
[322211]	4866862
[322220]	2844270
[322221]	5530525
[322222]	3128697
[330000]	37400
[331000]	238680
[331100]	763776
[331110]	1555840
[331111]	2187900
[332000]	680680
[332100]	2858856
[332110]	6417840
[332111]	9359350
[332200]	4343976
[332210]	12641200
[332211]	19909890
[332220]	12135552
[332221]	23891868
[332222]	13905320
[333000]	928200
[333100]	4331600
[333110]	10210200
[333111]	15193750
[333200]	8687952
[333210]	26546520
[333211]	42664050
[333220]	28316288
[333221]	56885400
[333222]	34763300
[333300]	7883512
[333310]	26546520
[333311]	44244200
[333320]	37165128
[333321]	77427350
[333322]	52144950
[333330]	23006984
[333331]	51765714
[333332]	45192290
[333333]	18076916
[400000]	1729
[410000]	16302
[411000]	73150
[411100]	203775
[411110]	389025
[411111]	532532
[420000]	77064
[421000]	456456
[421100]	1412840
[421110]	2832102
[421111]	3955952
[422000]	1043328
[422100]	4238520
[422110]	9363276
[422111]	13563264
[422200]	5977400
[422210]	17117100
[422211]	26778752
[422220]	15894450
[422221]	31082480
[422222]	17801784
[430000]	230945
[431000]	1526175
[431100]	4988412
[431110]	10287550
[431111]	14549535
[432000]	4618900
[432100]	19815081
[432110]	45034275
[432111]	66050270
[432200]	31177575
[432210]	91853125
[432211]	145495350
[432220]	90068550
[432221]	178335729
[432222]	105079975
[433000]	7054320
[433100]	33625592
[433110]	80242890
[433111]	120091400
[433200]	69837768
[433210]	216038550
[433211]	349188840
[433220]	235379144
[433221]	475561944
[433222]	294223930
[433300]	67251184
[433310]	229265400
[433311]	384292480
[433320]	327849522
[433321]	686922808
[433322]	468356460
[433330]	210159950
[433331]	475561944
[433332]	420319900
[433333]	171730702
[440000]	415701
[441000]	2909907
[441100]	9825660
[441110]	20633886
[441111]	29422393
[442000]	9848916
[442100]	43648605
[442110]	101015343
[442111]	149375226
[442200]	72747675
[442210]	218243025
[442211]	348542194
[442220]	221077350
[442221]	441335895
[442222]	264801537
[443000]	19953648
[443100]	98256600
[443110]	238763538
[443111]	360274200
[443200]	216164520
[443210]	680918238
[443211]	1109644536
[443220]	766401480
[443221]	1561188200
[443222]	983548566
[443300]	232792560
[443310]	808122744
[443311]	1365716352
[443320]	1193817690
[443321]	2521919400
[443322]	1750932612
[443330]	810616950
[443331]	1849407560
[443332]	1664466804
[443333]	702534690
[444000]	21744360
[444100]	113070672
[444110]	282676680
[444111]	431867150
[444200]	277537104
[444210]	899425800
[444211]	1484052570
[444220]	1069031808
[444221]	2204878104
[444222]	1429087660
[444300]	395747352
[444310]	1413383400
[444311]	2418456040
[444320]	2204878104
[444321]	4715989278
[444322]	3368563770
[444330]	1670362200
[444331]	3858536682
[444332]	3572719150
[444333]	1592411964
[444400]	287816256
[444410]	1079310960
[444411]	1884511200
[444420]	1870805664
[444421]	4083107600
[444422]	3062330700
[444430]	1870805664
[444431]	4409756208
[444432]	4287262980
[444433]	2123215952
[444440]	935402832
[444441]	2286540256
[444442]	2449864560
[444443]	1592411964
[444444]	530803988

A-26 Dimensions of the Irreducible Representations of R_{15}.

[0000000]	1
[1000000]	15
[1100000]	105
[1110000]	455
[1111000]	1365
[1111100]	3003
[1111110]	5005
[1111111]	6435
[2000000]	119
[2100000]	1105
[2110000]	5355
[2111000]	17017
[2111100]	38675
[2111110]	65637
[2111111]	85085
[2200000]	4080
[2210000]	26180
[2211000]	92820
[2211100]	222768
[2211110]	388960
[2211111]	510510
[2220000]	59500
[2221000]	278460
[2221100]	742560
[2221110]	1361360
[2221111]	1823250
[2222000]	433160
[2222100]	1516060
[2222110]	3063060
[2222111]	4254250
[2222200]	1689324
[2222210]	4424420
[2222211]	6636630
[2222220]	3539536
[2222221]	6636630
[2222222]	3476330

A-27 Dimensions of the Irreducible Representations of Sp_4.

<00>	1
<10>	4
<11>	5
<20>	10
<21>	16
<22>	14
<30>	20
<31>	35
<32>	40
<33>	30
<40>	35
<41>	64
<42>	81
<43>	80
<44>	55
<50>	56
<51>	105
<52>	140
<53>	154
<54>	140
<55>	91
<60>	84
<61>	160
<62>	220
<63>	256
<64>	260
<65>	224
<66>	140
<70>	120
<71>	231
<72>	324
<73>	390
<74>	420
<75>	405
<76>	336
<77>	204
<80>	165
<81>	320
<82>	455
<83>	560
<84>	625
<85>	640
<86>	595
<87>	480
<88>	285
<90>	220
<91>	429
<92>	616
<93>	770
<94>	880
<95>	935
<96>	924
<97>	836
<98>	660
<99>	385
<100>	286
<101>	560
<102>	810
<103>	1024
<104>	1190
<105>	1296
<106>	1330
<107>	1280
<108>	1134
<109>	880
<1010>	506
<110>	364
<111>	715
<112>	1040
<113>	1326
<114>	1560
<115>	1729
<116>	1820
<117>	1820
<118>	1716
<119>	1495
<1110>	1144
<1111>	650
<120>	455
<121>	896
<122>	1309
<123>	1680
<124>	1995
<125>	2240
<126>	2401
<127>	2464
<128>	2415
<129>	2240
<1210>	1925
<1211>	1456
<1212>	819
<130>	560
<131>	1105
<132>	1620
<133>	2090
<134>	2500
<135>	2835
<136>	3080
<137>	3220
<138>	3240
<139>	3125
<1310>	2860
<1311>	2430
<1312>	1820
<1313>	1015
<140>	680
<141>	1344
<142>	1976
<143>	2560
<144>	3080
<145>	3520
<146>	3864
<147>	4096
<148>	4200
<149>	4160
<1410>	3960
<1411>	3584
<1412>	3016
<1413>	2240
<1414>	1240
<150>	816
<151>	1615
<152>	2380
<153>	3094
<154>	3740
<155>	4301
<156>	4760
<157>	5100
<158>	5304
<159>	5355
<1510>	5236
<1511>	4930
<1512>	4420
<1513>	3689
<1514>	2720
<1515>	1496

Dimensions of the Irreducible Representations of Sp_6.

<000>	1	<521>	3072	<665>	12852	<821>	18018
		<522>	2464	<666>	5712	<822>	15092
<100>	6	<530>	3276			<830>	21120
<110>	14	<531>	5460	<700>	792	<831>	36036
<111>	14	<532>	5720	<710>	3003	<832>	39424
		<533>	3744	<711>	3640	<833>	27720
<200>	21	<540>	4116	<720>	6930	<840>	32725
<210>	64	<541>	7168	<721>	10752	<841>	58344
<211>	70	<542>	8316	<722>	8918	<842>	70686
<220>	90	<543>	7168	<730>	12375	<843>	65450
<221>	126	<544>	4004	<731>	21000	<844>	41140
<222>	84	<550>	3528	<732>	22750	<850>	44352
		<551>	6300	<733>	15750	<851>	81081
<300>	56	<552>	7700	<740>	18480	<852>	103488
<310>	189	<553>	7392	<741>	32768	<853>	106722
<311>	216	<554>	5460	<742>	39312	<854>	88704
<320>	350	<555>	2548	<743>	35840	<855>	51051
<321>	512			<744>	22000	<860>	52668
<322>	378	<600>	462	<750>	23562	<861>	97812
<330>	385	<610>	1728	<751>	42840	<862>	128744
<331>	616	<611>	2079	<752>	54145	<863>	140448
<332>	594	<620>	3900	<753>	54978	<864>	130416
<333>	330	<621>	6006	<754>	44625	<865>	99484
		<622>	4914	<755>	24752	<866>	52668
<400>	126	<630>	6720	<760>	24948	<870>	52800
<410>	448	<631>	11319	<761>	46080	<871>	99099
<411>	525	<632>	12096	<762>	60060	<872>	133056
<420>	924	<633>	8190	<763>	64512	<873>	150150
<421>	1386	<640>	9450	<764>	58500	<874>	147840
<422>	1078	<641>	16632	<765>	43008	<875>	126225
<430>	1344	<642>	19683	<766>	21420	<876>	88704
<431>	2205	<643>	17550	<770>	18810	<877>	42636
<432>	2240	<644>	10395	<771>	35112	<880>	38115
<433>	1386	<650>	10752	<772>	46683	<881>	72072
<440>	1274	<651>	19404	<773>	51870	<882>	98098
<441>	2184	<652>	24192	<774>	49875	<883>	113190
<442>	2457	<653>	24024	<775>	41040	<884>	115500
<443>	2002	<654>	18816	<776>	27132	<885>	104720
<444>	1001	<655>	9828	<777>	11628	<886>	82467
		<660>	8568			<887>	52668
<500>	252	<661>	15708	<800>	1287	<888>	21945
<510>	924	<662>	20196	<810>	4928		
<511>	1100	<663>	21216	<811>	6006		
<520>	2016	<664>	18564	<820>	11550		

A-29 Dimensions of the Irreducible Representations of Sp_8.

Representation	Dimension
<0000>	1
<1000>	8
<1100>	27
<1110>	48
<1111>	42
<2000>	36
<2100>	160
<2110>	315
<2111>	288
<2200>	308
<2210>	792
<2211>	792
<2220>	825
<2221>	1056
<2222>	594
<3000>	120
<3100>	594
<3110>	1232
<3111>	1155
<3200>	1512
<3210>	4096
<3211>	4200
<3220>	4752
<3221>	6237
<3222>	3696
<3300>	2184
<3310>	6552
<3311>	7020
<3320>	10010
<3321>	13728
<3322>	9009
<3330>	8008
<3331>	12012
<3332>	10296
<3333>	4719
<4000>	330
<4100>	1728
<4110>	3696
<4111>	3520
<4200>	4914
<4210>	13728
<4211>	14300
<4220>	16848
<4221>	22464
<4222>	13728
<4300>	9408
<4310>	29106
<4311>	31680
<4320>	47040
<4321>	65536
<4322>	44352
<4330>	42042
<4331>	64064
<4332>	56628
<4333>	27456
<4400>	11340
<4410>	36960
<4411>	41250
<4420>	66528
<4421>	95040
<4422>	67760
<4430>	78624
<4431>	122850
<4432>	114400
<4433>	61776
<4440>	53508
<4441>	87360
<4442>	90090
<4443>	64064
<4444>	26026
<5000>	792
<5100>	4290
<5110>	9360
<5111>	9009
<5200>	12936
<5210>	36864
<5211>	38808
<5220>	46800
<5221>	63063
<5222>	39312
<5300>	27720
<5310>	87480
<5311>	96228
<5320>	146250
<5321>	205920
<5322>	142155
<5330>	138600
<5331>	213444
<5332>	192456
<5333>	96525
<5400>	44352
<5410>	147456
<5411>	166320
<5420>	274560
<5421>	396396
<5422>	288288
<5430>	344064
<5431>	543312
<5432>	516096
<5433>	288288
<5440>	262080
<5441>	432432
<5442>	454896
<5443>	334620
<5444>	144144
<5500>	47124
<5510>	161568
<5511>	185130
<5520>	318240
<5521>	466752
<5522>	350064
<5530>	445536
<5531>	714714
<5532>	700128
<5533>	413712
<5540>	449820
<5541>	753984
<5542>	817938
<5543>	636480
<5544>	306306
<5550>	274176
<5551>	471240
<5552>	538560
<5553>	466752
<5554>	297024
<5555>	111384
<6000>	1716
<6100>	9504
<6110>	21021
<6111>	20384
<6200>	29700
<6210>	85800
<6211>	91000
<6220>	111375
<6221>	151200
<6222>	95550
<6300>	67584
<6310>	216216
<6311>	239616
<6320>	369600
<6321>	524288
<6322>	366912
<6330>	363000
<6331>	563200
<6332>	514800
<6333>	264000
<6400>	121176
<6410>	408408
<6411>	464100
<6420>	777546
<6421>	1130976
<6422>	833833
<6430>	1009800
<6431>	1606500
<6432>	1547000
<6433>	883575
<6440>	816816
<6441>	1357824
<6442>	1447992
<6443>	1089088
<6444>	486200
<6500>	171072
<6510>	594594
<6511>	686400
<6520>	1197504
<6521>	1769472
<6522>	1345344
<6530>	1737450
<6531>	2808000
<6532>	2788500
<6533>	1684800
<6540>	1862784
<6541>	3145728
<6542>	3459456
<6543>	2752512
<6544>	1372800
<6550>	1272348
<6551>	2203200
<6552>	2552550
<6553>	2261952
<6554>	1491750
<6555>	594048
<6600>	165528
<6610>	586872
<6611>	684684
<6620>	1222650
<6621>	1825824
<6622>	1416051
<6630>	1881000
<6631>	3072300
<6632>	3112200
<6633>	1945125
<6640>	2257200
<6641>	3852288
<6642>	4321512
<6643>	3556800
<6644>	1881000
<6650>	2046528
<6651>	3581424
<6652>	4232592
<6653>	3879876
<6654>	2713200
<6655>	1209312
<6660>	1151172
<6661>	2046528
<6662>	2494206
<6663>	2418624
<6664>	1889550
<6665>	1116288
<6666>	395352

Dimensions of the Irreducible Representations of Sp_{10}.

<00000>	1	<32200>	28028	<42100>	71500	<44200>	714714
		<32210>	66066	<42110>	131625	<44210>	1867008
<10000>	10	<32211>	62920	<42111>	115830	<44211>	1871870
<11000>	44	<32220>	60060	<42200>	118580	<44220>	2068560
<11100>	110	<32221>	73216	<42210>	285120	<44221>	2654652
<11110>	165	<32222>	37752	<42211>	274428	<44222>	1516944
<11111>	132	<33000>	8250	<42220>	268125	<44300>	1099560
		<33100>	35640	<42221>	330330	<44310>	3180870
<20000>	55	<33110>	67760	<42222>	173745	<44311>	3332340
<21000>	320	<33111>	60500	<43000>	42240	<44320>	4641000
<21100>	891	<33200>	73710	<43100>	188760	<44321>	6223360
<21110>	1408	<33210>	183040	<43110>	366080	<44322>	3938220
<21111>	1155	<33211>	178750	<43111>	330330	<44330>	3573570
<22000>	780	<33220>	185328	<43200>	413952	<44331>	5241236
<22100>	2860	<33221>	231660	<43210>	1048576	<44332>	4332042
<22110>	5005	<33222>	125840	<43211>	1034880	<44333>	1896180
<22111>	4290	<33300>	76440	<43220>	1098240	<44400>	942480
<22200>	4004	<33310>	210210	<43221>	1387386	<44410>	2872320
<22210>	9152	<33311>	214500	<43222>	768768	<44411>	3085500
<22211>	8580	<33320>	280280	<43300>	480480	<44420>	4667520
<22220>	7865	<33321>	366080	<43310>	1347840	<44421>	6417840
<22221>	9438	<33322>	220220	<43311>	1389960	<44422>	4278560
<22222>	4719	<33330>	182182	<43320>	1859000	<44430>	4752384
		<33331>	260260	<43321>	2453880	<44431>	7147140
<30000>	220	<33332>	204490	<43322>	1505790	<44432>	6223360
<31000>	1430	<33333>	81796	<43330>	1281280	<44433>	3033888
<31100>	4212			<43331>	1849848	<44440>	2598960
<31110>	6864	<40000>	715	<43332>	1482624	<44441>	4084080
<31111>	5720	<41000>	4928	<43333>	613470	<44442>	3938220
<32000>	4620	<41100>	15015	<44000>	61710	<44443>	2528240
<32100>	17920	<41110>	24960	<44100>	291720	<44444>	884884
<32110>	32340	<41111>	21021	<44110>	583440		
<32111>	28160	<42000>	17820	<44111>	534820		

A-31 Dimensions of the Irreducible Representations of Sp_{12}.

Representation	Dimension
<000000>	1
<100000>	12
<110000>	65
<111000>	208
<111100>	429
<111110>	572
<111111>	429
<200000>	78
<210000>	560
<211000>	2002
<211100>	4368
<211110>	6006
<211111>	4576
<220000>	1650
<221000>	7800
<221100>	18954
<221110>	27456
<221111>	21450
<222000>	13650
<222100>	43680
<222110>	70070
<222111>	57200
<222200>	49686
<222210>	104104
<222211>	92950
<222220>	78078
<222221>	89232
<222222>	40898
<300000>	364
<310000>	2925
<311000>	11088
<311100>	25025
<311110>	35100
<311111>	27027
<320000>	11440
<321000>	57344
<321100>	144144
<321110>	212992
<321111>	168168
<322000>	112320
<322100>	371800
<322110>	608400
<322111>	501930
<322200>	448448
<322210>	958464
<322211>	864864
<322220>	743600
<322221>	858858
<322222>	401544
<330000>	24310
<331000>	136136
<331100>	361998
<331110>	551616
<331111>	442442
<332000>	353430
<332100>	1237600
<332110>	2088450
<332111>	1750320
<332200>	1667666
<332210>	3675672
<332211>	3369366
<332220>	3016650
<332221>	3539536
<332222>	1706562
<333000>	448800
<333100>	1750320
<333110>	3111680
<333111>	2674100
<333200>	3118752
<333210>	7241728
<333211>	6806800
<333220>	6619392
<333221>	7963956
<333222>	4045184
<333300>	2598960
<333310>	6683040
<333311>	6563700
<333320>	8044400
<333321>	10112960
<333322>	5688540
<333330>	4504864
<333331>	6194188
<333332>	4550832
<333333>	1643356
<400000>	1365
<410000>	11648
<411000>	45760
<411100>	105600
<411110>	150150
<411111>	116480
<420000>	51051
<421000>	265200
<421100>	681615
<421110>	1021020
<421111>	812175
<422000>	551616
<422100>	1867008
<422110>	3097094
<422111>	2574208
<422200>	2333760
<422210>	5056480
<422211>	4596800
<422220>	4011150
<422221>	4667520
<422222>	2212210
<430000>	144144
<431000>	836550
<431100>	2274480
<431110>	3513510
<431111>	2839200
<432000>	2306304
<432100>	8257536
<432110>	14126112
<432111>	11927552
<432200>	11531520
<432210>	25765740
<432211>	23795200
<432220>	21621600
<432221>	25559040
<432222>	12492480
<433000>	3281850
<433100>	13087008
<433110>	23585562
<433111>	20420400
<433200>	24166350
<433210>	56885400
<433211>	53868750
<433220>	53165970
<433221>	64443600
<433222>	33183150
<433300>	21385728
<433310>	55747692
<433311>	55161600
<433320>	68612544
<433321>	86900736
<433322>	49553504
<433330>	39819780
<433331>	55161600
<433332>	41083900
<433333>	15169440
<440000>	247247
<441000>	1521520
<441100>	4279275
<441110>	6743100
<441111>	5506865
<442000>	4694976
<442100>	17388800
<442110>	30343950
<442111>	25893504
<442200>	25748800
<442210>	58687200
<442211>	54774720
<442220>	50943750
<442221>	60860800
<442222>	30343950
<443000>	8868288
<443100>	36581688
<443110>	67251184
<443111>	58844786
<443200>	71628480
<443210>	171991040
<443211>	164600800
<443220>	166280400
<443221>	203693490
<443222>	106990520
<443300>	70946304
<443310>	188652672
<443311>	188652672
<443320>	240182800
<443321>	307433984
<443322>	178827012
<443330>	147804800
<443331>	206926720
<443332>	157210560
<443333>	60045700
<444000>	9145422
<444100>	39907296
<444110>	75657582
<444111>	67251184
<444200>	87297210
<444210>	216164520
<444211>	210159950
<444220>	221077350
<444221>	275118480
<444222>	149022510
<444300>	114605568
<444310>	314269956
<444311>	319258368
<444320>	423259200
<444321>	550371328
<444322>	330142176
<444330>	290990700
<444331>	413853440
<444332>	324246780
<444333>	131008800
<444400>	78279696
<444410>	226141344
<444411>	235563900
<444420>	339212016
<444421>	452282688
<444422>	285817532
<444430>	308374560
<444431>	449712900
<444432>	371191600
<444433>	167036220
<444440>	143908128
<444441>	219288576
<444442>	200443464
<444443>	118781312
<444444>	37119160

A-32 Dimensions of the Irreducible Representations of Sp_{14}.

<0000000>	1	<2111000>	10752	<2221110>	452608
		<2111100>	20020	<2221111>	340340
<1000000>	14	<2111110>	24960	<2222000>	214200
<1100000>	90	<2111111>	18018	<2222100>	618800
<1110000>	350	<2200000>	3094	<2222110>	928200
<1111000>	910	<2210000>	17850	<2222111>	729300
<1111100>	1638	<2211000>	54978	<2222200>	606424
<1111110>	2002	<2211100>	108290	<2222210>	1188096
<1111111>	1430	<2211110>	139230	<2222211>	1021020
		<2211111>	102102	<2222220>	804440
<2000000>	105	<2220000>	37400	<2222221>	884884
<2100000>	896	<2221000>	152320	<2222222>	379236
<2110000>	3900	<2221100>	334152		

A-33 Dimensions of the Irreducible Representations of Sp_{16}.

<00000000>	1	<21111100>	88128	<22220000>	724812
		<21111110>	102102	<22221000>	2635680
<10000000>	16	<21111111>	70720	<22221100>	5337252
<11000000>	119	<22000000>	5320	<22221110>	6852768
<11100000>	544	<22100000>	36176	<22221111>	4996810
<11110000>	1700	<22110000>	134368	<22222000>	3162816
<11111000>	3808	<22111000>	330752	<22222100>	8434176
<11111100>	6188	<22111100>	568480	<22222110>	11992344
<11111110>	7072	<22111110>	671840	<22222111>	9137024
<11111111>	4862	<22111111>	470288	<22222200>	7379904
		<22200000>	88179	<22222210>	13705536
<20000000>	136	<22210000>	434112	<22222211>	11421280
<21000000>	1344	<22211000>	1193808	<22222220>	8565960
<21100000>	6885	<22211100>	2170560	<22222221>	9137024
<21110000>	22848	<22211110>	2645370	<22222222>	3711916
<21111000>	53040	<22211111>	1881152		

A-34 Dimensions of the Irreducible Representations of Sp_{18}.

<000000000>	1	<211111110>	413440	<222211000>	22682352
		<211111111>	277134	<222211100>	38663100
<100000000>	18	<220000000>	8568	<222211110>	45106950
<110000000>	152	<221000000>	67032	<222211111>	31274152
<111000000>	798	<221100000>	290871	<222220000>	12758823
<111100000>	2907	<221110000>	854658	<222221000>	42418944
<111110000>	7752	<221111000>	1813968	<222221100>	80529714
<111111000>	15504	<221111100>	2825604	<222221110>	98977536
<111111100>	23256	<221111110>	3133746	<222221111>	70366842
<111111110>	25194	<221111111>	2116296	<222222000>	45070128
<111111111>	16796	<222000000>	186200	<222222100>	112675320
		<222100000>	1072512	<222222110>	153363630
<200000000>	171	<222110000>	3527160	<222222111>	113927268
<210000000>	1920	<222111000>	7938048	<222222200>	90140256
<211000000>	11305	<222111100>	12790800	<222222210>	160249344
<211100000>	43776	<222111110>	14470400	<222222211>	130202592
<211110000>	121125	<222111111>	9876048	<222222220>	93896100
<211111000>	248064	<222200000>	2069613	<222222221>	97651944
<211111100>	377910	<222210000>	9021390	<222222222>	37975756

B-1 Outer Multiplication of S-Functions.

{0} x {0} = {0}

{1} x {0} = {1}
{1} x {1} = {11} + {2}

{2} x {0} = {2}
{2} x {1} = {21} + {3}
{2} x {2} = {22} + {31} + {4}
{2} x {11} = {211} + {31}
{11} x {0} = {11}
{11} x {1} = {111} + {21}
{11} x {11} = {1111} + {211} + {22}

{3} x {0} = {3}
{3} x {1} = {31} + {4}
{3} x {2} = {32} + {41} + {5}
{3} x {11} = {311} + {41}
{3} x {3} = {33} + {42} + {51} + {6}
{3} x {21} = {321} + {411} + {42} + {51}
{3} x {111} = {3111} + {411}
{21} x {0} = {21}
{21} x {1} = {211} + {22} + {31}
{21} x {2} = {221} + {311} + {32} + {41}
{21} x {11} = {2111} + {221} + {311} + {32}
{21} x {21} = {2211} + {222} + {3111} +2{321} + {33} + {411} + {42}
{21} x {111} = {21111} + {2211} + {3111} + {321}
{111} x {0} = {111}
{111} x {1} = {1111} + {211}
{111} x {2} = {2111} + {311}
{111} x {11} = {11111} + {2111} + {221}
{111} x {111} = {111111} + {21111} + {2211} + {222}

{4} x {0} = {4}
{4} x {1} = {41} + {5}
{4} x {2} = {42} + {51} + {6}
{4} x {11} = {411} + {51}
{4} x {3} = {43} + {52} + {61} + {7}
{4} x {21} = {421} + {511} + {52} + {61}
{4} x {111} = {4111} + {511}
{4} x {4} = {44} + {53} + {62} + {71} + {8}
{4} x {31} = {431} + {521} + {53} + {611} + {62} + {71}
{4} x {22} = {422} + {521} + {62}
{4} x {211} = {4211} + {5111} + {521} + {611}
{4} x {1111} = {41111} + {5111}
{31} x {0} = {31}
{31} x {1} = {311} + {32} + {41}
{31} x {2} = {321} + {33} + {411} + {42} + {51}
{31} x {11} = {3111} + {321} + {411} + {42}
{31} x {3} = {331} + {421} + {43} + {511} + {52} + {61}
{31} x {21} = {3211} + {322} + {331} + {4111} +2{421} + {43} + {511} + {52}
{31} x {111} = {31111} + {3211} + {4111} + {421}
{31} x {31} = {3311} + {332} + {4211} + {422} +2{431} + {44} + {5111} +2{521} + {53} + {611} + {62}
{31} x {22} = {3221} + {332} + {4211} + {422} + {431} + {521} + {53}
{31} x {211} = {32111} + {3221} + {3311} + {41111} +2{4211} + {422} + {431} + {5111} + {521}
{31} x {1111} = {311111} + {32111} + {41111} + {4211}
{22} x {0} = {22}
{22} x {1} = {221} + {32}
{22} x {2} = {222} + {321} + {42}
{22} x {11} = {2211} + {321} + {33}
{22} x {3} = {322} + {421} + {52}
{22} x {21} = {2221} + {3211} + {322} + {331} + {421} + {43}
{22} x {111} = {22111} + {3211} + {331}
{22} x {22} = {2222} + {3221} + {3311} + {422} + {431} + {44}
{22} x {211} = {22211} + {32111} + {3221} + {3311} + {332} + {4211} + {431}
{22} x {1111} = {221111} + {32111} + {3311}
{211} x {0} = {211}
{211} x {1} = {2111} + {221} + {311}
{211} x {2} = {2211} + {3111} + {321} + {411}
{211} x {11} = {21111} + {2211} + {222} + {3111} + {321}
{211} x {3} = {3211} + {4111} + {421} + {511}

{211} x {21} = {22111} + {2221} + {31111} +2{3211} + {322} + {331} + {4111} + {421}
{211} x {111} = {211111} + {22111} + {2221} + {31111} + {3211} + {322}
{211} x {211} = {221111} + {22211} + {2222} + {311111} +2{32111} +2{3221} + {3311} + {332} + {41111}
+ {4211} + {422}
{211} x {1111} = {2111111} + {221111} + {22211} + {311111} + {32111} + {3221}
{1111} x {0} = {1111}
{1111} x {1} = {11111} + {2111}
{1111} x {2} = {21111} + {3111}
{1111} x {11} = {111111} + {21111} + {2211}
{1111} x {3} = {31111} + {4111}
{1111} x {21} = {211111} + {22111} + {31111} + {3211}
{1111} x {111} = {1111111} + {211111} + {22111} + {2221}
{1111} x {1111} = {11111111} + {2111111} + {221111} + {22211} + {2222}

{41} x {0} = {41}
{41} x {1} = {411} + {42} + {51}
{41} x {2} = {421} + {43} + {511} + {52} + {61}
{41} x {11} = {4111} + {421} + {511} + {52}
{41} x {3} = {431} + {44} + {521} + {53} + {611} + {62} + {71}
{41} x {21} = {4211} + {422} + {431} + {5111} +2{521} + {53} + {611} + {62}
{41} x {111} = {41111} + {4211} + {5111} + {521}
{41} x {4} = {441} + {531} + {54} + {621} + {63} + {711} + {72} + {81}
{41} x {31} = {4311} + {432} + {441} + {5211} + {522} +2{531} + {54} + {6111} +2{621} + {63} + {711}
+ {72}
{41} x {22} = {4221} + {432} + {5211} + {522} + {531} + {621} + {63}
{41} x {211} = {42111} + {4221} + {4311} + {51111} +2{5211} + {522} + {531} + {6111} + {621}
{41} x {1111} = {411111} + {42111} + {51111} + {5211}
{41} x {41} = {4411} + {442} + {5311} + {532} +2{541} + {55} + {6211} + {622} +2{631} + {64}
+ {7111} +2{721} + {73} + {811} + {82}
{41} x {32} = {4321} + {433} + {442} + {5221} + {5311} +2{532} + {541} + {6211} + {622} +2{631}
+ {64} + {721} + {73}
{41} x {311} = {43111} + {4321} + {4411} + {52111} + {5221} +2{5311} + {532} + {541} + {61111}
+2{6211} + {622} + {631} + {7111} + {721}
{41} x {221} = {42211} + {4222} + {4321} + {52111} +2{5221} + {5311} + {532} + {6211} + {622}
+ {631}
{41} x {2111} = {421111} + {42211} + {43111} + {511111} +2{52111} + {5221} + {5311} + {61111}
+ {6211}
{41} x {11111} = {4111111} + {421111} + {511111} + {52111}
{32} x {0} = {32}
{32} x {1} = {321} + {33} + {42}
{32} x {2} = {322} + {331} + {421} + {43} + {52}
{32} x {11} = {3211} + {331} + {421} + {43}
{32} x {3} = {332} + {422} + {431} + {521} + {53} + {62}
{32} x {21} = {3221} + {3311} + {332} + {4211} + {422} +2{431} + {44} + {521} + {53}
{32} x {111} = {32111} + {3311} + {4211} + {431}
{32} x {4} = {432} + {522} + {531} + {621} + {63} + {72}
{32} x {31} = {3321} + {333} + {4221} + {4311} +2{432} + {441} + {5211} + {522} +2{531} + {54}
+ {621} + {63}
{32} x {22} = {3222} + {3321} + {4221} + {4311} + {432} + {441} + {522} + {531} + {54}
{32} x {211} = {32211} + {33111} + {3321} + {42111} + {4221} +2{4311} + {432} + {441} + {5211}
+ {531}
{32} x {1111} = {321111} + {33111} + {42111} + {4311}
{32} x {32} = {3322} + {3331} + {4222} +2{4321} + {433} + {4411} + {442} + {5221} + {5311} +2{532}
+2{541} + {55} + {622} + {631} + {64}
{32} x {311} = {33211} + {3331} + {42211} + {43111} +2{4321} + {433} + {4411} + {442} + {52111}
+ {5221} +2{5311} + {532} + {541} + {6211} + {631}
{32} x {221} = {32221} + {33211} + {3322} + {42211} + {4222} + {43111} +2{4321} + {4411} + {442}
+ {5221} + {5311} + {532} + {541}
{32} x {2111} = {322111} + {331111} + {33211} + {421111} + {42211} +2{43111} + {4321} + {4411}
+ {52111} + {5311}
{32} x {11111} = {3211111} + {331111} + {421111} + {43111}
{311} x {0} = {311}
{311} x {1} = {3111} + {321} + {411}
{311} x {2} = {3211} + {331} + {4111} + {421} + {511}
{311} x {11} = {31111} + {3211} + {322} + {4111} + {421}
{311} x {3} = {3311} + {4211} + {431} + {5111} + {521} + {611}
{311} x {21} = {32111} + {3221} + {3311} + {332} + {41111} +2{4211} + {422} + {431} + {5111} + {521}
{311} x {111} = {311111} + {32111} + {3221} + {41111} + {4211} + {422}
{311} x {4} = {4311} + {5211} + {531} + {6111} + {621} + {711}
{311} x {31} = {33111} + {3321} + {42111} + {4221} +2{4311} + {432} + {441} + {51111} +2{5211}
+ {522} + {531} + {6111} + {621}

{311} x {22} = {32211} + {3321} + {333} + {42111} + {4221} + {4311} + {432} + {5211} + {531}

{311} x {211} = {321111} + {32211} + {3222} + {33111} + {3321} + {411111} +2{42111} +2{4221} + {4311} + {432} + {51111} + {5211} + {522}

{311} x {1111} = {3111111} + {321111} + {32211} + {411111} + {42111} + {4221}

{311} x {311} = {331111} + {33211} + {3322} + {421111} + {42211} + {4222} +2{43111} +2{4321} + {4411} + {442} + {511111} +2{52111} +2{5221} + {5311} + {532} + {61111} + {6211} + {622}

{311} x {221} = {322111} + {32221} + {33211} + {3322} + {3331} + {421111} +2{42211} + {4222} + {43111} +2{4321} + {433} + {52111} + {5221} + {5311} + {532}

{311} x {2111} = {3211111} + {322111} + {32221} + {331111} + {33211} + {4111111} +2{421111} +2{42211} + {4222} + {43111} + {4321} + {511111} + {52111} + {5221}

{311} x {11111} = {31111111} + {3211111} + {322111} + {4111111} + {421111} + {42211}

{221} x {0} = {221}

{221} x {1} = {2211} + {222} + {321}

{221} x {2} = {2221} + {3211} + {322} + {421}

{221} x {11} = {22111} + {2221} + {3211} + {322} + {331}

{221} x {3} = {3221} + {4211} + {422} + {521}

{221} x {21} = {22211} + {2222} + {32111} +2{3221} + {3311} + {332} + {4211} + {422} + {431}

{221} x {111} = {221111} + {22211} + {32111} + {3221} + {3311} + {332}

{221} x {4} = {4221} + {5211} + {522} + {621}

{221} x {31} = {32211} + {3222} + {3321} + {42111} +2{4221} + {4311} + {432} + {5211} + {522} + {531}

{221} x {22} = {22221} + {32211} + {3222} + {33111} + {3321} + {4221} + {4311} + {432} + {441}

{221} x {211} = {222111} + {22221} + {321111} +2{32211} + {3222} + {33111} +2{3321} + {333} + {42111} + {4221} + {4311} + {432}

{221} x {1111} = {2211111} + {222111} + {321111} + {32211} + {33111} + {3321}

{221} x {221} = {222211} + {22222} + {322111} +2{32221} + {331111} +2{33211} + {3322} + {3331} + {42211} + {4222} + {43111} +2{4321} + {433} + {4411} + {442}

{221} x {2111} = {2221111} + {222211} + {3211111} +2{322111} + {32221} + {331111} +2{33211} + {3322} + {3331} + {421111} + {42211} + {43111} + {4321}

{221} x {11111} = {22111111} + {2221111} + {3211111} + {322111} + {331111} + {33211}

{2111} x {0} = {2111}

{2111} x {1} = {21111} + {2211} + {3111}

{2111} x {2} = {22111} + {31111} + {3211} + {4111}

{2111} x {11} = {211111} + {22111} + {2221} + {31111} + {3211}

{2111} x {3} = {32111} + {41111} + {4211} + {5111}

{2111} x {21} = {221111} + {22211} + {311111} +2{32111} + {3221} + {3311} + {41111} + {4211}

{2111} x {111} = {2111111} + {221111} + {22211} + {2222} + {311111} + {32111} + {3221}

{2111} x {4} = {42111} + {51111} + {5211} + {6111}

{2111} x {31} = {321111} + {32211} + {33111} + {411111} +2{42111} + {4221} + {4311} + {51111} + {5211}

{2111} x {22} = {222111} + {321111} + {32211} + {33111} + {3321} + {42111} + {4311}

{2111} x {211} = {2211111} + {222111} + {22221} + {3111111} +2{321111} +2{32211} + {3222} + {33111} + {3321} + {411111} + {42111} + {4221}

{2111} x {1111} = {21111111} + {2211111} + {222111} + {22221} + {3111111} + {321111} + {32211} + {3222}

{2111} x {2111} = {22111111} + {2221111} + {222211} + {22222} + {31111111} +2{3211111} +2{322111} +2{32221} + {331111} + {33211} + {3322} + {4111111} + {421111} + {42211} + {4222}

{2111} x {11111} = {211111111} + {22111111} + {2221111} + {222211} + {31111111} + {3211111} + {322111} + {32221}

{11111} x {0} = {11111}

{11111} x {1} = {111111} + {21111}

{11111} x {2} = {211111} + {31111}

{11111} x {11} = {1111111} + {211111} + {22111}

{11111} x {3} = {311111} + {41111}

{11111} x {21} = {2111111} + {221111} + {311111} + {32111}

{11111} x {111} = {11111111} + {2111111} + {221111} + {22211}

{11111} x {4} = {411111} + {51111}

{11111} x {31} = {3111111} + {321111} + {411111} + {42111}

{11111} x {22} = {2211111} + {321111} + {33111}

{11111} x {211} = {21111111} + {2211111} + {222111} + {3111111} + {321111} + {32211}

{11111} x {1111} = {111111111} + {21111111} + {2211111} + {222111} + {22221}

{11111} x {11111} = {1111111111} + {211111111} + {22111111} + {2221111} + {222211} + {22222}

{42} x {0} = {42}

{42} x {1} = {421} + {43} + {52}

{42} x {2} = {422} + {431} + {44} + {521} + {53} + {62}

{42} x {11} = {4211} + {431} + {521} + {53}

{42} x {3} = {432} + {441} + {522} + {531} + {54} + {621} + {63} + {72}

{42} x {21} = {4221} + {4311} + {432} + {441} + {5211} + {522} +2{531} + {54} + {621} + {63}

{42} x {111} = {42111} + {4311} + {5211} + {531}
{42} x {4} = {442} + {532} + {541} + {622} + {631} + {64} + {721} + {73} + {82}
{42} x {31} = {4321} + {433} + {4411} + {442} + {5221} + {5311} +2{532} +2{541} + {55} + {6211}
+ {622} +2{631} + {64} + {721} + {73}
{42} x {22} = {4222} + {4321} + {442} + {5221} + {5311} + {532} + {541} + {622} + {631} + {64}
{42} x {211} = {42211} + {43111} + {4321} + {4411} + {52111} + {5221} +2{5311} + {532} + {541}
+ {6211} + {631}
{42} x {1111} = {421111} + {43111} + {52111} + {5311}
{42} x {41} = {4421} + {443} + {5321} + {533} + {5411} +2{542} + {551} + {6221} + {6311} +2{632}
+2{641} + {65} + {7211} + {722} +2{731} + {74} + {821} + {83}
{42} x {32} = {4322} + {4331} + {4421} + {443} + {5222} +2{5321} + {533} + {5411} +2{542} + {551}
+ {6221} + {6311} +2{632} +2{641} + {65} + {722} + {731} + {74}
{42} x {311} = {43211} + {4331} + {44111} + {4421} + {52211} + {53111} +2{5321} + {533} +2{5411}
+ {542} + {551} + {62111} + {6221} +2{6311} + {632} + {641} + {7211} + {731}
{42} x {221} = {42221} + {43211} + {4322} + {4421} + {52211} + {5222} + {53111} +2{5321} + {5411}
+ {542} + {6221} + {6311} + {632} + {641}
{42} x {2111} = {422111} + {431111} + {43211} + {44111} + {521111} + {52211} +2{53111} + {5321}
+ {5411} + {62111} + {6311}
{42} x {11111} = {4211111} + {431111} + {521111} + {53111}
{42} x {42} = {4422} + {4431} + {444} + {5322} + {5331} +2{5421} +2{543} + {5511} + {552} + {6222}
+2{6321} + {633} + {6411} +3{642} +2{651} + {66} + {7221} + {7311} +2{732} +2{741}
+ {75} + {822} + {831} + {84}
{42} x {411} = {44211} + {4431} + {53211} + {5331} + {54111} +2{5421} + {543} + {5511} + {552}
+ {62211} + {63111} +2{6321} + {633} +2{6411} + {642} + {651} + {72111} + {7221}
+2{7311} + {732} + {741} + {8211} + {831}
{42} x {33} = {4332} + {4431} + {5322} + {5331} + {5421} + {543} + {552} + {6321} + {633} + {6411}
+ {642} + {651} + {732} + {741} + {75}
{42} x {321} = {43221} + {43311} + {4332} + {44211} + {4422} + {4431} + {52221} +2{53211} +2{5322}
+2{5331} + {54111} +3{5421} + {543} + {5511} + {552} + {62211} + {6222} + {63111}
+3{6321} + {633} +2{6411} +2{642} + {651} + {7221} + {7311} + {732} + {741}
{42} x {3111} = {432111} + {43311} + {441111} + {44211} + {522111} + {531111} +2{53211} + {5331}
+2{54111} + {5421} + {5511} + {621111} + {62211} +2{63111} + {6321} + {6411} + {72111}
+ {7311}
{42} x {222} = {42222} + {43221} + {4422} + {52221} + {53211} + {5322} + {5421} + {6222} + {6321}
+ {642}
{42} x {2211} = {422211} + {432111} + {43221} + {44211} + {522111} + {52221} + {531111} +2{53211}
+ {5322} + {54111} + {5421} + {62211} + {63111} + {6321} + {6411}
{42} x {21111} = {4221111} + {4311111} + {432111} + {441111} + {5211111} + {522111} +2{531111}
+ {53211} + {54111} + {621111} + {63111}
{42} x {111111} = {42111111} + {4311111} + {5211111} + {531111}
{411} x {0} = {411}
{411} x {1} = {4111} + {421} + {511}
{411} x {2} = {4211} + {431} + {5111} + {521} + {611}
{411} x {11} = {41111} + {4211} + {422} + {5111} + {521}
{411} x {3} = {4311} + {441} + {5211} + {531} + {6111} + {621} + {711}
{411} x {21} = {42111} + {4221} + {4311} + {432} + {51111} +2{5211} + {522} + {531} + {6111} + {621}
{411} x {111} = {411111} + {42111} + {4221} + {51111} + {5211} + {522}
{411} x {4} = {4411} + {5311} + {541} + {6211} + {631} + {7111} + {721} + {811}
{411} x {31} = {43111} + {4321} + {4411} + {442} + {52111} + {5221} +2{5311} + {532} + {541}
+ {61111} +2{6211} + {622} + {631} + {7111} + {721}
{411} x {22} = {42211} + {4321} + {433} + {52111} + {5221} + {5311} + {532} + {6211} + {631}
{411} x {211} = {421111} + {42211} + {4222} + {43111} + {4321} + {511111} +2{52111} +2{5221}
+ {5311} + {532} + {61111} + {6211} + {622}
{411} x {1111} = {4111111} + {421111} + {42211} + {511111} + {52111} + {5221}
{411} x {41} = {44111} + {4421} + {53111} + {5321} +2{5411} + {542} + {551} + {62111} + {6221}
+2{6311} + {632} + {641} + {71111} +2{7211} + {722} + {731} + {8111} + {821}
{411} x {32} = {43211} + {4331} + {4421} + {443} + {52211} + {53111} +2{5321} + {533} + {5411}
+ {542} + {62111} + {6221} +2{6311} + {632} + {641} + {7211} + {731}
{411} x {311} = {431111} + {43211} + {4322} + {44111} + {4421} + {521111} + {52211} + {5222}
+2{53111} +2{5321} + {5411} + {542} + {611111} +2{62111} +2{6221} + {6311} + {632}
+ {71111} + {7211} + {722}
{411} x {221} = {422111} + {42221} + {43211} + {4322} + {4331} + {521111} +2{52211} + {5222}
+ {53111} +2{5321} + {533} + {62111} + {6221} + {6311} + {632}
{411} x {2111} = {4211111} + {422111} + {42221} + {431111} + {43211} + {5111111} +2{521111}
+2{52211} + {5222} + {53111} + {5321} + {611111} + {62111} + {6221}
{411} x {11111} = {41111111} + {4211111} + {422111} + {5111111} + {521111} + {52211}
{411} x {411} = {441111} + {44211} + {4422} + {531111} + {53211} + {5322} +2{54111} +2{5421}
+ {5511} + {552} + {621111} + {62211} + {6222} +2{63111} +2{6321} + {6411} + {642}
+ {711111} +2{72111} +2{7221} + {7311} + {732} + {81111} + {8211} + {822}
{411} x {33} = {43311} + {4431} + {444} + {53211} + {5331} + {5421} + {543} + {63111} + {6321}
+ {6411} + {642} + {7311} + {741}

Outer Multiplication of S-Functions (contd).

{411} x {321} = {432111} + {43221} + {43311} + {4332} + {44211} + {4422} + {4431} + {522111}
+ {52221} + {531111} +3{53211} +2{5322} +2{5331} + {54111} +2{5421} + {543} + {621111}
+2{62211} + {6222} +2{63111} +3{6321} + {633} + {6411} + {642} + {72111} + {7221}
+ {7311} + {732}
{411} x {3111} = {4311111} + {432111} + {43221} + {441111} + {44211} + {5211111} + {522111}
+ {52221} +2{531111} +2{53211} + {5322} + {54111} + {5421} + {6111111} +2{621111}
+2{62211} + {6222} + {63111} + {6321} + {711111} + {72111} + {7221}
{411} x {222} = {422211} + {43221} + {4332} + {522111} + {52221} + {53211} + {5322} + {5331}
+ {62211} + {6321} + {633}
{411} x {2211} = {4221111} + {422211} + {42222} + {432111} + {43221} + {43311} + {5211111}
+2{522111} +2{52221} + {531111} +2{53211} + {5322} + {5331} + {621111} + {62211}
+ {6222} + {63111} + {6321}
{411} x {21111} = {42111111} + {4221111} + {422211} + {4311111} + {432111} + {51111111} +2{5211111}
+2{522111} + {52221} + {531111} + {53211} + {6111111} + {621111} + {62211}
{411} x {111111} = {411111111} + {42111111} + {4221111} + {51111111} + {5211111} + {522111}
{33} x {0} = {33}
{33} x {1} = {331} + {43}
{33} x {2} = {332} + {431} + {53}
{33} x {11} = {3311} + {431} + {44}
{33} x {3} = {333} + {432} + {531} + {63}
{33} x {21} = {3321} + {4311} + {432} + {441} + {531} + {54}
{33} x {111} = {33111} + {4311} + {441}
{33} x {4} = {433} + {532} + {631} + {73}
{33} x {31} = {3331} + {4321} + {433} + {442} + {5311} + {532} + {541} + {631} + {64}
{33} x {22} = {3322} + {4321} + {4411} + {532} + {541} + {55}
{33} x {211} = {33211} + {43111} + {4321} + {4411} + {442} + {5311} + {541}
{33} x {1111} = {331111} + {43111} + {4411}
{33} x {41} = {4331} + {443} + {5321} + {533} + {542} + {6311} + {632} + {641} + {731} + {74}
{33} x {32} = {3332} + {4322} + {4331} + {4421} + {5321} + {533} + {5411} + {542} + {551} + {632}
+ {641} + {65}
{33} x {311} = {33311} + {43211} + {4331} + {4421} + {443} + {53111} + {5321} + {5411} + {542}
+ {6311} + {641}
{33} x {221} = {33221} + {43211} + {4322} + {44111} + {4421} + {5321} + {5411} + {542} + {551}
{33} x {2111} = {332111} + {431111} + {43211} + {44111} + {4421} + {53111} + {5411}
{33} x {11111} = {3311111} + {431111} + {44111}
{33} x {33} = {3333} + {4332} + {4422} + {5331} + {5421} + {5511} + {633} + {642} + {651} + {66}
{33} x {321} = {33321} + {43221} + {43311} + {4332} + {44211} + {4422} + {4431} + {53211} + {5322}
+ {5331} + {54111} +2{5421} + {543} + {5511} + {552} + {6321} + {6411} + {642} + {651}
{33} x {3111} = {333111} + {432111} + {43311} + {44211} + {4431} + {531111} + {53211} + {54111}
+ {5421} + {63111} + {6411}
{33} x {222} = {33222} + {43221} + {44211} + {5322} + {5421} + {552}
{33} x {2211} = {332211} + {432111} + {43221} + {441111} + {44211} + {4422} + {53211} + {54111}
+ {5421} + {5511}
{33} x {21111} = {3321111} + {4311111} + {432111} + {441111} + {44211} + {531111} + {54111}
{33} x {111111} = {33111111} + {4311111} + {441111}
{321} x {0} = {321}
{321} x {1} = {3211} + {322} + {331} + {421}
{321} x {2} = {3221} + {3311} + {332} + {4211} + {422} + {431} + {521}
{321} x {11} = {32111} + {3221} + {3311} + {332} + {4211} + {422} + {431}
{321} x {3} = {3321} + {4221} + {4311} + {432} + {5211} + {522} + {531} + {621}
{321} x {21} = {32211} + {3222} + {33111} +2{3321} + {333} + {42111} +2{4221} +2{4311} +2{432}
+ {441} + {5211} + {522} + {531}
{321} x {111} = {321111} + {32211} + {33111} + {3321} + {42111} + {4221} + {4311} + {432}
{321} x {4} = {4321} + {5221} + {5311} + {532} + {6211} + {622} + {631} + {721}
{321} x {31} = {33211} + {3322} + {3331} + {42211} + {4222} + {43111} +3{4321} + {433} + {4411}
+ {442} + {52111} +2{5221} +2{5311} +2{532} + {541} + {6211} + {622} + {631}
{321} x {22} = {32221} + {33211} + {3322} + {3331} + {42211} + {4222} + {43111} +2{4321} + {433}
+ {4411} + {442} + {5221} + {5311} + {532} + {541}
{321} x {211} = {322111} + {32221} + {331111} +2{33211} + {3322} + {3331} + {421111} +2{42211}
+ {4222} +2{43111} +3{4321} + {433} + {4411} + {442} + {52111} + {5221} + {5311}
+ {532}
{321} x {1111} = {3211111} + {322111} + {331111} + {33211} + {421111} + {42211} + {43111} + {4321}
{321} x {41} = {43211} + {4322} + {4331} + {4421} + {52211} + {5222} + {53111} +3{5321} + {533}
+ {5411} + {542} + {62111} +2{6221} +2{6311} +2{632} + {641} + {7211} + {722} + {731}
{321} x {32} = {33221} + {33311} + {3332} + {42221} +2{43211} +2{4322} +2{4331} + {44111} +2{4421}
+ {443} + {52211} + {5222} + {53111} +3{5321} + {533} +2{5411} +2{542} + {551} + {6221}
+ {6311} + {632} + {641}
{321} x {311} = {332111} + {33221} + {33311} + {3332} + {422111} + {42221} + {431111} +3{43211}
+2{4322} +2{4331} + {44111} +2{4421} + {443} + {521111} +2{52211} + {5222} +2{53111}
+3{5321} + {533} + {5411} + {542} + {62111} + {6221} + {6311} + {632}

{321} x {221} = {322211} + {32222} + {332111} +2{33221} + {33311} + {3332} + {422111} +2{42221}
+ {431111} +3{43211} +2{4322} +2{4331} + {44111} +2{4421} + {443} + {52211} + {5222}
+ {53111} +2{5321} + {533} + {5411} + {542}

{321} x {2111} = {3221111} + {322211} + {3311111} +2{332111} + {33221} + {33311} + {4211111}
+2{422111} + {42221} +2{431111} +3{43211} + {4322} + {4331} + {44111} + {4421}
+ {521111} + {52211} + {53111} + {5321}

{321} x {11111} = {32111111} + {3221111} + {3311111} + {332111} + {4211111} + {422111} + {431111}
+ {43211}

{321} x {321} = {332211} + {33222} + {333111} +2{33321} + {3333} + {422211} + {42222} +2{432111}
+4{43221} +3{43311} +3{4332} + {441111} +3{44211} +2{4422} +3{4431} + {444} + {522111}
+2{52221} + {531111} +4{53211} +3{5322} +3{5331} +2{54111} +4{5421} +2{543} + {5511}
+ {552} + {62211} + {6222} + {63111} +2{6321} + {633} + {6411} + {642}

{321} x {3111} = {3321111} + {332211} + {333111} + {33321} + {4221111} + {422211} + {4311111}
+3{432111} +2{43221} +2{43311} + {4332} + {441111} +2{44211} + {4422} + {4431}
+ {5211111} +2{522111} + {52221} +2{531111} +3{53211} + {5322} + {5331} + {54111}
+ {5421} + {621111} + {62211} + {63111} + {6321}

{321} x {222} = {322221} + {332211} + {33222} + {33321} + {422211} + {42222} + {432111} +2{43221}
+ {43311} + {4332} + {44211} + {4422} + {4431} + {52221} + {53211} + {5322} + {5331}
+ {5421} + {543}

{321} x {2211} = {3222111} + {322221} + {3321111} +2{332211} + {33222} + {333111} + {33321}
+ {4221111} +2{422211} + {42222} + {4311111} +3{432111} +3{43221} +2{43311} + {4332}
+ {441111} +2{44211} + {4422} + {4431} + {522111} + {52221} + {531111} +2{53211}
+ {5322} + {5331} + {54111} + {5421}

{321} x {21111} = {32211111} + {3222111} + {33111111} +2{3321111} + {332211} + {333111} + {42111111}
+2{4221111} + {422211} +2{4311111} +3{432111} + {43221} + {43311} + {441111}
+ {44211} + {5211111} + {522111} + {531111} + {53211}

{321} x {111111} = {321111111} + {32211111} + {33111111} + {3321111} + {42111111} + {4221111}
+ {4311111} + {432111}

{3111} x {0} = {3111}

{3111} x {1} = {31111} + {3211} + {4111}

{3111} x {2} = {32111} + {3311} + {41111} + {4211} + {5111}

{3111} x {11} = {311111} + {32111} + {3221} + {41111} + {4211}

{3111} x {3} = {33111} + {42111} + {4311} + {51111} + {5211} + {6111}

{3111} x {21} = {321111} + {32211} + {33111} + {3321} + {411111} +2{42111} + {4221} + {4311}
+ {51111} + {5211}

{3111} x {111} = {3111111} + {321111} + {32211} + {3222} + {411111} + {42111} + {4221}

{3111} x {4} = {43111} + {52111} + {5311} + {61111} + {6211} + {7111}

{3111} x {31} = {331111} + {33211} + {421111} + {42211} +2{43111} + {4321} + {4411} + {511111}
+2{52111} + {5221} + {5311} + {61111} + {6211}

{3111} x {22} = {322111} + {33211} + {3331} + {421111} + {42211} + {43111} + {4321} + {52111}
+ {5311}

{3111} x {211} = {3211111} + {322111} + {32221} + {331111} + {33211} + {3322} + {4111111} +2{421111}
+2{42211} + {4222} + {43111} + {4321} + {511111} + {52111} + {5221}

{3111} x {1111} = {31111111} + {3211111} + {322111} + {32221} + {4111111} + {421111} + {42211}
+ {4222}

{3111} x {41} = {431111} + {43211} + {44111} + {521111} + {52211} +2{53111} + {5321} + {5411}
+ {611111} +2{62111} + {6221} + {6311} + {71111} + {7211}

{3111} x {32} = {332111} + {33311} + {422111} + {431111} +2{43211} + {4331} + {44111} + {4421}
+ {521111} + {52211} +2{53111} + {5321} + {5411} + {62111} + {6311}

{3111} x {311} = {3311111} + {332111} + {33221} + {4211111} + {422111} + {42221} +2{431111}
+2{43211} + {4322} + {44111} + {4421} + {5111111} +2{521111} +2{52211} + {5222}
+ {53111} + {5321} + {611111} + {62111} + {6221}

{3111} x {221} = {3221111} + {322211} + {332111} + {33221} + {33311} + {3332} + {4211111} +2{422111}
+ {42221} + {431111} +2{43211} + {4322} + {4331} + {521111} + {52211} + {53111}
+ {5321}

{3111} x {2111} = {32111111} + {3221111} + {322211} + {32222} + {3311111} + {332111} + {33221}
+ {41111111} +2{4211111} +2{422111} +2{42221} + {431111} + {43211} + {4322}
+ {5111111} + {521111} + {52211} + {5222}

{3111} x {11111} = {311111111} + {32111111} + {3221111} + {322211} + {41111111} + {4211111}
+ {422111} + {42221}

{3111} x {3111} = {33111111} + {3321111} + {332211} + {33222} + {42111111} + {4221111} + {422211}
+ {42222} +2{4311111} +2{432111} +2{43221} + {441111} + {44211} + {4422}
+ {51111111} +2{5211111} +2{522111} +2{52221} + {531111} + {53211} + {5322}
+ {6111111} + {621111} + {62211} + {6222}

{3111} x {222} = {3222111} + {332211} + {33321} + {3333} + {4221111} + {422211} + {432111} + {43221}
+ {43311} + {4332} + {522111} + {53211} + {5331}

{3111} x {2211} = {32211111} + {3222111} + {322221} + {3321111} + {332211} + {33222} + {333111}
+ {33321} + {42111111} +2{4221111} +2{422211} + {42222} + {4311111} +2{432111}
+2{43221} + {43311} + {4332} + {5211111} + {522111} + {52221} + {531111} + {53211}
+ {5322}

{3111} x {21111} = {321111111} + {32211111} + {3222111} + {322221} + {33111111} + {3321111}
+ {332211} + {411111111} +2{42111111} +2{4221111} +2{422211} + {42222} + {4311111}
+ {432111} + {43221} + {51111111} + {5211111} + {522111} + {52221}
{3111} x {111111} = {3111111111} + {321111111} + {32211111} + {3222111} + {411111111} + {42111111}
+ {4221111} + {422211}
{222} x {0} = {222}
{222} x {1} = {2221} + {322}
{222} x {2} = {2222} + {3221} + {422}
{222} x {11} = {22211} + {3221} + {332}
{222} x {3} = {3222} + {4221} + {522}
{222} x {21} = {22221} + {32211} + {3222} + {3321} + {4221} + {432}
{222} x {111} = {222111} + {32211} + {3321} + {333}
{222} x {4} = {4222} + {5221} + {622}
{222} x {31} = {32221} + {3322} + {42211} + {4222} + {4321} + {5221} + {532}
{222} x {22} = {22222} + {32221} + {33211} + {4222} + {4321} + {442}
{222} x {211} = {222211} + {322111} + {32221} + {33211} + {3322} + {3331} + {42211} + {4321} + {433}
{222} x {1111} = {2221111} + {322111} + {33211} + {3331}
{222} x {41} = {42221} + {4322} + {52211} + {5222} + {5321} + {6221} + {632}
{222} x {32} = {32222} + {33221} + {42221} + {43211} + {4322} + {4421} + {5222} + {5321} + {542}
{222} x {311} = {322211} + {33221} + {3332} + {422111} + {42221} + {43211} + {4322} + {4331}
+ {52211} + {5321} + {533}
{222} x {221} = {222221} + {322211} + {32222} + {332111} + {33221} + {33311} + {42221} + {43211}
+ {4322} + {4331} + {4421} + {443}
{222} x {2111} = {2222111} + {3221111} + {322211} + {332111} + {33221} + {33311} + {3332} + {422111}
+ {43211} + {4331}
{222} x {11111} = {22211111} + {3221111} + {332111} + {33311}
{222} x {222} = {222222} + {322221} + {332211} + {333111} + {42222} + {43221} + {43311} + {4422}
+ {4431} + {444}
{222} x {2211} = {2222211} + {3222111} + {322221} + {3321111} + {332211} + {33222} + {333111}
+ {33321} + {422211} + {432111} + {43221} + {43311} + {4332} + {44211} + {4431}
{222} x {21111} = {22221111} + {32211111} + {3222111} + {3321111} + {332211} + {333111} + {33321}
+ {4221111} + {432111} + {43311}
{222} x {111111} = {222111111} + {32211111} + {3321111} + {333111}
{2211} x {0} = {2211}
{2211} x {1} = {22111} + {2221} + {3211}
{2211} x {2} = {22211} + {32111} + {3221} + {4211}
{2211} x {11} = {221111} + {22211} + {2222} + {32111} + {3221} + {3311}
{2211} x {3} = {32211} + {42111} + {4221} + {5211}
{2211} x {21} = {222111} + {22221} + {321111} +2{32211} + {3222} + {33111} + {3321} + {42111}
+ {4221} + {4311}
{2211} x {111} = {2211111} + {222111} + {22221} + {321111} + {32211} + {3222} + {33111} + {3321}
{2211} x {4} = {42211} + {52111} + {5221} + {6211}
{2211} x {31} = {322111} + {32221} + {33211} + {421111} +2{42211} + {4222} + {43111} + {4321}
+ {52111} + {5221} + {5311}
{2211} x {22} = {222211} + {322111} + {32221} + {331111} + {33211} + {3322} + {42211} + {43111}
+ {4321} + {4411}
{2211} x {211} = {2221111} + {222211} + {22222} + {3211111} +2{322111} +2{32221} + {331111}
+2{33211} + {3322} + {3331} + {421111} + {42211} + {4222} + {43111} + {4321}
{2211} x {1111} = {22111111} + {2221111} + {222211} + {3211111} + {322111} + {32221} + {331111}
+ {33211} + {3322}
{2211} x {41} = {422111} + {42221} + {43211} + {521111} +2{52211} + {5222} + {53111} + {5321}
+ {62111} + {6221} + {6311}
{2211} x {32} = {322211} + {332111} + {33221} + {422111} + {42221} + {431111} +2{43211} + {4322}
+ {44111} + {4421} + {52211} + {53111} + {5321} + {5411}
{2211} x {311} = {3221111} + {322211} + {32222} + {332111} + {33221} + {33311} + {4211111}
+2{422111} +2{42221} + {431111} +2{43211} + {4322} + {4331} + {521111} + {52211}
+ {5222} + {53111} + {5321}
{2211} x {221} = {2222111} + {222221} + {3221111} +2{322211} + {32222} + {3311111} +2{332111}
+2{33221} + {33311} + {3332} + {422111} + {42221} + {431111} +2{43211} + {4322}
+ {4331} + {44111} + {4421}
{2211} x {2111} = {22211111} + {2222111} + {222221} + {32111111} +2{3221111} +2{322211} + {32222}
+ {3311111} +2{332111} +2{33221} + {33311} + {3332} + {4211111} + {422111} + {42221}
+ {431111} + {43211} + {4322}
{2211} x {11111} = {221111111} + {22211111} + {2222111} + {32111111} + {3221111} + {322211}
+ {3311111} + {332111} + {33221}
{2211} x {2211} = {22221111} + {2222211} + {222222} + {32211111} +2{3222111} +2{322221} + {33111111}
+2{3321111} +3{332211} + {33222} + {333111} +2{33321} + {3333} + {4221111}
+ {422211} + {42222} + {4311111} +2{432111} +2{43221} + {43311} + {4332} + {441111}
+ {44211} + {4422}
{2211} x {21111} = {222111111} + {22221111} + {2222211} + {321111111} +2{32211111} +2{3222111}
+ {322221} + {33111111} +2{3321111} +2{332211} + {33222} + {333111} + {33321}
+ {42111111} + {4221111} + {422211} + {4311111} + {432111} + {43221}

{2211} x {111111} = {2211111111} + {222111111} + {22221111} + {321111111} + {32211111} + {3222111}
+ {33111111} + {3321111} + {332211}
{21111} x {0} = {21111}
{21111} x {1} = {211111} + {22111} + {31111}
{21111} x {2} = {221111} + {311111} + {32111} + {41111}
{21111} x {11} = {2111111} + {221111} + {22211} + {311111} + {32111}
{21111} x {3} = {321111} + {411111} + {42111} + {51111}
{21111} x {21} = {2211111} + {222111} + {3111111} +2{321111} + {32211} + {33111} + {411111}
+ {42111}
{21111} x {111} = {21111111} + {2211111} + {222111} + {22221} + {3111111} + {321111} + {32211}
{21111} x {4} = {421111} + {511111} + {52111} + {61111}
{21111} x {31} = {3211111} + {322111} + {331111} + {4111111} +2{421111} + {42211} + {43111}
+ {511111} + {52111}
{21111} x {22} = {2221111} + {3211111} + {322111} + {331111} + {33211} + {421111} + {43111}
{21111} x {211} = {22111111} + {2221111} + {222211} + {31111111} +2{3211111} +2{322111} + {32221}
+ {331111} + {33211} + {4111111} + {421111} + {42211}
{21111} x {1111} = {211111111} + {22111111} + {2221111} + {222211} + {22222} + {31111111}
+ {3211111} + {322111} + {32221}
{21111} x {41} = {4211111} + {422111} + {431111} + {5111111} +2{521111} + {52211} + {53111}
+ {611111} + {62111}
{21111} x {32} = {3221111} + {3311111} + {332111} + {4211111} + {422111} +2{431111} + {43211}
+ {44111} + {521111} + {53111}
{21111} x {311} = {32111111} + {3221111} + {322211} + {3311111} + {332111} + {41111111} +2{4211111}
+2{422111} + {42221} + {431111} + {43211} + {5111111} + {521111} + {52211}
{21111} x {221} = {22211111} + {2222111} + {32111111} +2{3221111} + {322211} + {3311111} +2{332111}
+ {33221} + {33311} + {4211111} + {422111} + {431111} + {43211}
{21111} x {2111} = {221111111} + {22211111} + {2222111} + {222221} + {311111111} +2{32111111}
+2{3221111} +2{322211} + {32222} + {3311111} + {332111} + {33221} + {41111111}
+ {4211111} + {422111} + {42221}
{21111} x {11111} = {2111111111} + {221111111} + {22211111} + {2222111} + {222221} + {311111111}
+ {32111111} + {3221111} + {322211} + {32222}
{21111} x {21111} = {2211111111} + {222111111} + {22221111} + {2222211} + {222222} + {3111111111}
+2{321111111} +2{32211111} +2{3222111} +2{322221} + {33111111} + {3321111}
+ {332211} + {33222} + {411111111} + {42111111} + {4221111} + {422211} + {42222}
{21111} x {111111} = {21111111111} + {2211111111} + {222111111} + {22221111} + {2222211}
+ {3111111111} + {321111111} + {32211111} + {3222111} + {322221}
{111111} x {0} = {111111}
{111111} x {1} = {1111111} + {211111}
{111111} x {2} = {2111111} + {311111}
{111111} x {11} = {11111111} + {2111111} + {221111}
{111111} x {3} = {3111111} + {411111}
{111111} x {21} = {21111111} + {2211111} + {3111111} + {321111}
{111111} x {111} = {111111111} + {21111111} + {2211111} + {222111}
{111111} x {4} = {4111111} + {511111}
{111111} x {31} = {31111111} + {3211111} + {4111111} + {421111}
{111111} x {22} = {22111111} + {3211111} + {331111}
{111111} x {211} = {211111111} + {22111111} + {2221111} + {31111111} + {3211111} + {322111}
{111111} x {1111} = {1111111111} + {211111111} + {22111111} + {2221111} + {222211}
{111111} x {41} = {41111111} + {4211111} + {5111111} + {521111}
{111111} x {32} = {32111111} + {3311111} + {4211111} + {431111}
{111111} x {311} = {311111111} + {32111111} + {3221111} + {41111111} + {4211111} + {422111}
{111111} x {221} = {221111111} + {22211111} + {32111111} + {3221111} + {3311111} + {332111}
{111111} x {2111} = {2111111111} + {221111111} + {22211111} + {2222111} + {311111111} + {32111111}
+ {3221111} + {322211}
{111111} x {11111} = {11111111111} + {2111111111} + {221111111} + {22211111} + {2222111} + {222221}
{111111} x {111111} = {111111111111} + {21111111111} + {2211111111} + {222111111} + {22221111}
+ {2222211} + {222222}

{43} x {0} = {43}
{43} x {1} = {431} + {44} + {53}
{43} x {2} = {432} + {441} + {531} + {54} + {63}
{43} x {11} = {4311} + {441} + {531} + {54}
{43} x {3} = {433} + {442} + {532} + {541} + {631} + {64} + {73}
{43} x {21} = {4321} + {4411} + {442} + {5311} + {532} +2{541} + {55} + {631} + {64}
{43} x {111} = {43111} + {4411} + {5311} + {541}
{43} x {4} = {443} + {533} + {542} + {632} + {641} + {731} + {74} + {83}
{43} x {31} = {4331} + {4421} + {443} + {5321} + {533} + {5411} +2{542} + {551} + {6311} + {632}
+2{641} + {65} + {731} + {74}
{43} x {22} = {4322} + {4421} + {5321} + {5411} + {542} + {551} + {632} + {641} + {65}
{43} x {211} = {43211} + {44111} + {4421} + {53111} + {5321} +2{5411} + {542} + {551} + {6311}
+ {641}

{43} x {1111} = {431111} + {44111} + {53111} + {5411}
{43} x {41} = {4431} + {444} + {5331} + {5421} +2{543} + {552} + {6321} + {633} + {6411} +2{642}
 + {651} + {7311} + {732} +2{741} + {75} + {831} + {84}
{43} x {32} = {4332} + {4422} + {4431} + {5322} + {5331} +2{5421} + {543} + {5511} + {552} + {6321}
 + {633} + {6411} +2{642} +2{651} + {66} + {732} + {741} + {75}
{43} x {311} = {43311} + {44211} + {4431} + {53211} + {5331} + {54111} +2{5421} + {543} + {5511}
 + {552} + {63111} + {6321} +2{6411} + {642} + {651} + {7311} + {741}
{43} x {221} = {43221} + {44211} + {4422} + {53211} + {5322} + {54111} +2{5421} + {5511} + {552}
 + {6321} + {6411} + {642} + {651}
{43} x {2111} = {432111} + {441111} + {44211} + {531111} + {53211} +2{54111} + {5421} + {5511}
 + {63111} + {6411}
{43} x {11111} = {4311111} + {441111} + {531111} + {54111}
{43} x {42} = {4432} + {4441} + {5332} + {5422} +2{5431} + {544} + {5521} + {553} + {6322} + {6331}
 +2{6421} +2{643} + {6511} +2{652} + {661} + {7321} + {733} + {7411} +2{742} +2{751}
 + {76} + {832} + {841} + {85}
{43} x {411} = {44311} + {4441} + {53311} + {54211} +2{5431} + {544} + {5521} + {553} + {63211}
 + {6331} + {64111} +2{6421} + {643} + {6511} + {652} + {73111} + {7321} +2{7411}
 + {742} + {751} + {8311} + {841}
{43} x {33} = {4333} + {4432} + {5332} + {5422} + {5431} + {5521} + {6331} + {6421} + {643} + {6511}
 + {652} + {661} + {733} + {742} + {751} + {76}
{43} x {321} = {43321} + {44221} + {44311} + {4432} + {53221} + {53311} + {5332} +2{54211} +2{5422}
 +2{5431} + {55111} +2{5521} + {553} + {63211} + {6322} + {6331} + {64111} +3{6421}
 + {643} +2{6511} +2{652} + {661} + {7321} + {7411} + {742} + {751}
{43} x {3111} = {433111} + {442111} + {44311} + {532111} + {53311} + {541111} +2{54211} + {5431}
 + {55111} + {5521} + {631111} + {63211} +2{64111} + {6421} + {6511} + {73111} + {7411}
{43} x {222} = {43222} + {44221} + {53221} + {54211} + {5422} + {5521} + {6322} + {6421} + {652}
{43} x {2211} = {432211} + {442111} + {44221} + {532111} + {53221} + {541111} +2{54211} + {5422}
 + {55111} + {5521} + {63211} + {64111} + {6421} + {6511}
{43} x {21111} = {4321111} + {4411111} + {442111} + {5311111} + {532111} +2{541111} + {54211}
 + {55111} + {631111} + {64111}
{43} x {111111} = {43111111} + {4411111} + {5311111} + {541111}
{43} x {43} = {4433} + {4442} + {5333} +2{5432} + {5441} + {5522} + {5531} + {6332} + {6422}
 +2{6431} + {644} +2{6521} + {653} + {6611} + {662} + {7331} + {7421} +2{743} + {7511}
 +2{752} +2{761} + {77} + {833} + {842} + {851} + {86}
{43} x {421} = {44321} + {44411} + {4442} + {53321} + {54221} +2{54311} +2{5432} +2{5441} + {55211}
 + {5522} +2{5531} + {554} + {63221} + {63311} + {6332} +2{64211} +2{6422} +3{6431}
 + {644} + {65111} +3{6521} +2{653} + {6611} + {662} + {73211} + {7322} + {7331}
 + {74111} +3{7421} + {743} +2{7511} +2{752} + {761} + {8321} + {8411} + {842} + {851}
{43} x {4111} = {443111} + {44411} + {533111} + {542111} +2{54311} + {5441} + {55211} + {5531}
 + {632111} + {63311} + {641111} +2{64211} + {6431} + {65111} + {6521} + {731111}
 + {73211} +2{74111} + {7421} + {7511} + {83111} + {8411}
{43} x {331} = {43331} + {44321} + {4433} + {53321} + {5333} + {54221} + {54311} +2{5432} + {55211}
 + {5522} + {5531} + {63311} + {6332} + {64211} + {6422} +2{6431} + {65111} +2{6521}
 + {653} + {6611} + {662} + {7331} + {7421} + {743} + {7511} + {752} + {761}
{43} x {322} = {43322} + {44222} + {44321} + {53222} + {53321} +2{54221} + {54311} + {5432}
 + {55211} + {5522} + {5531} + {63221} + {6332} + {64211} +2{6422} + {6431} +2{6521}
 + {653} + {662} + {7322} + {7421} + {752}
{43} x {3211} = {433211} + {442211} + {443111} + {44321} + {532211} + {533111} + {53321} +2{542111}
 +2{54221} +2{54311} + {5432} + {551111} +2{55211} + {5522} + {5531} + {632111}
 + {63221} + {63311} + {641111} +3{64211} + {6422} + {6431} +2{65111} +2{6521} + {6611}
 + {73211} + {74111} + {7421} + {7511}
{43} x {31111} = {4331111} + {4421111} + {443111} + {5321111} + {533111} + {5411111} +2{542111}
 + {54311} + {551111} + {55211} + {6311111} + {632111} +2{641111} + {64211} + {65111}
 + {731111} + {74111}
{43} x {2221} = {432221} + {442211} + {44222} + {532211} + {53222} + {542111} +2{54221} + {55211}
 + {5522} + {63221} + {64211} + {6422} + {6521}
{43} x {22111} = {4322111} + {4421111} + {442211} + {5321111} + {532211} + {5411111} +2{542111}
 + {54221} + {551111} + {55211} + {632111} + {641111} + {64211} + {65111}
{43} x {211111} = {43211111} + {44111111} + {4421111} + {53111111} + {5321111} +2{5411111}
 + {542111} + {551111} + {6311111} + {641111}
{43} x {1111111} = {431111111} + {44111111} + {53111111} + {5411111}
{421} x {0} = {421}
{421} x {1} = {4211} + {422} + {431} + {521}
{421} x {2} = {4221} + {4311} + {432} + {441} + {5211} + {522} + {531} + {621}
{421} x {11} = {42111} + {4221} + {4311} + {432} + {5211} + {522} + {531}
{421} x {3} = {4321} + {4411} + {442} + {5221} + {5311} + {532} + {541} + {6211} + {622} + {631}
 + {721}
{421} x {21} = {42211} + {4222} + {43111} +2{4321} + {433} + {4411} + {442} + {52111} +2{5221}
 +2{5311} +2{532} + {541} + {6211} + {622} + {631}
{421} x {111} = {421111} + {42211} + {43111} + {4321} + {52111} + {5221} + {5311} + {532}

B-1 Outer Multiplication of S-Functions (contd).

{421} x {4} = {4421} + {5321} + {5411} + {542} + {6221} + {6311} + {632} + {641} + {7211} + {722}
+ {731} + {821}

{421} x {31} = {43211} + {4322} + {4331} + {44111} +2{4421} + {443} + {52211} + {5222} + {53111}
+3{5321} + {533} +2{5411} +2{542} + {551} + {62111} +2{6221} +2{6311} +2{632} + {641}
+ {7211} + {722} + {731}

{421} x {22} = {42221} + {43211} + {4322} + {4331} + {4421} + {443} + {52211} + {5222} + {53111}
+2{5321} + {533} + {5411} + {542} + {6221} + {6311} + {632} + {641}

{421} x {211} = {422111} + {42221} + {431111} +2{43211} + {4322} + {4331} + {44111} + {4421}
+ {521111} +2{52211} + {5222} +2{53111} +3{5321} + {533} + {5411} + {542} + {62111}
+ {6221} + {6311} + {632}

{421} x {1111} = {4211111} + {422111} + {431111} + {43211} + {521111} + {52211} + {53111} + {5321}

{421} x {41} = {44211} + {4422} + {4431} + {53211} + {5322} + {5331} + {54111} +3{5421} + {543}
+ {5511} + {552} + {62211} + {6222} + {63111} +3{6321} + {633} +2{6411} +2{642} + {651}
+ {72111} +2{7221} +2{7311} +2{732} + {741} + {8211} + {822} + {831}

{421} x {32} = {43221} + {43311} + {4332} + {44211} + {4422} +2{4431} + {444} + {52221} +2{53211}
+2{5322} +2{5331} + {54111} +3{5421} +2{543} + {5511} + {552} + {62211} + {6222}
+ {63111} +3{6321} + {633} +2{6411} +2{642} + {651} + {7221} + {7311} + {732} + {741}

{421} x {311} = {432111} + {43221} + {43311} + {4332} + {441111} +2{44211} + {4422} + {4431}
+ {522111} + {52221} + {531111} +3{53211} +2{5322} +2{5331} +2{54111} +3{5421} + {543}
+ {5511} + {552} + {621111} +2{62211} + {6222} +2{63111} +3{6321} + {633} + {6411}
+ {642} + {72111} + {7221} + {7311} + {732}

{421} x {221} = {422211} + {42222} + {432111} +2{43221} + {43311} + {4332} + {44211} + {4422}
+ {4431} + {522111} +2{52221} + {531111} +3{53211} +2{5322} +2{5331} + {54111}
+2{5421} + {543} + {62211} + {6222} + {63111} +2{6321} + {633} + {6411} + {642}

{421} x {2111} = {4221111} + {422211} + {4311111} +2{432111} + {43221} + {43311} + {441111}
+ {44211} + {5211111} +2{522111} + {52221} +2{531111} +3{53211} + {5322} + {5331}
+ {54111} + {5421} + {621111} + {62211} + {63111} + {6321}

{421} x {11111} = {42111111} + {4221111} + {4311111} + {432111} + {5211111} + {522111} + {531111}
+ {53211}

{421} x {42} = {44221} + {44311} + {4432} + {4441} + {53221} + {53311} + {5332} +2{54211} +2{5422}
+3{5431} + {544} + {55111} +2{5521} + {553} + {62221} +2{63211} +2{6322} +2{6331}
+ {64111} +4{6421} +2{643} +2{6511} +2{652} + {661} + {72211} + {7222} + {73111}
+3{7321} + {733} +2{7411} +2{742} + {751} + {8221} + {8311} + {832} + {841}

{421} x {411} = {442111} + {44221} + {44311} + {4432} + {532111} + {53221} + {53311} + {5332}
+ {541111} +3{54211} +2{5422} +2{5431} + {55111} +2{5521} + {553} + {622111} + {62221}
+ {631111} +3{63211} +2{6322} +2{6331} +2{64111} +3{6421} + {643} + {6511} + {652}
+ {721111} +2{72211} + {7222} +2{73111} +3{7321} + {733} + {7411} + {742} + {82111}
+ {8221} + {8311} + {832}

{421} x {33} = {43321} + {44311} + {4432} + {4441} + {53221} + {53311} + {5332} + {54211} + {5422}
+2{5431} + {544} + {5521} + {553} + {63211} + {6322} + {6331} + {64111} +2{6421}
+ {643} + {6511} + {652} + {7321} + {7411} + {742} + {751}

{421} x {321} = {432211} + {43222} + {433111} +2{43321} + {4333} + {442111} +2{44221} +2{44311}
+2{4432} + {4441} + {522211} + {52222} +2{532111} +4{53221} +3{53311} +3{5332}
+ {541111} +4{54211} +3{5422} +4{5431} + {544} + {55111} +2{5521} + {553} + {622111}
+2{62221} + {631111} +4{63211} +3{6322} +3{6331} +2{64111} +4{6421} +2{643} + {6511}
+ {652} + {72211} + {7222} + {73111} +2{7321} + {733} + {7411} + {742}

{421} x {3111} = {4321111} + {432211} + {433111} + {43321} + {4411111} +2{442111} + {44221}
+ {44311} + {5221111} + {522211} + {5311111} +3{532111} +2{53221} +2{53311} + {5332}
+2{541111} +3{54211} + {5422} + {5431} + {55111} + {5521} + {6211111} +2{622111}
+ {62221} +2{631111} +3{63211} + {6322} + {6331} + {64111} + {6421} + {721111}
+ {72211} + {73111} + {7321}

{421} x {222} = {422221} + {432211} + {43222} + {43321} + {44221} + {4432} + {522211} + {52222}
+ {532111} +2{53221} + {53311} + {5332} + {54211} + {5422} + {5431} + {62221}
+ {63211} + {6322} + {6331} + {6421} + {643}

{421} x {2211} = {4222111} + {422221} + {4321111} +2{432211} + {43222} + {433111} + {43321}
+ {442111} + {44221} + {44311} + {5221111} +2{522211} + {52222} + {5311111}
+3{532111} +3{53221} +2{53311} + {5332} + {541111} +2{54211} + {5422} + {5431}
+ {622111} + {62221} + {631111} +2{63211} + {6322} + {6331} + {64111} + {6421}

{421} x {21111} = {42211111} + {4222111} + {43111111} +2{4321111} + {432211} + {433111} + {4411111}
+ {442111} + {52111111} +2{5221111} + {522211} +2{5311111} +3{532111} + {53221}
+ {53311} + {541111} + {54211} + {6211111} + {622111} + {631111} + {63211}

{421} x {111111} = {421111111} + {42211111} + {43111111} + {4321111} + {52111111} + {5221111}
+ {5311111} + {532111}

{421} x {421} = {442211} + {44222} + {443111} +2{44321} + {4433} + {44411} + {4442} + {532211}
+ {53222} + {533111} +2{53321} + {5333} +2{542111} +4{54221} +4{54311} +4{5432}
+2{5441} + {551111} +3{55211} +2{5522} +3{5531} + {554} + {622211} + {62222}
+2{632111} +4{63221} +3{63311} +3{6332} + {641111} +5{64211} +4{6422} +5{6431} + {644}
+2{65111} +4{6521} +2{653} + {6611} + {662} + {722111} +2{72221} + {731111} +4{73211}
+3{7322} +3{7331} +2{74111} +4{7421} +2{743} + {7511} + {752} + {82211} + {8222}
+ {83111} +2{8321} + {833} + {8411} + {842}

{421} x {4111} = {4421111} + {442211} + {443111} + {44321} + {5321111} + {532211} + {533111}
+ {53321} + {5411111} +3{542111} +2{54221} +2{54311} + {5432} + {551111} +2{55211}
+ {5522} + {5531} + {6221111} + {622211} + {6311111} +3{632111} +2{63221} +2{63311}
+ {6332} +2{641111} +3{64211} + {6422} + {6431} + {65111} + {6521} + {7211111}
+2{722111} + {72221} +2{731111} +3{73211} + {7322} + {7331} + {74111} + {7421}
+ {821111} + {82211} + {83111} + {8321}

{421} x {331} = {433211} + {43322} + {43331} + {443111} +2{44321} + {4433} + {44411} + {4442}
+ {532211} + {53222} + {533111} +3{53321} + {5333} + {542111} +2{54221} +3{54311}
+3{5432} +2{5441} + {55211} + {5522} +2{5531} + {554} + {632111} +2{63221} +2{63311}
+2{6332} + {641111} +3{64211} +2{6422} +3{6431} + {644} + {65111} +2{6521} + {653}
+ {73211} + {7322} + {7331} + {74111} +2{7421} + {743} + {7511} + {752}

{421} x {322} = {432221} + {433211} + {43322} + {43331} + {442211} + {44222} +2{44321} + {4433}
+ {4442} + {522221} +2{532211} +2{53222} + {533111} +3{53321} + {5333} + {542111}
+3{54221} +2{54311} +3{5432} + {5441} + {55211} + {5522} + {5531} + {622211} + {62222}
+ {632111} +3{63221} +2{63311} +2{6332} +2{64211} +2{6422} +3{6431} + {644} + {6521}
+ {653} + {72221} + {73211} + {7322} + {7331} + {7421} + {743}

{421} x {3211} = {4322111} + {432221} + {4331111} +2{433211} + {43322} + {43331} + {4421111}
+2{442211} + {44222} +2{443111} +2{44321} + {44411} + {5222111} + {522221}
+2{5321111} +4{532211} +2{53222} +3{533111} +4{53321} + {5333} + {5411111} +4{542111}
+4{54221} +4{54311} +2{5432} + {5441} + {551111} +2{55211} + {5522} + {5531}
+ {6221111} +2{622211} + {62222} + {6311111} +4{632111} +4{63221} +3{63311} +2{6332}
+2{641111} +4{64211} +2{6422} +2{6431} + {65111} + {6521} + {722111} + {72221}
+ {731111} +2{73211} + {7322} + {7331} + {74111} + {7421}

{421} x {31111} = {43211111} + {4322111} + {4331111} + {433211} + {44111111} +2{4421111} + {442211}
+ {443111} + {52211111} + {5222111} + {53111111} +3{5321111} +2{532211} +2{533111}
+ {53321} +2{5411111} +3{542111} + {54221} + {54311} + {551111} + {55211}
+ {62111111} +2{6221111} + {622211} +2{6311111} +3{632111} + {63221} + {63311}
+ {641111} + {64211} + {7211111} + {722111} + {731111} + {73211}

{421} x {2221} = {4222211} + {422222} + {4322111} +2{432221} + {433211} + {43322} + {442211}
+ {44222} + {44321} + {5222111} +2{522221} + {5321111} +3{532211} +2{53222}
+ {533111} +2{53321} + {542111} +2{54221} + {54311} + {5432} + {622211} + {62222}
+ {632111} +2{63221} + {63311} + {6332} + {64211} + {6422} + {6431}

{421} x {22111} = {42221111} + {4222211} + {43211111} +2{4322111} + {432221} + {4331111} + {433211}
+ {4421111} + {442211} + {443111} + {52211111} +2{5222111} + {522221} + {53111111}
+3{5321111} +3{532211} + {53222} +2{533111} + {53321} + {5411111} +2{542111}
+ {54221} + {54311} + {6221111} + {622211} + {6311111} +2{632111} + {63221}
+ {63311} + {641111} + {64211}

{421} x {211111} = {422111111} + {42221111} + {431111111} +2{43211111} + {4322111} + {4331111}
+ {44111111} + {4421111} + {521111111} +2{52211111} + {5222111} +2{53111111}
+3{5321111} + {532211} + {533111} + {5411111} + {542111} + {62111111} + {6221111}
+ {6311111} + {632111}

{421} x {1111111} = {4211111111} + {422111111} + {431111111} + {43211111} + {521111111} + {52211111}
+ {53111111} + {5321111}

{4111} x {0} = {4111}

{4111} x {1} = {41111} + {4211} + {5111}

{4111} x {2} = {42111} + {4311} + {51111} + {5211} + {6111}

{4111} x {11} = {411111} + {42111} + {4221} + {51111} + {5211}

{4111} x {3} = {43111} + {4411} + {52111} + {5311} + {61111} + {6211} + {7111}

{4111} x {21} = {421111} + {42211} + {43111} + {4321} + {511111} +2{52111} + {5221} + {5311}
+ {61111} + {6211}

{4111} x {111} = {4111111} + {421111} + {42211} + {4222} + {511111} + {52111} + {5221}

{4111} x {4} = {44111} + {53111} + {5411} + {62111} + {6311} + {71111} + {7211} + {8111}

{4111} x {31} = {431111} + {43211} + {44111} + {4421} + {521111} + {52211} +2{53111} + {5321}
+ {5411} + {611111} +2{62111} + {6221} + {6311} + {71111} + {7211}

{4111} x {22} = {422111} + {43211} + {4331} + {521111} + {52211} + {53111} + {5321} + {62111}
+ {6311}

{4111} x {211} = {4211111} + {422111} + {42221} + {431111} + {43211} + {4322} + {5111111} +2{521111}
+2{52211} + {5222} + {53111} + {5321} + {611111} + {62111} + {6221}

{4111} x {1111} = {41111111} + {4211111} + {422111} + {42221} + {5111111} + {521111} + {52211}
+ {5222}

{4111} x {41} = {441111} + {44211} + {531111} + {53211} +2{54111} + {5421} + {5511} + {621111}
+ {62211} +2{63111} + {6321} + {6411} + {711111} +2{72111} + {7221} + {7311} + {81111}
+ {8211}

{4111} x {32} = {432111} + {43311} + {44211} + {4431} + {522111} + {531111} +2{53211} + {5331}
+ {54111} + {5421} + {621111} + {62211} +2{63111} + {6321} + {6411} + {72111} + {7311}

{4111} x {311} = {4311111} + {432111} + {43221} + {441111} + {44211} + {4422} + {5211111} + {522111}
+ {52221} +2{531111} +2{53211} + {5322} + {54111} + {5421} + {6111111} +2{621111}
+2{62211} + {6222} + {63111} + {6321} + {711111} + {72111} + {7221}

{4111} x {221} = {4221111} + {422211} + {432111} + {43221} + {43311} + {4332} + {5211111} +2{522111}
+ {52221} + {531111} +2{53211} + {5322} + {5331} + {621111} + {62211} + {63111}
+ {6321}

{4111} x {2111} = {42111111} + {4221111} + {422211} + {42222} + {4311111} + {432111} + {43221}
+ {5111111} +2{5211111} +2{522111} +2{52221} + {531111} + {53211} + {5322}
+ {6111111} + {621111} + {62211} + {6222}

{4111} x {11111} = {411111111} + {42111111} + {4221111} + {422211} + {5111111} + {5211111}
+ {522111} + {52221}

{4111} x {42} = {442111} + {44311} + {532111} + {53311} + {541111} +2{54211} + {5431} + {55111}
+ {5521} + {622111} + {631111} +2{63211} + {6331} +2{64111} + {6421} + {6511}
+ {721111} + {72211} +2{73111} + {7321} + {7411} + {82111} + {8311}

{4111} x {411} = {4411111} + {442111} + {44221} + {5311111} + {532111} + {53221} +2{541111}
+2{54211} + {5422} + {55111} + {5521} + {6211111} + {622111} + {62221} +2{631111}
+2{63211} + {6322} + {64111} + {6421} + {7111111} +2{721111} +2{72211} + {7222}
+ {73111} + {7321} + {811111} + {82111} + {8221}

{4111} x {33} = {433111} + {44311} + {4441} + {532111} + {53311} + {54211} + {5431} + {631111}
+ {63211} + {64111} + {6421} + {73111} + {7411}

{4111} x {321} = {4321111} + {432211} + {433111} + {43321} + {442111} + {44221} + {44311} + {4432}
+ {5221111} + {522211} + {5311111} +3{532111} +2{53221} +2{53311} + {5332} + {541111}
+2{54211} + {5422} + {5431} + {6211111} +2{622111} + {62221} +2{631111} +3{63211}
+ {6322} + {6331} + {64111} + {6421} + {721111} + {72211} + {73111} + {7321}

{4111} x {3111} = {43111111} + {4321111} + {432211} + {43222} + {4411111} + {442111} + {44221}
+ {52111111} + {5221111} + {522211} + {52222} +2{5311111} +2{532111} +2{53221}
+ {541111} + {54211} + {5422} + {61111111} +2{6211111} +2{622111} +2{62221}
+ {631111} + {63211} + {6322} + {7111111} + {721111} + {72211} + {7222}

{4111} x {222} = {4222111} + {432211} + {43321} + {4333} + {5221111} + {522211} + {532111} + {53221}
+ {53311} + {5332} + {622111} + {63211} + {6331}

{4111} x {2211} = {42211111} + {4222111} + {422221} + {4321111} + {432211} + {43222} + {433111}
+ {43321} + {52111111} +2{5221111} +2{522211} + {52222} + {5311111} +2{532111}
+2{53221} + {53311} + {5332} + {6211111} + {622111} + {62221} + {631111} + {63211}
+ {6322}

{4111} x {21111} = {421111111} + {42211111} + {4222111} + {422221} + {43111111} + {4321111}
+ {432211} + {511111111} +2{52111111} +2{5221111} +2{522211} + {52222} + {5311111}
+ {532111} + {53221} + {61111111} + {6211111} + {622111} + {62221}

{4111} x {111111} = {4111111111} + {421111111} + {42211111} + {4222111} + {511111111} + {52111111}
+ {5221111} + {522211}

{4111} x {4111} = {44111111} + {4421111} + {442211} + {44222} + {53111111} + {5321111} + {532211}
+ {53222} +2{5411111} +2{542111} +2{54221} + {551111} + {55211} + {5522}
+ {62111111} + {6221111} + {622211} + {62222} +2{6311111} +2{632111} +2{63221}
+ {641111} + {64211} + {6422} + {71111111} +2{7211111} +2{722111} +2{72221}
+ {731111} + {73211} + {7322} + {8111111} + {821111} + {82211} + {8222}

{4111} x {331} = {4331111} + {433211} + {443111} + {44321} + {44411} + {4442} + {5321111} + {532211}
+2{533111} + {53321} + {542111} + {54221} +2{54311} + {5432} + {5441} + {6311111}
+2{632111} + {63221} + {63311} + {641111} +2{64211} + {6422} + {6431} + {731111}
+ {73211} + {74111} + {7421}

{4111} x {322} = {4322111} + {433211} + {43331} + {442211} + {44321} + {4433} + {5222111}
+ {5321111} +2{532211} + {533111} +2{53321} + {5333} + {542111} + {54221} + {54311}
+ {5432} + {6221111} + {622211} +2{632111} + {63221} +2{63311} + {6332} + {64211}
+ {6431} + {722111} + {73211} + {7331}

{4111} x {3211} = {43211111} + {4322111} + {432221} + {4331111} + {433211} + {43322} + {4421111}
+ {442211} + {44222} + {443111} + {44321} + {52211111} + {5222111} + {522221}
+ {53111111} +3{5321111} +3{532211} +2{53222} +2{533111} +2{53321} + {5411111}
+2{542111} +2{54221} + {54311} + {5432} + {62111111} +2{6221111} +2{622211}
+ {62222} +2{6311111} +3{632111} +3{63221} + {63311} + {6332} + {641111} + {64211}
+ {6422} + {7211111} + {722111} + {72221} + {731111} + {73211} + {7322}

{4111} x {31111} = {431111111} + {43211111} + {4322111} + {432221} + {44111111} + {4421111}
+ {442211} + {521111111} + {52211111} + {5222111} + {522221} +2{53111111}
+2{5321111} +2{532211} + {53222} + {5411111} + {542111} + {54221} + {611111111}
+2{62111111} +2{6221111} +2{622211} + {62222} + {6311111} + {632111} + {63221}
+ {71111111} + {7211111} + {722111} + {72221}

{4111} x {2221} = {42221111} + {4222211} + {4322111} + {432221} + {433211} + {43322} + {43331}
+ {52211111} +2{5222111} + {522221} + {5321111} +2{532211} + {53222} + {533111}
+2{53321} + {5333} + {6221111} + {622211} + {632111} + {63221} + {63311} + {6332}

{4111} x {22111} = {422111111} + {42221111} + {4222211} + {422222} + {43211111} + {4322111}
+ {432221} + {4331111} + {433211} + {521111111} +2{52211111} +2{5222111} +2{522221}
+ {53111111} +2{5321111} +2{532211} + {53222} + {533111} + {53321} + {62111111}
+ {6221111} + {622211} + {62222} + {6311111} + {632111} + {63221}

{4111} x {211111} = {4211111111} + {422111111} + {42221111} + {4222211} + {431111111} + {43211111}
+ {4322111} + {5111111111} +2{521111111} +2{52211111} +2{5222111} + {522221}
+ {53111111} + {5321111} + {532211} + {611111111} + {62111111} + {6221111}
+ {622211}

{4111} x {1111111} = {41111111111} + {4211111111} + {422111111} + {42221111} + {5111111111}
+ {521111111} + {52211111} + {5222111}

Outer Multiplication of S-Functions (contd).

```
{331} x {0} = {331}
{331} x {1} = {3311} + {332} + {431}
{331} x {2} = {3321} + {333} + {4311} + {432} + {531}
{331} x {11} = {33111} + {3321} + {4311} + {432} + {441}
{331} x {3} = {3331} + {4321} + {433} + {5311} + {532} + {631}
{331} x {21} = {33211} + {3322} + {3331} + {43111} +2{4321} + {433} + {4411} + {442} + {5311}
         + {532} + {541}
{331} x {111} = {331111} + {33211} + {43111} + {4321} + {4411} + {442}
{331} x {4} = {4331} + {5321} + {533} + {6311} + {632} + {731}
{331} x {31} = {33311} + {3332} + {43211} + {4322} +2{4331} + {4421} + {443} + {53111} +2{5321}
         + {533} + {5411} + {542} + {6311} + {632} + {641}
{331} x {22} = {33221} + {3332} + {43211} + {4322} + {4331} + {44111} + {4421} + {5321} + {533}
         + {5411} + {542} + {551}
{331} x {211} = {332111} + {33221} + {33311} + {431111} +2{43211} + {4322} + {4331} + {44111}
          +2{4421} + {443} + {53111} + {5321} + {5411} + {542}
{331} x {1111} = {3311111} + {332111} + {431111} + {43211} + {44111} + {4421}
{331} x {41} = {43311} + {4332} + {4431} + {53211} + {5322} +2{5331} + {5421} + {543} + {63111}
         +2{6321} + {633} + {6411} + {642} + {7311} + {732} + {741}
{331} x {32} = {33321} + {3333} + {43221} + {43311} +2{4332} + {44211} + {4422} + {4431} + {53211}
         + {5322} +2{5331} + {54111} +2{5421} + {543} + {5511} + {552} + {6321} + {633} + {6411}
         + {642} + {651}
{331} x {311} = {333111} + {33321} + {432111} + {43221} +2{43311} + {4332} + {44211} + {4422}
          +2{4431} + {444} + {531111} +2{53211} + {5322} + {5331} + {54111} +2{5421} + {543}
          + {63111} + {6321} + {6411} + {642}
{331} x {221} = {332211} + {33222} + {33321} + {432111} +2{43221} + {43311} + {4332} + {441111}
          +2{44211} + {4422} + {4431} + {53211} + {5322} + {5331} + {54111} +2{5421} + {543}
          + {5511} + {552}
{331} x {2111} = {3321111} + {332211} + {333111} + {4311111} +2{432111} + {43221} + {43311}
           + {441111} +2{44211} + {4422} + {4431} + {531111} + {53211} + {54111} + {5421}
{331} x {11111} = {33111111} + {3321111} + {4311111} + {432111} + {441111} + {44211}
{331} x {42} = {43321} + {4333} + {44311} + {4432} + {53221} + {53311} +2{5332} + {54211} + {5422}
         +2{5431} + {5521} + {553} + {63211} + {6322} +2{6331} + {64111} +2{6421} + {643}
         + {6511} + {652} + {7321} + {733} + {7411} + {742} + {751}
{331} x {411} = {433111} + {43321} + {44311} + {4432} + {4441} + {532111} + {53221} +2{53311}
          + {5332} + {54211} + {5422} +2{5431} + {544} + {631111} +2{63211} + {6322} + {6331}
          + {64111} +2{6421} + {643} + {73111} + {7321} + {7411} + {742}
{331} x {33} = {33331} + {43321} + {4333} + {44221} + {4432} + {53311} + {5332} + {54211} + {5422}
         + {5431} + {55111} + {5521} + {6331} + {6421} + {643} + {6511} + {652} + {661}
{331} x {321} = {333211} + {33322} + {33331} + {432211} + {43222} + {433111} +3{43321} + {4333}
          + {442111} +2{44221} +2{44311} +2{4432} + {4441} + {532111} +2{53221} +2{53311}
          +2{5332} + {541111} +3{54211} +2{5422} +3{5431} + {544} + {55111} +2{5521} + {553}
          + {63211} + {6322} + {6331} + {64111} +2{6421} + {643} + {6511} + {652}
{331} x {3111} = {3331111} + {333211} + {4321111} + {432211} +2{433111} + {43321} + {442111}
           + {44221} +2{44311} + {4432} + {4441} + {5311111} +2{532111} + {53221} + {53311}
           + {541111} +2{54211} + {5422} + {5431} + {631111} + {63211} + {64111} + {6421}
{331} x {222} = {332221} + {33322} + {432211} + {43222} + {43321} + {442111} + {44221} + {44311}
          + {53221} + {5332} + {54211} + {5422} + {5431} + {5521} + {553}
{331} x {2211} = {3322111} + {332221} + {333211} + {4321111} +2{432211} + {43222} + {433111}
           + {43321} + {4411111} +2{442111} +2{44221} + {44311} + {4432} + {532111} + {53221}
           + {53311} + {541111} +2{54211} + {5422} + {5431} + {55111} + {5521}
{331} x {21111} = {33211111} + {3322111} + {3331111} + {43111111} +2{4321111} + {432211} + {433111}
            + {4411111} +2{442111} + {44221} + {44311} + {5311111} + {532111} + {541111}
            + {54211}
{331} x {111111} = {331111111} + {33211111} + {43111111} + {4321111} + {4411111} + {442111}
{331} x {331} = {333311} + {33332} + {433211} + {43322} +2{43331} + {442211} + {44222} +2{44321}
          + {4433} + {4442} + {533111} +2{53321} + {5333} + {542111} +2{54221} +2{54311}
          +2{5432} + {5441} + {551111} +2{55211} + {5522} + {5531} + {63311} + {6332} + {64211}
          + {6422} +2{6431} + {644} + {65111} +2{6521} + {653} + {6611} + {662}
{331} x {322} = {333221} + {33332} + {432221} + {433211} +2{43322} + {43331} + {442211} + {44222}
          + {443111} +2{44321} + {44411} + {532211} + {53222} +2{53321} + {5333} + {542111}
          +2{54221} +2{54311} +2{5432} + {5441} + {55211} + {5522} +2{5531} + {554} + {63221}
          + {6332} + {64211} + {6422} + {6431} + {6521} + {653}
{331} x {3211} = {3332111} + {333221} + {333311} + {4322111} + {432221} + {4331111} +3{433211}
           + {43322} + {43331} + {4421111} +2{442211} + {44222} +2{443111} +3{44321} + {4433}
           + {44411} + {4442} + {5321111} +2{532211} + {53222} +2{533111} +2{53321} + {5411111}
           +3{542111} +3{54221} +3{54311} +2{5432} + {5441} + {551111} +2{55211} + {5522}
           + {5531} + {632111} + {63221} + {63311} + {641111} +2{64211} + {6422} + {6431}
           + {65111} + {6521}
{331} x {31111} = {33311111} + {3332111} + {43211111} + {4322111} +2{4331111} + {433211} + {4421111}
            + {442211} +2{443111} + {44321} + {44411} + {53111111} +2{5321111} + {532211}
            + {533111} + {5411111} +2{542111} + {54221} + {54311} + {6311111} + {632111}
            + {641111} + {64211}
```

{331} x {2221} = {3322211} + {332222} + {333221} + {4322111} +2{432221} + {433211} + {43322}
+ {4421111} +2{442211} + {44222} + {443111} + {44321} + {532211} + {53222} + {53321}
+ {542111} +2{54221} + {54311} + {5432} + {55211} + {5522} + {5531}
{331} x {22111} = {33221111} + {3322211} + {3332111} + {43211111} +2{4322111} + {432221} + {4331111}
+ {433211} + {44111111} +2{4421111} +2{442211} + {44222} + {443111} + {44321}
+ {5321111} + {532211} + {533111} + {5411111} +2{542111} + {54221} + {54311}
+ {551111} + {55211}
{331} x {211111} = {332111111} + {33221111} + {33311111} + {431111111} +2{43211111} + {4322111}
+ {4331111} + {44111111} +2{4421111} + {442211} + {443111} + {53111111} + {5321111}
+ {5411111} + {542111}
{331} x {1111111} = {3311111111} + {332111111} + {431111111} + {43211111} + {44111111} + {4421111}
{322} x {0} = {322}
{322} x {1} = {3221} + {332} + {422}
{322} x {2} = {3222} + {3321} + {4221} + {432} + {522}
{322} x {11} = {32211} + {3321} + {333} + {4221} + {432}
{322} x {3} = {3322} + {4222} + {4321} + {5221} + {532} + {622}
{322} x {21} = {32221} + {33211} + {3322} + {3331} + {42211} + {4222} +2{4321} + {433} + {442}
+ {5221} + {532}
{322} x {111} = {322111} + {33211} + {3331} + {42211} + {4321} + {433}
{322} x {4} = {4322} + {5222} + {5321} + {6221} + {632} + {722}
{322} x {31} = {33221} + {3332} + {42221} + {43211} +2{4322} + {4331} + {4421} + {52211} + {5222}
+2{5321} + {533} + {542} + {6221} + {632}
{322} x {22} = {32222} + {33221} + {33311} + {42221} + {43211} + {4322} + {4331} + {4421} + {443}
+ {5222} + {5321} + {542}
{322} x {211} = {322211} + {332111} + {33221} + {33311} + {3332} + {422111} + {42221} +2{43211}
+ {4322} +2{4331} + {4421} + {443} + {52211} + {5321} + {533}
{322} x {1111} = {3221111} + {332111} + {33311} + {422111} + {43211} + {4331}
{322} x {41} = {43221} + {4332} + {4422} + {52221} + {53211} +2{5322} + {5331} + {5421} + {62211}
+ {6222} +2{6321} + {633} + {642} + {7221} + {732}
{322} x {32} = {33222} + {33321} + {42222} +2{43221} + {43311} + {4332} + {44211} + {4422} + {4431}
+ {52221} + {53211} +2{5322} + {5331} +2{5421} + {543} + {552} + {6222} + {6321}
+ {642}
{322} x {311} = {332211} + {33321} + {3333} + {422211} + {432111} +2{43221} + {43311} +2{4332}
+ {44211} + {4422} + {4431} + {522111} + {52221} +2{53211} + {5322} +2{5331} + {5421}
+ {543} + {62211} + {6321} + {633}
{322} x {221} = {322221} + {332211} + {33222} + {333111} + {33321} + {422211} + {42222} + {432111}
+2{43221} +2{43311} + {4332} + {44211} + {4422} +2{4431} + {444} + {52221} + {53211}
+ {5322} + {5331} + {5421} + {543}
{322} x {2111} = {3222111} + {3321111} + {332211} + {333111} + {33321} + {4221111} + {422211}
+2{432111} + {43221} +2{43311} + {4332} + {44211} + {4431} + {522111} + {53211}
+ {5331}
{322} x {11111} = {32211111} + {3321111} + {333111} + {4221111} + {432111} + {43311}
{322} x {42} = {43222} + {43321} + {44221} + {4432} + {52222} +2{53221} + {53311} + {5332} + {54211}
+2{5422} + {5431} + {5521} + {62221} + {63211} +2{6322} + {6331} +2{6421} + {643}
+ {652} + {7222} + {7321} + {742}
{322} x {411} = {432211} + {43321} + {4333} + {44221} + {4432} + {522211} + {532111} +2{53221}
+ {53311} +2{5332} + {54211} + {5422} + {5431} + {622111} + {62221} +2{63211} + {6322}
+2{6331} + {6421} + {643} + {72211} + {7321} + {733}
{322} x {33} = {33322} + {43222} + {43321} + {44221} + {44311} + {53221} + {5332} + {54211} + {5422}
+ {5431} + {5521} + {553} + {6322} + {6421} + {652}
{322} x {321} = {332221} + {333211} + {33322} + {33331} + {422221} +2{432211} +2{43222} + {433111}
+3{43321} + {4333} + {442111} +2{44221} +2{44311} +2{4432} + {4441} + {522211}
+ {52222} + {532111} +3{53221} +2{53311} +2{5332} +2{54211} +2{5422} +3{5431} + {544}
+ {5521} + {553} + {62221} + {63211} + {6322} + {6331} + {6421} + {643}
{322} x {3111} = {3322111} + {333211} + {33331} + {4222111} + {4321111} +2{432211} + {433111}
+2{43321} + {4333} + {442111} + {44221} + {44311} + {4432} + {5221111} + {522211}
+2{532111} + {53221} +2{53311} + {5332} + {54211} + {5431} + {622111} + {63211}
+ {6331}
{322} x {222} = {322222} + {332221} + {333211} + {422221} + {432211} + {43222} + {433111} + {43321}
+ {44221} + {44311} + {4432} + {4441} + {52222} + {53221} + {53311} + {5422} + {5431}
+ {544}
{322} x {2211} = {3222211} + {3322111} + {332221} + {3331111} + {333211} + {33322} + {4222111}
+ {422221} + {4321111} +2{432211} + {43222} +2{433111} +2{43321} + {442111} + {44221}
+2{44311} + {4432} + {4441} + {522211} + {532111} + {53221} + {53311} + {5332}
+ {54211} + {5431}
{322} x {21111} = {32221111} + {33211111} + {3322111} + {3331111} + {333211} + {42211111}
+ {4222111} +2{4321111} + {432211} +2{433111} + {43321} + {442111} + {44311}
+ {5221111} + {532111} + {53311}
{322} x {111111} = {322111111} + {33211111} + {3331111} + {42211111} + {4321111} + {433111}
{322} x {322} = {332222} + {333221} + {333311} + {422222} +2{432221} +2{433211} + {43322} + {43331}
+ {442211} + {44222} + {443111} +2{44321} + {4433} + {44411} + {4442} + {522221}
+ {532211} +2{53222} + {533111} +2{53321} +2{54221} +2{54311} +2{5432} +2{5441}
+ {5522} + {5531} + {554} + {62222} + {63221} + {63311} + {6422} + {6431} + {644}

```
{31111} x {211111} = {32111111111} + {3221111111} + {322211111} + {32222111} + {3222221}
                   + {3311111111} + {332111111} + {33221111} + {3322211} + {41111111111}
                   +2{4211111111} +2{422111111} +2{42221111} +2{4222211} + {422222} + {431111111}
                   + {43211111} + {4322111} + {432221} + {5111111111} + {521111111} + {52211111}
                   + {5222111} + {522221}
{31111} x {1111111} = {311111111111} + {32111111111} + {3221111111} + {322211111} + {32222111}
                    + {41111111111} + {4211111111} + {422111111} + {42221111} + {4222211}
{2221} x {0} = {2221}
{2221} x {1} = {22211} + {2222} + {3221}
{2221} x {2} = {22221} + {32211} + {3222} + {4221}
{2221} x {11} = {222111} + {22221} + {32211} + {3222} + {3321}
{2221} x {3} = {32221} + {42211} + {4222} + {5221}
{2221} x {21} = {222211} + {22222} + {322111} +2{32221} + {33211} + {3322} + {42211} + {4222}
             + {4321}
{2221} x {111} = {2221111} + {222211} + {322111} + {32221} + {33211} + {3322} + {3331}
{2221} x {4} = {42221} + {52211} + {5222} + {6221}
{2221} x {31} = {322211} + {32222} + {33221} + {422111} +2{42221} + {43211} + {4322} + {52211}
             + {5222} + {5321}
{2221} x {22} = {222221} + {322211} + {32222} + {332111} + {33221} + {42221} + {43211} + {4322}
             + {4421}
{2221} x {211} = {2222111} + {222221} + {3221111} +2{322211} + {32222} + {332111} +2{33221}
              + {33311} + {3332} + {422111} + {42221} + {43211} + {4322} + {4331}
{2221} x {1111} = {22211111} + {2222111} + {3221111} + {322211} + {332111} + {33221} + {33311}
               + {3332}
{2221} x {41} = {422211} + {42222} + {43221} + {522111} +2{52221} + {53211} + {5322} + {62211}
             + {6222} + {6321}
{2221} x {32} = {322221} + {332211} + {33222} + {422211} + {42222} + {432111} +2{43221} + {44211}
             + {4422} + {52221} + {53211} + {5322} + {5421}
{2221} x {311} = {3222111} + {322221} + {332211} + {33222} + {33321} + {4221111} +2{422211}
              + {42222} + {432111} +2{43221} + {43311} + {4332} + {522111} + {52221} + {53211}
              + {5322} + {5331}
{2221} x {221} = {2222211} + {222222} + {3222111} +2{322221} + {3321111} +2{332211} + {33222}
              + {333111} + {33321} + {422211} + {42222} + {432111} +2{43221} + {43311} + {4332}
              + {44211} + {4422} + {4431}
{2221} x {2111} = {22221111} + {2222211} + {32211111} +2{3222111} + {322221} + {3321111} +2{332211}
               + {33222} + {333111} +2{33321} + {3333} + {4221111} + {422211} + {432111} + {43221}
               + {43311} + {4332}
{2221} x {11111} = {222111111} + {22221111} + {32211111} + {3222111} + {3321111} + {332211}
                + {333111} + {33321}
{2221} x {42} = {422221} + {432211} + {43222} + {44221} + {522211} + {52222} + {532111} +2{53221}
             + {54211} + {5422} + {62221} + {63211} + {6322} + {6421}
{2221} x {411} = {4222111} + {422221} + {432211} + {43222} + {43321} + {5221111} +2{522211}
              + {52222} + {532111} +2{53221} + {53311} + {5332} + {622111} + {62221} + {63211}
              + {6322} + {6331}
{2221} x {33} = {332221} + {432211} + {43222} + {442111} + {44221} + {53221} + {54211} + {5422}
             + {5521}
{2221} x {321} = {3222211} + {322222} + {3322111} +2{332221} + {333211} + {33322} + {4222111}
              +2{422221} + {4321111} +3{432211} +2{43222} + {433111} +2{43321} + {442111} +2{44221}
              + {44311} + {4432} + {522211} + {52222} + {532111} +2{53221} + {53311} + {5332}
              + {54211} + {5422} + {5431}
{2221} x {3111} = {32221111} + {3222211} + {3322111} + {332221} + {333211} + {33322} + {33331}
               + {42211111} +2{4222111} + {422221} + {4321111} +2{432211} + {43222} + {433111}
               +2{43321} + {4333} + {5221111} + {522211} + {532111} + {53221} + {53311} + {5332}
{2221} x {222} = {2222221} + {3222211} + {322222} + {3322111} + {332221} + {3331111} + {333211}
              + {422221} + {432211} + {43222} + {433111} + {43321} + {44221} + {44311} + {4432}
              + {4441}
{2221} x {2211} = {22222111} + {2222221} + {32221111} +2{3222211} + {322222} + {33211111}
               +2{3322111} +2{332221} + {3331111} +2{333211} + {33322} + {33331} + {4222111}
               + {422221} + {4321111} +2{432211} + {43222} + {433111} +2{43321} + {4333} + {442111}
               + {44221} + {44311} + {4432}
{2221} x {21111} = {222211111} + {22222111} + {322111111} +2{32221111} + {3222211} + {33211111}
                +2{3322111} + {332221} + {3331111} +2{333211} + {33322} + {33331} + {42211111}
                + {4222111} + {4321111} + {432211} + {433111} + {43321}
{2221} x {111111} = {2221111111} + {222211111} + {322111111} + {32221111} + {33211111} + {3322111}
                 + {3331111} + {333211}
{2221} x {2221} = {22222211} + {2222222} + {32222111} +2{3222221} + {33221111} +2{3322211}
               + {332222} + {33311111} +2{3332111} + {333221} + {333311} + {4222211} + {422222}
               + {4322111} +2{432221} + {4331111} +2{433211} + {43322} + {43331} + {442211}
               + {44222} + {443111} +2{44321} + {4433} + {44411} + {4442}
{2221} x {22111} = {222221111} + {22222211} + {322211111} +2{32222111} + {3222221} + {332111111}
                +2{33221111} +2{3322211} + {332222} + {33311111} +2{3332111} +2{333221} + {333311}
                + {33332} + {42221111} + {4222211} + {43211111} +2{4322111} + {432221} + {4331111}
                +2{433211} + {43322} + {43331} + {4421111} + {442211} + {443111} + {44321}
```

Outer Multiplication of S-Functions (contd).

{3211} x {33} = {333211} + {432211} + {433111} + {43321} + {442111} + {44221} + {44311} + {4432}
+ {532111} + {53221} + {53311} + {541111} +2{54211} + {5422} + {5431} + {55111}
+ {5521} + {63211} + {64111} + {6421} + {6511}

{3211} x {321} = {3322111} + {332221} + {3331111} +2{333211} + {33322} + {33331} + {422211}
+ {422221} +2{4321111} +4{432211} +2{43222} +3{433111} +4{43321} + {4333} + {4411111}
+3{442111} +3{44221} +3{44311} +2{4432} + {4441} + {5221111} +2{522211} + {52222}
+ {5311111} +4{532111} +4{53221} +3{53311} +2{5332} +2{541111} +4{54211} +2{5422}
+2{5431} + {55111} + {5521} + {622111} + {62221} + {631111} +2{63211} + {6322}
+ {6331} + {64111} + {6421}

{3211} x {3111} = {33211111} + {3322111} + {332221} + {3331111} + {333211} + {33322} + {42211111}
+ {4222111} + {422221} + {43111111} +3{4321111} +3{432211} +2{43222} +2{433111}
+2{43321} + {4411111} +2{442111} +2{44221} + {44311} + {4432} + {52111111}
+2{5221111} +2{522211} + {52222} +2{5311111} +3{532111} +3{53221} + {53311} + {5332}
+ {541111} + {54211} + {5422} + {6211111} + {622111} + {62221} + {631111} + {63211}
+ {6322}

{3211} x {222} = {3222211} + {3322111} + {332221} + {333211} + {33322} + {33331} + {4222111}
+ {422221} + {4321111} +2{432211} + {43222} + {433111} +2{43321} + {4333} + {442111}
+ {44221} + {44311} + {4432} + {522211} + {532111} + {53221} + {53311} + {5332}
+ {54211} + {5431}

{3211} x {2211} = {32221111} + {3222211} + {322222} + {33211111} +2{3322111} +2{332221} + {3331111}
+2{333211} + {33322} + {33331} + {42211111} +2{4222111} +2{422221} + {43111111}
+3{4321111} +4{432211} +2{43222} +2{433111} +3{43321} + {4333} + {4411111}
+2{442111} +2{44221} + {44311} + {4432} + {5221111} + {522211} + {52222} + {5311111}
+2{532111} +2{53221} + {53311} + {5332} + {541111} + {54211} + {5422}

{3211} x {21111} = {322111111} + {32221111} + {3222211} + {331111111} +2{33211111} +2{3322111}
+ {332221} + {3331111} + {333211} + {421111111} +2{42211111} +2{4222111} + {422221}
+2{43111111} +3{4321111} +3{432211} + {43222} + {433111} + {43321} + {4411111}
+ {442111} + {44221} + {52111111} + {5221111} + {522211} + {5311111} + {532111}
+ {53221}

{3211} x {111111} = {3211111111} + {322111111} + {32221111} + {331111111} + {33211111} + {3322111}
+ {421111111} + {42211111} + {4222111} + {43111111} + {4321111} + {432211}

{3211} x {3211} = {33221111} + {3322211} + {332222} + {33311111} +2{3332111} +2{333221} + {333311}
+ {33332} + {42221111} + {4222211} + {422222} +2{43211111} +4{4322111} +4{432221}
+3{4331111} +5{433211} +3{43322} +2{43331} + {44111111} +3{4421111} +4{442211}
+2{44222} +3{443111} +4{44321} + {4433} + {44411} + {4442} + {52211111} +2{5222111}
+2{522221} + {53111111} +4{5321111} +5{532211} +3{53222} +3{533111} +4{53321}
+ {5333} +2{5411111} +4{542111} +4{54221} +2{54311} +2{5432} + {551111} + {55211}
+ {5522} + {6221111} + {622211} + {62222} + {6311111} +2{632111} +2{63221} + {63311}
+ {6332} + {641111} + {64211} + {6422}

{3211} x {31111} = {332111111} + {33221111} + {3322211} + {33311111} + {3332111} + {333221}
+ {422111111} + {42221111} + {4222211} + {431111111} +3{43211111} +3{4322111}
+2{432221} +2{4331111} +2{433211} + {43322} + {44111111} +2{4421111} +2{442211}
+ {44222} + {443111} + {44321} + {521111111} +2{52211111} +2{5222111} + {522221}
+2{53111111} +3{5321111} +3{532211} + {53222} + {533111} + {53321} + {5411111}
+ {542111} + {54221} + {62111111} + {6221111} + {622211} + {6311111} + {632111}
+ {63221}

{3211} x {2221} = {32222111} + {3222221} + {33221111} +2{3322211} + {332222} + {3332111} +2{333221}
+ {333311} + {33332} + {42221111} +2{4222211} + {422222} + {43211111} +3{4322111}
+3{432221} + {4331111} +3{433211} +2{43322} +2{43331} + {4421111} +2{442211}
+ {44222} + {443111} +2{44321} + {4433} + {5222111} + {522221} + {5321111}
+2{532211} + {53222} + {533111} +2{53321} + {5333} + {542111} + {54221} + {54311}
+ {5432}

{3211} x {22111} = {322211111} + {32222111} + {3222221} + {332111111} +2{33221111} +2{3322211}
+ {332222} + {33311111} +2{3332111} + {333221} + {333311} + {422111111}
+2{42221111} +2{4222211} + {422222} + {431111111} +3{43211111} +4{4322111}
+3{432221} +2{4331111} +3{433211} + {43322} + {43331} + {44111111} +2{4421111}
+2{442211} + {44222} + {443111} + {44321} + {52211111} + {5222111} + {522221}
+ {53111111} +2{5321111} +2{532211} + {53222} + {533111} + {53321} + {5411111}
+ {542111} + {54221}

{3211} x {211111} = {3221111111} + {322211111} + {32222111} + {3311111111} +2{332111111}
+2{33221111} + {3322211} + {33311111} + {3332111} + {4211111111} +2{422111111}
+2{42221111} + {4222211} +2{431111111} +3{43211111} +3{4322111} + {432221}
+ {4331111} + {433211} + {44111111} + {4421111} + {442211} + {521111111}
+ {52211111} + {5222111} + {53111111} + {5321111} + {532211}

{3211} x {1111111} = {32111111111} + {3221111111} + {322211111} + {3311111111} + {332111111}
+ {33221111} + {4211111111} + {422111111} + {42221111} + {431111111} + {43211111}
+ {4322111}

{31111} x {0} = {31111}

{31111} x {1} = {311111} + {32111} + {41111}

{31111} x {2} = {321111} + {33111} + {411111} + {42111} + {51111}

Outer Multiplication of S-Functions (contd).

{31111} x {11} = {3111111} + {321111} + {32211} + {411111} + {42111}

{31111} x {3} = {331111} + {421111} + {43111} + {511111} + {52111} + {61111}

{31111} x {21} = {3211111} + {322111} + {331111} + {33211} + {4111111} +2{421111} + {42211}
+ {43111} + {511111} + {52111}

{31111} x {111} = {31111111} + {3211111} + {322111} + {32221} + {4111111} + {421111} + {42211}

{31111} x {4} = {431111} + {521111} + {53111} + {611111} + {62111} + {71111}

{31111} x {31} = {3311111} + {332111} + {4211111} + {422111} +2{431111} + {43211} + {44111}
+ {5111111} +2{521111} + {52211} + {53111} + {611111} + {62111}

{31111} x {22} = {3221111} + {332111} + {33311} + {4211111} + {422111} + {431111} + {43211}
+ {521111} + {53111}

{31111} x {211} = {32111111} + {3221111} + {322211} + {3311111} + {332111} + {33221} + {41111111}
+2{4211111} +2{422111} + {42221} + {431111} + {43211} + {5111111} + {521111}
+ {52211}

{31111} x {1111} = {311111111} + {32111111} + {3221111} + {322211} + {32222} + {41111111}
+ {4211111} + {422111} + {42221}

{31111} x {41} = {4311111} + {432111} + {441111} + {5211111} + {522111} +2{531111} + {53211}
+ {54111} + {6111111} +2{621111} + {62211} + {63111} + {711111} + {72111}

{31111} x {32} = {3321111} + {333111} + {4221111} + {4311111} +2{432111} + {43311} + {441111}
+ {44211} + {5211111} + {522111} +2{531111} + {53211} + {54111} + {621111} + {63111}

{31111} x {311} = {33111111} + {3321111} + {332211} + {42111111} + {4221111} + {422211} +2{4311111}
+2{432111} + {43221} + {441111} + {44211} + {51111111} +2{5211111} +2{522111}
+ {52221} + {531111} + {53211} + {6111111} + {621111} + {62211}

{31111} x {221} = {32211111} + {3222111} + {3321111} + {332211} + {333111} + {33321} + {42111111}
+2{4221111} + {422211} + {4311111} +2{432111} + {43221} + {43311} + {5211111}
+ {522111} + {531111} + {53211}

{31111} x {2111} = {321111111} + {32211111} + {3222111} + {322221} + {33111111} + {3321111}
+ {332211} + {33222} + {411111111} +2{42111111} +2{4221111} +2{422211} + {42222}
+ {4311111} + {432111} + {43221} + {51111111} + {5211111} + {522111} + {52221}

{31111} x {11111} = {3111111111} + {321111111} + {32211111} + {3222111} + {322221} + {411111111}
+ {42111111} + {4221111} + {422211} + {42222}

{31111} x {42} = {4321111} + {433111} + {4411111} + {442111} + {5221111} + {5311111} +2{532111}
+ {53311} +2{541111} + {54211} + {55111} + {6211111} + {622111} +2{631111} + {63211}
+ {64111} + {721111} + {73111}

{31111} x {411} = {43111111} + {4321111} + {432211} + {4411111} + {442111} + {52111111} + {5221111}
+ {522211} +2{5311111} +2{532111} + {53221} + {541111} + {54211} + {61111111}
+2{6211111} +2{622111} + {62221} + {631111} + {63211} + {7111111} + {721111}
+ {72211}

{31111} x {33} = {3331111} + {4321111} + {433111} + {442111} + {44311} + {5311111} + {532111}
+ {541111} + {54211} + {631111} + {64111}

{31111} x {321} = {33211111} + {3322111} + {3331111} + {333211} + {42211111} + {4222111}
+ {43111111} +3{4321111} +2{432211} +2{433111} + {43321} + {4411111} +2{442111}
+ {44221} + {44311} + {52111111} +2{5221111} + {522211} +2{5311111} +3{532111}
+ {53221} + {53311} + {541111} + {54211} + {6211111} + {622111} + {631111} + {63211}

{31111} x {3111} = {331111111} + {33211111} + {3322111} + {332221} + {421111111} + {42211111}
+ {4222111} + {422221} +2{43111111} +2{4321111} +2{432211} + {43222} + {4411111}
+ {442111} + {44221} + {511111111} +2{52111111} +2{5221111} +2{522211} + {52222}
+ {5311111} + {532111} + {53221} + {61111111} + {6211111} + {622111} + {62221}

{31111} x {222} = {3222111} + {3322111} + {333211} + {33331} + {42211111} + {4222111} + {4321111}
+ {432211} + {433111} + {43321} + {5221111} + {532111} + {53311}

{31111} x {2211} = {322111111} + {32221111} + {3222211} + {33211111} + {3322111} + {332221}
+ {3331111} + {333211} + {33322} + {421111111} +2{42211111} +2{4222111} + {422221}
+ {43111111} +2{4321111} +2{432211} + {43222} + {433111} + {43321} + {52111111}
+ {5221111} + {522211} + {5311111} + {532111} + {53221}

{31111} x {21111} = {3211111111} + {322111111} + {32221111} + {3222211} + {322222} + {331111111}
+ {33211111} + {3322111} + {332221} + {4111111111} +2{421111111} +2{42211111}
+2{4222111} +2{422221} + {43111111} + {4321111} + {432211} + {43222} + {511111111}
+ {52111111} + {5221111} + {522211} + {52222}

{31111} x {111111} = {31111111111} + {3211111111} + {322111111} + {32221111} + {3222211}
+ {4111111111} + {421111111} + {42211111} + {4222111} + {422221}

{31111} x {31111} = {3311111111} + {332111111} + {33221111} + {3322211} + {332222} + {4211111111}
+ {422111111} + {42221111} + {4222211} + {422222} +2{431111111} +2{43211111}
+2{4322111} +2{432221} + {44111111} + {4421111} + {442211} + {44222}
+ {5111111111} +2{521111111} +2{52211111} +2{5222111} +2{522221} + {53111111}
+ {5321111} + {532211} + {53222} + {611111111} + {62111111} + {6221111} + {622211}
+ {62222}

{31111} x {2221} = {322211111} + {32222111} + {33221111} + {3322211} + {3332111} + {333221}
+ {333311} + {33332} + {422111111} +2{42221111} + {4222211} + {43211111}
+2{4322111} + {432221} + {4331111} +2{433211} + {43322} + {43331} + {52211111}
+ {5222111} + {5321111} + {532211} + {533111} + {53321}

{31111} x {22111} = {3221111111} + {322211111} + {32222111} + {3222221} + {332111111} + {33221111}
+ {3322211} + {332222} + {33311111} + {3332111} + {333221} + {4211111111}
+2{422111111} +2{42221111} +2{4222211} + {422222} + {431111111} +2{43211111}
+2{4322111} +2{432221} + {4331111} + {433211} + {43322} + {521111111} + {52211111}
+ {5222111} + {522221} + {53111111} + {5321111} + {532211} + {53222}

{2221} x {211111} = {2222111111} + {222221111} + {3221111111} +2{322211111} + {32222111}
+ {332111111} +2{33221111} + {3322211} + {33311111} +2{3332111} + {333221}
+ {333311} + {422111111} + {42221111} + {43211111} + {4322111} + {4331111}
+ {433211}

{2221} x {1111111} = {22211111111} + {2222111111} + {3221111111} + {322211111} + {332111111}
+ {33221111} + {33311111} + {3332111}

{22111} x {0} = {22111}

{22111} x {1} = {221111} + {22211} + {32111}

{22111} x {2} = {222111} + {321111} + {32211} + {42111}

{22111} x {11} = {2211111} + {222111} + {22221} + {321111} + {32211} + {33111}

{22111} x {3} = {322111} + {421111} + {42211} + {52111}

{22111} x {21} = {2221111} + {222211} + {3211111} +2{322111} + {32221} + {331111} + {33211}
+ {421111} + {42211} + {43111}

{22111} x {111} = {22111111} + {2221111} + {222211} + {22222} + {3211111} + {322111} + {32221}
+ {331111} + {33211}

{22111} x {4} = {422111} + {521111} + {52211} + {62111}

{22111} x {31} = {3221111} + {322211} + {332111} + {4211111} +2{422111} + {42221} + {431111}
+ {43211} + {521111} + {52211} + {53111}

{22111} x {22} = {2222111} + {3221111} + {322211} + {3311111} + {332111} + {33221} + {422111}
+ {431111} + {43211} + {44111}

{22111} x {211} = {22211111} + {2222111} + {222221} + {32111111} +2{3221111} +2{322211} + {32222}
+ {3311111} +2{332111} + {33221} + {33311} + {4211111} + {422111} + {42221}
+ {431111} + {43211}

{22111} x {1111} = {221111111} + {22211111} + {2222111} + {222221} + {32111111} + {3221111}
+ {322211} + {32222} + {3311111} + {332111} + {33221}

{22111} x {41} = {4221111} + {422211} + {432111} + {5211111} +2{522111} + {52221} + {531111}
+ {53211} + {621111} + {62211} + {63111}

{22111} x {32} = {3222111} + {3321111} + {332211} + {4221111} + {422211} + {4311111} +2{432111}
+ {43221} + {441111} + {44211} + {522111} + {531111} + {53211} + {54111}

{22111} x {311} = {32211111} + {3222111} + {322221} + {3321111} + {332211} + {333111} + {42111111}
+2{4221111} +2{422211} + {42222} + {4311111} +2{432111} + {43221} + {43311}
+ {5211111} + {522111} + {52221} + {531111} + {53211}

{22111} x {221} = {22221111} + {2222211} + {32211111} +2{3222111} + {322221} + {33111111}
+2{3321111} +2{332211} + {33222} + {333111} + {33321} + {4221111} + {422211}
+ {4311111} +2{432111} + {43221} + {43311} + {441111} + {44211}

{22111} x {2111} = {222111111} + {22221111} + {2222211} + {222222} + {321111111} +2{32211111}
+2{3222111} +2{322221} + {33111111} +2{3321111} +2{332211} + {33222} + {333111}
+ {33321} + {42111111} + {4221111} + {422211} + {42222} + {4311111} + {432111}
+ {43221}

{22111} x {11111} = {2211111111} + {222111111} + {22221111} + {2222211} + {321111111} + {32211111}
+ {3222111} + {322221} + {33111111} + {3321111} + {332211} + {33222}

{22111} x {42} = {4222111} + {4321111} + {432211} + {442111} + {5221111} + {522211} + {5311111}
+2{532111} + {53221} + {541111} + {54211} + {622111} + {631111} + {63211} + {64111}

{22111} x {411} = {42211111} + {4222111} + {422221} + {4321111} + {432211} + {433111} + {52111111}
+2{5221111} +2{522211} + {52222} + {5311111} +2{532111} + {53221} + {53311}
+ {6211111} + {622111} + {62221} + {631111} + {63211}

{22111} x {33} = {3322111} + {4321111} + {432211} + {4411111} + {442111} + {44221} + {532111}
+ {541111} + {54211} + {55111}

{22111} x {321} = {32221111} + {3222211} + {33211111} +2{3322111} + {332221} + {3331111} + {333211}
+ {42211111} +2{4222111} + {422221} + {43111111} +3{4321111} +3{432211} + {43222}
+2{433111} + {43321} + {4411111} +2{442111} + {44221} + {44311} + {5221111}
+ {522211} + {5311111} +2{532111} + {53221} + {53311} + {541111} + {54211}

{22111} x {3111} = {322111111} + {32221111} + {3222211} + {322222} + {33211111} + {3322111}
+ {332221} + {3331111} + {333211} + {421111111} +2{42211111} +2{4222111} +2{422221}
+ {43111111} +2{4321111} +2{432211} + {43222} + {433111} + {43321} + {52111111}
+ {5221111} + {522211} + {52222} + {5311111} + {532111} + {53221}

{22111} x {222} = {22222111} + {32221111} + {3222211} + {33211111} + {3322111} + {332221}
+ {3331111} + {333211} + {33322} + {4222111} + {4321111} + {432211} + {433111}
+ {43321} + {442111} + {44311}

{22111} x {2211} = {222211111} + {22222111} + {2222221} + {322111111} +2{32221111} +2{3222211}
+ {322222} + {331111111} +2{33211111} +3{3322111} +2{332221} + {3331111} +2{333211}
+ {33322} + {33331} + {42211111} + {4222111} + {422221} + {43111111} +2{4321111}
+2{432211} + {43222} + {433111} + {43321} + {4411111} + {442111} + {44221}

{22111} x {21111} = {2221111111} + {222211111} + {22222111} + {2222221} + {3211111111} +2{322111111}
+2{32221111} +2{3222211} + {322222} + {331111111} +2{33211111} +2{3322111}
+2{332221} + {3331111} + {333211} + {33322} + {421111111} + {42211111} + {4222111}
+ {422221} + {43111111} + {4321111} + {432211} + {43222}

{22111} x {111111} = {22111111111} + {2221111111} + {222211111} + {22222111} + {3211111111}
+ {322111111} + {32221111} + {3222211} + {331111111} + {33211111} + {3322111}
+ {332221}

{322} x {3211} = {3322211} + {3332111} + {333221} + {333311} + {33332} + {4222211} +2{4322111}
+2{432221} + {4331111} +3{433211} +2{43322} +2{43331} + {4421111} +2{442211}
+ {44222} +2{443111} +3{44321} + {4433} + {44411} + {4442} + {5222111} + {522221}
+ {5321111} +3{532211} + {53222} +2{533111} +3{53321} + {5333} +2{542111} +2{54221}
+3{54311} +2{5432} + {5441} + {55211} + {5531} + {622211} + {632111} + {63221}
+ {63311} + {6332} + {64211} + {6431}

{322} x {31111} = {33221111} + {3332111} + {333311} + {42221111} + {43211111} +2{4322111}
+ {4331111} +2{433211} + {43331} + {4421111} + {442211} + {443111} + {44321}
+ {52211111} + {5222111} +2{5321111} + {532211} +2{533111} + {53321} + {542111}
+ {54311} + {6221111} + {632111} + {63311}

{322} x {2221} = {3222221} + {3322211} + {332222} + {3332111} + {333221} + {4222211} + {422222}
+ {4322111} +2{432221} + {4331111} +2{433211} + {43322} + {442211} + {44222}
+ {443111} +2{44321} + {44411} + {4442} + {522221} + {532211} + {53222} + {533111}
+ {53321} + {54221} + {54311} + {5432} + {5441}

{322} x {22111} = {32222111} + {33221111} + {3322211} + {33311111} + {3332111} + {333221}
+ {42221111} + {4222211} + {43211111} +2{4322111} + {432221} +2{4331111} +2{433211}
+ {43322} + {4421111} + {442211} +2{443111} + {44321} + {44411} + {5222111}
+ {5321111} + {532211} + {533111} + {53321} + {542111} + {54311}

{322} x {211111} = {322211111} + {332111111} + {33221111} + {33311111} + {3332111} + {422111111}
+ {42221111} +2{43211111} + {4322111} +2{4331111} + {433211} + {4421111} + {443111}
+ {52211111} + {5321111} + {533111}

{322} x {1111111} = {3221111111} + {332111111} + {33311111} + {422111111} + {43211111} + {4331111}

{3211} x {0} = {3211}

{3211} x {1} = {32111} + {3221} + {3311} + {4211}

{3211} x {2} = {32211} + {33111} + {3321} + {42111} + {4221} + {4311} + {5211}

{3211} x {11} = {321111} + {32211} + {3222} + {33111} + {3321} + {42111} + {4221} + {4311}

{3211} x {3} = {33211} + {42211} + {43111} + {4321} + {52111} + {5221} + {5311} + {6211}

{3211} x {21} = {322111} + {32221} + {331111} +2{33211} + {3322} + {3331} + {421111} +2{42211}
+ {4222} +2{43111} +2{4321} + {4411} + {52111} + {5221} + {5311}

{3211} x {111} = {3211111} + {322111} + {32221} + {331111} + {33211} + {3322} + {421111} + {42211}
+ {4222} + {43111} + {4321}

{3211} x {4} = {43211} + {52211} + {53111} + {5321} + {62111} + {6221} + {6311} + {7211}

{3211} x {31} = {332111} + {33221} + {33311} + {422111} + {42221} + {431111} +3{43211} + {4322}
+ {4331} + {44111} + {4421} + {521111} +2{52211} + {5222} +2{53111} +2{5321} + {5411}
+ {62111} + {6221} + {6311}

{3211} x {22} = {322211} + {332111} + {33221} + {33311} + {3332} + {422111} + {42221} + {431111}
+2{43211} + {4322} + {4331} + {44111} + {4421} + {52211} + {53111} + {5321} + {5411}

{3211} x {211} = {3221111} + {322211} + {32222} + {3311111} +2{332111} +2{33221} + {33311} + {3332}
+ {4211111} +2{422111} +2{42221} +2{431111} +3{43211} +2{4322} + {4331} + {44111}
+ {4421} + {521111} + {52211} + {5222} + {53111} + {5321}

{3211} x {1111} = {32111111} + {3221111} + {322211} + {3311111} + {332111} + {33221} + {4211111}
+ {422111} + {42221} + {431111} + {43211} + {4322}

{3211} x {41} = {432111} + {43221} + {43311} + {44211} + {522111} + {52221} + {531111} +3{53211}
+ {5322} + {5331} + {54111} + {5421} + {621111} +2{62211} + {6222} +2{63111} +2{6321}
+ {6411} + {72111} + {7221} + {7311}

{3211} x {32} = {332211} + {333111} + {33321} + {422211} +2{432111} +2{43221} +2{43311} + {4332}
+ {441111} +2{44211} + {4422} + {4431} + {522111} + {52221} + {531111} +3{53211}
+ {5322} + {5331} +2{54111} +2{5421} + {5511} + {62211} + {63111} + {6321} + {6411}

{3211} x {311} = {3321111} + {332211} + {33222} + {333111} + {33321} + {4221111} + {422211}
+ {42222} + {4311111} +3{432111} +3{43221} +2{43311} + {4332} + {441111} +2{44211}
+ {4422} + {4431} + {5211111} +2{522111} +2{52221} +2{531111} +3{53211} +2{5322}
+ {5331} + {54111} + {5421} + {621111} + {62211} + {6222} + {63111} + {6321}

{3211} x {221} = {3222111} + {322221} + {3321111} +2{332211} + {33222} + {333111} +2{33321} + {3333}
+ {4221111} +2{422211} + {42222} + {4311111} +3{432111} +3{43221} +2{43311} +2{4332}
+ {441111} +2{44211} + {4422} + {4431} + {522111} + {52221} + {531111} +2{53211}
+ {5322} + {5331} + {54111} + {5421}

{3211} x {2111} = {32211111} + {3222111} + {322221} + {33111111} +2{3321111} +2{332211} + {33222}
+ {333111} + {33321} + {42111111} +2{4221111} +2{422211} + {42222} +2{4311111}
+3{432111} +3{43221} + {43311} + {4332} + {441111} + {44211} + {4422} + {5211111}
+ {522111} + {52221} + {531111} + {53211} + {5322}

{3211} x {11111} = {321111111} + {32211111} + {3222111} + {33111111} + {3321111} + {332211}
+ {42111111} + {4221111} + {422211} + {4311111} + {432111} + {43221}

{3211} x {42} = {432211} + {433111} + {43321} + {442111} + {44221} + {44311} + {522211} +2{532111}
+2{53221} +2{53311} + {5332} + {541111} +3{54211} + {5422} + {5431} + {55111} + {5521}
+ {622111} + {62221} + {631111} +3{63211} + {6322} + {6331} +2{64111} +2{6421}
+ {6511} + {72211} + {73111} + {7321} + {7411}

{3211} x {411} = {4321111} + {432211} + {43222} + {433111} + {43321} + {442111} + {44221} + {44311}
+ {5221111} + {522211} + {52222} + {5311111} +3{532111} +3{53221} +2{53311} + {5332}
+ {541111} +2{54211} + {5422} + {5431} + {6211111} +2{622111} +2{62221} +2{631111}
+3{63211} +2{6322} + {6331} + {64111} + {6421} + {721111} + {72211} + {7222}
+ {73111} + {7321}

B-1 Outer Multiplication of S-Functions (contd).

{22111} x {22111} = {2222111111} + {222221111} + {22222211} + {2222222} + {3221111111} +2{322211111}
+2{32222111} +2{3222221} + {3311111111} +2{332111111} +3{33221111} +3{3322211}
+ {332222} + {33311111} +2{3332111} +2{333221} + {333311} + {33332} + {422111111}
+ {42221111} + {4222211} + {422222} + {431111111} +2{43211111} +2{4322111}
+2{432221} + {4331111} + {433211} + {43322} + {44111111} + {4421111} + {442211}
+ {44222}

{22111} x {211111} = {22211111111} + {2222111111} + {222221111} + {22222211} + {32111111111}
+2{3221111111} +2{322211111} +2{32222111} + {3222221} + {3311111111}
+2{332111111} +2{33221111} +2{3322211} + {332222} + {33311111} + {3332111}
+ {333221} + {4211111111} + {422111111} + {42221111} + {4222211} + {431111111}
+ {43211111} + {4322111} + {432221}

{22111} x {1111111} = {221111111111} + {22211111111} + {2222111111} + {222221111} + {32111111111}
+ {3221111111} + {322211111} + {32222111} + {3311111111} + {332111111}
+ {33221111} + {3322211}

{211111} x {0} = {211111}

{211111} x {1} = {2111111} + {221111} + {311111}

{211111} x {2} = {2211111} + {3111111} + {321111} + {411111}

{211111} x {11} = {21111111} + {2211111} + {222111} + {3111111} + {321111}

{211111} x {3} = {3211111} + {4111111} + {421111} + {511111}

{211111} x {21} = {22111111} + {2221111} + {31111111} +2{3211111} + {322111} + {331111} + {4111111}
+ {421111}

{211111} x {111} = {211111111} + {22111111} + {2221111} + {222211} + {31111111} + {3211111}
+ {322111}

{211111} x {4} = {4211111} + {5111111} + {521111} + {611111}

{211111} x {31} = {32111111} + {3221111} + {3311111} + {41111111} +2{4211111} + {422111} + {431111}
+ {5111111} + {521111}

{211111} x {22} = {22211111} + {32111111} + {3221111} + {3311111} + {332111} + {4211111} + {431111}

{211111} x {211} = {221111111} + {22211111} + {2222111} + {311111111} +2{32111111} +2{3221111}
+ {322211} + {3311111} + {332111} + {41111111} + {4211111} + {422111}

{211111} x {1111} = {2111111111} + {221111111} + {22211111} + {2222111} + {222221} + {311111111}
+ {32111111} + {3221111} + {322211}

{211111} x {41} = {42111111} + {4221111} + {4311111} + {51111111} +2{5211111} + {522111} + {531111}
+ {6111111} + {621111}

{211111} x {32} = {32211111} + {33111111} + {3321111} + {42111111} + {4221111} +2{4311111}
+ {432111} + {441111} + {5211111} + {531111}

{211111} x {311} = {321111111} + {32211111} + {3222111} + {33111111} + {3321111} + {411111111}
+2{42111111} +2{4221111} + {422211} + {4311111} + {432111} + {51111111} + {5211111}
+ {522111}

{211111} x {221} = {222111111} + {22221111} + {321111111} +2{32211111} + {3222111} + {33111111}
+2{3321111} + {332211} + {333111} + {42111111} + {4221111} + {4311111} + {432111}

{211111} x {2111} = {2211111111} + {222111111} + {22221111} + {2222211} + {3111111111} +2{321111111}
+2{32211111} +2{3222111} + {322221} + {33111111} + {3321111} + {332211}
+ {411111111} + {42111111} + {4221111} + {422211}

{211111} x {11111} = {21111111111} + {2211111111} + {222111111} + {22221111} + {2222211} + {222222}
+ {3111111111} + {321111111} + {32211111} + {3222111} + {322221}

{211111} x {42} = {42211111} + {43111111} + {4321111} + {4411111} + {52111111} + {5221111}
+2{5311111} + {532111} + {541111} + {6211111} + {631111}

{211111} x {411} = {421111111} + {42211111} + {4222111} + {43111111} + {4321111} + {511111111}
+2{52111111} +2{5221111} + {522211} + {5311111} + {532111} + {61111111} + {6211111}
+ {622111}

{211111} x {33} = {33211111} + {43111111} + {4321111} + {4411111} + {442111} + {5311111} + {541111}

{211111} x {321} = {322111111} + {32221111} + {331111111} +2{33211111} + {3322111} + {3331111}
+ {421111111} +2{42211111} + {4222111} +2{43111111} +3{4321111} + {432211}
+ {433111} + {4411111} + {442111} + {52111111} + {5221111} + {5311111} + {532111}

{211111} x {3111} = {3211111111} + {322111111} + {32221111} + {3222211} + {331111111} + {33211111}
+ {3322111} + {4111111111} +2{421111111} +2{42211111} +2{4222111} + {422221}
+ {43111111} + {4321111} + {432211} + {511111111} + {52111111} + {5221111}
+ {522211}

{211111} x {222} = {222211111} + {322111111} + {32221111} + {33211111} + {3322111} + {3331111}
+ {333211} + {42211111} + {4321111} + {433111}

{211111} x {2211} = {2221111111} + {222211111} + {22222111} + {3211111111} +2{322111111}
+2{32221111} + {3222211} + {331111111} +2{33211111} +2{3322111} + {332221}
+ {3331111} + {333211} + {421111111} + {42211111} + {4222111} + {43111111}
+ {4321111} + {432211}

{211111} x {21111} = {22111111111} + {2221111111} + {222211111} + {22222111} + {2222221}
+ {31111111111} +2{3211111111} +2{322111111} +2{32221111} +2{3222211} + {322222}
+ {331111111} + {33211111} + {3322111} + {332221} + {4111111111} + {421111111}
+ {42211111} + {4222111} + {422221}

{211111} x {111111} = {211111111111} + {22111111111} + {2221111111} + {222211111} + {22222111}
+ {2222221} + {31111111111} + {3211111111} + {322111111} + {32221111}
+ {3222211} + {322222}

{211111} x {211111} = {221111111111} + {22211111111} + {2222111111} + {222221111} + {22222211}
+ {2222222} + {311111111111} +2{32111111111} +2{3221111111} +2{322211111}
+2{32222111} +2{3222221} + {3311111111} + {332111111} + {33221111} + {3322211}
+ {332222} + {41111111111} + {4211111111} + {422111111} + {42221111} + {4222211}
+ {422222}

{211111} x {1111111} = {2111111111111} + {221111111111} + {22211111111} + {2222111111} + {222221111}
+ {22222211} + {311111111111} + {32111111111} + {3221111111} + {322211111}
+ {32222111} + {3222221}

{1111111} x {0} = {1111111}

{1111111} x {1} = {11111111} + {2111111}

{1111111} x {2} = {21111111} + {3111111}

{1111111} x {11} = {111111111} + {21111111} + {2211111}

{1111111} x {3} = {31111111} + {4111111}

{1111111} x {21} = {211111111} + {22111111} + {31111111} + {3211111}

{1111111} x {111} = {1111111111} + {211111111} + {22111111} + {2221111}

{1111111} x {4} = {41111111} + {5111111}

{1111111} x {31} = {311111111} + {32111111} + {41111111} + {4211111}

{1111111} x {22} = {221111111} + {32111111} + {3311111}

{1111111} x {211} = {2111111111} + {221111111} + {22211111} + {311111111} + {32111111} + {3221111}

{1111111} x {1111} = {11111111111} + {2111111111} + {221111111} + {22211111} + {2222111}

{1111111} x {41} = {411111111} + {42111111} + {51111111} + {5211111}

{1111111} x {32} = {321111111} + {33111111} + {42111111} + {4311111}

{1111111} x {311} = {3111111111} + {321111111} + {32211111} + {411111111} + {42111111} + {4221111}

{1111111} x {221} = {2211111111} + {222111111} + {321111111} + {32211111} + {33111111} + {3321111}

{1111111} x {2111} = {21111111111} + {2211111111} + {222111111} + {22221111} + {3111111111}
+ {321111111} + {32211111} + {3222111}

{1111111} x {11111} = {111111111111} + {21111111111} + {2211111111} + {222111111} + {22221111}
+ {2222211}

{1111111} x {42} = {421111111} + {43111111} + {52111111} + {5311111}

{1111111} x {411} = {4111111111} + {421111111} + {42211111} + {511111111} + {52111111} + {5221111}

{1111111} x {33} = {331111111} + {43111111} + {4411111}

{1111111} x {321} = {3211111111} + {322111111} + {331111111} + {33211111} + {421111111} + {42211111}
+ {43111111} + {4321111}

{1111111} x {3111} = {31111111111} + {3211111111} + {322111111} + {32221111} + {4111111111}
+ {421111111} + {42211111} + {4222111}

{1111111} x {222} = {2221111111} + {322111111} + {33211111} + {3331111}

{1111111} x {2211} = {22111111111} + {2221111111} + {222211111} + {3211111111} + {322111111}
+ {32221111} + {331111111} + {33211111} + {3322111}

{1111111} x {21111} = {211111111111} + {22111111111} + {2221111111} + {222211111} + {22222111}
+ {31111111111} + {3211111111} + {322111111} + {32221111} + {3222211}

{1111111} x {111111} = {1111111111111} + {211111111111} + {22111111111} + {2221111111} + {222211111}
+ {22222111} + {2222221}

{1111111} x {1111111} = {11111111111111} + {2111111111111} + {221111111111} + {22211111111}
+ {2222111111} + {222221111} + {22222211} + {2222222}

{21} o {21} = {111} + {21} + {3}

{31} o {31} = {211} + {22} + {31} + {4}
{31} o {22} = {211} + {31}
{22} o {22} = {1111} + {22} + {4}

{41} o {41} = {311} + {32} + {41} + {5}
{41} o {32} = {221} + {311} + {32} + {41}
{41} o {311} = {2111} + {221} + {311} + {32} + {41}
{32} o {32} = {2111} + {221} + {311} + {32} + {41} + {5}
{32} o {311} = {2111} + {221} +2{311} + {32} + {41}
{311} o {311} = {11111} + {2111} +2{221} + {311} +2{32} + {41} + {5}

{51} o {51} = {411} + {42} + {51} + {6}
{51} o {42} = {321} + {33} + {411} + {42} + {51}
{51} o {411} = {3111} + {321} + {411} + {42} + {51}
{51} o {33} = {321} + {42}
{51} o {321} = {2211} + {222} + {3111} +2{321} + {33} + {411} + {42}
{42} o {42} = {222} + {3111} +2{321} + {411} +2{42} + {51} + {6}
{42} o {411} = {2211} + {3111} +2{321} + {33} +2{411} + {42} + {51}
{42} o {33} = {2211} + {321} + {33} + {411} + {51}
{42} o {321} = {21111} +2{2211} + {222} +2{3111} +3{321} + {33} +2{411} +2{42} + {51}
{411} o {411} = {21111} + {2211} + {222} + {3111} +2{321} + {33} + {411} +2{42} + {51} + {6}
{411} o {33} = {222} + {3111} + {321} + {411} + {42}
{411} o {321} = {21111} +2{2211} + {222} +2{3111} +4{321} + {33} +2{411} +2{42} + {51}
{33} o {33} = {222} + {3111} + {42} + {6}
{33} o {321} = {21111} + {2211} + {3111} +2{321} + {411} + {42} + {51}
{321} o {321} = {111111} +2{21111} +3{2211} +2{222} +4{3111} +5{321} +2{33} +4{411} +3{42} +2{51}
+ {6}

{61} o {61} = {511} + {52} + {61} + {7}
{61} o {52} = {421} + {43} + {511} + {52} + {61}
{61} o {511} = {4111} + {421} + {511} + {52} + {61}
{61} o {43} = {331} + {421} + {43} + {52}
{61} o {421} = {3211} + {322} + {331} + {4111} +2{421} + {43} + {511} + {52}
{61} o {4111} = {31111} + {3211} + {4111} + {421} + {511}
{61} o {331} = {3211} + {322} + {331} + {421} + {43}
{52} o {52} = {322} + {331} + {4111} +2{421} + {43} + {511} +2{52} + {61} + {7}
{52} o {511} = {3211} + {331} + {4111} +2{421} + {43} +2{511} + {52} + {61}
{52} o {43} = {3211} + {322} + {331} +2{421} + {43} + {511} + {52} + {61}
{52} o {421} = {2221} + {31111} +3{3211} +2{322} +2{331} +2{4111} +4{421} +2{43} +2{511} +2{52}
+ {61}
{52} o {4111} = {22111} + {31111} +2{3211} + {322} + {331} +2{4111} +2{421} + {511} + {52}
{52} o {331} = {22111} + {2221} +2{3211} + {322} +2{331} + {4111} +2{421} + {43} + {511} + {52}
{511} o {511} = {31111} + {3211} + {322} + {4111} +2{421} + {43} + {511} +2{52} + {61} + {7}
{511} o {43} = {3211} + {322} + {331} + {4111} +2{421} + {43} + {511} + {52}
{511} o {421} = {22111} + {2221} + {31111} +3{3211} +2{322} +3{331} +2{4111} +4{421} +2{43} +2{511}
+2{52} + {61}
{511} o {4111} = {211111} + {22111} + {2221} + {31111} +2{3211} + {322} + {331} + {4111} +2{421}
+ {43} + {511} + {52} + {61}
{511} o {331} = {2221} + {31111} +2{3211} +2{322} + {331} + {4111} +3{421} + {43} + {52}
{43} o {43} = {2221} + {3211} + {322} + {331} + {4111} + {421} + {43} + {511} + {52} + {61} + {7}
{43} o {421} = {22111} + {2221} + {31111} +3{3211} +2{322} +2{331} +2{4111} +4{421} + {43} +2{511}
+2{52} + {61}
{43} o {4111} = .{2221} + {31111} +2{3211} + {322} + {331} +2{4111} +2{421} + {43} + {511}
{43} o {331} = {22111} + {2221} + {31111} +2{3211} + {322} + {331} + {4111} +2{421} + {43} + {511}
+ {52} + {61}
{421} o {421} = {211111} +3{22111} +3{2221} +3{31111} +8{3211} +5{322} +5{331} +5{4111} +9{421}
+4{43} +4{511} +4{52} +2{61} + {7}
{421} o {4111} = {211111} +2{22111} +2{2221} +2{31111} +5{3211} +3{322} +3{331} +2{4111} +5{421}
+2{43} +2{511} +2{52} + {61}
{421} o {331} = {211111} +2{22111} +2{2221} +2{31111} +5{3211} +3{322} +3{331} +3{4111} +5{421}
+2{43} +3{511} +2{52} + {61}
{4111} o {4111} = {1111111} + {211111} +2{22111} +2{2221} + {31111} +2{3211} +2{322} +2{331}
+ {4111} +2{421} +2{43} + {511} +2{52} + {61} + {7}
{4111} o {331} = {22111} + {2221} + {31111} +3{3211} +2{322} +2{331} +2{4111} +3{421} + {43} + {511}
+ {52}
{331} o {331} = {211111} + {22111} + {2221} +2{31111} +3{3211} +2{322} + {331} +2{4111} +3{421}
+ {43} + {511} +2{52} + {61} + {7}

{71} o {71} = {611} + {62} + {71} + {8}

{71} o {62} = {521} + {53} + {611} + {62} + {71}

{71} o {611} = {5111} + {521} + {611} + {62} + {71}

{71} o {53} = {431} + {44} + {521} + {53} + {62}

{71} o {521} = {4211} + {422} + {431} + {5111} +2{521} + {53} + {611} + {62}

{71} o {5111} = {41111} + {4211} + {5111} + {521} + {611}

{71} o {44} = {431} + {53}

{71} o {431} = {3311} + {332} + {4211} + {422} +2{431} + {44} + {521} + {53}

{71} o {422} = {3221} + {332} + {4211} + {422} + {431} + {521}

{71} o {4211} = {32111} + {3221} + {3311} + {41111} +2{4211} + {422} + {431} + {5111} + {521}

{71} o {332} = {3221} + {3311} + {332} + {422} + {431}

{62} o {62} = {422} + {431} + {44} + {5111} +2{521} + {53} + {611} +2{62} + {71} + {8}

{62} o {611} = {4211} + {431} + {5111} +2{521} + {53} +2{611} + {62} + {71}

{62} o {53} = {332} + {4211} + {422} +2{431} +2{521} +2{53} + {611} + {62} + {71}

{62} o {521} = {3221} + {3311} + {332} + {41111} +3{4211} +2{422} +3{431} + {44} +2{5111} +4{521}
+2{53} +2{611} +2{62} + {71}

{62} o {5111} = {32111} + {3311} + {41111} +2{4211} + {422} + {431} +2{5111} +2{521} + {611} + {62}

{62} o {44} = {3311} + {422} + {431} + {44} + {521} + {62}

{62} o {431} = {32111} +2{3221} +2{3311} +2{332} +3{4211} +2{422} +4{431} + {44} + {5111} +3{521}
+2{53} + {611} + {62}

{62} o {422} = {2222} + {32111} +2{3221} +2{3311} + {332} + {41111} +2{4211} +3{422} +2{431} + {44}
+ {5111} +2{521} + {53} + {62}

{62} o {4211} = {22211} + {311111} +3{32111} +3{3221} +2{3311} +2{332} +2{41111} +5{4211} +2{422}
+3{431} +2{5111} +3{521} + {53} + {611}

{62} o {332} = {22211} + {32111} +2{3221} + {3311} +2{332} +2{4211} + {422} +2{431} + {521} + {53}

{611} o {611} = {41111} + {4211} + {422} + {5111} +2{521} + {53} + {611} +2{62} + {71} + {8}

{611} o {53} = {3311} + {4211} + {422} +2{431} + {44} + {5111} +2{521} + {53} + {611} + {62}

{611} o {521} = {32111} + {3221} + {3311} + {332} + {41111} +3{4211} +2{422} +3{431} + {44} +2{5111}
+4{521} +2{53} +2{611} +2{62} + {71}

{611} o {5111} = {311111} + {32111} + {3221} + {41111} +2{4211} + {422} + {431} + {5111} +2{521}
+ {53} + {611} + {62} + {71}

{611} o {44} = {332} + {4211} + {431} + {521} + {53}

{611} o {431} = {32111} +2{3221} +2{3311} +2{332} + {41111} +3{4211} +3{422} +4{431} + {44} + {5111}
+3{521} +2{53} + {62}

{611} o {422} = {22211} + {32111} +2{3221} +2{3311} +2{332} +3{4211} + {422} +3{431} + {5111}
+2{521} + {53} + {611}

{611} o {4211} = {221111} + {22211} + {2222} + {311111} +3{32111} +3{3221} +3{3311} +2{332}
+2{41111} +4{4211} +3{422} +3{431} + {44} +2{5111} +3{521} + {53} + {611} + {62}

{611} o {332} = {2222} + {32111} +2{3221} +2{3311} + {332} +2{4211} +2{422} +2{431} + {44} + {521}

{53} o {53} = {3221} + {3311} + {332} + {4211} +2{422} +2{431} + {44} + {5111} +2{521} + {53}
+ {611} +2{62} + {71} + {8}

{53} o {521} = {32111} +2{3221} +2{3311} +2{332} + {41111} +4{4211} +3{422} +4{431} + {44} +2{5111}
+5{521} +2{53} +2{611} +2{62} + {71}

{53} o {5111} = {32111} + {3221} + {3311} + {332} + {41111} +3{4211} + {422} +2{431} +2{5111}
+2{521} + {53} + {611}

{53} o {44} = {3221} + {332} + {4211} + {431} + {521} + {53} + {611} + {71}

{53} o {431} = {22211} + {2222} +2{32111} +3{3221} +3{3311} +2{332} + {41111} +4{4211} +3{422}
+4{431} + {44} +2{5111} +4{521} +2{53} +2{611} +2{62} + {71}

{53} o {422} = {22211} + {311111} +2{32111} +3{3221} + {3311} +2{332} + {41111} +4{4211} +2{422}
+3{431} + {5111} +3{521} +2{53} + {611} + {62}

{53} o {4211} = {221111} + {22211} + {2222} + {311111} +4{32111} +4{3221} +4{3311} +2{332} +3{41111}
+6{4211} +4{422} +4{431} + {44} +3{5111} +4{521} + {53} + {611} + {62}

{53} o {332} = {221111} + {22211} + {2222} +2{32111} +2{3221} +2{3311} + {332} + {41111} +2{4211}
+2{422} +2{431} + {44} + {5111} +2{521} + {53} + {62}

{521} o {521} = {22211} + {2222} + {311111} +4{32111} +5{3221} +5{3311} +4{332} +3{41111} +9{4211}
+6{422} +9{431} +2{44} +5{5111} +9{521} +5{53} +4{611} +4{62} +2{71} + {8}

{521} o {5111} = {221111} + {22211} + {311111} +3{32111} +3{3221} +3{3311} +2{332} +2{41111}
+5{4211} +3{422} +4{431} + {44} +2{5111} +5{521} +2{53} +2{611} +2{62} + {71}

{521} o {44} = {32111} + {3221} + {3311} + {332} +2{4211} +2{422} +2{431} + {5111} +2{521} + {53}
+ {611} + {62}

{521} o {431} = {221111} +2{22211} + {2222} + {311111} +5{32111} +7{3221} +6{3311} +5{332} +3{41111}
+10{4211} +7{422} +9{431} +2{44} +4{5111} +9{521} +4{53} +3{611} +3{62} + {71}

{521} o {422} = {221111} +2{22211} + {2222} + {311111} +5{32111} +6{3221} +5{3311} +4{332} +3{41111}
+8{4211} +5{422} +7{431} +2{44} +3{5111} +6{521} +3{53} +2{611} +2{62} + {71}

{521} o {4211} = {2111111} +3{221111} +4{22211} +2{2222} +3{311111} +9{32111} +10{3221} +8{3311}
+6{332} +5{41111} +13{4211} +8{422} +10{431} +2{44} +5{5111} +9{521} +4{53} +3{611}
+3{62} + {71}

{521} o {332} = {221111} +2{22211} + {2222} + {311111} +4{32111} +5{3221} +4{3311} +3{332} +2{41111}
+6{4211} +4{422} +5{431} + {44} +2{5111} +4{521} +2{53} + {611} + {62}

{5111} o {5111} = {2111111} + {221111} + {22211} + {2222} + {311111} +2{32111} +2{3221} + {3311}
+ {332} + {41111} +2{4211} +2{422} +2{431} + {44} + {5111} +2{521} +2{53} + {611}
+2{62} + {71} + {8}

```
{5111} o {44} = {3221} + {3311} + {41111} + {4211} + {422} + {431} + {44} + {5111} + {521}
{5111} o {431} = {22211} + {2222} + {311111} +3{32111} +4{3221} +3{3311} +3{332} +2{41111} +6{4211}
         +4{422} +5{431} + {44} +2{5111} +4{521} +2{53} + {611} + {62}
{5111} o {422} = {221111} + {22211} + {2222} +3{32111} +3{3221} +4{3311} +2{332} + {41111} +4{4211}
         +3{422} +4{431} + {44} +2{5111} +3{521} + {53} + {611} + {62}
{5111} o {4211} = {2111111} +2{221111} +3{22211} + {2222} +2{311111} +5{32111} +6{3221} +4{3311}
         +4{332} +2{41111} +6{4211} +4{422} +6{431} + {44} +2{5111} +5{521} +3{53} +2{611}
         +2{62} + {71}
{5111} o {332} = {22211} +2{32111} +3{3221} +2{3311} +2{332} + {41111} +4{4211} +2{422} +3{431}
         + {5111} +2{521} + {53}
{44} o {44} = {2222} + {3311} + {422} + {44} + {5111} + {62} + {8}
{44} o {431} = {22211} + {32111} +2{3221} + {3311} + {332} + {41111} +2{4211} + {422} +2{431}
         + {5111} +2{521} + {53} + {611} + {62} + {71}
{44} o {422} = {221111} + {2222} + {32111} + {3221} +2{3311} + {41111} + {4211} +2{422} + {431}
         + {44} + {5111} +2{521} + {62}
{44} o {4211} = {22211} + {311111} +2{32111} +2{3221} + {3311} +2{332} + {41111} +4{4211} + {422}
         +2{431} + {5111} +2{521} + {53} + {611}
{44} o {332} = {22211} + {311111} + {32111} + {3221} + {332} +2{4211} + {431} + {521} + {53} + {611}
{431} o {431} =2{221111} +3{22211} +2{2222} +2{311111} +7{32111} +8{3221} +6{3311} +5{332} +4{41111}
         +11{4211} +7{422} +8{431} +2{44} +5{5111} +9{521} +4{53} +4{611} +4{62} +2{71} + {8}
{431} o {422} = {2111111} +2{221111} +3{22211} + {2222} +2{311111} +6{32111} +6{3221} +5{3311}
         +4{332} +3{41111} +9{4211} +5{422} +7{431} + {44} +4{5111} +7{521} +3{53} +3{611}
         +2{62} + {71}
{431} o {4211} = {2111111} +3{221111} +4{22211} +2{2222} +3{311111} +10{32111} +11{3221} +9{3311}
         +6{332} +6{41111} +14{4211} +9{422} +11{431} +2{44} +6{5111} +10{521} +4{53} +3{611}
         +3{62} + {71}
{431} o {332} = {2111111} +2{221111} +2{22211} + {2222} +2{311111} +5{32111} +5{3221} +4{3311}
         +2{332} +3{41111} +6{4211} +4{422} +5{431} + {44} +3{5111} +5{521} +2{53} +2{611}
         +2{62} + {71}
{422} o {422} =2{221111} + {22211} +2{2222} +2{311111} +5{32111} +5{3221} +5{3311} +2{332} +4{41111}
         +6{4211} +6{422} +5{431} +2{44} +3{5111} +5{521} +2{53} + {611} +3{62} + {71} + {8}
{422} o {4211} = {2111111} +2{221111} +4{22211} + {2222} +3{311111} +8{32111} +9{3221} +6{3311}
         +6{332} +4{41111} +12{4211} +6{422} +9{431} + {44} +4{5111} +8{521} +4{53} +3{611}
         +2{62} + {71}
{422} o {332} = {2111111} + {221111} +2{22211} +2{311111} +4{32111} +4{3221} +2{3311} +3{332}
         +2{41111} +6{4211} +2{422} +4{431} +2{5111} +4{521} +2{53} +2{611} + {62} + {71}
{4211} o {4211} = {11111111} +2{2111111} +5{221111} +6{22211} +4{2222} +4{311111} +13{32111}
         +14{3221} +12{3311} +8{332} +6{41111} +17{4211} +12{422} +14{431} +4{44} +6{5111}
         +13{521} +6{53} +4{611} +5{62} +2{71} + {8}
{4211} o {332} =2{221111} +2{22211} +2{2222} +2{311111} +6{32111} +6{3221} +6{3311} +4{332}
         +4{41111} +8{4211} +6{422} +6{431} +2{44} +4{5111} +6{521} +2{53} +2{611} +2{62}
{332} o {332} = {11111111} + {2111111} +2{221111} + {22211} + {2222} + {311111} +3{32111} +2{3221}
         +3{3311} + {332} +2{41111} +4{4211} +3{422} +2{431} + {44} +2{5111} +3{521} + {53}
         + {611} +2{62} + {71} + {8}

{81} o {81} = {711} + {72} + {81} + {9}
{81} o {72} = {621} + {63} + {711} + {72} + {81}
{81} o {711} = {6111} + {621} + {711} + {72} + {81}
{81} o {63} = {531} + {54} + {621} + {63} + {72}
{81} o {621} = {5211} + {522} + {531} + {6111} +2{621} + {63} + {711} + {72}
{81} o {6111} = {51111} + {5211} + {6111} + {621} + {711}
{81} o {54} = {441} + {531} + {54} + {63}
{81} o {531} = {4311} + {432} + {441} + {5211} + {522} +2{531} + {54} + {621} + {63}
{81} o {522} = {4221} + {432} + {5211} + {522} + {531} + {621}
{81} o {5211} = {42111} + {4221} + {4311} + {51111} +2{5211} + {522} + {531} + {6111} + {621}
{81} o {51111} = {411111} + {42111} + {51111} + {5211} + {6111}
{81} o {441} = {4311} + {432} + {441} + {531} + {54}
{81} o {432} = {3321} + {333} + {4221} + {4311} +2{432} + {441} + {522} + {531}
{81} o {4311} = {33111} + {3321} + {42111} + {4221} +2{4311} + {432} + {441} + {5211} + {531}
{81} o {333} = {3321} + {432}
{72} o {72} = {522} + {531} + {54} + {6111} +2{621} + {63} + {711} +2{72} + {81} + {9}
{72} o {711} = {5211} + {531} + {6111} +2{621} + {63} +2{711} + {72} + {81}
{72} o {63} = {432} + {441} + {5211} + {522} +2{531} + {54} +2{621} +2{63} + {711} + {72} + {81}
{72} o {621} = {4221} + {4311} + {432} + {441} + {51111} +3{5211} +2{522} +3{531} + {54} +2{6111}
         +4{621} +2{63} +2{711} +2{72} + {81}
{72} o {6111} = {42111} + {4311} + {51111} +2{5211} + {522} + {531} +2{6111} +2{621} + {711} + {72}
{72} o {54} = {4311} + {432} + {441} + {522} +2{531} + {54} + {621} + {63} + {72}
{72} o {531} = {3321} + {333} + {42111} +2{4221} +3{4311} +3{432} +2{441} +3{5211} +2{522} +5{531}
         +2{54} + {6111} +3{621} +2{63} + {711} + {72}
{72} o {522} = {3222} + {3321} + {42111} +2{4221} +2{4311} +2{432} + {441} + {51111} +2{5211}
         +3{522} +2{531} + {54} + {6111} +2{621} + {63} + {72}
```

{72} o {5211} = {32211} + {33111} + {3321} + {411111} +3{42111} +3{4221} +3{4311} +2{432} + {441}
+2{51111} +5{5211} +2{522} +3{531} +2{6111} +3{621} + {63} + {711}

{72} o {51111} = {321111} + {33111} + {411111} +2{42111} + {4221} + {4311} +2{51111} +2{5211}
+ {522} + {6111} + {621}

{72} o {441} = {33111} + {3321} + {4221} +2{4311} +2{432} +2{441} + {5211} + {522} +2{531} + {54}
+ {621} + {63}

{72} o {432} = {32211} + {3222} + {33111} +3{3321} + {333} + {42111} +3{4221} +3{4311} +4{432}
+2{441} +2{5211} +2{522} +3{531} + {54} + {621} + {63}

{72} o {4311} = {321111} +2{32211} + {3222} +2{33111} +3{3321} + {333} +3{42111} +3{4221} +5{4311}
+3{432} +2{441} + {51111} +3{5211} +2{522} +3{531} + {54} + {6111} + {621}

{72} o {333} = {32211} + {3321} + {333} + {4221} + {4311} + {432} + {531}

{711} o {711} = {51111} + {5211} + {522} + {6111} +2{621} + {63} + {711} +2{72} + {81} + {9}

{711} o {63} = {4311} + {441} + {5211} + {522} +2{531} + {54} + {6111} +2{621} + {63} + {711} + {72}

{711} o {621} = {42111} + {4221} + {4311} + {432} + {51111} +3{5211} +2{522} +3{531} + {54} +2{6111}
+4{621} +2{63} +2{711} +2{72} + {81}

{711} o {6111} = {411111} + {42111} + {4221} + {51111} +2{5211} + {522} + {531} + {6111} +2{621}
+ {63} + {711} + {72} + {81}

{711} o {54} = {4311} + {432} + {441} + {5211} +2{531} + {54} + {621} + {63}

{711} o {531} = {33111} + {3321} + {42111} +2{4221} +3{4311} +3{432} +3{441} + {51111} +3{5211}
+3{522} +4{531} +2{54} + {6111} +3{621} +2{63} + {72}

{711} o {522} = {32211} + {3321} + {333} + {42111} +2{4221} +2{4311} +2{432} + {441} +3{5211}
+ {522} +3{531} + {6111} +2{621} + {63} + {711}

{711} o {5211} = {321111} + {32211} + {3222} + {33111} + {3321} + {411111} +3{42111} +3{4221}
+3{4311} +2{432} + {441} +2{51111} +4{5211} +3{522} +3{531} + {54} +2{6111} +3{621}
+ {63} + {711} + {72}

{711} o {51111} = {3111111} + {321111} + {32211} + {411111} +2{42111} + {4221} + {4311} + {51111}
+2{5211} + {531} + {6111} + {621} + {711}

{711} o {441} = {3321} + {333} + {42111} + {4221} +2{4311} +2{432} + {441} + {5211} + {522} +3{531}
+ {54} + {63}

{711} o {432} = {32211} + {3222} + {33111} +3{3321} + {333} + {42111} +3{4221} +4{4311} +4{432}
+2{441} +2{5211} +2{522} +3{531} + {54} + {621}

{711} o {4311} = {321111} +2{32211} + {3222} +2{33111} +3{3321} + {333} + {411111} +3{42111}
+4{4221} +4{4311} +4{432} +2{441} + {51111} +3{5211} +2{522} +3{531} + {54} + {621}
+ {63}

{711} o {333} = {3222} + {33111} + {3321} + {4221} + {4311} + {432} + {441} + {522}

{63} o {63} = {333} + {4221} + {4311} +2{432} + {441} + {5211} +2{522} +3{531} + {54} + {6111}
+2{621} +2{63} + {711} +2{72} + {81} + {9}

{63} o {621} = {3321} + {42111} +2{4221} +3{4311} +3{432} +2{441} + {51111} +4{5211} +3{522} +5{531}
+2{54} +2{6111} +5{621} +2{63} +2{711} +2{72} + {81}

{63} o {6111} = {33111} + {42111} + {4221} +2{4311} + {432} + {441} + {51111} +3{5211} + {522}
+2{531} +2{6111} +2{621} + {63} + {711}

{63} o {54} = {3321} + {4221} + {4311} +2{432} + {441} + {5211} + {522} +2{531} + {54} +2{621}
+ {63} + {711} + {72} + {81}

{63} o {531} = {32211} + {3222} + {33111} +3{3321} + {333} +2{42111} +4{4221} +5{4311} +5{432}
+3{441} + {51111} +5{5211} +4{522} +6{531} +2{54} +2{6111} +5{621} +3{63} +2{711}
+2{72} + {81}

{63} o {522} = {32211} + {3222} + {33111} +2{3321} + {333} + {411111} +2{42111} +4{4221} +3{4311}
+3{432} +2{441} + {51111} +4{5211} +3{522} +4{531} + {54} + {6111} +3{621} +2{63}
+ {711} + {72}

{63} o {5211} = {321111} +2{32211} + {3222} +2{33111} +3{3321} + {333} + {411111} +5{42111} +5{4221}
+6{4311} +4{432} +2{441} +3{51111} +7{5211} +4{522} +5{531} + {54} +3{6111} +4{621}
+ {63} + {711} + {72}

{63} o {51111} = {321111} + {32211} + {33111} + {3321} + {411111} +3{42111} +2{4221} +2{4311}
+ {432} +2{51111} +3{5211} + {522} + {531} + {6111} + {621}

{63} o {441} = {32211} + {3222} + {33111} +2{3321} + {333} + {42111} +2{4221} +3{4311} +2{432}
+2{441} +2{5211} +2{522} +3{531} + {54} + {6111} +2{621} + {63} + {711} + {72}

{63} o {432} = {22221} + {321111} +3{32211} +2{3222} +2{33111} +4{3321} + {333} +3{42111} +5{4221}
+5{4311} +5{432} +2{441} + {51111} +4{5211} +3{522} +5{531} +2{54} + {6111} +3{621}
+2{63} + {72}

{63} o {4311} = {222111} + {22221} +2{321111} +4{32211} +2{3222} +4{33111} +5{3321} + {333}
+ {411111} +5{42111} +6{4221} +7{4311} +5{432} +3{441} +2{51111} +6{5211} +3{522}
+5{531} + {54} +2{6111} +3{621} + {63} + {711}

{63} o {333} = {222111} + {32211} + {3222} + {33111} + {3321} + {333} + {42111} + {4221} + {4311}
+ {432} + {441} + {5211} + {522} + {531} + {63}

{621} o {621} = {32211} + {3222} + {33111} +2{3321} + {333} + {411111} +4{42111} +5{4221} +6{4311}
+5{432} +4{441} +3{51111} +9{5211} +6{522} +9{531} +3{54} +5{6111} +9{621} +5{63}
+4{711} +4{72} +2{81} + {9}

{621} o {6111} = {321111} + {32211} + {33111} + {3321} + {411111} +3{42111} +3{4221} +3{4311}
+2{432} + {441} +2{51111} +5{5211} +3{522} +4{531} + {54} +2{6111} +5{621} +2{63}
+2{711} +2{72} + {81}

{621} o {54} = {33111} + {3321} + {333} + {42111} +2{4221} +3{4311} +3{432} +2{441} +3{5211} +3{522}
+4{531} + {54} + {6111} +3{621} +2{63} + {711} + {72}

{621} o {531} = {321111} +3{32211} +2{3222} +3{33111} +6{3321} +2{333} + {411111} +6{42111} +9{4221}
+11{4311} +10{432} +6{441} +3{51111} +11{5211} +8{522} +13{531} +4{54} +4{6111}
+9{621} +5{63} +3{711} +3{72} + {81}

{621} o {522} = {22221} + {321111} +3{32211} +2{3222} +3{33111} +5{3321} + {333} + {411111}
+5{42111} +7{4221} +8{4311} +7{432} +4{441} +3{51111} +8{5211} +5{522} +8{531} +3{54}
+3{6111} +6{621} +3{63} +2{711} +2{72} + {81}

{621} o {5211} = {222111} + {22221} + {3111111} +4{321111} +6{32211} +3{3222} +5{33111} +7{3321}
+2{333} +3{411111} +10{42111} +11{4221} +12{4311} +9{432} +5{441} +5{51111} +13{5211}
+8{522} +11{531} +3{54} +5{6111} +9{621} +4{63} +3{711} +3{72} + {81}

{621} o {51111} = {2211111} + {222111} + {3111111} +3{321111} +3{32211} + {3222} +3{33111} +2{3321}
+2{411111} +5{42111} +4{4221} +4{4311} +2{432} + {441} +2{51111} +5{5211} +3{522}
+3{531} +2{6111} +3{621} + {63} + {711} + {72}

{621} o {441} = {321111} +2{32211} + {3222} +2{33111} +4{3321} + {333} +3{42111} +5{4221} +6{4311}
+6{432} +3{441} + {51111} +5{5211} +4{522} +6{531} +2{54} + {6111} +4{621} +2{63}
+ {72}

{621} o {432} = {222111} + {22221} +2{321111} +6{32211} +4{3222} +5{33111} +9{3321} +3{333}
+ {411111} +7{42111} +11{4221} +12{4311} +10{432} +6{441} +2{51111} +9{5211} +7{522}
+10{531} +3{54} +2{6111} +5{621} +3{63} + {711} + {72}

{621} o {4311} = {2211111} +2{222111} +2{22221} + {3111111} +5{321111} +9{32211} +5{3222} +7{33111}
+11{3321} +3{333} +3{411111} +11{42111} +14{4221} +14{4311} +12{432} +6{441}
+4{51111} +12{5211} +8{522} +11{531} +3{54} +3{6111} +6{621} +3{63} + {711} + {72}

{621} o {333} = {22221} + {321111} +2{32211} + {3222} + {33111} +3{3321} +2{42111} +3{4221} +3{4311}
+3{432} + {441} +2{5211} + {522} +2{531} + {54} + {621}

{6111} o {6111} = {3111111} + {321111} + {32211} + {3222} + {411111} +2{42111} +2{4221} + {4311}
+ {432} + {51111} +2{5211} +2{522} +2{531} + {54} + {6111} +2{621} +2{63} + {711}
+2{72} + {81} + {9}

{6111} o {54} = {3321} + {42111} + {4221} +2{4311} + {432} + {441} + {51111} +2{5211} + {522}
+2{531} + {54} + {6111} + {621}

{6111} o {531} = {321111} +2{32211} + {3222} +2{33111} +3{3321} + {333} + {411111} +4{42111}
+5{4221} +6{4311} +5{432} +3{441} +2{51111} +6{5211} +4{522} +6{531} +2{54} +2{6111}
+4{621} +2{63} + {711} + {72}

{6111} o {522} = {222111} + {321111} +2{32211} + {3222} +2{33111} +3{3321} + {333} +3{42111}
+3{4221} +5{4311} +3{432} +2{441} + {51111} +4{5211} +3{522} +4{531} + {54} +2{6111}
+3{621} + {63} + {711} + {72}

{6111} o {5211} = {2211111} + {222111} + {22221} + {3111111} +3{321111} +4{32211} +2{3222} +3{33111}
+4{3321} + {333} +2{411111} +5{42111} +6{4221} +5{4311} +5{432} +3{441} +2{51111}
+6{5211} +4{522} +6{531} +2{54} +2{6111} +5{621} +3{63} +2{711} +2{72} + {81}

{6111} o {51111} = {21111111} + {2211111} + {222111} + {22221} + {3111111} +2{321111} +2{32211}
+ {3222} + {33111} + {3321} + {411111} +2{42111} +2{4221} +2{4311} + {432} + {441}
+ {51111} +2{5211} + {522} +2{531} + {54} + {6111} +2{621} + {63} + {711} + {72}
+ {81}

{6111} o {441} = {32211} + {3222} + {33111} +2{3321} + {333} + {411111} +2{42111} +3{4221} +3{4311}
+3{432} +2{441} + {51111} +3{5211} +2{522} +3{531} + {54} + {621} + {63}

{6111} o {432} = {22221} + {321111} +3{32211} +2{3222} +3{33111} +5{3321} + {333} +4{42111} +6{4221}
+7{4311} +6{432} +3{441} + {51111} +5{5211} +3{522} +5{531} + {54} + {6111} +2{621}
+ {63}

{6111} o {4311} = {222111} + {22221} + {3111111} +3{321111} +5{32211} +3{3222} +3{33111} +6{3321}
+2{333} +2{411111} +6{42111} +8{4221} +7{4311} +7{432} +3{441} +2{51111} +5{5211}
+5{522} +6{531} +2{54} + {6111} +3{621} +2{63} + {72}

{6111} o {333} = {32211} + {3222} + {33111} + {3321} + {333} + {42111} +2{4221} +2{4311} + {432}
+ {441} + {5211} + {522} + {531}

{54} o {54} = {3222} + {3321} + {333} + {4221} + {4311} + {432} + {441} + {5211} + {522} + {531}
+ {54} + {6111} + {621} + {63} + {711} + {72} + {81} + {9}

{54} o {531} =2{32211} + {3222} + {33111} +3{3321} + {333} +2{42111} +4{4221} +4{4311} +4{432}
+2{441} + {51111} +4{5211} +3{522} +5{531} + {54} +2{6111} +4{621} +2{63} +2{711}
+2{72} + {81}

{54} o {522} = {321111} + {32211} + {3222} + {33111} +2{3321} +2{42111} +3{4221} +3{4311} +3{432}
+ {441} + {51111} +3{5211} +3{522} +3{531} + {54} + {6111} +3{621} + {63} + {72}

{54} o {5211} = {321111} +2{32211} + {3222} +2{33111} +3{3321} + {333} + {411111} +4{42111} +5{4221}
+5{4311} +4{432} +2{441} +2{51111} +6{5211} +3{522} +4{531} + {54} +2{6111} +3{621}
+ {63} + {711}

{54} o {51111} = {32211} + {3222} + {33111} + {3321} + {411111} +2{42111} +2{4221} +2{4311} + {432}
+ {441} +2{51111} +2{5211} + {522} + {531} + {6111}

{54} o {441} = {22221} + {32211} + {3222} + {33111} +2{3321} + {42111} +2{4221} +2{4311} +2{432}
+ {441} + {51111} +2{5211} + {522} +2{531} + {54} + {6111} +2{621} + {63} + {711}
+ {72} + {81}

{54} o {432} = {222111} + {22221} + {321111} +3{32211} +2{3222} +2{33111} +3{3321} + {333}
+ {411111} +3{42111} +4{4221} +4{4311} +3{432} +2{441} + {51111} +4{5211} +3{522}
+4{531} + {54} + {6111} +3{621} +2{63} + {711} + {72}

{54} o {4311} = {222111} + {22221} +2{321111} +4{32211} +2{3222} +3{33111} +4{3321} + {333}
+ {411111} +5{42111} +5{4221} +6{4311} +4{432} +2{441} +2{51111} +5{5211} +3{522}
+4{531} + {54} +2{6111} +3{621} + {63} + {711} + {72}

{54} o {333} = {22221} + {321111} + {32211} + {3321} + {42111} + {4221} + {4311} + {432} + {5211}
+ {531} + {54} + {621}

{531} o {531} = {222111} +2{22221} +3{321111} +7{32211} +5{3222} +7{33111} +11{3321} +3{333}
+2{411111} +10{42111} +15{4221} +16{4311} +14{432} +8{441} +5{51111} +16{5211}
+12{522} +15{531} +5{54} +6{6111} +13{621} +6{63} +4{711} +5{72} +2{81} + {9}

{531} o {522} = {222111} + {22221} + {3111111} +3{321111} +7{32211} +3{3222} +4{33111} +8{3321}
+3{333} +2{411111} +9{42111} +11{4221} +12{4311} +10{432} +5{441} +3{51111} +12{5211}
+7{522} +12{531} +3{54} +4{6111} +8{621} +4{63} +3{711} +2{72} + {81}

{531} o {5211} = {2211111} +2{222111} +2{22221} + {3111111} +6{321111} +10{32211} +6{3222} +9{33111}
+13{3321} +3{333} +4{411111} +15{42111} +18{4221} +19{4311} +15{432} +8{441}
+7{51111} +18{5211} +12{522} +16{531} +4{54} +6{6111} +11{621} +5{63} +3{711} +3{72}
+ {81}

{531} o {51111} = {222111} + {22221} + {3111111} +3{321111} +5{32211} +2{3222} +3{33111} +5{3321}
+2{333} +2{411111} +7{42111} +7{4221} +7{4311} +5{432} +2{441} +2{51111} +7{5211}
+3{522} +5{531} + {54} +2{6111} +3{621} + {63} + {711}

{531} o {441} = {222111} + {22221} +2{321111} +5{32211} +3{3222} +3{33111} +6{3321} +2{333}
+ {411111} +6{42111} +8{4221} +8{4311} +7{432} +3{441} +2{51111} +8{5211} +5{522}
+8{531} +2{54} +3{6111} +6{621} +3{63} +3{711} +2{72} + {81}

{531} o {432} = {2211111} +3{222111} +3{22221} + {3111111} +6{321111} +11{32211} +6{3222} +8{33111}
+12{3321} +3{333} +3{411111} +13{42111} +16{4221} +16{4311} +13{432} +7{441} +5{51111}
+15{5211} +10{522} +14{531} +4{54} +5{6111} +10{621} +5{63} +3{711} +3{72} + {81}

{531} o {4311} =2{2211111} +4{222111} +4{22221} +2{3111111} +9{321111} +15{32211} +8{3222}
+11{33111} +16{3321} +4{333} +5{411111} +18{42111} +21{4221} +21{4311} +16{432}
+8{441} +7{51111} +19{5211} +12{522} +16{531} +4{54} +6{6111} +11{621} +5{63} +3{711}
+3{72} + {81}

{531} o {333} = {2211111} + {222111} + {22221} +2{321111} +3{32211} +2{3222} +3{33111} +3{3321}
+ {411111} +3{42111} +4{4221} +4{4311} +3{432} +2{441} +2{51111} +3{5211} +3{522}
+3{531} + {54} + {6111} +2{621} + {63} + {72}

{522} o {522} = {222111} + {22221} +3{321111} +4{32211} +4{3222} +5{33111} +6{3321} + {333}
+2{411111} +7{42111} +8{4221} +9{4311} +7{432} +5{441} +4{51111} +7{5211} +7{522}
+7{531} +3{54} +3{6111} +5{621} +3{63} + {711} +3{72} + {81} + {9}

{522} o {5211} = {2211111} +2{222111} +2{22221} + {3111111} +5{321111} +9{32211} +4{3222} +7{33111}
+10{3321} +3{333} +3{411111} +11{42111} +13{4221} +14{4311} +11{432} +6{441}
+4{51111} +13{5211} +7{522} +12{531} +3{54} +4{6111} +8{621} +4{63} +3{711} +2{72}
+ {81}

{522} o {51111} = {2211111} + {222111} + {22221} +3{321111} +3{32211} +2{3222} +4{33111} +4{3321}
+ {411111} +4{42111} +5{4221} +5{4311} +4{432} +2{441} +2{51111} +4{5211} +4{522}
+3{531} + {54} + {6111} +3{621} + {63} + {72}

{522} o {441} = {2211111} + {222111} + {22221} +2{321111} +3{32211} +2{3222} +4{33111} +4{3321}
+ {333} + {411111} +4{42111} +6{4221} +6{4311} +5{432} +3{441} +2{51111} +6{5211}
+5{522} +5{531} + {54} +2{6111} +4{621} +2{63} + {711} + {72}

{522} o {432} = {2211111} +2{222111} +2{22221} + {3111111} +5{321111} +8{32211} +4{3222} +6{33111}
+9{3321} +2{333} +3{411111} +10{42111} +12{4221} +12{4311} +10{432} +5{441} +4{51111}
+11{5211} +7{522} +10{531} +3{54} +3{6111} +7{621} +3{63} +2{711} +2{72} + {81}

{522} o {4311} = {21111111} +2{2211111} +4{222111} +3{22221} +2{3111111} +7{321111} +11{32211}
+6{3222} +8{33111} +12{3321} +3{333} +3{411111} +13{42111} +15{4221} +16{4311}
+12{432} +6{441} +5{51111} +14{5211} +9{522} +12{531} +3{54} +5{6111} +8{621} +3{63}
+2{711} +2{72}

{522} o {333} = {222111} + {3111111} + {321111} +3{32211} + {3222} + {33111} +2{3321} +2{333}
+ {411111} +3{42111} +3{4221} +3{4311} +2{432} + {441} +3{5211} + {522} +3{531}
+ {6111} + {621} + {63} + {711}

{5211} o {5211} = {21111111} +3{2211111} +5{222111} +4{22221} +3{3111111} +10{321111} +15{32211}
+8{3222} +11{33111} +16{3321} +4{333} +5{411111} +17{42111} +20{4221} +21{4311}
+17{432} +9{441} +6{51111} +18{5211} +13{522} +18{531} +6{54} +6{6111} +13{621}
+7{63} +4{711} +5{72} +2{81} + {9}

{5211} o {51111} = {21111111} +2{2211111} +3{222111} +2{22221} +2{3111111} +5{321111} +7{32211}
+3{3222} +4{33111} +6{3321} +2{333} +2{411111} +6{42111} +7{4221} +7{4311} +6{432}
+3{441} +2{51111} +6{5211} +4{522} +7{531} +2{54} +2{6111} +5{621} +3{63} +2{711}
+2{72} + {81}

{5211} o {441} = {222111} + {22221} + {3111111} +3{321111} +6{32211} +3{3222} +4{33111} +7{3321}
+2{333} +2{411111} +8{42111} +10{4221} +10{4311} +8{432} +4{441} +3{51111} +9{5211}
+6{522} +8{531} +2{54} +3{6111} +5{621} +2{63} + {711} + {72}

{5211} o {432} = {2211111} +3{222111} +3{22221} + {3111111} +7{321111} +13{32211} +7{3222}
+10{33111} +15{3321} +4{333} +4{411111} +16{42111} +19{4221} +20{4311} +16{432}
+8{441} +6{51111} +17{5211} +11{522} +15{531} +4{54} +5{6111} +9{621} +4{63} +2{711}
+2{72}

{5211} o {4311} = {21111111} +3{2211111} +5{222111} +5{22221} +3{3111111} +11{321111} +18{32211}
+10{3222} +13{33111} +19{3321} +5{333} +6{411111} +20{42111} +25{4221} +24{4311}
+20{432} +10{441} +7{51111} +21{5211} +14{522} +19{531} +5{54} +5{6111} +12{621}
+6{63} +3{711} +3{72} + {81}

Inner Multiplication of S-Functions (contd).

{5211} o {333} = {222111} + {22221} +2{321111} +3{32211} +2{3222} +3{33111} +4{3321} + {333}
+ {411111} +4{42111} +5{4221} +5{4311} +4{432} +2{441} +2{51111} +4{5211} +3{522}
+3{531} + {54} + {6111} +2{621} + {63}

{51111} o {51111} = {111111111} + {21111111} +2{2211111} +2{222111} +2{22221} + {3111111} +2{321111}
+2{32211} +2{3222} +2{33111} +2{3321} + {411111} +2{42111} +2{4221} +2{4311}
+2{432} +2{441} + {51111} +2{5211} +2{522} +2{531} +2{54} + {6111} +2{621} +2{63}
+ {711} +2{72} + {81} + {9}

{51111} o {441} = {22221} + {321111} +2{32211} +2{3222} +2{33111} +3{3321} + {411111} +3{42111}
+4{4221} +4{4311} +3{432} +2{441} +2{51111} +3{5211} +2{522} +2{531} + {54} + {6111}
+ {621}

{51111} o {432} = {222111} + {22221} +2{321111} +5{32211} +3{3222} +4{33111} +6{3321} +2{333}
+ {411111} +6{42111} +8{4221} +8{4311} +6{432} +3{441} +2{51111} +6{5211} +4{522}
+5{531} + {54} + {6111} +2{621} + {63}

{51111} o {4311} = {2211111} +2{222111} +2{22221} + {3111111} +4{321111} +7{32211} +4{3222}
+5{33111} +8{3321} +2{333} +2{411111} +7{42111} +9{4221} +9{4311} +8{432} +4{441}
+2{51111} +7{5211} +5{522} +7{531} +2{54} +2{6111} +4{621} +2{63} + {711} + {72}

{51111} o {333} =2{32211} +2{3321} +2{42111} +2{4221} +2{4311} +2{432} +2{5211} +2{531}

{441} o {441} = {222111} + {22221} + {321111} +3{32211} +2{3222} +2{33111} +3{3321} + {333}
+ {411111} +3{42111} +4{4221} +4{4311} +3{432} +2{441} +2{51111} +4{5211} +3{522}
+3{531} + {54} +2{6111} +3{621} +2{63} + {711} +2{72} + {81} + {9}

{441} o {432} = {2211111} +2{222111} +2{22221} + {3111111} +4{321111} +6{32211} +3{3222} +4{33111}
+6{3321} + {333} +2{411111} +7{42111} +8{4221} +8{4311} +6{432} +3{441} +3{51111}
+8{5211} +5{522} +7{531} +2{54} +3{6111} +6{621} +2{63} +2{711} +2{72} + {81}

{441} o {4311} = {2211111} +2{222111} +2{22221} + {3111111} +5{321111} +8{32211} +4{3222} +6{33111}
+8{3321} +2{333} +3{411111} +10{42111} +11{4221} +10{4311} +8{432} +4{441} +4{51111}
+10{5211} +6{522} +8{531} +2{54} +3{6111} +6{621} +3{63} +2{711} +2{72} + {81}

{441} o {333} = {222111} + {3111111} + {321111} +2{32211} + {3222} + {33111} + {3321} + {333}
+ {411111} +2{42111} +2{4221} +2{4311} + {432} + {441} +2{5211} + {522} +2{531}
+ {6111} + {621} + {63} + {711}

{432} o {432} = {21111111} +3{2211111} +4{222111} +3{22221} +3{3111111} +9{321111} +12{32211}
+6{3222} +9{33111} +11{3321} +3{333} +5{411111} +15{42111} +16{4221} +16{4311}
+12{432} +6{441} +6{51111} +16{5211} +10{522} +13{531} +3{54} +6{6111} +10{621} +5{63}
+4{711} +4{72} +2{81} + {9}

{432} o {4311} = {21111111} +3{2211111} +5{222111} +4{22221} +3{3111111} +11{321111} +16{32211}
+8{3222} +12{33111} +16{3321} +4{333} +6{411111} +19{42111} +21{4221} +21{4311}
+16{432} +8{441} +8{51111} +20{5211} +12{522} +16{531} +4{54} +7{6111} +12{621}
+5{63} +4{711} +3{72} + {81}

{432} o {333} = {21111111} + {2211111} + {222111} + {22221} + {3111111} +3{321111} +3{32211}
+ {3222} +2{33111} +3{3321} + {411111} +4{42111} +4{4221} +4{4311} +3{432} + {441}
+2{51111} +4{5211} +2{522} +3{531} + {54} + {6111} +3{621} + {63} + {711} + {72}
+ {81}

{4311} o {4311} = {21111111} +3{2211111} +6{222111} +5{22221} +4{3111111} +14{321111} +21{32211}
+11{3222} +15{33111} +21{3321} +5{333} +8{411111} +25{42111} +28{4221} +27{4311}
+21{432} +10{441} +9{51111} +24{5211} +16{522} +21{531} +6{54} +7{6111} +14{621}
+7{63} +4{711} +5{72} +2{81} + {9}

{4311} o {333} = {2211111} + {222111} + {22221} + {3111111} +3{321111} +4{32211} +2{3222} +3{33111}
+4{3321} + {333} +2{411111} +5{42111} +5{4221} +5{4311} +4{432} +2{441} +2{51111}
+5{5211} +3{522} +4{531} + {54} +2{6111} +3{621} + {63} + {711} + {72}

{333} o {333} = {111111111} + {2211111} + {222111} + {3222} +2{33111} + {333} + {411111} + {42111}
+ {4221} + {4311} + {441} + {5211} +2{522} + {6111} + {63} + {72} + {9}

B-3 Expansion of S-Functions into Products of the Symmetric Functions h_r.

{0} = (0)

{1} = (1)

{2} = (2)
{11}=- (2) + (11)

{3} = (3)
{21}=- (3) + (21)
{111} = (3) -2(21) + (111)

{4} = (4)
{31}=- (4) + (31)
{22}=- (31) + (22)
{211} = (4) - (31) - (22) + (211)
{1111}=- (4) +2(31) + (22) -3(211) + (1111)

{5} = (5)
{41}=- (5) + (41)
{32}=- (41) + (32)
{311} = (5) - (41) - (32) + (311)
{221} = (41) - (32) - (311) + (221)
{2111}=- (5) + (41) +2(32) - (311) -2(221) + (2111)
{11111} = (5) -2(41) -2(32) +3(311) +3(221) -4(2111) + (11111)

{6} = (6)
{51}=- (6) + (51)
{42}=- (51) + (42)
{411} = (6) - (51) - (42) + (411)
{33}=- (42) + (33)
{321} = (51) - (411) - (33) + (321)
{3111}=- (6) + (51) + (42) - (411) + (33) -2(321) + (3111)
{222}=- (42) + (411) + (33) -2(321) + (222)
{2211}=- (51) + (42) + (411) - (3111) - (222) + (2211)
{21111} = (6) - (51) -2(42) + (411) - (33) +4(321) - (3111) + (222) -3(2211) + (21111)
{111111}=- (6) +2(51) +2(42) -3(411) + (33) -6(321) +4(3111) - (222) +6(2211) -5(21111) + (111111)

{7} = (7)
{61}=- (7) + (61)
{52}=- (61) + (52)
{511} = (7) - (61) - (52) + (511)
{43}=- (52) + (43)
{421} = (61) - (511) - (43) + (421)
{4111}=- (7) + (61) + (52) - (511) + (43) -2(421) + (4111)
{331} = (52) - (43) - (421) + (331)
{322}=- (52) + (511) + (43) - (421) - (331) + (322)
{3211}=- (61) + (511) + (43) + (421) - (4111) - (331) - (322) + (3211)
{31111} = (7) - (61) - (52) + (511) -2(43) +2(421) - (4111) +2(331) + (322) -3(3211) + (31111)
{2221} = (52) - (511) - (43) + (4111) +2(331) - (322) -2(3211) + (2221)
{22111} = (61) - (52) - (511) + (4111) - (331) +2(322) + (3211) - (31111) -2(2221) + (22111)
{211111}=- (7) + (61) +2(52) - (511) +2(43) -4(421) + (4111) -2(331) -3(322) +6(3211) - (31111)
+3(2221) -4(22111) + (211111)
{1111111} = (7) -2(61) -2(52) +3(511) -2(43) +6(421) -4(4111) +3(331) +3(322) -12(3211) +5(31111)
-4(2221) +10(22111) -6(211111) + (1111111)

{8} = (8)
{71}=- (8) + (71)
{62}=- (71) + (62)
{611} = (8) - (71) - (62) + (611)
{53}=- (62) + (53)
{521} = (71) - (611) - (53) + (521)
{5111}=- (8) + (71) + (62) - (611) + (53) -2(521) + (5111)
{44}=- (53) + (44)
{431} = (62) - (521) - (44) + (431)
{422}=- (62) + (611) + (53) - (521) - (431) + (422)
{4211}=- (71) + (611) + (521) - (5111) + (44) - (431) - (422) + (4211)
{41111} = (8) - (71) - (62) + (611) - (53) +2(521) - (5111) - (44) +2(431) + (422) -3(4211)
+ (41111)
{332}=- (53) + (521) + (44) - (431) - (422) + (332)
{3311}=- (62) + (53) + (521) - (431) + (422) - (4211) - (332) + (3311)

Expansion of S-Functions into Products of the Symmetric Functions h_r (contd).

{3221} = (62) - (611) - (521) + (5111) - (44) +3(431) - (4211) - (332) - (3311) + (3221)
{32111} = (71) - (611) - (53) - (521) + (5111) +2(4211) - (41111) +2(332) - (3311) -2(3221)
+ (32111)
{311111}=- (8) + (71) + (62) - (611) +2(53) -2(521) + (5111) + (44) -4(431) - (422) +3(4211)
- (41111) -2(332) +3(3311) +3(3221) -4(32111) + (311111)
{2222}=- (53) +2(521) - (5111) + (44) -2(431) -2(422) +2(4211) +2(332) + (3311) -3(3221) + (2222)
{22211}=- (62) + (611) + (53) - (5111) -2(431) +2(422) - (4211) + (41111) - (332) +2(3311) + (3221)
-2(32111) - (2222) + (22211)
{221111}=- (71) + (62) + (611) - (5111) +2(431) -2(422) - (4211) + (41111) - (332) -2(3311) +3(3221)
+2(32111) - (311111) + (2222) -3(22211) + (221111)
{2111111} = (8) - (71) -2(62) + (611) -2(53) +4(521) - (5111) - (44) +4(431) +3(422) -6(4211)
+ (41111) +3(332) -3(3311) -9(3221) +8(32111) - (311111) - (2222) +6(22211) -5(221111)
+ (2111111)
{11111111}=- (8) +2(71) +2(62) -3(611) +2(53) -6(521) +4(5111) + (44) -6(431) -3(422) +12(4211)
-5(41111) -3(332) +6(3311) +12(3221) -20(32111) +6(311111) + (2222) -10(22211)
+15(221111) -7(2111111) + (11111111)

{9} = (9)
{81}=- (9) + (81)
{72}=- (81) + (72)
{711} = (9) - (81) - (72) + (711)
{63}=- (72) + (63)
{621} = (81) - (711) - (63) + (621)
{6111}=- (9) + (81) + (72) - (711) + (63) -2(621) + (6111)
{54}=- (63) + (54)
{531} = (72) - (621) - (54) + (531)
{522}=- (72) + (711) + (63) - (621) - (531) + (522)
{5211}=- (81) + (711) + (621) - (6111) + (54) - (531) - (522) + (5211)
{51111} = (9) - (81) - (72) + (711) - (63) +2(621) - (6111) - (54) +2(531) + (522) -3(5211)
+ (51111)
{441} = (63) - (54) - (531) + (441)
{432}=- (63) + (621) + (54) - (522) - (441) + (432)
{4311}=- (72) + (621) + (54) + (522) - (5211) - (441) - (432) + (4311)
{4221} = (72) - (711) - (621) + (6111) - (54) +2(531) - (5211) + (441) - (432) - (4311) + (4221)
{42111} = (81) - (711) - (621) + (6111) - (54) - (531) +2(5211) - (51111) + (441) +2(432) - (4311)
-2(4221) + (42111)
{411111}=- (9) + (81) + (72) - (711) + (63) -2(621) + (6111) +2(54) -2(531) - (522) +3(5211)
- (51111) -2(441) -2(432) +3(4311) +3(4221) -4(42111) + (411111)
{333}=- (531) + (522) + (441) -2(432) + (333)
{3321} = (63) - (621) - (54) + (5211) + (441) + (432) - (4311) - (4221) - (333) + (3321)
{33111} = (72) - (63) - (621) + (531) - (522) + (5211) - (4311) +2(4221) - (42111) + (333) -2(3321)
+ (33111)
{3222}=- (63) +2(621) - (6111) + (54) - (531) - (522) + (5211) - (441) +2(4311) - (4221) + (333)
-2(3321) + (3222)
{32211}=- (72) + (711) + (621) - (6111) + (54) - (531) + (522) -2(5211) + (51111) -2(441) +3(4311)
+ (4221) - (42111) - (33111) - (3222) + (32211)
{321111}=- (81) + (711) + (63) + (621) - (6111) -2(5211) + (51111) + (441) -2(432) - (4311) - (4221)
+3(42111) - (411111) - (333) +4(3321) - (33111) + (3222) -3(32211) + (321111)
{3111111} = (9) - (81) - (72) + (711) -2(63) +2(621) - (6111) -2(54) +4(531) + (522) -3(5211)
+ (51111) +2(441) +4(432) -6(4311) -3(4221) +4(42111) - (411111) + (333) -6(3321)
+4(33111) - (3222) +6(32211) -5(321111) + (3111111)
{22221} = (63) -2(621) + (6111) - (54) + (522) + (5211) - (51111) +2(441) -4(4311) - (4221)
+2(42111) - (333) +4(3321) + (33111) - (3222) -3(32211) + (22221)
{222111} = (72) - (711) - (63) + (6111) +2(531) -2(522) + (5211) - (51111) - (4311) +2(4221)
-2(42111) + (411111) + (333) -5(3321) +2(33111) +2(3222) +3(32211) -2(321111) -2(22221)
+ (222111)
{2211111} = (81) - (72) - (711) + (6111) -2(531) +2(522) + (5211) - (51111) - (441) +2(432) +4(4311)
-3(4221) -2(42111) + (411111) -3(33111) -3(3222) +3(32211) +3(321111) - (3111111)
+3(22221) -4(222111) + (2211111)
{21111111}=- (9) + (81) +2(72) - (711) +2(63) -4(621) + (6111) +2(54) -4(531) -3(522) +6(5211)
- (51111) -2(441) -6(432) +6(4311) +9(4221) -8(42111) + (411111) - (333) +9(3321)
-4(33111) +4(3222) -18(32211) +10(321111) - (3111111) -4(22221) +10(222111) -6(2211111)
+ (21111111)
{111111111} = (9) -2(81) -2(72) +3(711) -2(63) +6(621) -4(6111) -2(54) +6(531) +3(522) -12(5211)
+5(51111) +3(441) +6(432) -12(4311) -12(4221) +20(42111) -6(411111) + (333) -12(3321)
+10(33111) -4(3222) +30(32211) -30(321111) +7(3111111) +5(22221) -20(222111)
+21(2211111) -8(21111111) + (111111111)

{10} = (10)
{91}=- (10) + (91)

B-3 Expansion of S-Functions into Products of the Symmetric Functions h_r (contd).

```
{82}=- (91) + (82)
{811} = (10) - (91) - (82) + (811)
{73}=- (82) + (73)
{721} = (91) - (811) - (73) + (721)
{7111}=- (10) + (91) + (82) - (811) + (73) -2(721) + (7111)
{64}=- (73) + (64)
{631} = (82) - (721) - (64) + (631)
{622}=- (82) + (811) + (73) - (721) - (631) + (622)
{6211}=- (91) + (811) + (721) - (7111) + (64) - (631) - (622) + (6211)
{61111} = (10) - (91) - (82) + (811) - (73) +2(721) - (7111) - (64) +2(631) + (622) -3(6211)
        + (61111)
{55}=- (64) + (55)
{541} = (73) - (631) - (55) + (541)
{532}=- (73) + (721) + (64) - (622) - (541) + (532)
{5311}=- (82) + (721) + (622) - (6211) + (55) - (541) - (532) + (5311)
{5221} = (82) - (811) - (721) + (7111) - (64) +2(631) - (6211) + (541) - (532) - (5311) + (5221)
{52111} = (91) - (811) - (721) + (7111) - (631) +2(6211) - (61111) - (55) + (541) +2(532) - (5311)
        -2(5221) + (52111)
{511111}=- (10) + (91) + (82) - (811) + (73) -2(721) + (7111) + (64) -2(631) - (622) +3(6211)
         - (61111) + (55) -2(541) -2(532) +3(5311) +3(5221) -4(52111) + (511111)
{442}=- (64) + (631) + (55) - (541) - (532) + (442)
{4411}=- (73) + (64) + (631) - (541) + (532) - (5311) - (442) + (4411)
{433}=- (631) + (622) + (541) - (532) - (442) + (433)
{4321} = (73) - (721) - (631) + (6211) - (55) +2(541) + (532) - (5221) - (4411) - (433) + (4321)
{43111} = (82) - (721) - (64) - (622) + (6211) + (541) - (532) +2(5221) - (52111) + (442) - (4411)
        + (433) -2(4321) + (43111)
{4222}=- (73) +2(721) - (7111) + (64) - (631) - (622) + (6211) - (541) + (532) + (5311) - (5221)
       - (442) + (4411) + (433) -2(4321) + (4222)
{42211}=- (82) + (811) + (721) - (7111) + (622) -2(6211) + (61111) + (55) -3(541) - (532) +2(5311)
        + (5221) - (52111) + (442) + (4411) - (43111) - (4222) + (42211)
{421111}=- (91) + (811) + (721) - (7111) + (64) + (631) -2(6211) + (61111) -2(5311) - (5221)
         +3(52111) - (511111) -2(442) + (4411) - (433) +4(4321) - (43111) + (4222) -3(42211)
         + (421111)
{4111111} = (10) - (91) - (82) + (811) - (73) +2(721) - (7111) -2(64) +2(631) + (622) -3(6211)
          + (61111) - (55) +4(541) +2(532) -3(5311) -3(5221) +4(52111) - (511111) +2(442) -3(4411)
          + (433) -6(4321) +4(43111) - (4222) +6(42211) -5(421111) + (4111111)
{3331} = (631) - (622) - (541) + (532) - (5311) + (5221) + (442) + (4411) - (433) -2(4321) + (3331)
{3322}=- (64) + (631) + (622) - (6211) + (55) - (541) -3(532) + (5311) + (5221) + (442) + (433)
       - (4222) - (3331) + (3322)
{33211}=- (73) + (721) + (64) - (6211) - (541) + (532) - (5221) + (52111) -2(442) + (4411) + (433)
        +2(4321) - (43111) + (4222) - (42211) - (3331) - (3322) + (33211)
{331111}=- (82) + (73) + (721) - (631) + (622) - (6211) + (5311) -2(5221) + (52111) + (442) -2(433)
         - (43111) - (4222) +3(42211) - (421111) +2(3331) + (3322) -3(33211) + (331111)
{32221} = (73) -2(721) + (7111) - (631) +2(6211) - (61111) - (55) +3(541) +2(532) -3(5311) - (5221)
        + (52111) -2(4411) -2(433) +2(4321) +2(43111) - (42211) +2(3331) - (3322) -2(33211)
        + (32221)
{322111} = (82) - (811) - (721) + (7111) - (64) + (631) - (622) +2(6211) - (61111) +2(541) -2(532)
         +2(5221) -3(52111) + (511111) + (442) -2(4411) + (433) -4(4321) +3(43111) +2(42211)
         - (421111) - (3331) +2(3322) + (33211) - (331111) -2(32221) + (322111)
{3211111} = (91) - (811) - (73) - (721) + (7111) +2(6211) - (61111) -2(541) +2(532) + (5311)
          + (5221) -3(52111) + (511111) +2(4411) +2(433) -2(4321) -2(43111) -3(42211) +4(421111)
          - (4111111) -2(3331) -3(3322) +6(33211) - (331111) +3(32221) -4(322111) + (3211111)
{31111111}=- (10) + (91) + (82) - (811) +2(73) -2(721) + (7111) +2(64) -4(631) - (622) +3(6211)
           - (61111) + (55) -4(541) -4(532) +6(5311) +3(5221) -4(52111) + (511111) -2(442) +3(4411)
           -3(433) +12(4321) -8(43111) + (4222) -6(42211) +5(421111) - (4111111) +3(3331) +3(3322)
           -12(33211) +5(331111) -4(32221) +10(322111) -6(3211111) + (31111111)
{22222}=- (64) +2(631) + (622) -3(6211) + (61111) + (55) -2(541) -4(532) +2(5311) +4(5221) -2(52111)
        +2(442) + (4411) + (433) -2(4321) -2(43111) -3(4222) +3(42211) -2(3331) +3(3322) +3(33211)
        -4(32221) + (22222)
{222211}=- (73) +2(721) - (7111) + (64) - (622) - (6211) + (61111) -2(541) +2(532) + (5311) -2(5221)
         +2(52111) - (511111) -2(442) +3(4411) + (433) -4(43111) +3(4222) -3(42211) +2(421111)
         - (3331) -2(3322) +3(33211) + (331111) +2(32221) -3(322111) - (22222) + (222211)
{2221111}=- (82) + (811) + (73) - (7111) -2(631) +2(622) - (6211) + (61111) + (5311) -2(5221)
          +2(52111) - (511111) + (442) - (4411) -2(433) +4(4321) -3(4222) +3(42211) -3(421111)
          + (4111111) +3(3331) -9(33211) +2(331111) +2(32221) +5(322111) -2(3211111) + (22222)
          -3(222211) + (2221111)
{22111111}=- (91) + (82) + (811) - (7111) +2(631) -2(622) - (6211) + (61111) +2(541) -2(532)
           -4(5311) +3(5221) +2(52111) - (511111) - (442) -2(4411) +6(43111) +3(4222) -3(42211)
           -3(421111) + (4111111) - (3331) +3(3322) +3(33211) -4(331111) -8(32221) +2(322111)
           +4(3211111) - (31111111) - (22222) +6(222211) -5(2221111) + (22111111)
```

{211111111} = (10) - (91) -2(82) + (811) -2(73) +4(721) - (7111) -2(64) +4(631) +3(622) -6(6211)
+ (61111) - (55) +4(541) +6(532) -6(5311) -9(5221) +8(52111) - (511111) +3(442) -3(4411)
+3(433) -18(4321) +8(43111) -4(4222) +18(42211) -10(421111) + (4111111) -3(3331)
-6(3322) +18(33211) -5(331111) +16(32221) -30(322111) +12(3211111) - (31111111)
+ (22222) -10(222211) +15(2221111) -7(22111111) + (211111111)

{1111111111}=- (10) +2(91) +2(82) -3(811) +2(73) -6(721) +4(7111) +2(64) -6(631) -3(622) +12(6211)
-5(61111) + (55) -6(541) -6(532) +12(5311) +12(5221) -20(52111) +6(511111) -3(442)
+6(4411) -3(433) +24(4321) -20(43111) +4(4222) -30(42211) +30(421111) -7(4111111)
+4(3331) +6(3322) -30(33211) +15(331111) -20(32221) +60(322111) -42(3211111)
+8(31111111) - (22222) +15(222211) -35(2221111) +28(22111111) -9(211111111)
+ (1111111111)

{11} = (11)

{101}=- (11) + (101)

{92}=- (101) + (92)

{911} = (11) - (101) - (92) + (911)

{83}=- (92) + (83)

{821} = (101) - (911) - (83) + (821)

{8111}=- (11) + (101) + (92) - (911) + (83) -2(821) + (8111)

{74}=- (83) + (74)

{731} = (92) - (821) - (74) + (731)

{722}=- (92) + (911) + (83) - (821) - (731) + (722)

{7211}=- (101) + (911) + (821) - (8111) + (74) - (731) - (722) + (7211)

{71111} = (11) - (101) - (92) + (911) - (83) +2(821) - (8111) - (74) +2(731) + (722) -3(7211)
+ (71111)

{65}=- (74) + (65)

{641} = (83) - (731) - (65) + (641)

{632}=- (83) + (821) + (74) - (722) - (641) + (632)

{6311}=- (92) + (821) + (722) - (7211) + (65) - (641) - (632) + (6311)

{6221} = (92) - (911) - (821) + (8111) - (74) +2(731) - (7211) + (641) - (632) - (6311) + (6221)

{62111} = (101) - (911) - (821) + (8111) - (731) +2(7211) - (71111) - (65) + (641) +2(632) - (6311)
-2(6221) + (62111)

{611111}=- (11) + (101) + (92) - (911) + (83) -2(821) + (8111) + (74) -2(731) - (722) +3(7211)
- (71111) + (65) -2(641) -2(632) +3(6311) +3(6221) -4(62111) + (611111)

{551} = (74) - (65) - (641) + (551)

{542}=- (74) + (731) + (65) - (632) - (551) + (542)

{5411}=- (83) + (731) + (65) + (632) - (6311) - (551) - (542) + (5411)

{533}=- (731) + (722) + (641) - (632) - (542) + (533)

{5321} = (83) - (821) - (731) + (7211) - (65) + (641) + (632) - (6221) + (551) - (5411) - (533)
+ (5321)

{53111} = (92) - (821) - (722) + (7211) - (65) - (632) +2(6221) - (62111) + (551) + (542) - (5411)
+ (533) -2(5321) + (53111)

{5222}=- (83) +2(821) - (8111) + (74) - (731) - (722) + (7211) - (641) + (632) + (6311) - (6221)
- (542) + (5411) + (533) -2(5321) + (5222)

{52211}=- (92) + (911) + (821) - (8111) + (722) -2(7211) + (71111) + (65) -2(641) - (632) +2(6311)
+ (6221) - (62111) - (551) + (542) + (5411) - (53111) - (5222) + (52211)

{521111}=- (101) + (911) + (821) - (8111) + (731) -2(7211) + (71111) + (65) + (641) -2(6311)
- (6221) +3(62111) - (611111) - (551) -2(542) + (5411) - (533) +4(5321) - (53111) + (5222)
-3(52211) + (521111)

{5111111} = (11) - (101) - (92) + (911) - (83) +2(821) - (8111) - (74) +2(731) + (722) -3(7211)
+ (71111) -2(65) +2(641) +2(632) -3(6311) -3(6221) +4(62111) - (611111) +2(551) +2(542)
-3(5411) + (533) -6(5321) +4(53111) - (5222) +6(52211) -5(521111) + (5111111)

{443}=- (641) + (632) + (551) - (542) - (533) + (443)

{4421} = (74) - (731) - (65) + (6311) + (551) - (5411) + (533) - (5321) - (443) + (4421)

{44111} = (83) - (74) - (731) + (641) - (632) + (6311) + (542) - (5411) - (533) +2(5321) - (53111)
+ (443) -2(4421) + (44111)

{4331} = (731) - (722) - (6311) + (6221) - (551) +2(542) + (5411) - (5321) - (443) - (4421) + (4331)

{4322}=- (74) + (731) + (722) - (7211) + (65) -2(632) + (6221) - (551) + (5411) + (5321) - (5222)
+ (443) - (4421) - (4331) + (4322)

{43211}=- (83) + (821) + (731) - (7211) + (65) - (641) + (632) - (6311) - (6221) + (62111) - (551)
-2(542) +2(5411) + (5321) + (5222) - (52211) + (443) + (4421) - (44111) - (4331) - (4322)
+ (43211)

{431111}=- (92) + (821) + (74) + (722) - (7211) - (641) + (632) -2(6221) + (62111) + (5411) -2(5321)
- (5222) +3(52211) - (521111) -2(443) +2(4421) - (44111) +2(4331) + (4322) -3(43211)
+ (431111)

{42221} = (83) -2(821) + (8111) - (731) +2(7211) - (71111) - (65) +2(641) + (632) -2(6311) - (6221)
+ (62111) + (551) + (542) -3(5411) - (533) +2(5321) + (53111) - (52211) - (443) + (44111)
+2(4331) - (4322) -2(43211) + (42221)

{422111} = (92) - (911) - (821) + (8111) - (722) +2(7211) - (71111) - (65) + (641) - (632) + (6311)
+2(6221) -3(62111) + (611111) +2(551) -3(5411) + (533) -4(5321) +2(53111) +2(52211)
- (521111) + (44111) - (4331) +2(4322) + (43211) - (431111) -2(42221) + (422111)

{4211111} = (101) - (911) - (821) + (8111) - (74) - (731) +2(7211) - (71111) +2(6311) + (6221) -3(62111) + (611111) - (551) +2(542) + (5411) +2(5321) -3(53111) -3(52211) +4(521111) - (5111111) +2(443) -4(4421) + (44111) -2(4331) -3(4322) +6(43211) - (431111) +3(42221) -4(422111) + (4211111)

{41111111}=- (11) + (101) + (92) - (911) + (83) -2(821) + (8111) +2(74) -2(731) - (722) +3(7211) - (71111) +2(65) -4(641) -2(632) +3(6311) +3(6221) -4(62111) + (611111) -2(551) -4(542) +6(5411) - (533) +6(5321) -4(53111) + (5222) -6(52211) +5(521111) - (5111111) -2(443) +6(4421) -4(44111) +3(4331) +3(4322) -12(43211) +5(431111) -4(42221) +10(422111) -6(4211111) + (41111111)

{3332}=- (641) + (632) + (6311) - (6221) + (551) - (542) - (5411) - (533) + (5222) + (443) +2(4421) - (4331) -2(4322) + (3332)

{33311}=- (731) + (722) + (641) - (632) + (6311) - (6221) - (542) - (5411) + (533) +2(5321) - (53111) - (5222) + (52211) + (44111) - (4331) +2(4322) -2(43211) - (3332) + (33311)

{33221} = (74) - (731) - (722) + (7211) - (65) + (632) + (6221) - (62111) + (551) + (542) - (5411) + (533) -4(5321) + (53111) + (52211) -2(443) +3(4331) + (4322) - (42221) - (3332) - (33311) + (33221)

{332111} = (83) - (821) - (74) + (7211) + (641) - (632) + (6221) - (62111) +2(542) - (5411) -2(533) +2(5321) -2(52211) + (521111) + (443) -3(4421) + (44111) -2(4322) +3(43211) - (431111) +2(42221) - (422111) +2(3332) - (33311) -2(33221) + (332111)

{3311111} = (92) - (83) - (821) + (731) - (722) + (7211) - (6311) +2(6221) - (62111) -2(542) +2(533) + (53111) + (5222) -3(52211) + (521111) + (443) +2(4421) -4(4331) + (4322) - (431111) -3(42221) +4(422111) - (4211111) -2(3332) +3(33311) +3(33221) -4(332111) + (3311111)

{32222}=- (74) +2(731) + (722) -3(7211) + (71111) + (65) - (641) -2(632) + (6311) +2(6221) - (62111) - (551) +2(5411) - (533) +2(5321) -2(53111) - (5222) + (52211) +2(443) - (44111) -4(4331) - (4322) +4(43211) - (42221) +2(3332) + (33311) -3(33221) + (32222)

{322211}=- (83) +2(821) - (8111) + (731) -2(7211) + (71111) + (65) - (641) -2(6221) +3(62111) - (611111) -2(551) -2(542) +5(5411) + (533) +2(5321) -3(53111) + (5222) -2(52211) + (521111) + (443) -2(44111) -3(4331) + (4322) +2(431111) + (42221) - (422111) - (3332) +2(33311) + (33221) -2(332111) - (32222) + (322211)

{3221111}=- (92) + (911) + (821) - (8111) + (74) - (731) + (722) -2(7211) + (71111) -2(641) +2(632) -2(6221) +3(62111) - (611111) + (5411) -2(5321) + (53111) - (5222) +4(52211) -4(521111) + (5111111) -2(443) +4(4421) -2(44111) +5(4331) - (4322) -8(43211) +3(431111) - (42221) +3(422111) - (4211111) - (3332) -2(33311) +3(33221) +2(332111) - (3311111) + (32222) -3(322211) + (3221111)

{32111111}=- (101) + (911) + (83) + (821) - (8111) -2(7211) + (71111) +2(641) -2(632) - (6311) - (6221) +3(62111) - (611111) + (551) -4(5411) -2(533) +2(5321) +2(53111) +3(52211) -4(521111) + (5111111) - (443) -2(4421) +3(44111) +3(4331) +3(4322) -3(431111) + (42221) -6(422111) +5(4211111) - (41111111) +3(3332) -3(33311) -9(33221) +8(332111) - (3311111) - (32222) +6(322211) -5(3221111) + (32111111)

{311111111} = (11) - (101) - (92) + (911) -2(83) +2(821) - (8111) -2(74) +4(731) + (722) -3(7211) + (71111) -2(65) +4(641) +4(632) -6(6311) -3(6221) +4(62111) - (611111) +2(551) +4(542) -6(5411) +3(533) -12(5321) +8(53111) - (5222) +6(52211) -5(521111) + (5111111) +3(443) -6(4421) +4(44111) -9(4331) -6(4322) +24(43211) -10(431111) +4(42221) -10(422111) +6(4211111) - (41111111) -3(3332) +6(33311) +12(33221) -20(332111) +6(3311111) + (32222) -10(322211) +15(3221111) -7(32111111) + (311111111)

{222221} = (74) -2(731) - (722) +3(7211) - (71111) - (65) +2(632) + (6311) - (6221) -2(62111) + (611111) +2(551) -4(5411) + (533) -6(5321) +4(53111) + (5222) +3(52211) -2(521111) -2(443) +2(4421) +2(44111) +5(4331) + (4322) -6(43211) -2(431111) -2(42221) +3(422111) -2(3332) -3(33311) +6(33221) +3(332111) - (32222) -4(322211) + (222221)

{2222111} = (83) -2(821) + (8111) - (74) + (722) + (7211) - (71111) +2(641) -2(632) - (6311) +2(6221) -2(62111) + (611111) +2(542) -3(5411) -2(533) +6(5321) -2(5222) -3(52211) +3(521111) - (5111111) + (443) -4(4421) +3(44111) - (4331) -2(4322) +6(43211) -4(431111) +4(42221) -5(422111) +2(4211111) +3(3332) -9(33221) +2(332111) + (3311111) +2(32222) +5(322211) -3(3221111) -2(222221) + (2222111)

{22211111} = (92) - (911) - (83) + (8111) +2(731) -2(722) + (7211) - (71111) - (6311) +2(6221) -2(62111) + (611111) -2(542) +2(5411) +2(533) -4(5321) +3(5222) -3(52211) +3(521111) - (5111111) + (443) + (4421) -2(44111) -6(4331) +3(4322) +6(43211) + (431111) -6(42221) +5(422111) -4(4211111) + (41111111) -2(3332) +5(33311) +6(33221) -13(332111) +2(3311111) -3(32222) +7(3221111) -2(32111111) +3(222221) -4(2222111) + (22211111)

{221111111} = (101) - (92) - (911) + (8111) -2(731) +2(722) + (7211) - (71111) -2(641) +2(632) +4(6311) -3(6221) -2(62111) + (611111) - (551) +2(542) +4(5411) -6(53111) -3(5222) +3(52211) +3(521111) - (5111111) -3(44111) +3(4331) -6(4322) -6(43211) +8(431111) +8(42221) -2(422111) -4(4211111) + (41111111) - (3332) -3(33311) +6(33221) +8(332111) -5(3311111) +4(32222) -14(322211) +5(32111111) - (311111111) -4(222221) +10(2222111) -6(22211111) + (221111111)

{2111111111}=- (11) + (101) +2(92) - (911) +2(83) -4(821) + (8111) +2(74) -4(731) -3(722) +6(7211) - (71111) +2(65) -4(641) -6(632) +6(6311) +9(6221) -8(62111) + (611111) -2(551) -6(542) +6(5411) -3(533) +18(5321) -8(53111) +4(5222) -18(52211) +10(521111) - (5111111) -3(443) +9(4421) -4(44111) +9(4331) +12(4322) -36(43211) +10(431111) -16(42221) +30(422111) -12(4211111) + (41111111) +4(3332) -6(33311) -24(33221) +30(332111) -6(3311111) -5(32222) +40(322211) -45(3221111) +14(32111111) - (311111111) +5(222221) -20(2222111) +21(22211111) -8(221111111) + (2111111111)

Expansion of S-Functions into Products of the Symmetric Functions h_r (contd).

{11111111111} = (11) -2(101) -2(92) +3(911) -2(83) +6(821) -4(8111) -2(74) +6(731) +3(722) -12(7211)
+5(71111) -2(65) +6(641) +6(632) -12(6311) -12(6221) +20(62111) -6(611111) +3(551)
+6(542) -12(5411) +3(533) -24(5321) +20(53111) -4(5222) +30(52211) -30(521111)
+7(5111111) +3(443) -12(4421) +10(44111) -12(4331) -12(4322) +60(43211) -30(431111)
+20(42221) -60(422111) +42(4211111) -8(41111111) -4(3332) +10(33311) +30(33221)
-60(332111) +21(3311111) +5(32222) -60(322211) +105(3221111) -56(32111111)
+9(311111111) -6(222221) +35(2222111) -56(22211111) +36(221111111) -10(2111111111)
+ (11111111111)

{12} = (12)
{111}=- (12) + (111)
{102}=- (111) + (102)
{1011} = (12) - (111) - (102) + (1011)
{93}=- (102) + (93)
{921} = (111) - (1011) - (93) + (921)
{9111}=- (12) + (111) + (102) - (1011) + (93) -2(921) + (9111)
{84}=- (93) + (84)
{831} = (102) - (921) - (84) + (831)
{822}=- (102) + (1011) + (93) - (921) - (831) + (822)
{8211}=- (111) + (1011) + (921) - (9111) + (84) - (831) - (822) + (8211)
{81111} = (12) - (111) - (102) + (1011) - (93) +2(921) - (9111) - (84) +2(831) + (822) -3(8211)
+ (81111)
{75}=- (84) + (75)
{741} = (93) - (831) - (75) + (741)
{732}=- (93) + (921) + (84) - (822) - (741) + (732)
{7311}=- (102) + (921) + (822) - (8211) + (75) - (741) - (732) + (7311)
{7221} = (102) - (1011) - (921) + (9111) - (84) +2(831) - (8211) + (741) - (732) - (7311) + (7221)
{72111} = (111) - (1011) - (921) + (9111) - (831) +2(8211) - (81111) - (75) + (741) +2(732) - (7311)
-2(7221) + (72111)
{711111}=- (12) + (111) + (102) - (1011) + (93) -2(921) + (9111) + (84) -2(831) - (822) +3(8211)
- (81111) + (75) -2(741) -2(732) +3(7311) +3(7221) -4(72111) + (711111)
{66}=- (75) + (66)
{651} = (84) - (741) - (66) + (651)
{642}=- (84) + (831) + (75) - (732) - (651) + (642)
{6411}=- (93) + (831) + (732) - (7311) + (66) - (651) - (642) + (6411)
{633}=- (831) + (822) + (741) - (732) - (642) + (633)
{6321} = (93) - (921) - (831) + (8211) - (75) + (741) + (732) - (7221) + (651) - (6411) - (633)
+ (6321)
{63111} = (102) - (921) - (822) + (8211) - (732) +2(7221) - (72111) - (66) + (651) + (642) - (6411)
+ (633) -2(6321) + (63111)
{6222}=- (93) +2(921) - (9111) + (84) - (831) - (822) + (8211) - (741) + (732) + (7311) - (7221)
- (642) + (6411) + (633) -2(6321) + (6222)
{62211}=- (102) + (1011) + (921) - (9111) + (822) -2(8211) + (81111) + (75) -2(741) - (732) +2(7311)
+ (7221) - (72111) - (651) + (642) + (6411) - (63111) - (6222) + (62211)
{621111}=- (111) + (1011) + (921) - (9111) + (831) -2(8211) + (81111) + (741) -2(7311) - (7221)
+3(72111) - (711111) + (66) - (651) -2(642) + (6411) - (633) +4(6321) - (63111) + (6222)
-3(62211) + (621111)
{6111111} = (12) - (111) - (102) + (1011) - (93) +2(921) - (9111) - (84) +2(831) + (822) -3(8211)
+ (81111) - (75) +2(741) +2(732) -3(7311) -3(7221) +4(72111) - (711111) - (66) +2(651)
+2(642) -3(6411) + (633) -6(6321) +4(63111) - (6222) +6(62211) -5(621111) + (6111111)
{552}=- (75) + (741) + (66) - (651) - (642) + (552)
{5511}=- (84) + (75) + (741) - (651) + (642) - (6411) - (552) + (5511)
{543}=- (741) + (732) + (651) - (633) - (552) + (543)
{5421} = (84) - (831) - (741) + (7311) - (66) +2(651) + (633) - (6321) - (5511) - (543) + (5421)
{54111} = (93) - (831) - (75) - (732) + (7311) + (651) - (633) +2(6321) - (63111) + (552) - (5511)
+ (543) -2(5421) + (54111)
{5331} = (831) - (822) - (7311) + (7221) - (651) + (642) + (6411) - (6321) + (552) - (543) - (5421)
+ (5331)
{5322}=- (84) + (831) + (822) - (8211) + (75) -2(732) + (7221) - (651) + (642) + (6321) - (6222)
- (552) + (5511) + (543) - (5421) - (5331) + (5322)
{53211}=- (93) + (921) + (831) - (8211) + (732) - (7311) - (7221) + (72111) + (66) -2(651) -2(642)
+ (6411) + (6321) + (6222) - (62211) + (5511) + (543) + (5421) - (54111) - (5331) - (5322)
+ (53211)
{531111}=- (102) + (921) + (822) - (8211) + (75) + (732) -2(7221) + (72111) - (651) + (642) -2(6321)
- (6222) +3(62211) - (621111) - (552) + (5511) -2(543) +2(5421) - (54111) +2(5331) + (5322)
-3(53211) + (531111)
{52221} = (93) -2(921) + (9111) - (831) +2(8211) - (81111) - (75) +2(741) + (732) -2(7311) - (7221)
+ (72111) + (651) -2(6411) - (633) +2(6321) + (63111) - (62211) + (552) - (5511) - (543)
+ (54111) +2(5331) - (5322) -2(53211) + (52221)
{522111} = (102) - (1011) - (921) + (9111) - (822) +2(8211) - (81111) - (732) + (7311) +2(7221)
-3(72111) + (711111) - (66) +3(651) + (642) -2(6411) + (633) -4(6321) +2(63111) +2(62211)
- (621111) - (552) - (5511) + (54111) - (5331) +2(5322) + (53211) - (531111) -2(52221)
+ (522111)

Expansion of Orthogonal Group Characters into S-Functions.

```
[0] = {0}

[1] = {1}

[2]=- {0} + {2}
[11] = {11}

[3]=- {1} + {3}
[21]=- {1} + {21}
[111] = {111}

[4]=- {2} + {4}
[31] = {0} - {11} - {2} + {31}
[22]=- {2} + {22}
[211]=- {11} + {211}
[1111] = {1111}

[41] = {1} - {21} - {3} + {41}
[32] = {1} - {21} - {3} + {32}
[311] = {1} - {111} - {21} + {311}
[221]=- {21} + {221}
[2111]=- {111} + {2111}
[11111] = {11111}

[42] = {11} + {2} - {22} - {31} - {4} + {42}
[411]=- {0} + {11} + {2} - {211} - {31} + {411}
[33]=- {0} + {2} - {31} + {33}
[321] = {11} + {2} - {211} - {22} - {31} + {321}
[3111] = {11} - {1111} - {211} + {3111}
[222]=- {22} + {222}
[2211]=- {211} + {2211}
[21111]=- {1111} + {21111}
[111111] = {111111}

[43]=- {1} + {21} + {3} - {32} - {41} + {43}
[421]=- {1} + {111} +2{21} - {221} + {3} - {311} - {32} - {41} + {421}
[4111]=- {1} + {111} + {21} - {2111} - {311} + {4111}
[331]=- {1} + {21} + {3} - {311} - {32} + {331}
[322] = {21} - {221} - {32} + {322}
[3211] = {111} + {21} - {2111} - {221} - {311} + {3211}
[31111] = {111} - {11111} - {2111} + {31111}
[2221]=- {221} + {2221}
[22111]=- {2111} + {22111}
[211111]=- {11111} + {211111}
[1111111] = {1111111}

[44]=- {11} + {31} - {42} + {44}
[431] = {0} - {11} -2{2} + {211} + {22} +2{31} - {321} - {33} + {4} - {411} - {42} + {431}
[422]=- {11} + {211} + {22} - {222} + {31} - {321} - {42} + {422}
[4211]=- {11} + {1111} - {2} +2{211} + {22} - {2211} + {31} - {3111} - {321} - {411} + {4211}
[41111]=- {11} + {1111} + {211} - {21111} - {3111} + {41111}
[332]=- {2} + {22} + {31} - {321} - {33} + {332}
[3311]=- {11} + {211} + {31} - {3111} - {321} + {3311}
[3221] = {211} + {22} - {2211} - {222} - {321} + {3221}
[32111] = {1111} + {211} - {21111} - {2211} - {3111} + {32111}
[311111] = {1111} - {111111} - {21111} + {311111}
[2222]=- {222} + {2222}
[22211]=- {2211} + {22211}
[221111]=- {21111} + {221111}
[2111111]=- {111111} + {2111111}
[11111111] = {11111111}

[441] = {1} - {111} - {21} - {3} + {311} + {32} + {41} - {421} - {43} + {441}
[432] = {1} -2{21} + {221} - {3} + {311} +2{32} - {322} - {331} + {41} - {421} - {43} + {432}
[4311] = {1} - {111} -2{21} + {2111} + {221} - {3} +2{311} + {32} - {3211} - {331} + {41} - {4111}
        - {421} + {4311}
[4221]=- {111} - {21} + {2111} +2{221} - {2221} + {311} + {32} - {3211} - {322} - {421} + {4221}
[42111]=- {111} + {11111} - {21} +2{2111} + {221} - {22111} + {311} - {31111} - {3211} - {4111}
         + {42111}
[411111]=- {111} + {11111} + {2111} - {211111} - {31111} + {411111}
```

Expansion of Orthogonal Group Characters into S-Functions (contd).

[333]=- {3} + {32} - {331} + {333}
[3321]=- {21} + {221} + {311} + {32} - {3211} - {322} - {331} + {3321}
[33111]=- {111} + {2111} + {311} - {31111} - {3211} + {33111}
[3222] = {221} - {2221} - {322} + {3222}
[32211] = {2111} + {221} - {22111} - {2221} - {3211} + {32211}
[321111] = {11111} + {2111} - {211111} - {22111} - {31111} + {321111}
[3111111] = {11111} - {1111111} - {211111} + {3111111}
[22221]=- {2221} + {22221}
[222111]=- {22111} + {222111}
[2211111]=- {211111} + {2211111}
[21111111]=- {1111111} + {21111111}
[111111111] = {111111111}

[442]=- {0} + {11} + {2} - {211} -2{31} + {321} + {33} + {411} + {42} - {422} - {431} - {44} + {442}
[4411] = {11} - {1111} + {2} - {211} - {22} - {31} + {3111} + {321} - {4} + {411} + {42} - {4211}
 - {431} + {4411}
[433] = {2} - {22} - {31} + {321} + {33} - {332} - {4} + {42} - {431} + {433}
[4321] = {11} + {2} -2{211} -2{22} + {2211} + {222} -2{31} + {3111} +3{321} - {3221} + {33} - {3311}
 - {332} + {411} + {42} - {4211} - {422} - {431} + {4321}
[43111] = {11} - {1111} -2{211} + {21111} + {2211} - {31} +2{3111} + {321} - {32111} - {3311}
 + {411} - {41111} - {4211} + {43111}
[4222]=- {211} + {2211} + {222} - {2222} + {321} - {3221} - {422} + {4222}
[42211]=- {1111} - {211} + {21111} - {22} +2{2211} + {222} - {22211} + {3111} + {321} - {32111}
 - {3221} - {4211} + {42211}
[421111]=- {1111} + {111111} - {211} +2{21111} + {2211} - {221111} + {3111} - {311111} - {32111}
 - {41111} + {421111}
[4111111]=- {1111} + {111111} + {21111} - {2111111} - {311111} + {4111111}
[3331]=- {31} + {321} + {33} - {3311} - {332} + {3331}
[3322]=- {22} + {222} + {321} - {3221} - {332} + {3322}
[33211]=- {211} + {2211} + {3111} + {321} - {32111} - {3221} - {3311} + {33211}
[331111]=- {1111} + {21111} + {3111} - {311111} - {32111} + {331111}
[32221] = {2211} + {222} - {22211} - {2222} - {3221} + {32221}
[322111] = {21111} + {2211} - {221111} - {22211} - {32111} + {322111}
[3211111] = {111111} + {21111} - {2111111} - {221111} - {311111} + {3211111}
[31111111] = {111111} - {11111111} - {2111111} + {31111111}
[22222]=- {2222} + {22222}
[222211]=- {22211} + {222211}
[2221111]=- {221111} + {2221111}
[22111111]=- {2111111} + {22111111}
[211111111]=- {11111111} + {211111111}
[1111111111] = {1111111111}

[443]=- {1} + {21} + {3} - {311} - {32} + {331} - {41} + {421} + {43} - {432} - {441} + {443}
[4421]=- {1} + {111} +2{21} - {2111} - {221} + {3} -2{311} -2{32} + {3211} + {322} + {331} - {41}
 + {4111} +2{421} - {4221} + {43} - {4311} - {432} - {441} + {4421}
[44111] = {111} - {11111} + {21} - {2111} - {221} - {311} + {31111} + {3211} - {41} + {4111} + {421}
 - {42111} - {4311} + {44111}
[4331] = {21} - {221} + {3} - {311} -2{32} + {3211} + {322} +2{331} - {3321} - {333} - {41} + {421}
 + {43} - {4311} - {432} + {4331}
[4322] = {21} -2{221} + {2221} - {311} - {32} + {3211} +2{322} - {3222} + {331} - {3321} + {421}
 - {4221} - {432} + {4322}
[43211] = {111} + {21} -2{2111} -2{221} + {22111} + {2221} -2{311} + {31111} - {32} +3{3211} + {322}
 - {32211} + {331} - {33111} - {3321} + {4111} + {421} - {42111} - {4221} - {4311} + {43211}
[431111] = {111} - {11111} -2{2111} + {211111} + {22111} - {311} +2{31111} + {3211} - {321111}
 - {33111} + {4111} - {411111} - {42111} + {431111}
[42221]=- {2111} - {221} + {22111} +2{2221} - {22221} + {3211} + {322} - {32211} - {3222} - {4221}
 + {42221}
[422111]=- {11111} - {2111} + {211111} - {221} +2{22111} + {2221} - {222111} + {31111} + {3211}
 - {321111} - {32211} - {42111} + {422111}
[4211111]=- {11111} + {1111111} - {2111} +2{211111} + {22111} - {2211111} + {31111} - {3111111}
 - {321111} - {411111} + {4211111}
[41111111]=- {11111} + {1111111} + {211111} - {21111111} - {3111111} + {41111111}
[3332]=- {32} + {322} + {331} - {3321} - {333} + {3332}
[33311]=- {311} + {3211} + {331} - {33111} - {3321} + {33311}
[33221]=- {221} + {2221} + {3211} + {322} - {32211} - {3222} - {3321} + {33221}
[332111]=- {2111} + {22111} + {31111} + {3211} - {321111} - {32211} - {33111} + {332111}
[3311111]=- {11111} + {211111} + {31111} - {3111111} - {321111} + {3311111}
[32222] = {2221} - {22221} - {3222} + {32222}
[322211] = {22111} + {2221} - {222111} - {22221} - {32211} + {322211}
[3221111] = {211111} + {22111} - {2211111} - {222111} - {321111} + {3221111}

[32111111] = {1111111} + {211111} - {21111111} - {2211111} - {3111111} + {32111111}

[311111111] = {1111111} - {111111111} - {21111111} + {311111111}

[222221]=- {22221} + {222221}

[2222111]=- {222111} + {2222111}

[22211111]=- {2211111} + {22211111}

[221111111]=- {21111111} + {221111111}

[2111111111]=- {111111111} + {2111111111}

[11111111111] = {11111111111}

[444] = {0} - {2} + {31} - {33} - {411} + {431} - {442} + {444}

[4431]=- {11} - {2} + {211} + {22} +2{31} - {3111} -2{321} - {33} + {3311} + {332} + {4} - {411} -2{42} + {4211} + {422} +2{431} - {4321} - {433} + {44} - {4411} - {442} + {4431}

[4422]=- {2} + {211} + {22} - {2211} + {31} -2{321} + {3221} - {33} + {332} - {411} + {4211} + {422} - {4222} + {431} - {4321} - {442} + {4422}

[44211]=- {11} + {1111} +2{211} - {21111} + {22} - {2211} - {222} + {31} -2{3111} -2{321} + {32111} + {3221} + {3311} - {411} + {41111} - {42} +2{4211} + {422} - {42211} + {431} - {43111} - {4321} - {4411} + {44211}

[441111] = {1111} - {111111} + {211} - {21111} - {2211} - {3111} + {311111} + {32111} - {411} + {41111} + {4211} - {421111} - {43111} + {441111}

[4332] = {22} - {222} + {31} -2{321} + {3221} - {33} + {3311} +2{332} - {3322} - {3331} - {42} + {422} + {431} - {4321} - {433} + {4332}

[43311] = {211} - {2211} + {31} - {3111} -2{321} + {32111} + {3221} - {33} +2{3311} + {332} - {33211} - {3331} - {411} + {4211} + {431} - {43111} - {4321} + {43311}

[43221] = {211} + {22} -2{2211} -2{222} + {22211} + {2222} - {3111} -2{321} + {32111} +3{3221} - {32221} + {3311} + {332} - {33211} - {3322} + {4211} + {422} - {42211} - {4222} - {4321} + {43221}

[432111] = {1111} + {211} -2{21111} -2{2211} + {221111} + {22211} -2{3111} + {311111} - {321} +3{32111} + {3221} - {322111} + {3311} - {331111} - {33211} + {41111} + {4211} - {421111} - {42211} - {43111} + {432111}

[4311111] = {1111} - {111111} -2{21111} + {2111111} + {221111} - {3111} +2{311111} + {32111} - {3211111} - {331111} + {41111} - {4111111} - {421111} + {4311111}

[42222]=- {2211} + {22211} + {2222} - {22222} + {3221} - {32221} - {4222} + {42222}

[422211]=- {21111} - {2211} + {221111} - {222} +2{22211} + {2222} - {222211} + {32111} + {3221} - {322111} - {32221} - {42211} + {422211}

[4221111]=- {111111} - {21111} + {2111111} - {2211} +2{221111} + {22211} - {2221111} + {311111} + {32111} - {3211111} - {322111} - {421111} + {4221111}

[42111111]=- {111111} + {11111111} - {21111} +2{2111111} + {221111} - {22111111} + {311111} - {31111111} - {3211111} - {4111111} + {42111111}

[411111111]=- {111111} + {11111111} + {2111111} - {211111111} - {31111111} + {411111111}

[3333]=- {33} + {332} - {3331} + {3333}

[33321]=- {321} + {3221} + {3311} + {332} - {33211} - {3322} - {3331} + {33321}

[333111]=- {3111} + {32111} + {3311} - {331111} - {33211} + {333111}

[33222]=- {222} + {2222} + {3221} - {32221} - {3322} + {33222}

[332211]=- {2211} + {22211} + {32111} + {3221} - {322111} - {32221} - {33211} + {332211}

[3321111]=- {21111} + {221111} + {311111} + {32111} - {3211111} - {322111} - {331111} + {3321111}

[33111111]=- {111111} + {2111111} + {311111} - {31111111} - {3211111} + {33111111}

[322221] = {22211} + {2222} - {222211} - {22222} - {32221} + {322221}

[3222111] = {221111} + {22211} - {2221111} - {222211} - {322111} + {3222111}

[32211111] = {2111111} + {221111} - {22111111} - {2221111} - {3211111} + {32211111}

[321111111] = {11111111} + {2111111} - {211111111} - {22111111} - {31111111} + {321111111}

[3111111111] = {11111111} - {1111111111} - {211111111} + {3111111111}

[222222]=- {22222} + {222222}

[2222211]=- {222211} + {2222211}

[22221111]=- {2221111} + {22221111}

[222111111]=- {22111111} + {222111111}

[2211111111]=- {211111111} + {2211111111}

[21111111111]=- {1111111111} + {21111111111}

[111111111111] = {111111111111}

[4441] = {1} - {21} - {3} + {311} + {32} - {331} + {41} - {4111} - {421} - {43} + {4311} + {432} + {441} - {4421} - {443} + {4441}

[4432]=- {21} + {221} - {3} + {311} +2{32} - {3211} - {322} -2{331} + {3321} + {333} + {41} -2{421} + {4221} - {43} + {4311} +2{432} - {4322} - {4331} + {441} - {4421} - {443} + {4432}

[44311]=- {111} - {21} + {2111} + {221} +2{311} - {31111} + {32} -2{3211} - {322} - {331} + {33111} + {3321} + {41} - {4111} -2{421} + {42111} + {4221} - {43} +2{4311} + {432} - {43211} - {4331} + {441} - {44111} - {4421} + {44311}

[44221]=- {21} + {2111} +2{221} - {22111} - {2221} + {311} + {32} -2{3211} -2{322} + {32211} + {3222} - {331} + {3321} - {4111} - {421} + {42111} +2{4221} - {42221} + {4311} + {432} - {43211} - {4322} - {4421} + {44221}

[442111]=- {111} + {11111} +2{2111} - {211111} + {221} - {22111} - {2221} + {311} -2{31111} -2{3211} + {321111} + {32211} + {33111} - {4111} + {411111} - {421} +2{42111} + {4221} - {422111} + {4311} - {431111} - {43211} - {44111} + {442111}

[4411111] = {11111} - {1111111} + {2111} - {211111} - {22111} - {31111} + {3111111} + {321111}
- {4111} + {411111} + {42111} - {4211111} - {431111} + {4411111}
[4333] = {32} - {322} - {331} + {3321} + {333} - {3332} - {43} + {432} - {4331} + {4333}
[43321] = {221} - {2221} + {311} + {32} -2{3211} -2{322} + {32211} + {3222} -2{331} + {33111}
+3{3321} - {33221} + {333} - {33311} - {3332} - {421} + {4221} + {4311} + {432} - {43211}
- {4322} - {4331} + {43321}
[433111] = {2111} - {22111} + {311} - {31111} -2{3211} + {321111} + {32211} - {331} +2{33111}
+ {3321} - {332111} - {33311} - {4111} + {42111} + {4311} - {431111} - {43211} + {433111}
[43222] = {221} -2{2221} + {22221} - {3211} - {322} + {32211} +2{3222} - {32222} + {3321} - {33221}
+ {4221} - {42221} - {4322} + {43222}
[432211] = {2111} + {221} -2{22111} -2{2221} + {222111} + {22221} - {31111} -2{3211} + {321111}
- {322} +3{32211} + {3222} - {322211} + {33111} + {3321} - {332111} - {33221} + {42111}
+ {4221} - {422111} - {42221} - {43211} + {432211}
[4321111] = {11111} + {2111} -2{211111} -2{22111} + {2211111} + {222111} -2{31111} + {3111111}
- {3211} +3{321111} + {32211} - {3221111} + {33111} - {3311111} - {332111} + {411111}
+ {42111} - {4211111} - {422111} - {431111} + {4321111}
[43111111] = {11111} - {1111111} -2{211111} + {21111111} + {2211111} - {31111} +2{3111111}
+ {321111} - {32111111} - {3311111} + {411111} - {41111111} - {4211111} + {43111111}
[422221]=- {22111} - {2221} + {222111} +2{22221} - {222221} + {32211} + {3222} - {322211} - {32222}
- {42221} + {422221}
[4222111]=- {211111} - {22111} + {2211111} - {2221} +2{222111} + {22221} - {2222111} + {321111}
+ {32211} - {3221111} - {322211} - {422111} + {4222111}
[42211111]=- {1111111} - {211111} + {21111111} - {22111} +2{2211111} + {222111} - {22211111}
+ {3111111} + {321111} - {32111111} - {3221111} - {4211111} + {42211111}
[421111111]=- {1111111} + {111111111} - {211111} +2{21111111} + {2211111} - {221111111} + {3111111}
- {311111111} - {32111111} - {41111111} + {421111111}
[4111111111]=- {1111111} + {111111111} + {21111111} - {2111111111} - {311111111} + {4111111111}
[33331]=- {331} + {3321} + {333} - {33311} - {3332} + {33331}
[33322]=- {322} + {3222} + {3321} - {33221} - {3332} + {33322}
[333211]=- {3211} + {32211} + {33111} + {3321} - {332111} - {33221} - {33311} + {333211}
[3331111]=- {31111} + {321111} + {33111} - {3311111} - {332111} + {3331111}
[332221]=- {2221} + {22221} + {32211} + {3222} - {322211} - {32222} - {33221} + {332221}
[3322111]=- {22111} + {222111} + {321111} + {32211} - {3221111} - {322211} - {332111} + {3322111}
[33211111]=- {211111} + {2211111} + {3111111} + {321111} - {32111111} - {3221111} - {3311111}
+ {33211111}
[331111111]=- {1111111} + {21111111} + {3111111} - {311111111} - {32111111} + {331111111}
[322222] = {22221} - {222221} - {32222} + {322222}
[3222211] = {222111} + {22221} - {2222111} - {222221} - {322211} + {3222211}
[32221111] = {2211111} + {222111} - {22211111} - {2222111} - {3221111} + {32221111}
[322111111] = {21111111} + {2211111} - {221111111} - {22211111} - {32111111} + {322111111}
[3211111111] = {111111111} + {21111111} - {2111111111} - {221111111} - {311111111} + {3211111111}
[31111111111] = {111111111} - {11111111111} - {2111111111} + {31111111111}
[2222221]=- {222221} + {2222221}
[22222111]=- {2222111} + {22222111}
[222211111]=- {22211111} + {222211111}
[2221111111]=- {221111111} + {2221111111}
[22111111111]=- {2111111111} + {22111111111}
[211111111111]=- {11111111111} + {211111111111}
[1111111111111] = {1111111111111}

[4442] = {2} - {22} - {31} + {321} + {33} - {332} - {4} + {411} + {42} - {4211} -2{431} + {4321}
+ {433} + {4411} + {442} - {4422} - {4431} - {444} + {4442}
[44411] = {11} - {211} - {31} + {3111} + {321} - {3311} + {411} - {41111} + {42} - {4211} - {422}
- {431} + {43111} + {4321} - {44} + {4411} + {442} - {44211} - {4431} + {44411}
[4433]=- {31} + {321} + {33} - {3311} - {332} + {3331} + {42} - {422} - {431} + {4321} + {433}
- {4332} - {44} + {442} - {4431} + {4433}
[44321]=- {211} - {22} + {2211} + {222} - {31} + {3111} +3{321} - {32111} -2{3221} + {33} -2{3311}
-2{332} + {33211} + {3322} + {3331} + {411} + {42} -2{4211} -2{422} + {42211} + {4222}
-2{431} + {43111} +3{4321} - {43221} + {433} - {43311} - {4332} + {4411} + {442} - {44211}
- {4422} - {4431} + {44321}
[443111]=- {1111} - {211} + {21111} + {2211} +2{3111} - {311111} + {321} -2{32111} - {3221} - {3311}
+ {331111} + {33211} + {411} - {41111} -2{4211} + {421111} + {42211} - {431} +2{43111}
+ {4321} - {432111} - {43311} + {4411} - {441111} - {44211} + {443111}
[44222]=- {22} + {2211} + {222} - {22211} + {321} -2{3221} + {32221} - {332} + {3322} - {4211}
+ {42211} + {4222} - {42222} + {4321} - {43221} - {4422} + {44222}
[442211]=- {211} + {21111} +2{2211} - {221111} + {222} - {22211} - {2222} + {3111} + {321} -2{32111}
-2{3221} + {322111} + {32221} - {3311} + {33211} - {41111} - {4211} + {421111} - {422}
+2{42211} + {4222} - {422211} + {43111} + {4321} - {432111} - {43221} - {44211} + {442211}
[4421111]=- {1111} + {111111} +2{21111} - {2111111} + {2211} - {221111} - {22211} + {3111}
-2{311111} -2{32111} + {3211111} + {322111} + {331111} - {41111} + {4111111} - {4211}
+2{421111} + {42211} - {4221111} + {43111} - {4311111} - {432111} - {441111} + {4421111}

B-4 Expansion of Orthogonal Group Characters into S-Functions (contd).

[44111111] = {111111} - {11111111} + {21111} - {2111111} - {221111} - {311111} + {31111111}
+ {3211111} - {41111} + {4111111} + {421111} - {42111111} - {4311111} + {44111111}
[43331] = {321} - {3221} + {33} - {3311} -2{332} + {33211} + {3322} +2{3331} - {33321} - {3333}
- {431} + {4321} + {433} - {43311} - {4332} + {43331}
[43322] = {222} - {2222} + {321} -2{3221} + {32221} - {3311} - {332} + {33211} +2{3322} - {33222}
+ {3331} - {33321} - {422} + {4222} + {4321} - {43221} - {4332} + {43322}
[433211] = {2211} - {22211} + {3111} + {321} -2{32111} -2{3221} + {322111} + {32221} -2{3311}
+ {331111} - {332} +3{33211} + {3322} - {332211} + {3331} - {333111} - {33321} - {4211}
+ {42211} + {43111} + {4321} - {432111} - {43221} - {43311} + {433211}
[4331111] = {21111} - {221111} + {3111} - {311111} -2{32111} + {3211111} + {322111} - {3311}
+2{331111} + {33211} - {3321111} - {333111} - {41111} + {421111} + {43111} - {4311111}
- {432111} + {4331111}
[432221] = {2211} + {222} -2{22211} -2{2222} + {222211} + {22222} - {32111} -2{3221} + {322111}
+3{32221} - {322221} + {33211} + {3322} - {332211} - {33222} + {42211} + {4222} - {422211}
- {42222} - {43221} + {432221}
[4322111] = {21111} + {2211} -2{221111} -2{22211} + {2221111} + {222211} - {311111} -2{32111}
+ {3211111} - {3221} +3{322111} + {32221} - {3222111} + {331111} + {33211} - {3321111}
- {332211} + {421111} + {42211} - {4221111} - {422211} - {432111} + {4322111}
[43211111] = {111111} + {21111} -2{2111111} -2{221111} + {22111111} + {2221111} -2{311111}
+ {31111111} - {32111} +3{3211111} + {322111} - {32211111} + {331111} - {33111111}
- {3321111} + {4111111} + {421111} - {42111111} - {4221111} - {4311111} + {43211111}
[431111111] = {111111} - {11111111} -2{2111111} + {211111111} + {22111111} - {311111} +2{31111111}
+ {3211111} - {321111111} - {33111111} + {4111111} - {411111111} - {42111111}
+ {431111111}
[422222]=- {22211} + {222211} + {22222} - {222222} + {32221} - {322221} - {42222} + {422222}
[4222211]=- {221111} - {22211} + {2221111} - {2222} +2{222211} + {22222} - {2222211} + {322111}
+ {32221} - {3222111} - {322221} - {422211} + {4222211}
[42221111]=- {2111111} - {221111} + {22111111} - {22211} +2{2221111} + {222211} - {22221111}
+ {3211111} + {322111} - {32211111} - {3222111} - {4221111} + {42221111}
[422111111]=- {11111111} - {2111111} + {211111111} - {221111} +2{22111111} + {2221111} - {222111111}
+ {31111111} + {3211111} - {321111111} - {32211111} - {42111111} + {422111111}
[4211111111]=- {11111111} + {1111111111} - {2111111} +2{211111111} + {22111111} - {2211111111}
+ {31111111} - {3111111111} - {321111111} - {411111111} + {4211111111}
[41111111111]=- {11111111} + {1111111111} + {211111111} - {21111111111} - {3111111111}
+ {41111111111}
[33332]=- {332} + {3322} + {3331} - {33321} - {3333} + {33332}
[333311]=- {3311} + {33211} + {3331} - {333111} - {33321} + {333311}
[333221]=- {3221} + {32221} + {33211} + {3322} - {332211} - {33222} - {33321} + {333221}
[3332111]=- {32111} + {322111} + {331111} + {33211} - {3321111} - {332211} - {333111} + {3332111}
[33311111]=- {311111} + {3211111} + {331111} - {33111111} - {3321111} + {33311111}
[332222]=- {2222} + {22222} + {32221} - {322221} - {33222} + {332222}
[3322211]=- {22211} + {222211} + {322111} + {32221} - {3222111} - {322221} - {332211} + {3322211}
[33221111]=- {221111} + {2221111} + {3211111} + {322111} - {32211111} - {3222111} - {3321111}
+ {33221111}
[332111111]=- {2111111} + {22111111} + {31111111} + {3211111} - {321111111} - {32211111}
- {33111111} + {332111111}
[3311111111]=- {11111111} + {211111111} + {31111111} - {3111111111} - {321111111} + {3311111111}
[3222221] = {222211} + {22222} - {2222211} - {222222} - {322221} + {3222221}
[32222111] = {2221111} + {222211} - {22221111} - {2222211} - {3222111} + {32222111}
[322211111] = {22111111} + {2221111} - {222111111} - {22221111} - {32211111} + {322211111}
[3221111111] = {211111111} + {22111111} - {2211111111} - {222111111} - {321111111} + {3221111111}
[32111111111] = {1111111111} + {211111111} - {21111111111} - {2211111111} - {3111111111}
+ {32111111111}
[311111111111] = {1111111111} - {111111111111} - {21111111111} + {311111111111}
[2222222]=- {222222} + {2222222}
[22222211]=- {2222211} + {22222211}
[222221111]=- {22221111} + {222221111}
[2222111111]=- {222111111} + {2222111111}
[22211111111]=- {2211111111} + {22211111111}
[221111111111]=- {21111111111} + {221111111111}
[2111111111111]=- {111111111111} + {2111111111111}
[11111111111111] = {11111111111111}

[4443] = {3} - {32} + {331} - {333} - {41} + {421} + {43} - {4311} - {432} + {4331} - {441} + {4421}
+ {443} - {4432} - {4441} + {4443}
[44421] = {21} - {221} - {311} - {32} + {3211} + {322} + {331} - {3321} - {41} + {4111} +2{421}
- {42111} - {4221} + {43} -2{4311} -2{432} + {43211} + {4322} + {4331} - {441} + {44111}
+2{4421} - {44221} + {443} - {44311} - {4432} - {4441} + {44421}
[444111] = {111} - {2111} - {311} + {31111} + {3211} - {33111} + {4111} - {411111} + {421} - {42111}
- {4221} - {4311} + {431111} + {43211} - {441} + {44111} + {4421} - {442111} - {44311}
+ {444111}

Expansion of Orthogonal Group Characters into S-Functions (contd).

[44331]=- {311} - {32} + {3211} + {322} +2{331} - {33111} -2{3321} - {333} + {33311} + {3332}
+ {421} - {4221} + {43} - {4311} -2{432} + {43211} + {4322} +2{4331} - {43321} - {4333}
- {441} + {4421} + {443} - {44311} - {4432} + {44331}

[44322]=- {221} + {2221} - {32} + {3211} +2{322} - {32211} - {3222} + {331} -2{3321} + {33221}
- {333} + {3332} + {421} -2{4221} + {42221} - {4311} - {432} + {43211} +2{4322} - {43222}
+ {4331} - {43321} + {4421} - {44221} - {4432} + {44322}

[443211]=- {2111} - {221} + {22111} + {2221} - {311} + {31111} +3{3211} - {321111} + {322} -2{32211}
- {3222} + {331} -2{33111} -2{3321} + {332111} + {33221} + {33311} + {4111} + {421}
-2{42111} -2{4221} + {422111} + {42221} -2{4311} + {431111} - {432} +3{43211} + {4322}
- {432211} + {4331} - {433111} - {43321} + {44111} + {4421} - {442111} - {44221} - {44311}
+ {443211}

[4431111]=- {11111} - {2111} + {211111} + {22111} +2{31111} - {3111111} + {3211} -2{321111}
- {32211} - {33111} + {3311111} + {332111} + {4111} - {411111} -2{42111} + {4211111}
+ {422111} - {4311} +2{431111} + {43211} - {4321111} - {433111} + {44111} - {4411111}
- {442111} + {4431111}

[442221]=- {221} + {22111} +2{2221} - {222111} - {22221} + {3211} + {322} -2{32211} -2{3222}
+ {322211} + {32222} - {3321} + {33221} - {42111} - {4221} + {422111} +2{42221} - {422221}
+ {43211} + {4322} - {432211} - {43222} - {44221} + {442221}

[4422111]=- {2111} + {211111} +2{22111} - {2211111} + {2221} - {222111} - {22221} + {31111} + {3211}
-2{321111} -2{32211} + {3221111} + {322211} - {33111} + {332111} - {411111} - {42111}
+ {4211111} - {4221} +2{422111} + {42221} - {4222111} + {431111} + {43211} - {4321111}
- {432211} - {442111} + {4422111}

[44211111]=- {11111} + {1111111} +2{211111} - {21111111} + {22111} - {2211111} - {222111} + {31111}
-2{3111111} -2{321111} + {32111111} + {3221111} + {3311111} - {411111} + {41111111}
- {42111} +2{4211111} + {422111} - {42211111} + {431111} - {43111111} - {4321111}
- {4411111} + {44211111}

[441111111] = {1111111} - {111111111} + {211111} - {21111111} - {2211111} - {3111111} + {311111111}
+ {32111111} - {411111} + {41111111} + {4211111} - {421111111} - {43111111}
+ {441111111}

[43332] = {322} - {3222} + {331} -2{3321} + {33221} - {333} + {33311} +2{3332} - {33322} - {33331}
- {432} + {4322} + {4331} - {43321} - {4333} + {43332}

[433311] = {3211} - {32211} + {331} - {33111} -2{3321} + {332111} + {33221} - {333} +2{33311}
+ {3332} - {333211} - {33331} - {4311} + {43211} + {4331} - {433111} - {43321} + {433311}

[433221] = {2221} - {22221} + {3211} + {322} -2{32211} -2{3222} + {322211} + {32222} - {33111}
-2{3321} + {332111} +3{33221} - {332221} + {33311} + {3332} - {333211} - {33322} - {4221}
+ {42221} + {43211} + {4322} - {432211} - {43222} - {43321} + {433221}

[4332111] = {22111} - {222111} + {31111} + {3211} -2{321111} -2{32211} + {3221111} + {322211}
-2{33111} + {3311111} - {3321} +3{332111} + {33221} - {3322111} + {33311} - {3331111}
- {333211} - {42111} + {422111} + {431111} + {43211} - {4321111} - {432211} - {433111}
+ {4332111}

[43311111] = {211111} - {2211111} + {31111} - {3111111} -2{321111} + {32111111} + {3221111}
- {33111} +2{3311111} + {332111} - {33211111} - {3331111} - {411111} + {4211111}
+ {431111} - {43111111} - {4321111} + {43311111}

[432222] = {2221} -2{22221} + {222221} - {32211} - {3222} + {322211} +2{32222} - {322222} + {33221}
- {332221} + {42221} - {422221} - {43222} + {432222}

[4322211] = {22111} + {2221} -2{222111} -2{22221} + {2222111} + {222221} - {321111} -2{32211}
+ {3221111} - {3222} +3{322211} + {32222} - {3222211} + {332111} + {33221} - {3322111}
- {332221} + {422111} + {42221} - {4222111} - {422221} - {432211} + {4322211}

[43221111] = {211111} + {22111} -2{2211111} -2{222111} + {22211111} + {2222111} - {3111111}
-2{321111} + {32111111} - {32211} +3{3221111} + {322211} - {32221111} + {3311111}
+ {332111} - {33211111} - {3322111} + {4211111} + {422111} - {42211111} - {4222111}
- {4321111} + {43221111}

[432111111] = {1111111} + {211111} -2{21111111} -2{2211111} + {221111111} + {22211111} -2{3111111}
+ {311111111} - {321111} +3{32111111} + {3221111} - {322111111} + {3311111}
- {331111111} - {33211111} + {41111111} + {4211111} - {421111111} - {42211111}
- {43111111} + {432111111}

[4311111111] = {1111111} - {111111111} -2{21111111} + {2111111111} + {221111111} - {3111111}
+2{311111111} + {32111111} - {3211111111} - {331111111} + {41111111} - {4111111111}
- {421111111} + {4311111111}

[4222221]=- {222111} - {22221} + {2222111} +2{222221} - {2222221} + {322211} + {32222} - {3222211}
- {322222} - {422221} + {4222221}

[42222111]=- {2211111} - {222111} + {22211111} - {22221} +2{2222111} + {222221} - {22222111}
+ {3221111} + {322211} - {32221111} - {3222211} - {4222111} + {42222111}

[422211111]=- {21111111} - {2211111} + {221111111} - {222111} +2{22211111} + {2222111} - {222211111}
+ {32111111} + {3221111} - {322111111} - {32221111} - {42211111} + {422211111}

[4221111111]=- {111111111} - {21111111} + {2111111111} - {2211111} +2{221111111} + {22211111}
- {2221111111} + {311111111} + {32111111} - {3211111111} - {322111111} - {421111111}
+ {4221111111}

[42111111111]=- {111111111} + {11111111111} - {21111111} +2{2111111111} + {221111111}
- {22111111111} + {311111111} - {31111111111} - {3211111111} - {4111111111}
+ {42111111111}

Expansion of Orthogonal Group Characters into S-Functions (contd).

[411111111111]=- {111111111} + {11111111111} + {2111111111} - {211111111111} - {31111111111}
+ {411111111111}
[33333]=- {333} + {3332} - {33331} + {33333}
[333321]=- {3321} + {33221} + {33311} + {3332} - {333211} - {33322} - {33331} + {333321}
[3333111]=- {33111} + {332111} + {33311} - {3331111} - {333211} + {3333111}
[333222]=- {3222} + {32222} + {33221} - {332221} - {33322} + {333222}
[3332211]=- {32211} + {322211} + {332111} + {33221} - {3322111} - {332221} - {333211} + {3332211}
[33321111]=- {321111} + {3221111} + {3311111} + {332111} - {33211111} - {3322111} - {3331111}
+ {33321111}
[333111111]=- {3111111} + {32111111} + {3311111} - {331111111} - {33211111} + {333111111}
[3322221]=- {22221} + {222221} + {322211} + {32222} - {3222211} - {322222} - {332221} + {3322221}
[33222111]=- {222111} + {2222111} + {3221111} + {322211} - {32221111} - {3222211} - {3322111}
+ {33222111}
[332211111]=- {2211111} + {22211111} + {32111111} + {3221111} - {322111111} - {32221111}
- {33211111} + {332211111}
[3321111111]=- {21111111} + {221111111} + {311111111} + {32111111} - {3211111111} - {322111111}
- {331111111} + {3321111111}
[33111111111]=- {111111111} + {2111111111} + {311111111} - {31111111111} - {3211111111}
+ {33111111111}
[3222222] = {222221} - {2222221} - {322222} + {3222222}
[32222211] = {2222111} + {222221} - {22222111} - {2222221} - {3222211} + {32222211}
[322221111] = {22211111} + {2222111} - {222211111} - {22222111} - {32221111} + {322221111}
[3222111111] = {221111111} + {22211111} - {2221111111} - {222211111} - {322111111} + {3222111111}
[32211111111] = {2111111111} + {221111111} - {22111111111} - {2221111111} - {3211111111}
+ {32211111111}
[321111111111] = {11111111111} + {2111111111} - {211111111111} - {22111111111} - {31111111111}
+ {321111111111}
[3111111111111] = {11111111111} - {1111111111111} - {211111111111} + {3111111111111}
[22222221]=- {2222221} + {22222221}
[222222111]=- {22222111} + {222222111}
[2222211111]=- {222211111} + {2222211111}
[22221111111]=- {2221111111} + {22221111111}
[222111111111]=- {22111111111} + {222111111111}
[2211111111111]=- {211111111111} + {2211111111111}
[21111111111111]=- {1111111111111} + {21111111111111}
[111111111111111] = {111111111111111}

[4444] = {4} - {42} + {431} - {433} - {4411} + {4431} - {4442} + {4444}
[44431] = {31} - {321} - {33} + {3311} + {332} - {3331} - {411} - {42} + {4211} + {422} +2{431}
- {43111} -2{4321} - {433} + {43311} + {4332} + {44} - {4411} -2{442} + {44211} + {4422}
+2{4431} - {44321} - {4433} + {444} - {44411} - {4442} + {44431}
[44422] = {22} - {222} - {321} + {3221} + {332} - {3322} - {42} + {4211} + {422} - {42211} + {431}
-2{4321} + {43221} - {433} + {4332} - {4411} + {44211} + {4422} - {44222} + {4431} - {44321}
- {4442} + {44422}
[444211] = {211} - {2211} - {3111} - {321} + {32111} + {3221} + {3311} - {33211} - {411} + {41111}
+2{4211} - {421111} + {422} - {42211} - {4222} + {431} -2{43111} -2{4321} + {432111}
+ {43221} + {43311} - {4411} + {441111} - {442} +2{44211} + {4422} - {442211} + {4431}
- {443111} - {44321} - {44411} + {444211}
[4441111] = {1111} - {21111} - {3111} + {311111} + {32111} - {331111} + {41111} - {4111111} + {4211}
- {421111} - {42211} - {43111} + {4311111} + {432111} - {4411} + {441111} + {44211}
- {4421111} - {443111} + {4441111}
[44332]=- {321} + {3221} - {33} + {3311} +2{332} - {33211} - {3322} -2{3331} + {33321} + {3333}
+ {422} - {4222} + {431} -2{4321} + {43221} - {433} + {43311} +2{4332} - {43322} - {43331}
- {442} + {4422} + {4431} - {44321} - {4433} + {44332}
[443311]=- {3111} - {321} + {32111} + {3221} +2{3311} - {331111} + {332} -2{33211} - {3322} - {3331}
+ {333111} + {33321} + {4211} - {42211} + {431} - {43111} -2{4321} + {432111} + {43221}
- {433} +2{43311} + {4332} - {433211} - {43331} - {4411} + {44211} + {4431} - {443111}
- {44321} + {443311}
[443221]=- {2211} - {222} + {22211} + {2222} - {321} + {32111} +3{3221} - {322111} -2{32221}
+ {3311} + {332} -2{33211} -2{3322} + {332211} + {33222} - {3331} + {33321} + {4211}
+ {422} -2{42211} -2{4222} + {422211} + {42222} - {43111} -2{4321} + {432111} +3{43221}
- {432221} + {43311} + {4332} - {433211} - {43322} + {44211} + {4422} - {442211} - {44222}
- {44321} + {443221}
[4432111]=- {21111} - {2211} + {221111} + {22211} - {3111} + {311111} +3{32111} - {3211111} + {3221}
-2{322111} - {32221} + {3311} -2{331111} -2{33211} + {3321111} + {332211} + {333111}
+ {41111} + {4211} -2{421111} -2{42211} + {4221111} + {422211} -2{43111} + {4311111}
- {4321} +3{432111} + {43221} - {4322111} + {43311} - {4331111} - {433211} + {441111}
+ {44211} - {4421111} - {442211} - {443111} + {4432111}
[44311111]=- {111111} - {21111} + {2111111} + {221111} +2{311111} - {31111111} + {32111} -2{3211111}
- {322111} - {331111} + {33111111} + {3321111} + {41111} - {4111111} -2{421111}
+ {42111111} + {4221111} - {43111} +2{4311111} + {432111} - {43211111} - {4331111}
+ {441111} - {44111111} - {4421111} + {44311111}

[442222]=- {222} + {22211} + {2222} - {222211} + {3221} -2{32221} + {322221} - {3322} + {33222}
- {42211} + {422211} + {42222} - {422222} + {43221} - {432221} - {44222} + {442222}

[4422211]=- {2211} + {221111} +2{22211} - {2221111} + {2222} - {222211} - {22222} + {32111} + {3221}
-2{322111} -2{32221} + {3222111} + {322221} - {33211} + {332211} - {421111} - {42211}
+ {4221111} - {4222} +2{422211} + {42222} - {4222211} + {432111} + {43221} - {4322111}
- {432221} - {442211} + {4422211}

[44221111]=- {21111} + {2111111} +2{221111} - {22111111} + {22211} - {2221111} - {222211} + {311111}
+ {32111} -2{3211111} -2{322111} + {32211111} + {3222111} - {331111} + {3321111}
- {4111111} - {421111} + {42111111} - {42211} +2{4221111} + {422211} - {42221111}
+ {4311111} + {432111} - {43211111} - {4322111} - {4421111} + {44221111}

[442111111]=- {111111} + {11111111} +2{2111111} - {211111111} + {221111} - {22111111} - {2221111}
+ {311111} -2{31111111} -2{3211111} + {321111111} + {32211111} + {33111111} - {4111111}
+ {411111111} - {421111} +2{42111111} + {4221111} - {422111111} + {4311111}
- {431111111} - {43211111} - {44111111} + {442111111}

[4411111111] = {11111111} - {1111111111} + {2111111} - {211111111} - {22111111} - {31111111}
+ {3111111111} + {321111111} - {4111111} + {411111111} + {42111111} - {4211111111}
- {431111111} + {4411111111}

[43333] = {332} - {3322} - {3331} + {33321} + {3333} - {33332} - {433} + {4332} - {43331} + {43333}

[433321] = {3221} - {32221} + {3311} + {332} -2{33211} -2{3322} + {332211} + {33222} -2{3331}
+ {333111} +3{33321} - {333221} + {3333} - {333311} - {33332} - {4321} + {43221} + {43311}
+ {4332} - {433211} - {43322} - {43331} + {433321}

[4333111] = {32111} - {322111} + {3311} - {331111} -2{33211} + {3321111} + {332211} - {3331}
+2{333111} + {33321} - {3332111} - {333311} - {43111} + {432111} + {43311} - {4331111}
- {433211} + {4333111}

[433222] = {2222} - {22222} + {3221} -2{32221} + {322221} - {33211} - {3322} + {332211} +2{33222}
- {332222} + {33321} - {333221} - {4222} + {42222} + {43221} - {432221} - {43322}
+ {433222}

[4332211] = {22211} - {222211} + {32111} + {3221} -2{322111} -2{32221} + {3222111} + {322221}
- {331111} -2{33211} + {3321111} - {3322} +3{332211} + {33222} - {3322211} + {333111}
+ {33321} - {3332111} - {333221} - {42211} + {422211} + {432111} + {43221} - {4322111}
- {432221} - {433211} + {4332211}

[43321111] = {221111} - {2221111} + {311111} + {32111} -2{3211111} -2{322111} + {32211111}
+ {3222111} -2{331111} + {33111111} - {33211} +3{3321111} + {332211} - {33221111}
+ {333111} - {33311111} - {3332111} - {421111} + {4221111} + {4311111} + {432111}
- {43211111} - {4322111} - {4331111} + {43321111}

[433111111] = {2111111} - {22111111} + {311111} - {31111111} -2{3211111} + {321111111} + {32211111}
- {331111} +2{33111111} + {3321111} - {332111111} - {33311111} - {4111111} + {42111111}
+ {4311111} - {431111111} - {43211111} + {433111111}

[4322221] = {22211} + {2222} -2{222211} -2{22222} + {2222211} + {222222} - {322111} -2{32221}
+ {3222111} +3{322221} - {3222221} + {332211} + {33222} - {3322211} - {332222} + {422211}
+ {42222} - {4222211} - {422222} - {432221} + {4322221}

[43222111] = {221111} + {22211} -2{2221111} -2{222211} + {22221111} + {2222211} - {3211111}
-2{322111} + {32211111} - {32221} +3{3222111} + {322221} - {32222111} + {3321111}
+ {332211} - {33221111} - {3322211} + {4221111} + {422211} - {42221111} - {4222211}
- {4322111} + {43222111}

[432211111] = {2111111} + {221111} -2{22111111} -2{2221111} + {222111111} + {22221111} - {31111111}
-2{3211111} + {321111111} - {322111} +3{32211111} + {3222111} - {322211111} + {33111111}
+ {3321111} - {332111111} - {33221111} + {42111111} + {4221111} - {422111111}
- {42221111} - {43211111} + {432211111}

[4321111111] = {11111111} + {2111111} -2{211111111} -2{22111111} + {2211111111} + {222111111}
-2{31111111} + {3111111111} - {3211111} +3{321111111} + {32211111} - {3221111111}
+ {33111111} - {3311111111} - {332111111} + {411111111} + {42111111} - {4211111111}
- {422111111} - {431111111} + {4321111111}

[43111111111] = {11111111} - {1111111111} -2{211111111} + {21111111111} + {2211111111} - {31111111}
+2{3111111111} + {321111111} - {32111111111} - {3311111111} + {411111111}
- {41111111111} - {4211111111} + {43111111111}

[4222222]=- {222211} + {2222211} + {222222} - {2222222} + {322221} - {3222221} - {422222}
+ {4222222}

[42222211]=- {2221111} - {222211} + {22221111} - {22222} +2{2222211} + {222222} - {22222211}
+ {3222211} + {322221} - {32222111} - {3222221} - {4222211} + {42222211}

[422221111]=- {22111111} - {2221111} + {222111111} - {222211} +2{22221111} + {2222211} - {222221111}
+ {32211111} + {3222111} - {322211111} - {32222111} - {42221111} + {422221111}

[4222111111]=- {211111111} - {22111111} + {2211111111} - {2221111} +2{222111111} + {22221111}
- {2222111111} + {321111111} + {32211111} - {3221111111} - {322211111} - {422111111}
+ {4222111111}

[42211111111]=- {1111111111} - {211111111} + {21111111111} - {22111111} +2{2211111111} + {222111111}
- {22211111111} + {3111111111} + {321111111} - {32111111111} - {3221111111}
- {4211111111} + {42211111111}

[421111111111]=- {1111111111} + {111111111111} - {211111111} +2{21111111111} + {2211111111}
- {221111111111} + {3111111111} - {311111111111} - {32111111111} - {41111111111}
+ {421111111111}

Expansion of Orthogonal Group Characters into S-Functions (contd).

[4111111111111]=- {1111111111} + {111111111111} + {21111111111} - {2111111111111} - {311111111111}
+ {4111111111111}
[333331]=- {3331} + {33321} + {3333} - {333311} - {33332} + {333331}
[333322]=- {3322} + {33222} + {33321} - {333221} - {33332} + {333322}
[3333211]=- {33211} + {332211} + {333111} + {33321} - {3332111} - {333221} - {333311} + {3333211}
[33331111]=- {331111} + {3321111} + {333111} - {33311111} - {3332111} + {33331111}
[3332221]=- {32221} + {322221} + {332211} + {33222} - {3322211} - {332222} - {333221} + {3332221}
[33322111]=- {322111} + {3222111} + {3321111} + {332211} - {33221111} - {3322211} - {3332111}
+ {33322111}
[333211111]=- {3211111} + {32211111} + {33111111} + {3321111} - {332111111} - {33221111}
- {33311111} + {333211111}
[3331111111]=- {31111111} + {321111111} + {33111111} - {3311111111} - {332111111} + {3331111111}
[3322222]=- {22222} + {222222} + {322221} - {3222221} - {332222} + {3322222}
[33222211]=- {222211} + {2222211} + {3222111} + {322221} - {32222111} - {3222221} - {3322211}
+ {33222211}
[332221111]=- {2221111} + {22221111} + {32211111} + {3222111} - {322211111} - {32222111}
- {33221111} + {332221111}
[3322111111]=- {22111111} + {222111111} + {321111111} + {32211111} - {3221111111} - {322211111}
- {332111111} + {3322111111}
[33211111111]=- {211111111} + {2211111111} + {3111111111} + {321111111} - {32111111111}
- {3221111111} - {3311111111} + {33211111111}
[331111111111]=- {1111111111} + {21111111111} + {3111111111} - {311111111111} - {32111111111}
+ {331111111111}
[32222221] = {2222211} + {222222} - {22222211} - {2222222} - {3222221} + {32222221}
[322222111] = {22221111} + {2222211} - {222221111} - {22222211} - {32222111} + {322222111}
[3222211111] = {222111111} + {22221111} - {2222111111} - {222221111} - {322211111} + {3222211111}
[32221111111] = {2211111111} + {222111111} - {22211111111} - {2222111111} - {3221111111}
+ {32221111111}
[322111111111] = {21111111111} + {2211111111} - {221111111111} - {22211111111} - {32111111111}
+ {322111111111}
[3211111111111] = {111111111111} + {21111111111} - {2111111111111} - {221111111111} - {311111111111}
+ {3211111111111}
[31111111111111] = {111111111111} - {11111111111111} - {2111111111111} + {31111111111111}
[22222222]=- {2222222} + {22222222}
[222222211]=- {22222211} + {222222211}
[2222221111]=- {222221111} + {2222221111}
[22222111111]=- {2222111111} + {22222111111}
[222211111111]=- {22211111111} + {222211111111}
[2221111111111]=- {221111111111} + {2221111111111}
[22111111111111]=- {2111111111111} + {22111111111111}
[211111111111111]=- {11111111111111} + {211111111111111}
[1111111111111111] = {1111111111111111-11111111111111}

Expansion of Symplectic Group Characters into S-Functions.

<0> = {0}

<1> = {1}

<2> = {2}
<11>=- {0} + {11}

<3> = {3}
<21>=- {1} + {21}
<111>=- {1} + {111}

<4> = {4}
<31>=- {2} + {31}
<22>=- {11} + {22}
<211> = {0} - {11} - {2} + {211}
<1111>=- {11} + {1111}

<41>=- {3} + {41}
<32>=- {21} + {32}
<311> = {1} - {21} - {3} + {311}
<221> = {1} - {111} - {21} + {221}
<2111> = {1} - {111} - {21} + {2111}
<11111>=- {111} + {11111}

<42>=- {31} + {42}
<411> = {2} - {31} - {4} + {411}
<33>=- {22} + {33}
<321> = {11} + {2} - {211} - {22} - {31} + {321}
<3111>=- {0} + {11} + {2} - {211} - {31} + {3111}
<222>=- {0} + {11} - {211} + {222}
<2211> = {11} - {1111} + {2} - {211} - {22} + {2211}
<21111> = {11} - {1111} - {211} + {21111}
<111111>=- {1111} + {111111}

<43>=- {32} + {43}
<421> = {21} + {3} - {311} - {32} - {41} + {421}
<4111>=- {1} + {21} + {3} - {311} - {41} + {4111}
<331> = {21} - {221} - {32} + {331}
<322>=- {1} + {111} + {21} - {221} - {311} + {322}
<3211>=- {1} + {111} +2{21} - {2111} - {221} + {3} - {311} - {32} + {3211}
<31111>=- {1} + {111} + {21} - {2111} - {311} + {31111}
<2221>=- {1} + {111} + {21} - {2111} - {221} + {2221}
<22111> = {111} - {11111} + {21} - {2111} - {221} + {22111}
<211111> = {111} - {11111} - {2111} + {211111}
<1111111>=- {11111} + {1111111}

<44>=- {33} + {44}
<431> = {22} + {31} - {321} - {33} - {42} + {431}
<422>=- {2} + {211} + {31} - {321} - {411} + {422}
<4211>=- {11} - {2} + {211} + {22} +2{31} - {3111} - {321} + {4} - {411} - {42} + {4211}
<41111>=- {11} - {2} + {211} + {31} - {3111} - {411} + {41111}
<332>=- {11} + {211} + {22} - {222} - {321} + {332}
<3311>=- {2} + {211} + {22} - {2211} + {31} - {321} - {33} + {3311}
<3221> = {0} -2{11} + {1111} - {2} +2{211} + {22} - {2211} - {222} + {31} - {3111} - {321} + {3221}
<32111>=- {11} + {1111} - {2} +2{211} - {21111} + {22} - {2211} + {31} - {3111} - {321} + {32111}
<311111>=- {11} + {1111} + {211} - {21111} - {3111} + {311111}
<2222>=- {2} + {211} - {2211} + {2222}
<22211>=- {11} + {1111} + {211} - {21111} + {22} - {2211} - {222} + {22211}
<221111> = {1111} - {111111} + {211} - {21111} - {2211} + {221111}
<2111111> = {1111} - {111111} - {21111} + {2111111}
<11111111>=- {111111} + {11111111}

<441> = {32} - {331} - {43} + {441}
<432>=- {21} + {221} + {311} + {32} - {322} - {331} - {421} + {432}
<4311>=- {21} + {221} - {3} + {311} +2{32} - {3211} - {331} + {41} - {421} - {43} + {4311}
<4221> = {1} - {111} -2{21} + {2111} + {221} - {3} +2{311} + {32} - {3211} - {322} + {41} - {4111}
- {421} + {4221}
<42111>=- {111} -2{21} + {2111} + {221} - {3} +2{311} - {31111} + {32} - {3211} + {41} - {4111}
- {421} + {42111}
<411111>=- {111} - {21} + {2111} + {311} - {31111} - {4111} + {411111}

Expansion of Symplectic Group Characters into S-Functions (contd).

<333>=- {111} + {221} - {322} + {333}
<3321> = {1} - {111} -2{21} + {2111} +2{221} - {2221} + {311} + {32} - {3211} - {322} - {331}
+ {3321}
<33111>=- {21} + {2111} + {221} - {22111} - {3} + {311} + {32} - {3211} - {331} + {33111}
<3222> = {1} - {111} - {21} + {2111} + {221} - {2221} - {3} + {311} - {3211} + {3222}
<32211> = {1} -2{111} + {11111} -2{21} +2{2111} +2{221} - {22111} - {2221} + {311} - {31111} + {32}
- {3211} - {322} + {32211}
<321111>=- {111} + {11111} - {21} +2{2111} - {211111} + {221} - {22111} + {311} - {31111} - {3211}
+ {321111}
<3111111>=- {111} + {11111} + {2111} - {211111} - {31111} + {3111111}
<22221>=- {21} + {2111} + {221} - {22111} - {2221} + {22221}
<222111>=- {111} + {11111} + {2111} - {211111} + {221} - {22111} - {2221} + {222111}
<2211111> = {11111} - {1111111} + {2111} - {211111} - {22111} + {2211111}
<21111111> = {11111} - {1111111} - {211111} + {21111111}
<111111111>=- {1111111} + {111111111}

<442>=- {22} + {321} + {33} - {332} - {431} + {442}
<4411>=- {31} + {321} + {33} - {3311} + {42} - {431} - {44} + {4411}
<433>=- {211} + {222} + {321} - {332} - {422} + {433}
<4321> = {11} + {2} -2{211} -2{22} + {2211} + {222} -2{31} + {3111} +3{321} - {3221} + {33} - {3311}
- {332} + {411} + {42} - {4211} - {422} - {431} + {4321}
<43111>=- {211} - {22} + {2211} -2{31} + {3111} +2{321} - {32111} + {33} - {3311} - {4} + {411}
+ {42} - {4211} - {431} + {43111}
<4222> = {11} - {1111} + {2} - {211} - {22} + {2211} - {31} + {3111} + {321} - {3221} - {4} + {411}
- {4211} + {4222}
<42211> = {11} - {1111} + {2} -3{211} + {21111} - {22} + {2211} + {222} -2{31} +2{3111} +2{321}
- {32111} - {3221} + {411} - {41111} + {42} - {4211} - {422} + {42211}
<421111>=- {1111} -2{211} + {21111} - {22} + {2211} - {31} +2{3111} - {311111} + {321} - {32111}
+ {411} - {41111} - {4211} + {421111}
<4111111>=- {1111} - {211} + {21111} + {3111} - {311111} - {41111} + {4111111}
<3331> = {11} - {1111} - {211} - {22} + {2211} + {222} + {321} - {3221} - {332} + {3331}
<3322>=- {0} + {11} + {2} -2{211} + {2211} + {222} - {2222} - {31} + {3111} + {321} - {3221}
- {3311} + {3322}
<33211> = {11} - {1111} + {2} -2{211} + {21111} -2{22} +2{2211} + {222} - {22211} - {31} + {3111}
+2{321} - {32111} - {3221} + {33} - {3311} - {332} + {33211}
<331111>=- {211} + {21111} + {2211} - {221111} - {31} + {3111} + {321} - {32111} - {3311} + {331111}
<32221> = {11} - {1111} + {2} -2{211} + {21111} - {22} +2{2211} + {222} - {22211} - {2222} - {31}
+ {3111} + {321} - {32111} - {3221} + {32221}
<322111> = {11} -2{1111} + {111111} -2{211} +2{21111} - {22} +2{2211} - {221111} + {222} - {22211}
+ {3111} - {311111} + {321} - {32111} - {3221} + {322111}
<3211111>=- {1111} + {111111} - {211} +2{21111} - {2111111} + {2211} - {221111} + {3111} - {311111}
- {32111} + {3211111}
<31111111>=- {1111} + {111111} + {21111} - {2111111} - {311111} + {31111111}
<22222>=- {22} + {2211} - {22211} + {22222}
<222211>=- {211} + {21111} + {2211} - {221111} + {222} - {22211} - {2222} + {222211}
<2221111>=- {1111} + {111111} + {21111} - {2111111} + {2211} - {221111} - {22211} + {2221111}
<22111111> = {111111} - {11111111} + {21111} - {2111111} - {221111} + {22111111}
<211111111> = {111111} - {11111111} - {2111111} + {211111111}
<1111111111>=- {11111111} + {1111111111}

<443>=- {221} + {322} + {331} - {333} - {432} + {443}
<4421> = {21} - {221} - {311} -2{32} + {3211} + {322} +2{331} - {3321} + {421} + {43} - {4311}
- {432} - {441} + {4421}
<44111>=- {311} - {32} + {3211} + {331} - {33111} - {41} + {421} + {43} - {4311} - {441} + {44111}
<4331> = {111} + {21} - {2111} -2{221} + {2221} - {311} - {32} + {3211} +2{322} + {331} - {3321}
- {333} + {421} - {4221} - {432} + {4331}
<4322>=- {1} + {111} +2{21} - {2111} -2{221} + {2221} + {3} -2{311} - {32} +2{3211} + {322} - {3222}
+ {331} - {3321} - {41} + {4111} + {421} - {4221} - {4311} + {4322}
<43211> = {111} +2{21} -2{2111} -3{221} + {22111} + {2221} + {3} -3{311} + {31111} -3{32} +3{3211}
+2{322} - {32211} +2{331} - {33111} - {3321} - {41} + {4111} +2{421} - {42111} - {4221}
+ {43} - {4311} - {432} + {43211}
<431111>=- {2111} - {221} + {22111} -2{311} + {31111} - {32} +2{3211} - {321111} + {331} - {33111}
- {41} + {4111} + {421} - {42111} - {4311} + {431111}
<42221> = {111} - {11111} +2{21} -2{2111} -2{221} + {22111} + {2221} + {3} -2{311} + {31111} - {32}
+2{3211} + {322} - {32211} - {3222} - {41} + {4111} + {421} - {42111} - {4221} + {42221}
<422111> = {111} - {11111} + {21} -3{2111} + {211111} -2{221} + {22111} + {2221} -2{311} +2{31111}
- {32} +2{3211} - {321111} + {322} - {32211} + {4111} - {411111} + {421} - {42111} - {4221}
+ {422111}
<4211111>=- {11111} -2{2111} + {211111} - {221} + {22111} - {311} +2{31111} - {3111111} + {3211}
- {321111} + {4111} - {411111} - {42111} + {4211111}

B-5 Expansion of Symplectic Group Characters into S-Functions (contd).

<41111111>=- {11111} - {2111} + {211111} + {31111} - {3111111} - {411111} + {41111111}
<3332>=- {1} + {111} + {21} - {2111} - {221} + {2221} - {311} + {3211} + {322} - {3222} - {3321}
+ {3332}
<33311> = {111} - {11111} + {21} - {2111} -2{221} + {22111} + {2221} - {32} + {3211} + {322}
- {32211} + {331} - {3321} - {333} + {33311}
<33221>=- {1} + {111} +2{21} -2{2111} -2{221} + {22111} +2{2221} - {22221} + {3} -2{311} + {31111}
- {32} +2{3211} + {322} - {32211} - {3222} + {331} - {33111} - {3321} + {33221}
<332111> = {111} - {11111} + {21} -2{2111} + {211111} -2{221} +2{22111} + {2221} - {222111} - {311}
+ {31111} - {32} +2{3211} - {321111} + {322} - {32211} + {331} - {33111} - {3321}
+ {332111}
<3311111>=- {2111} + {211111} + {22111} - {2211111} - {311} + {31111} + {3211} - {321111} - {33111}
+ {3311111}
<32222> = {21} - {2111} - {221} + {22111} + {2221} - {22221} - {32} + {3211} - {32211} + {32222}
<322211> = {111} - {11111} + {21} -2{2111} + {211111} -2{221} +2{22111} +2{2221} - {222111}
- {22221} - {311} + {31111} + {3211} - {321111} + {322} - {32211} - {3222} + {322211}
<3221111> = {111} -2{11111} + {1111111} -2{2111} +2{211111} - {221} +2{22111} - {2211111} + {2221}
- {222111} + {31111} - {3111111} + {3211} - {321111} - {32211} + {3221111}
<32111111>=- {11111} + {1111111} - {2111} +2{211111} - {21111111} + {22111} - {2211111} + {31111}
- {3111111} - {321111} + {32111111}
<311111111>=- {11111} + {1111111} + {211111} - {21111111} - {3111111} + {311111111}
<222221>=- {221} + {22111} + {2221} - {222111} - {22221} + {222221}
<2222111>=- {2111} + {211111} + {22111} - {2211111} + {2221} - {222111} - {22221} + {2222111}
<22211111>=- {11111} + {1111111} + {211111} - {21111111} + {22111} - {2211111} - {222111}
+ {22211111}
<221111111> = {1111111} - {111111111} + {211111} - {21111111} - {2211111} + {221111111}
<2111111111> = {1111111} - {111111111} - {21111111} + {2111111111}
<11111111111>=- {111111111} + {11111111111}

<444>=- {222} + {332} - {433} + {444}
<4431> = {211} + {22} - {2211} - {222} -2{321} + {3221} - {33} + {3311} +2{332} - {3331} + {422}
+ {431} - {4321} - {433} - {442} + {4431}
<4422>=- {11} + {211} + {22} - {222} + {31} - {3111} -2{321} + {3221} + {3311} + {332} - {3322}
- {42} + {4211} + {431} - {4321} - {4411} + {4422}
<44211> = {211} + {22} - {2211} + {31} - {3111} -3{321} + {32111} + {3221} -2{33} +2{3311} + {332}
- {33211} - {411} - {42} + {4211} + {422} +2{431} - {43111} - {4321} + {44} - {4411} - {442}
+ {44211}
<441111>=- {3111} - {321} + {32111} + {3311} - {331111} - {411} - {42} + {4211} + {431} - {43111}
- {4411} + {441111}
<4332>=- {11} + {1111} - {2} +2{211} + {22} -2{2211} - {222} + {2222} + {31} - {3111} -2{321}
+2{3221} + {3311} + {332} - {3322} - {3331} - {411} + {4211} + {422} - {4222} - {4321}
+ {4332}
<43311> = {1111} +2{211} - {21111} + {22} -2{2211} -2{222} + {22211} + {31} - {3111} -3{321}
+ {32111} +2{3221} - {33} + {3311} +2{332} - {33211} - {3331} - {42} + {4211} + {422}
- {42211} + {431} - {4321} - {433} + {43311}
<43221>=- {11} + {1111} - {2} +3{211} - {21111} +2{22} -3{2211} -2{222} + {22211} + {2222} +3{31}
-3{3111} -4{321} +2{32111} +3{3221} - {32221} - {33} +2{3311} + {332} - {33211} - {3322}
+ {4} -2{411} + {41111} - {42} +2{4211} + {422} - {42211} - {4222} + {431} - {43111}
- {4321} + {43221}
<432111> = {1111} +2{211} -2{21111} + {22} -3{2211} + {221111} - {222} + {22211} + {31} -3{3111}
+ {311111} -4{321} +3{32111} +2{3221} - {322111} - {33} +2{3311} - {331111} + {332}
- {33211} - {411} + {41111} - {42} +2{4211} - {421111} + {422} - {42211} + {431} - {43111}
- {4321} + {432111}
<4311111>=- {21111} - {2211} + {221111} -2{3111} + {311111} - {321} +2{32111} - {3211111} + {3311}
- {331111} - {411} + {41111} + {4211} - {421111} - {43111} + {4311111}
<42222> = {211} - {21111} + {22} - {2211} - {222} + {22211} + {31} - {3111} - {321} + {32111}
+ {3221} - {32221} - {42} + {4211} - {42211} + {42222}
<422211> = {1111} - {111111} +2{211} -2{21111} + {22} -3{2211} + {221111} - {222} + {22211} + {2222}
+ {31} -2{3111} + {311111} -2{321} +2{32111} +2{3221} - {322111} - {32221} - {411}
+ {41111} + {4211} - {421111} + {422} - {42211} - {4222} + {422211}
<4221111> = {1111} - {111111} + {211} -3{21111} + {2111111} -2{2211} + {221111} - {222} + {22211}
-2{3111} +2{311111} - {321} +2{32111} - {3211111} + {3221} - {322111} + {41111}
- {4111111} + {4211} - {421111} - {42211} + {4221111}
<42111111>=- {111111} -2{21111} + {2111111} - {2211} + {221111} - {3111} +2{311111} - {31111111}
+ {32111} - {3211111} + {41111} - {4111111} - {421111} + {42111111}
<411111111>=- {111111} - {21111} + {2111111} + {311111} - {31111111} - {4111111} + {411111111}
<3333> = {0} - {11} + {211} - {222} - {3111} + {3221} - {3322} + {3333}
<33321>=- {11} + {1111} - {2} +2{211} - {21111} + {22} -2{2211} - {222} + {22211} + {2222} + {31}
- {3111} -2{321} + {32111} +2{3221} - {32221} + {3311} + {332} - {33211} - {3322} - {3331}
+ {33321}
<333111> = {1111} - {111111} + {211} - {21111} + {22} -2{2211} + {221111} - {222} + {22211} - {321}
+ {32111} + {3221} - {322111} - {33} + {3311} + {332} - {33211} - {3331} + {333111}

B-5 Expansion of Symplectic Group Characters into S-Functions (contd).

```
<33222>=- {2} + {211} + {22} -2{2211} + {22211} + {2222} - {22222} + {31} - {3111} - {321} + {32111}
        + {3221} - {32221} - {33} + {3311} - {33211} + {33222}
<332211>=- {11} + {1111} +2{211} -2{21111} + {22} -2{2211} + {221111} -2{222} +2{22211} + {2222}
          - {222211} + {31} -2{3111} + {311111} -2{321} +2{32111} +2{3221} - {322111} - {32221}
          + {3311} - {331111} + {332} - {33211} - {3322} + {332211}
<3321111> = {1111} - {111111} + {211} -2{21111} + {2111111} -2{2211} +2{221111} + {22211}
          - {2221111} - {3111} + {311111} - {321} +2{32111} - {3211111} + {3221} - {322111} + {3311}
          - {331111} - {33211} + {3321111}
<33111111>=- {21111} + {2111111} + {221111} - {22111111} - {3111} + {311111} + {32111} - {3211111}
          - {331111} + {33111111}
<322221> = {211} - {21111} + {22} -2{2211} + {221111} - {222} +2{22211} + {2222} - {222211}
          - {22222} - {321} + {32111} + {3221} - {322111} - {32221} + {322221}
<3222111> = {1111} - {111111} + {211} -2{21111} + {2111111} -2{2211} +2{221111} - {222} +2{22211}
          - {2221111} + {2222} - {222211} - {3111} + {311111} + {32111} - {3211111} + {3221}
          - {322111} - {32221} + {3222111}
<32211111> = {1111} -2{111111} + {11111111} -2{21111} +2{2111111} - {2211} +2{221111} - {22111111}
          + {22211} - {2221111} + {311111} - {31111111} + {32111} - {3211111} - {322111}
          + {32211111}
<321111111>=- {111111} + {11111111} - {21111} +2{2111111} - {211111111} + {221111} - {22111111}
          + {311111} - {31111111} - {3211111} + {321111111}
<3111111111>=- {111111} + {11111111} + {2111111} - {211111111} - {31111111} + {3111111111}
<222222>=- {222} + {22211} - {222211} + {222222}
<2222211>=- {2211} + {221111} + {22211} - {2221111} + {2222} - {222211} - {22222} + {2222211}
<22221111>=- {21111} + {2111111} + {221111} - {22111111} + {22211} - {2221111} - {222211}
          + {22221111}
<222111111>=- {111111} + {11111111} + {2111111} - {211111111} + {221111} - {22111111} - {2221111}
          + {222111111}
<2211111111> = {11111111} - {1111111111} + {2111111} - {211111111} - {22111111} + {2211111111}
<21111111111> = {11111111} - {1111111111} - {211111111} + {21111111111}
<111111111111>=- {1111111111} + {111111111111}

<4441> = {221} - {2221} - {322} - {331} + {3321} + {333} + {432} - {4331} - {443} + {4441}
<4432>=- {111} - {21} + {2111} +2{221} - {2221} + {311} + {32} -2{3211} -2{322} + {3222} - {331}
        +2{3321} + {333} - {3332} - {421} + {4221} + {4311} + {432} - {4322} - {4331} - {4421}
        + {4432}
<44311> = {2111} +2{221} - {22111} - {2221} + {311} + {32} -2{3211} -2{322} + {32211} -3{331}
        + {33111} +2{3321} + {333} - {33311} - {421} + {4221} - {43} + {4311} +2{432} - {43211}
        - {4331} + {441} - {4421} - {443} + {44311}
<44221>=- {111} - {21} + {2111} +2{221} - {2221} +2{311} - {31111} +2{32} -3{3211} -2{322} + {32211}
        + {3222} -2{331} + {33111} +2{3321} - {33221} + {41} - {4111} -2{421} + {42111} + {4221}
        - {43} +2{4311} + {432} - {43211} - {4322} + {441} - {44111} - {4421} + {44221}
<442111> = {2111} + {221} - {22111} + {311} - {31111} + {32} -3{3211} + {321111} - {322} + {32211}
         -2{331} +2{33111} + {3321} - {332111} - {4111} -2{421} + {42111} + {4221} - {43} +2{4311}
         - {431111} + {432} - {43211} + {441} - {44111} - {4421} + {442111}
<4411111>=- {31111} - {3211} + {321111} + {33111} - {3311111} - {4111} - {421} + {42111} + {4311}
          - {431111} - {44111} + {4411111}
<4333> = {1} - {111} - {21} + {2111} + {221} - {2221} + {311} - {3211} - {322} + {3222} + {3321}
        - {3332} - {4111} + {4221} - {4322} + {4333}
<43321>=- {111} + {11111} -2{21} +3{2111} +3{221} -2{22111} -3{2221} + {22221} - {3} +3{311}
        - {31111} +2{32} -4{3211} -3{322} +2{32211} +2{3222} -2{331} + {33111} +3{3321} - {33221}
        + {333} - {33311} - {3332} + {41} - {4111} -2{421} + {42111} +2{4221} - {42221} + {4311}
        + {432} - {43211} - {4322} - {4331} + {43321}
<433111> = {11111} +2{2111} - {211111} +2{221} -2{22111} -2{2221} + {222111} + {311} - {31111}
         + {32} -3{3211} + {321111} -2{322} +2{32211} -2{331} + {33111} +2{3321} - {332111} + {333}
         - {33311} - {421} + {42111} + {4221} - {422111} - {43} + {4311} + {432} - {43211} - {4331}
         + {433111}
<43222>=- {21} + {2111} +2{221} - {22111} -2{2221} + {22221} - {3} +2{311} - {31111} +2{32} -3{3211}
        - {322} +2{32211} + {3222} - {32222} - {331} + {33111} + {3321} - {33221} + {41} - {4111}
        - {421} + {42111} + {4221} - {42221} - {43} + {4311} - {43211} + {43222}
<432211>=- {111} + {11111} - {21} +3{2111} - {211111} +3{221} -3{22111} -3{2221} + {222111}
         + {22221} +3{311} -3{31111} +2{32} -5{3211} +2{321111} -3{322} +3{32211} +2{3222}
         - {322211} -2{331} +2{33111} +2{3321} - {332111} - {33221} + {41} -2{4111} + {411111}
         -2{421} +2{42111} +2{4221} - {422111} - {42221} + {4311} - {431111} + {432} - {43211}
         - {4322} + {432211}
<4321111> = {11111} +2{2111} -2{211111} + {221} -3{22111} + {2211111} - {2221} + {222111} + {311}
          -3{31111} + {3111111} -4{3211} +3{321111} - {322} +2{32211} - {3221111} - {331} +2{33111}
          - {3311111} + {3321} - {332111} - {4111} + {411111} - {421} +2{42111} - {4211111} + {4221}
          - {422111} + {4311} - {431111} - {43211} + {4321111}
<43111111>=- {211111} - {22111} + {2211111} -2{31111} + {3111111} - {3211} +2{321111} - {32111111}
          + {33111} - {3311111} - {4111} + {411111} + {42111} - {4211111} - {431111} + {43111111}
```

 Expansion of Symplectic Group Characters into S-Functions (contd).

<422221> = {2111} - {211111} +2{221} -2{22111} -2{2221} + {222111} + {22221} + {311} - {31111}
+ {32} -2{3211} + {321111} - {322} +2{32211} + {3222} - {322211} - {32222} - {421}
+ {42111} + {4221} - {422111} - {42221} + {422221}
<4222111> = {11111} - {1111111} +2{2111} -2{211111} + {221} -3{22111} + {2211111} -2{2221}
+ {222111} + {22221} + {311} -2{31111} + {3111111} -2{3211} +2{321111} - {322} +2{32211}
- {3221111} + {3222} - {322211} - {4111} + {411111} + {42111} - {4211111} + {4221}
- {422111} - {42221} + {4222111}
<42211111> = {11111} - {1111111} + {2111} -3{211111} + {21111111} -2{22111} + {2211111} - {2221}
+ {222111} -2{31111} +2{3111111} - {3211} +2{321111} - {32111111} + {32211} - {3221111}
+ {411111} - {41111111} + {42111} - {4211111} - {422111} + {42211111}
<421111111>=- {1111111} -2{211111} + {21111111} - {22111} + {2211111} - {31111} +2{3111111}
- {311111111} + {321111} - {32111111} + {411111} - {41111111} - {4211111} + {421111111}
<4111111111>=- {1111111} - {211111} + {21111111} + {3111111} - {311111111} - {41111111}
+ {4111111111}
<33331> = {1} - {111} - {21} + {2111} + {221} - {2221} + {311} - {31111} - {3211} - {322} + {32211}
+ {3222} + {3321} - {33221} - {3332} + {33331}
<33322>=- {21} + {2111} + {221} - {22111} - {2221} + {22221} - {3} + {311} + {32} -2{3211} + {32211}
+ {3222} - {32222} - {331} + {33111} + {3321} - {33221} - {33311} + {33322}
<333211>=- {111} + {11111} - {21} +2{2111} - {211111} +2{221} -2{22111} -2{2221} + {222111}
+ {22221} + {311} - {31111} + {32} -2{3211} + {321111} -2{322} +2{32211} + {3222}
- {322211} - {331} + {33111} +2{3321} - {332111} - {33221} + {333} - {33311} - {3332}
+ {333211}
<3331111> = {11111} - {1111111} + {2111} - {211111} + {221} -2{22111} + {2211111} - {2221}
+ {222111} - {3211} + {321111} + {32211} - {3221111} - {331} + {33111} + {3321} - {332111}
- {33311} + {3331111}
<332221>=- {21} + {2111} +2{221} -2{22111} -2{2221} + {222111} +2{22221} - {222221} + {311}
- {31111} + {32} -2{3211} + {321111} - {322} +2{32211} + {3222} - {322211} - {32222}
- {331} + {33111} + {3321} - {332111} - {33221} + {332221}
<3322111>=- {111} + {11111} +2{2111} -2{211111} + {221} -2{22111} + {2211111} -2{2221} +2{222111}
+ {22221} - {2222111} + {311} -2{31111} + {3111111} -2{3211} +2{321111} - {322} +2{32211}
- {3221111} + {3222} - {322211} + {33111} - {3311111} + {3321} - {332111} - {33221}
+ {3322111}
<33211111> = {11111} - {1111111} + {2111} -2{211111} + {21111111} -2{22111} +2{2211111} + {222111}
- {22211111} - {31111} + {3111111} - {3211} +2{321111} - {32111111} + {32211} - {3221111}
+ {33111} - {3311111} - {332111} + {33211111}
<331111111>=- {211111} + {21111111} + {2211111} - {221111111} - {31111} + {3111111} + {321111}
- {32111111} - {3311111} + {331111111}
<322222> = {221} - {22111} - {2221} + {222111} + {22221} - {222221} - {322} + {32211} - {322211}
+ {322222}
<3222211> = {2111} - {211111} + {221} -2{22111} + {2211111} -2{2221} +2{222111} +2{22221}
- {2222111} - {222221} - {3211} + {321111} + {32211} - {3221111} + {3222} - {322211}
- {32222} + {3222211}
<32221111> = {11111} - {1111111} + {2111} -2{211111} + {21111111} -2{22111} +2{2211111} - {2221}
+2{222111} - {22211111} + {22221} - {2222111} - {31111} + {3111111} + {321111}
- {32111111} + {32211} - {3221111} - {322211} + {32221111}
<322111111> = {11111} -2{1111111} + {111111111} -2{211111} +2{21111111} - {22111} +2{2211111}
- {221111111} + {222111} - {22211111} + {3111111} - {311111111} + {321111} - {32111111}
- {3221111} + {322111111}
<3211111111>=- {1111111} + {111111111} - {211111} +2{21111111} - {2111111111} + {2211111}
- {221111111} + {3111111} - {311111111} - {32111111} + {3211111111}
<31111111111>=- {1111111} + {111111111} + {21111111} - {2111111111} - {311111111} + {31111111111}
<2222221>=- {2221} + {222111} + {22221} - {2222111} - {222221} + {2222221}
<22222111>=- {22111} + {2211111} + {222111} - {22211111} + {22221} - {2222111} - {222221}
+ {22222111}
<222211111>=- {211111} + {21111111} + {2211111} - {221111111} + {222111} - {22211111} - {2222111}
+ {222211111}
<2221111111>=- {1111111} + {111111111} + {21111111} - {2111111111} + {2211111} - {221111111}
- {22211111} + {2221111111}
<22111111111> = {111111111} - {11111111111} + {21111111} - {2111111111} - {221111111}
+ {22111111111}
<211111111111> = {111111111} - {11111111111} - {2111111111} + {211111111111}
<1111111111111>=- {11111111111} + {1111111111111}

<4442>=- {211} + {2211} + {222} - {2222} + {321} - {3221} - {3311} - {332} + {3322} + {3331} - {422}
+ {4321} + {433} - {4332} - {4431} + {4442}
<44411> = {2211} + {222} - {22211} + {321} - {3221} - {3311} -2{332} + {33211} + {3331} - {431}
+ {4321} + {433} - {43311} + {442} - {4431} - {444} + {44411}
<4433> = {11} - {1111} - {211} - {22} + {2211} + {222} + {3111} + {321} -2{3221} - {332} + {3322}
+ {3331} - {3333} - {4211} + {4222} + {4321} - {4332} - {4422} + {4433}
<44321>=- {1111} -2{211} + {21111} - {22} +3{2211} +2{222} - {22211} - {2222} - {31} +2{3111}
+4{321} -2{32111} -4{3221} + {32221} + {33} -3{3311} -3{332} +2{33211} +2{3322} +2{3331}
- {33321} + {411} + {42} -2{4211} -2{422} + {42211} + {4222} -2{431} + {43111} +3{4321}
- {43221} + {433} - {43311} - {4332} + {4411} + {442} - {44211} - {4422} - {4431} + {44321}

Expansion of Symplectic Group Characters into S-Functions (contd).

<443111> = {21111} +2{2211} - {221111} + {222} - {22211} + {3111} +2{321} -2{32111} -2{3221}
+ {322111} + {33} -3{3311} + {331111} -2{332} +2{33211} + {3331} - {333111} - {4211}
- {422} + {42211} -2{431} + {43111} +2{4321} - {432111} + {433} - {43311} - {44} + {4411}
+ {442} - {44211} - {4431} + {443111}
<44222>=- {211} + {2211} + {222} - {2222} - {31} + {3111} +2{321} - {32111} -2{3221} + {32221}
+ {33} -2{3311} + {33211} + {3322} - {33222} + {411} - {41111} + {42} - {4211} - {422}
+ {42211} - {431} + {43111} + {4321} - {43221} - {44} + {4411} - {44211} + {44222}
<442211>=- {1111} - {211} + {21111} - {22} +2{2211} + {222} - {22211} +2{3111} - {311111} +3{321}
-3{32111} -3{3221} + {322111} + {32221} + {33} -2{3311} + {331111} -2{332} +2{33211}
+ {3322} - {332211} + {411} - {41111} + {42} -3{4211} + {421111} - {422} + {42211} + {4222}
-2{431} +2{43111} +2{4321} - {432111} - {43221} + {4411} - {441111} + {442} - {44211}
- {4422} + {442211}
<4421111> = {21111} + {2211} - {221111} + {3111} - {311111} + {321} -3{32111} + {3211111} - {3221}
+ {322111} -2{3311} +2{331111} + {33211} - {3321111} - {41111} -2{4211} + {421111} - {422}
+ {42211} - {431} +2{43111} - {4311111} + {4321} - {432111} + {4411} - {441111} - {44211}
+ {4421111}
<44111111>=- {311111} - {32111} + {3211111} + {331111} - {33111111} - {41111} - {4211} + {421111}
+ {43111} - {4311111} - {441111} + {44111111}
<43331> = {11} - {1111} + {2} -2{211} + {21111} - {22} +2{2211} + {222} - {22211} - {2222} - {31}
+2{3111} +2{321} - {32111} -3{3221} + {32221} - {3311} - {332} + {33211} +2{3322} + {3331}
- {33321} - {3333} + {411} - {41111} - {4211} - {422} + {42211} + {4222} + {4321} - {43221}
- {4332} + {43331}
<43322>=- {211} + {21111} - {22} +2{2211} + {222} -2{22211} - {2222} + {22222} -2{31} +2{3111}
+3{321} -2{32111} -3{3221} +2{32221} + {33} -2{3311} - {332} +2{33211} + {3322} - {33222}
+ {3331} - {33321} - {4} + {411} + {42} -2{4211} + {42211} + {4222} - {42222} - {431}
+ {43111} + {4321} - {43221} - {43311} + {43322}
<433211>=- {1111} + {111111} -2{211} +3{21111} - {22} +4{2211} -2{221111} +2{222} -3{22211} -2{2222}
+ {222211} - {31} +3{3111} - {311111} +4{321} -4{32111} -5{3221} +2{322111} +2{32221}
+ {33} -3{3311} + {331111} -3{332} +3{33211} +2{3322} - {332211} +2{3331} - {333111}
- {33321} + {411} - {41111} + {42} -2{4211} + {421111} -2{422} +2{42211} + {4222}
- {422211} - {431} + {43111} +2{4321} - {432111} - {43221} + {433} - {43311} - {4332}
+ {433211}
<4331111> = {111111} +2{21111} - {2111111} +2{2211} -2{221111} + {222} -2{22211} + {2221111}
+ {3111} - {311111} + {321} -3{32111} + {3211111} -2{3221} +2{322111} -2{3311} + {331111}
- {332} +2{33211} - {3321111} + {3331} - {333111} - {4211} + {421111} + {42211}
- {4221111} - {431} + {43111} + {4321} - {432111} - {43311} + {4331111}
<432221>=- {211} + {21111} - {22} +3{2211} - {221111} +2{222} -3{22211} -2{2222} + {222211}
+ {22222} - {31} +2{3111} - {311111} +4{321} -4{32111} -4{3221} +2{322111} +3{32221}
- {322221} + {33} -2{3311} + {331111} - {332} +2{33211} + {3322} - {332211} - {33222}
+ {411} - {41111} + {42} -2{4211} + {421111} - {422} +2{42211} + {4222} - {422211}
- {42222} - {431} + {43111} + {4321} - {432111} - {43221} + {432221}
<4322111>=- {1111} + {111111} - {211} +3{21111} - {2111111} +3{2211} -3{221111} + {222} -3{22211}
+ {2221111} - {2222} + {222211} +3{3111} -3{311111} +2{321} -5{32111} +2{3211111} -4{3221}
+3{322111} +2{32221} - {3222111} -2{3311} +2{331111} - {332} +2{33211} - {3321111}
+ {3322} - {332211} + {411} -2{41111} + {4111111} -2{4211} +2{421111} - {422} +2{42211}
- {4221111} + {4222} - {422211} + {43111} - {4311111} + {4321} - {432111} - {43221}
+ {4322111}
<43211111> = {111111} +2{21111} -2{2111111} + {2211} -3{221111} + {22111111} - {22211} + {2221111}
+ {3111} -3{311111} + {31111111} -4{32111} +3{3211111} - {3221} +2{322111} - {32211111}
- {3311} +2{331111} - {33111111} + {33211} - {3321111} - {41111} + {4111111} - {4211}
+2{421111} - {42111111} + {42211} - {4221111} + {43111} - {4311111} - {432111}
+ {43211111}
<431111111>=- {2111111} - {221111} + {22111111} -2{311111} + {31111111} - {32111} +2{3211111}
- {321111111} + {331111} - {33111111} - {41111} + {4111111} + {421111} - {42111111}
- {4311111} + {431111111}
<422222> = {2211} - {221111} + {222} - {22211} - {2222} + {222211} + {321} - {32111} - {3221}
+ {322111} + {32221} - {322221} - {422} + {42211} - {422211} + {422222}
<4222211> = {21111} - {2111111} +2{2211} -2{221111} + {222} -3{22211} + {2221111} - {2222}
+ {222211} + {22222} + {3111} - {311111} + {321} -2{32111} + {3211111} -2{3221} +2{322111}
+2{32221} - {3222111} - {322221} - {4211} + {421111} + {42211} - {4221111} + {4222}
- {422211} - {42222} + {4222211}
<42221111> = {111111} - {11111111} +2{21111} -2{2111111} + {2211} -3{221111} + {22111111} -2{22211}
+ {2221111} - {2222} + {222211} + {3111} -2{311111} + {31111111} -2{32111} +2{3211111}
- {3221} +2{322111} - {32211111} + {32221} - {3222111} - {41111} + {4111111} + {421111}
- {42111111} + {42211} - {4221111} - {422211} + {42221111}
<422111111> = {111111} - {11111111} + {21111} -3{2111111} + {211111111} -2{221111} + {22111111}
- {22211} + {2221111} -2{311111} +2{31111111} - {32111} +2{3211111} - {321111111}
+ {322111} - {32211111} + {4111111} - {411111111} + {421111} - {42111111} - {4221111}
+ {422111111}
<4211111111>=- {11111111} -2{2111111} + {211111111} - {221111} + {22111111} - {311111} +2{31111111}
- {3111111111} + {3211111} - {321111111} + {4111111} - {411111111} - {42111111}
+ {4211111111}

Expansion of Symplectic Group Characters into S-Functions (contd).

<41111111111>=- {11111111} - {2111111} + {211111111} + {31111111} - {3111111111} - {411111111}
+ {41111111111}
<33332> = {2} - {211} + {2211} - {2222} - {31} + {3111} + {321} - {32111} - {3221} + {32221}
- {3311} + {33211} + {3322} - {33222} - {33321} + {33332}
<333311> = {11} - {1111} - {211} + {21111} - {22} + {2211} + {222} - {22211} + {3111} - {311111}
+ {321} - {32111} -2{3221} + {322111} + {32221} - {332} + {33211} + {3322} - {332211}
+ {3331} - {33321} - {3333} + {333311}
<333221>=- {211} + {21111} - {22} +2{2211} - {221111} + {222} -2{22211} - {2222} + {222211}
+ {22222} - {31} + {3111} +2{321} -2{32111} -2{3221} + {322111} +2{32221} - {322221} + {33}
-2{3311} + {331111} - {332} +2{33211} + {3322} - {332211} - {33222} + {3331} - {333111}
- {33321} + {333221}
<3332111>=- {1111} + {111111} - {211} +2{21111} - {2111111} +2{2211} -2{221111} + {222} -2{22211}
+ {2221111} - {2222} + {222211} + {3111} - {311111} + {321} -2{32111} + {3211111} -2{3221}
+2{322111} + {32221} - {3222111} - {3311} + {331111} - {332} +2{33211} - {3321111}
+ {3322} - {332211} + {3331} - {333111} - {33321} + {3332111}
<33311111> = {111111} - {11111111} + {21111} - {2111111} + {2211} -2{221111} + {22111111} - {22211}
+ {2221111} - {32111} + {3211111} + {322111} - {32211111} - {3311} + {331111} + {33211}
- {3321111} - {333111} + {33311111}
<332222>=- {22} + {2211} + {222} -2{22211} + {222211} + {22222} - {222222} + {321} - {32111}
- {3221} + {322111} + {32221} - {322221} - {332} + {33211} - {332211} + {332222}
<3322211>=- {211} + {21111} +2{2211} -2{221111} + {222} -2{22211} + {2221111} -2{2222} +2{222211}
+ {22222} - {2222211} + {3111} - {311111} + {321} -2{32111} + {3211111} -2{3221}
+2{322111} +2{32221} - {3222111} - {322221} - {3311} + {331111} + {33211} - {3321111}
+ {3322} - {332211} - {33222} + {3322211}
<33221111>=- {1111} + {111111} +2{21111} -2{2111111} + {2211} -2{221111} + {22111111} -2{22211}
+2{2221111} + {222211} - {22221111} + {3111} -2{311111} + {31111111} -2{32111}
+2{3211111} - {3221} +2{322111} - {32211111} + {32221} - {3222111} + {331111}
- {33111111} + {33211} - {3321111} - {332211} + {33221111}
<332111111> = {111111} - {11111111} + {21111} -2{2111111} + {211111111} -2{221111} +2{22111111}
+ {2221111} - {222111111} - {311111} + {31111111} - {32111} +2{3211111} - {321111111}
+ {322111} - {32211111} + {331111} - {33111111} - {3321111} + {332111111}
<3311111111>=- {2111111} + {211111111} + {22111111} - {2211111111} - {311111} + {31111111}
+ {3211111} - {321111111} - {33111111} + {3311111111}
<3222221> = {2211} - {221111} + {222} -2{22211} + {2221111} - {2222} +2{222211} + {22222}
- {2222211} - {222222} - {3221} + {322111} + {32221} - {3222111} - {322221} + {3222221}
<32222111> = {21111} - {2111111} + {2211} -2{221111} + {22111111} -2{22211} +2{2221111} - {2222}
+2{222211} - {22221111} + {22222} - {2222211} - {32111} + {3211111} + {322111}
- {32211111} + {32221} - {3222111} - {322221} + {32222111}
<322211111> = {111111} - {11111111} + {21111} -2{2111111} + {211111111} -2{221111} +2{22111111}
- {22211} +2{2221111} - {222111111} + {222211} - {22221111} - {311111} + {31111111}
+ {3211111} - {321111111} + {322111} - {32211111} - {3222111} + {322211111}
<3221111111> = {111111} -2{11111111} + {1111111111} -2{2111111} +2{211111111} - {221111}
+2{22111111} - {2211111111} + {2221111} - {222111111} + {31111111} - {3111111111}
+ {3211111} - {321111111} - {32211111} + {3221111111}
<32111111111>=- {11111111} + {1111111111} - {2111111} +2{211111111} - {21111111111} + {22111111}
- {2211111111} + {31111111} - {3111111111} - {321111111} + {32111111111}
<311111111111>=- {11111111} + {1111111111} + {211111111} - {21111111111} - {3111111111}
+ {311111111111}
<2222222>=- {2222} + {222211} - {2222211} + {2222222}
<22222211>=- {22211} + {2221111} + {222211} - {22221111} + {22222} - {2222211} - {222222}
+ {22222211}
<222221111>=- {221111} + {22111111} + {2221111} - {222111111} + {222211} - {22221111} - {2222211}
+ {222221111}
<2222111111>=- {2111111} + {211111111} + {22111111} - {2211111111} + {2221111} - {222111111}
- {22221111} + {2222111111}
<22211111111>=- {11111111} + {1111111111} + {211111111} - {21111111111} + {22111111} - {2211111111}
- {222111111} + {22211111111}
<221111111111> = {1111111111} - {111111111111} + {211111111} - {21111111111} - {2211111111}
+ {221111111111}
<2111111111111> = {1111111111} - {111111111111} - {21111111111} + {2111111111111}
<11111111111111>=- {111111111111} + {11111111111111}

Plethysms within the Group GL_2.

{0} ⊗ {0} = {0}

{1} ⊗ {0} = {0}
{1} ⊗ {1} = {1}

{2} ⊗ {0} = {0}
{2} ⊗ {1} = {2}
{2} ⊗ {2} = {0} + {4}

{3} ⊗ {0} = {0}
{3} ⊗ {1} = {3}
{3} ⊗ {2} = {2} + {6}
{3} ⊗ {3} = {3} + {5} + {9}

{4} ⊗ {0} = {0}
{4} ⊗ {1} = {4}
{4} ⊗ {2} = {0} + {4} + {8}
{4} ⊗ {3} = {0} + {4} + {6} + {8} + {12}
{4} ⊗ {4} = {0} +2{4} +2{8} + {10} + {12} + {16}

{5} ⊗ {0} = {0}
{5} ⊗ {1} = {5}
{5} ⊗ {2} = {2} + {6} + {10}
{5} ⊗ {3} = {3} + {5} + {7} + {9} + {11} + {15}
{5} ⊗ {4} = {0} +2{4} + {6} +2{8} + {10} +2{12} + {14} + {16} + {20}
{5} ⊗ {5} = {1} + {3} +2{5} +2{7} +3{9} +2{11} +2{13} +2{15} +2{17} + {19} + {21} + {25}

{6} ⊗ {0} = {0}
{6} ⊗ {1} = {6}
{6} ⊗ {2} = {0} + {4} + {8} + {12}
{6} ⊗ {3} = {2} +2{6} + {8} + {10} + {12} + {14} + {18}
{6} ⊗ {4} =2{0} +2{4} + {6} +3{8} + {10} +3{12} + {14} +2{16} + {18} + {20} + {24}
{6} ⊗ {5} =2{2} + {4} +4{6} +2{8} +4{10} +3{12} +4{14} +2{16} +3{18} +2{20} +2{22} + {24} + {26}
+ {30}
{6} ⊗ {6} =3{0} +4{4} +3{6} +6{8} +3{10} +7{12} +4{14} +6{16} +4{18} +5{20} +2{22} +4{24} +2{26}
+2{28} + {30} + {32} + {36}

{7} ⊗ {0} = {0}
{7} ⊗ {1} = {7}
{7} ⊗ {2} = {2} + {6} + {10} + {14}
{7} ⊗ {3} = {3} + {5} + {7} +2{9} + {11} + {13} + {15} + {17} + {21}
{7} ⊗ {4} = {0} +3{4} + {6} +3{8} +2{10} +3{12} +2{14} +3{16} + {18} +2{20} + {22} + {24} + {28}
{7} ⊗ {5} = {1} +2{3} +3{5} +4{7} +4{9} +5{11} +4{13} +5{15} +4{17} +4{19} +3{21} +3{23} +2{25}
+2{27} + {29} + {31} + {35}
{7} ⊗ {6} =4{2} +2{4} +7{6} +5{8} +8{10} +7{12} +9{14} +6{16} +9{18} +6{20} +7{22} +5{24} +5{26}
+3{28} +4{30} +2{32} +2{34} + {36} + {38} + {42}
{7} ⊗ {7} =3{1} +4{3} +7{5} +9{7} +10{9} +11{11} +13{13} +12{15} +13{17} +12{19} +12{21} +10{23}
+11{25} +8{27} +8{29} +6{31} +5{33} +4{35} +4{37} +2{39} +2{41} + {43} + {45} + {49}

{8} ⊗ {0} = {0}
{8} ⊗ {1} = {8}
{8} ⊗ {2} = {0} + {4} + {8} + {12} + {16}
{8} ⊗ {3} = {0} + {4} + {6} +2{8} + {10} +2{12} + {14} + {16} + {18} + {20} + {24}
{8} ⊗ {4} =2{0} +3{4} + {6} +4{8} +2{10} +4{12} +2{14} +4{16} +2{18} +3{20} + {22} +2{24} + {26}
+ {28} + {32}
{8} ⊗ {5} =2{0} + {2} +4{4} +3{6} +6{8} +5{10} +7{12} +5{14} +7{16} +5{18} +6{20} +4{22} +5{24}
+3{26} +3{28} +2{30} +2{32} + {34} + {36} + {40}
{8} ⊗ {6} =4{0} + {2} +7{4} +5{6} +11{8} +7{10} +13{12} +9{14} +13{16} +10{18} +12{20} +8{22}
+11{24} +7{26} +8{28} +5{30} +6{32} +3{34} +4{36} +2{38} +2{40} + {42} + {44} + {48}
{8} ⊗ {7} =4{0} +3{2} +10{4} +9{6} +16{8} +14{10} +19{12} +17{14} +21{16} +18{18} +21{20} +17{22}
+19{24} +15{26} +16{28} +12{30} +13{32} +9{34} +9{36} +6{38} +6{40} +4{42} +4{44} +2{46}
+2{48} + {50} + {52} + {56}
{8} ⊗ {8} =7{0} +4{2} +16{4} +13{6} +25{8} +21{10} +31{12} +26{14} +35{16} +29{18} +35{20} +29{22}
+34{24} +27{26} +30{28} +23{30} +25{32} +19{34} +20{36} +14{38} +15{40} +10{42} +10{44}
+6{46} +7{48} +4{50} +4{52} +2{54} +2{56} + {58} + {60} + {64}

{9} ⊗ {0} = {0}
{9} ⊗ {1} = {9}
{9} ⊗ {2} = {2} + {6} + {10} + {14} + {18}
{9} ⊗ {3} = {3} + {5} + {7} +2{9} +2{11} + {13} +2{15} + {17} + {19} + {21} + {23} + {27}

Plethysms within the Group GL_2 (contd).

{9} ⊗ {4} =2{0} +3{4} +2{6} +4{8} +2{10} +5{12} +3{14} +4{16} +3{18} +4{20} +2{22} +3{24} + {26} +2{28} + {30} + {32} + {36}

{9} ⊗ {5} = {1} +3{3} +5{5} +5{7} +7{9} +7{11} +8{13} +8{15} +8{17} +7{19} +8{21} +6{23} +6{25} +5{27} +5{29} +3{31} +3{33} +2{35} +2{37} + {39} + {41} + {45}

{9} ⊗ {6} =6{2} +4{4} +12{6} +9{8} +14{10} +13{12} +17{14} +13{16} +18{18} +14{20} +16{22} +13{24} +14{26} +10{28} +12{30} +8{32} +8{34} +6{36} +6{38} +3{40} +4{42} +2{44} +2{46} + {48} + {50} + {54}

{9} ⊗ {7} =4{1} +10{3} +13{5} +17{7} +21{9} +24{11} +25{13} +29{15} +28{17} +29{19} +29{21} +29{23} +26{25} +27{27} +23{29} +22{31} +19{33} +18{35} +14{37} +14{39} +10{41} +9{43} +7{45} +6{47} +4{49} +4{51} +2{53} +2{55} + {57} + {59} + {63}

{9} ⊗ {8} =8{0} +8{2} +21{4} +22{6} +35{8} +33{10} +45{12} +42{14} +51{16} +48{18} +54{20} +48{22} +54{24} +47{26} +49{28} +43{30} +44{32} +36{34} +37{36} +29{38} +29{40} +23{42} +22{44} +16{46} +16{48} +11{50} +10{52} +7{54} +7{56} +4{58} +4{60} +2{62} +2{64} + {66} + {68} + {72}

{9} ⊗ {9} =11{1} +24{3} +34{5} +43{7} +56{9} +62{11} +69{13} +76{15} +82{17} +82{19} +88{21} +86{23} +87{25} +85{27} +83{29} +77{31} +77{33} +69{35} +65{37} +59{39} +55{41} +46{43} +44{45} +36{47} +33{49} +27{51} +24{53} +18{55} +17{57} +12{59} +10{61} +8{63} +7{65} +4{67} +4{69} +2{71} +2{73} + {75} + {77} + {81}

{10} ⊗ {0} = {0}

{10} ⊗ {1} = {10}

{10} ⊗ {2} = {0} + {4} + {8} + {12} + {16} + {20}

{10} ⊗ {3} = {2} +2{6} + {8} +2{10} +2{12} +2{14} + {16} +2{18} + {20} + {22} + {24} + {26} + {30}

{10} ⊗ {4} =2{0} +4{4} + {6} +5{8} +3{10} +5{12} +3{14} +6{16} +3{18} +5{20} +3{22} +4{24} +2{26} +3{28} + {30} +2{32} + {34} + {36} + {40}

{10} ⊗ {5} =4{2} +3{4} +7{6} +6{8} +10{10} +8{12} +11{14} +9{16} +11{18} +9{20} +10{22} +8{24} +9{26} +6{28} +7{30} +5{32} +5{34} +3{36} +3{38} +2{40} +2{42} + {44} + {46} + {50}

{10} ⊗ {6} =6{0} +2{2} +11{4} +9{6} +17{8} +13{10} +22{12} +17{14} +23{16} +19{18} +24{20} +18{22} +23{24} +17{26} +19{28} +15{30} +16{32} +11{34} +13{36} +8{38} +9{40} +6{42} +6{44} +3{46} +4{48} +2{50} +2{52} + {54} + {56} + {60}

{10} ⊗ {7} =12{2} +13{4} +23{6} +23{8} +34{10} +31{12} +40{14} +38{16} +43{18} +40{20} +45{22} +39{24} +42{26} +37{28} +38{30} +32{32} +33{34} +26{36} +26{38} +21{40} +20{42} +15{44} +15{46} +10{48} +10{50} +7{52} +6{54} +4{56} +4{58} +2{60} +2{62} + {64} + {66} + {70}

{10} ⊗ {8} =12{0} +10{2} +32{4} +30{6} +51{8} +48{10} +66{12} +61{14} +77{16} +70{18} +83{20} +74{22} +84{24} +74{26} +80{28} +69{30} +74{32} +62{34} +64{36} +53{38} +54{40} +43{42} +43{44} +33{46} +33{48} +25{50} +24{52} +17{54} +17{56} +11{58} +11{60} +7{62} +7{64} +4{66} +4{68} +2{70} +2{72} + {74} + {76} + {80}

{10} ⊗ {9} =5{0} +32{2} +37{4} +66{6} +70{8} +94{10} +98{12} +118{14} +117{16} +136{18} +131{20} +143{22} +138{24} +146{26} +135{28} +142{30} +129{32} +130{34} +118{36} +116{38} +102{40} +101{42} +86{44} +82{46} +70{48} +66{50} +53{52} +51{54} +40{56} +37{58} +29{60} +26{62} +19{64} +18{66} +12{68} +11{70} +8{72} +7{74} +4{76} +4{78} +2{80} +2{82} + {84} + {86} + {90}

{10} ⊗ {10} =24{0} +32{2} +81{4} +85{6} +131{8} +136{10} +175{12} +175{14} +212{16} +205{18} +236{20} +227{22} +248{24} +235{26} +252{28} +233{30} +244{32} +224{34} +228{36} +205{38} +208{40} +182{42} +181{44} +158{46} +153{48} +131{50} +127{52} +105{54} +100{56} +82{58} +77{60} +61{62} +58{64} +44{66} +41{68} +31{70} +28{72} +20{74} +19{76} +12{78} +12{80} +8{82} +7{84} +4{86} +4{88} +2{90} +2{92} + {94} + {96} + {100}

{11} ⊗ {0} = {0}

{11} ⊗ {1} = {11}

{11} ⊗ {2} = {2} + {6} + {10} + {14} + {18} + {22}

{11} ⊗ {3} = {3} + {5} + {7} +2{9} +2{11} +2{13} +2{15} +2{17} + {19} +2{21} + {23} + {25} + {27} + {29} + {33}

{11} ⊗ {4} =2{0} +4{4} +2{6} +5{8} +3{10} +6{12} +4{14} +6{16} +4{18} +6{20} +4{22} +5{24} +3{26} +4{28} +2{30} +3{32} + {34} +2{36} + {38} + {40} + {44}

{11} ⊗ {5} =2{1} +4{3} +6{5} +8{7} +9{9} +11{11} +11{13} +13{15} +12{17} +13{19} +12{21} +12{23} +11{25} +11{27} +9{29} +9{31} +7{33} +7{35} +5{37} +5{39} +3{41} +3{43} +2{45} +2{47} + {49} + {51} + {55}

{11} ⊗ {6} =9{2} +7{4} +18{6} +15{8} +23{10} +21{12} +28{14} +24{16} +31{18} +26{20} +30{22} +26{24} +29{26} +23{28} +26{30} +20{32} +21{34} +17{36} +17{38} +12{40} +13{42} +9{44} +9{46} +6{48} +6{50} +3{52} +4{54} +2{56} +2{58} + {60} + {62} + {66}

{11} ⊗ {7} =8{1} +16{3} +24{5} +31{7} +38{9} +43{11} +49{13} +52{15} +57{17} +57{19} +60{21} +59{23} +60{25} +57{27} +57{29} +52{31} +51{33} +46{35} +44{37} +38{39} +36{41} +30{43} +28{45} +23{47} +21{49} +16{51} +15{53} +11{55} +10{57} +7{59} +6{61} +4{63} +4{65} +2{67} +2{69} + {71} + {73} + {77}

{11} ⊗ {8} =13{0} +17{2} +42{4} +45{6} +69{8} +70{10} +91{12} +90{14} +108{16} +105{18} +118{20} +112{22} +123{24} +114{26} +121{28} +110{30} +114{32} +102{34} +103{36} +90{38} +90{40} +77{42} +75{44} +63{46} +61{48} +49{50} +47{52} +37{54} +35{56} +27{58} +25{60} +18{62} +17{64} +12{66} +11{68} +7{70} +7{72} +4{74} +4{76} +2{78} +2{80} + {82} + {84} + {88}

{11} ⊗ {9} =25{1} +53{3} +76{5} +100{7} +123{9} +144{11} +160{13} +180{15} +191{17} +204{19} +212{21} +220{23} +220{25} +226{27} +221{29} +220{31} +213{33} +209{35} +196{37} +191{39} +176{41} +167{43} +153{45} +143{47} +127{49} +119{51} +103{53} +94{55} +81{57} +73{59} +60{61} +55{63} +44{65} +39{67} +31{69} +27{71} +20{73} +18{75} +13{77} +11{79} +8{81} +7{83} +4{85} +4{87} +2{89} +2{91} + {93} + {95} + {99}

{11} ⊗ {10} =13{0} +75{2} +99{4} +159{6} +180{8} +236{10} +252{12} +301{14} +311{16} +352{18}
+356{20} +388{22} +384{24} +408{26} +396{28} +412{30} +394{32} +401{34} +378{36}
+379{38} +352{40} +347{42} +318{44} +309{46} +278{48} +268{50} +237{52} +225{54}
+197{56} +184{58} +158{60} +147{62} +123{64} +113{66} +93{68} +85{70} +68{72} +62{74}
+48{76} +43{78} +33{80} +29{82} +21{84} +19{86} +13{88} +12{90} +8{92} +7{94} +4{96}
+4{98} +2{100} +2{102} + {104} + {106} + {110}
{11} ⊗ {11} =72{1} +142{3} +212{5} +277{7} +343{9} +401{11} +459{13} +507{15} +555{17} +592{19}
+630{21} +653{23} +679{25} +690{27} +704{29} +703{31} +706{33} +693{35} +686{37}
+664{39} +649{41} +619{43} +598{45} +562{47} +537{49} +499{51} +470{53} +431{55}
+402{57} +363{59} +336{61} +299{63} +273{65} +240{67} +217{69} +187{71} +168{73}
+142{75} +126{77} +105{79} +93{81} +75{83} +66{85} +52{87} +45{89} +35{91} +30{93}
+22{95} +19{97} +14{99} +12{101} +8{103} +7{105} +4{107} +4{109} +2{111} +2{113} + {115}
+ {117} + {121}

{12} ⊗ {0} = {0}
{12} ⊗ {1} = {12}
{12} ⊗ {2} = {0} + {4} + {8} + {12} + {16} + {20} + {24}
{12} ⊗ {3} = {0} + {4} + {6} +2{8} + {10} +3{12} +2{14} +2{16} +2{18} +2{20} + {22} +2{24} + {26}
+ {28} + {30} + {32} + {36}
{12} ⊗ {4} =3{0} +4{4} +2{6} +6{8} +3{10} +7{12} +4{14} +7{16} +5{18} +7{20} +4{22} +7{24} +4{26}
+5{28} +3{30} +4{32} +2{34} +3{36} + {38} +2{40} + {42} + {44} + {48}
{12} ⊗ {5} =3{0} +2{2} +7{4} +7{6} +11{8} +10{10} +15{12} +13{14} +16{16} +14{18} +17{20} +14{22}
+16{24} +13{26} +14{28} +12{30} +12{32} +9{34} +10{36} +7{38} +7{40} +5{42} +5{44} +3{46}
+3{48} +2{50} +2{52} + {54} + {56} + {60}
{12} ⊗ {6} =8{0} +4{2} +16{4} +14{6} +26{8} +21{10} +34{12} +28{14} +37{16} +33{18} +40{20} +33{22}
+41{24} +33{26} +37{28} +31{30} +34{32} +26{34} +29{36} +22{38} +23{40} +18{42} +18{44}
+12{46} +14{48} +9{50} +9{52} +6{54} +6{56} +3{58} +4{60} +2{62} +2{64} + {66} + {68}
+ {72}
{12} ⊗ {7} =10{0} +12{2} +29{4} +33{6} +48{8} +49{10} +65{12} +64{14} +75{16} +74{18} +83{20}
+78{22} +86{24} +79{26} +83{28} +77{30} +78{32} +69{34} +71{36} +61{38} +60{40} +52{42}
+50{44} +41{46} +40{48} +32{50} +30{52} +24{54} +22{56} +16{58} +16{60} +11{62} +10{64}
+7{66} +6{68} +4{70} +4{72} +2{74} +2{76} + {78} + {80} + {84}
{12} ⊗ {8} =20{0} +22{2} +58{4} +61{6} +96{8} +95{10} +128{12} +124{14} +152{16} +147{18} +169{20}
+159{22} +179{24} +166{26} +178{28} +164{30} +173{32} +155{34} +161{36} +142{38} +144{40}
+126{42} +125{44} +106{46} +106{48} +88{50} +85{52} +70{54} +67{56} +53{58} +51{60}
+39{62} +37{64} +28{66} +26{68} +18{70} +18{72} +12{74} +11{76} +7{78} +7{80} +4{82}
+4{84} +2{86} +2{88} + {90} + {92} + {96}
{12} ⊗ {9} =28{0} +48{2} +99{4} +122{6} +169{8} +184{10} +232{12} +242{14} +278{16} +287{18}
+317{20} +315{22} +341{24} +333{26} +348{28} +338{30} +345{32} +326{34} +332{36} +309{38}
+305{40} +283{42} +276{44} +249{46} +242{48} +215{50} +204{52} +181{54} +169{56} +145{58}
+137{60} +115{62} +105{64} +88{66} +80{68} +64{70} +59{72} +46{74} +41{76} +32{78}
+28{80} +20{82} +19{84} +13{86} +11{88} +8{90} +7{92} +4{94} +4{96} +2{98} +2{100}
+ {102} + {104} + {108}
{12} ⊗ {10} =52{0} +83{2} +179{4} +212{6} +304{8} +328{10} +416{12} +432{14} +507{16} +518{18}
+581{20} +578{22} +633{24} +620{26} +657{28} +639{30} +664{32} +633{34} +650{36}
+611{38} +616{40} +575{42} +570{44} +523{46} +516{48} +467{50} +452{52} +406{54}
+389{56} +342{58} +327{60} +283{62} +266{64} +229{66} +212{68} +178{70} +166{72}
+136{74} +124{76} +101{78} +92{80} +72{82} +66{84} +50{86} +45{88} +34{90} +30{92}
+21{94} +20{96} +13{98} +12{100} +8{102} +7{104} +4{106} +4{108} +2{110} +2{112} + {114}
+ {116} + {120}
{12} ⊗ {11} =73{0} +155{2} +295{4} +376{6} +509{8} +578{10} +705{12} +762{14} +869{16} +916{18}
+1007{20} +1034{22} +1109{24} +1118{26} +1171{28} +1167{30} +1200{32} +1176{34}
+1195{36} +1157{38} +1156{40} +1109{42} +1095{44} +1035{46} +1013{48} +947{50} +915{52}
+848{54} +810{56} +740{58} +703{60} +635{62} +594{64} +532{66} +494{68} +435{70}
+402{72} +349{74} +318{76} +274{78} +247{80} +208{82} +188{84} +155{86} +138{88}
+113{90} +100{92} +79{94} +70{96} +54{98} +47{100} +36{102} +31{104} +22{106} +20{108}
+14{110} +12{112} +8{114} +7{116} +4{118} +4{120} +2{122} +2{124} + {126} + {128}
+ {132}
{12} ⊗ {12} =127{0} +253{2} +493{4} +621{6} +852{8} +961{10} +1182{12} +1273{14} +1466{16} +1543{18}
+1709{20} +1754{22} +1900{24} +1919{26} +2025{28} +2025{30} +2102{32} +2069{34}
+2122{36} +2065{38} +2086{40} +2015{42} +2010{44} +1916{46} +1898{48} +1792{50}
+1750{52} +1641{54} +1589{56} +1470{58} +1415{60} +1296{62} +1233{64} +1122{66}
+1057{68} +948{70} +891{72} +790{74} +732{76} +645{78} +594{80} +514{82} +472{84}
+403{86} +365{88} +310{90} +278{92} +230{94} +208{96} +169{98} +150{100} +121{102}
+107{104} +83{106} +74{108} +56{110} +49{112} +37{114} +32{116} +22{118} +21{120}
+14{122} +12{124} +8{126} +7{128} +4{130} +4{132} +2{134} +2{136} + {138} + {140}
+ {144}

B-7 S-Function Division.

{0} / {0} = {0}

{1} / {0} = {1}
{1} / {1} = {0}

{2} / {0} = {2}
{2} / {1} = {1}
{2} / {2} = {0}
{11} / {0} = {11}
{11} / {1} = {1}
{11} / {11} = {0}

{3} / {0} = {3}
{3} / {1} = {2}
{3} / {2} = {1}
{3} / {3} = {0}
{21} / {0} = {21}
{21} / {1} = {11} + {2}
{21} / {2} = {1}
{21} / {11} = {1}
{21} / {21} = {0}
{111} / {0} = {111}
{111} / {1} = {11}
{111} / {11} = {1}
{111} / {111} = {0}

{4} / {0} = {4}
{4} / {1} = {3}
{4} / {2} = {2}
{4} / {3} = {1}
{4} / {4} = {0}
{31} / {0} = {31}
{31} / {1} = {21} + {3}
{31} / {2} = {11} + {2}
{31} / {11} = {2}
{31} / {3} = {1}
{31} / {21} = {1}
{31} / {31} = {0}
{22} / {0} = {22}
{22} / {1} = {21}
{22} / {2} = {2}
{22} / {11} = {11}
{22} / {21} = {1}
{22} / {22} = {0}
{211} / {0} = {211}
{211} / {1} = {111} + {21}
{211} / {2} = {11}
{211} / {11} = {11} + {2}
{211} / {21} = {1}
{211} / {111} = {1}
{211} / {211} = {0}
{1111} / {0} = {1111}
{1111} / {1} = {111}
{1111} / {11} = {11}
{1111} / {111} = {1}
{1111} / {1111} = {0}

{41} / {0} = {41}
{41} / {1} = {31} + {4}
{41} / {2} = {21} + {3}
{41} / {11} = {3}
{41} / {3} = {11} + {2}
{41} / {21} = {2}
{41} / {4} = {1}
{41} / {31} = {1}
{41} / {41} = {0}
{32} / {0} = {32}
{32} / {1} = {22} + {31}
{32} / {2} = {21} + {3}
{32} / {11} = {21}
{32} / {3} = {2}
{32} / {21} = {11} + {2}
{32} / {31} = {1}
{32} / {22} = {1}
{32} / {32} = {0}
{311} / {0} = {311}
{311} / {1} = {211} + {31}
{311} / {2} = {111} + {21}
{311} / {11} = {21} + {3}
{311} / {3} = {11}
{311} / {21} = {11} + {2}
{311} / {111} = {2}
{311} / {31} = {1}
{311} / {211} = {1}
{311} / {311} = {0}
{221} / {0} = {221}
{221} / {1} = {211} + {22}
{221} / {2} = {21}
{221} / {11} = {111} + {21}
{221} / {21} = {11} + {2}
{221} / {111} = {11}
{221} / {22} = {1}
{221} / {211} = {1}
{221} / {221} = {0}
{2111} / {0} = {2111}
{2111} / {1} = {1111} + {211}
{2111} / {2} = {111}
{2111} / {11} = {111} + {21}
{2111} / {21} = {11}
{2111} / {111} = {11} + {2}
{2111} / {211} = {1}
{2111} / {1111} = {1}
{2111} / {2111} = {0}
{11111} / {0} = {11111}
{11111} / {1} = {1111}
{11111} / {11} = {111}
{11111} / {111} = {11}
{11111} / {1111} = {1}
{11111} / {11111} = {0}

{42} / {0} = {42}
{42} / {1} = {32} + {41}
{42} / {2} = {22} + {31} + {4}
{42} / {11} = {31}
{42} / {3} = {21} + {3}
{42} / {21} = {21} + {3}
{42} / {4} = {2}
{42} / {31} = {11} + {2}
{42} / {22} = {2}
{42} / {41} = {1}
{42} / {32} = {1}
{42} / {42} = {0}
{411} / {0} = {411}
{411} / {1} = {311} + {41}
{411} / {2} = {211} + {31}
{411} / {11} = {31} + {4}
{411} / {3} = {111} + {21}
{411} / {21} = {21} + {3}
{411} / {111} = {3}
{411} / {4} = {11}
{411} / {31} = {11} + {2}
{411} / {211} = {2}
{411} / {41} = {1}
{411} / {311} = {1}
{411} / {411} = {0}
{33} / {0} = {33}
{33} / {1} = {32}
{33} / {2} = {31}
{33} / {11} = {22}
{33} / {3} = {3}

{33} / {21} = {21}
{33} / {31} = {2}
{33} / {22} = {11}
{33} / {32} = {1}
{33} / {33} = {0}
{321} / {0} = {321}
{321} / {1} = {221} + {311} + {32}
{321} / {2} = {211} + {22} + {31}
{321} / {11} = {211} + {22} + {31}
{321} / {3} = {21}
{321} / {21} = {111} +2{21} + {3}
{321} / {111} = {21}
{321} / {31} = {11} + {2}
{321} / {22} = {11} + {2}
{321} / {211} = {11} + {2}
{321} / {32} = {1}
{321} / {311} = {1}
{321} / {221} = {1}
{321} / {321} = {0}
{3111} / {0} = {3111}
{3111} / {1} = {2111} + {311}
{3111} / {2} = {1111} + {211}
{3111} / {11} = {211} + {31}
{3111} / {3} = {111}
{3111} / {21} = {111} + {21}
{3111} / {111} = {21} + {3}
{3111} / {31} = {11}
{3111} / {211} = {11} + {2}
{3111} / {1111} = {2}
{3111} / {311} = {1}
{3111} / {2111} = {1}
{3111} / {3111} = {0}
{222} / {0} = {222}
{222} / {1} = {221}
{222} / {2} = {22}
{222} / {11} = {211}
{222} / {21} = {21}
{222} / {111} = {111}
{222} / {22} = {2}
{222} / {211} = {11}
{222} / {221} = {1}
{222} / {222} = {0}
{2211} / {0} = {2211}
{2211} / {1} = {2111} + {221}
{2211} / {2} = {211}
{2211} / {11} = {1111} + {211} + {22}
{2211} / {21} = {111} + {21}
{2211} / {111} = {111} + {21}
{2211} / {22} = {11}
{2211} / {211} = {11} + {2}
{2211} / {1111} = {11}
{2211} / {221} = {1}
{2211} / {2111} = {1}
{2211} / {2211} = {0}
{21111} / {0} = {21111}
{21111} / {1} = {11111} + {2111}
{21111} / {2} = {1111}
{21111} / {11} = {1111} + {211}
{21111} / {21} = {111}
{21111} / {111} = {111} + {21}
{21111} / {211} = {11}
{21111} / {1111} = {11} + {2}
{21111} / {2111} = {1}
{21111} / {11111} = {1}
{21111} / {21111} = {0}
{111111} / {0} = {111111}
{111111} / {1} = {11111}
{111111} / {11} = {1111}
{111111} / {111} = {111}
{111111} / {1111} = {11}
{111111} / {11111} = {1}
{111111} / {111111} = {0}

{43} / {0} = {43}
{43} / {1} = {33} + {42}
{43} / {2} = {32} + {41}
{43} / {11} = {32}
{43} / {3} = {31} + {4}
{43} / {21} = {22} + {31}
{43} / {4} = {3}
{43} / {31} = {21} + {3}
{43} / {22} = {21}
{43} / {41} = {2}
{43} / {32} = {11} + {2}
{43} / {42} = {1}
{43} / {33} = {1}
{43} / {43} = {0}
{421} / {0} = {421}
{421} / {1} = {321} + {411} + {42}
{421} / {2} = {221} + {311} + {32} + {41}
{421} / {11} = {311} + {32} + {41}
{421} / {3} = {211} + {22} + {31}
{421} / {21} = {211} + {22} +2{31} + {4}
{421} / {111} = {31}
{421} / {4} = {21}
{421} / {31} = {111} +2{21} + {3}
{421} / {22} = {21} + {3}
{421} / {211} = {21} + {3}
{421} / {41} = {11} + {2}
{421} / {32} = {11} + {2}
{421} / {311} = {11} + {2}
{421} / {221} = {2}
{421} / {42} = {1}
{421} / {411} = {1}
{421} / {321} = {1}
{421} / {421} = {0}
{4111} / {0} = {4111}
{4111} / {1} = {3111} + {411}
{4111} / {2} = {2111} + {311}
{4111} / {11} = {311} + {41}
{4111} / {3} = {1111} + {211}
{4111} / {21} = {211} + {31}
{4111} / {111} = {31} + {4}
{4111} / {4} = {111}
{4111} / {31} = {111} + {21}
{4111} / {211} = {21} + {3}
{4111} / {1111} = {3}
{4111} / {41} = {11}
{4111} / {311} = {11} + {2}
{4111} / {2111} = {2}
{4111} / {411} = {1}
{4111} / {3111} = {1}
{4111} / {4111} = {0}
{331} / {0} = {331}
{331} / {1} = {321} + {33}
{331} / {2} = {311} + {32}
{331} / {11} = {221} + {32}
{331} / {3} = {31}
{331} / {21} = {211} + {22} + {31}
{331} / {111} = {22}
{331} / {31} = {21} + {3}
{331} / {22} = {111} + {21}
{331} / {211} = {21}
{331} / {32} = {11} + {2}
{331} / {311} = {2}
{331} / {221} = {11}
{331} / {33} = {1}
{331} / {321} = {1}
{331} / {331} = {0}
{322} / {0} = {322}

{322} / {1} = {222} + {321}
{322} / {2} = {221} + {32}
{322} / {11} = {221} + {311}
{322} / {3} = {22}
{322} / {21} = {211} + {22} + {31}
{322} / {111} = {211}
{322} / {31} = {21}
{322} / {22} = {21} + {3}
{322} / {211} = {111} + {21}
{322} / {32} = {2}
{322} / {311} = {11}
{322} / {221} = {11} + {2}
{322} / {321} = {1}
{322} / {222} = {1}
{322} / {322} = {0}
{3211} / {0} = {3211}
{3211} / {1} = {2211} + {3111} + {321}
{3211} / {2} = {2111} + {221} + {311}
{3211} / {11} = {2111} + {221} + {311} + {32}
{3211} / {3} = {211}
{3211} / {21} = {1111} +2{211} + {22} + {31}
{3211} / {111} = {211} + {22} + {31}
{3211} / {31} = {111} + {21}
{3211} / {22} = {111} + {21}
{3211} / {211} = {111} +2{21} + {3}
{3211} / {1111} = {21}
{3211} / {32} = {11}
{3211} / {311} = {11} + {2}
{3211} / {221} = {11} + {2}
{3211} / {2111} = {11} + {2}
{3211} / {321} = {1}
{3211} / {3111} = {1}
{3211} / {2211} = {1}
{3211} / {3211} = {0}
{31111} / {0} = {31111}
{31111} / {1} = {21111} + {3111}
{31111} / {2} = {11111} + {2111}
{31111} / {11} = {2111} + {311}
{31111} / {3} = {1111}
{31111} / {21} = {1111} + {211}
{31111} / {111} = {211} + {31}
{31111} / {31} = {111}
{31111} / {211} = {111} + {21}
{31111} / {1111} = {21} + {3}
{31111} / {311} = {11}
{31111} / {2111} = {11} + {2}
{31111} / {11111} = {2}
{31111} / {3111} = {1}
{31111} / {21111} = {1}
{31111} / {31111} = {0}
{2221} / {0} = {2221}
{2221} / {1} = {2211} + {222}
{2221} / {2} = {221}
{2221} / {11} = {2111} + {221}
{2221} / {21} = {211} + {22}
{2221} / {111} = {1111} + {211}
{2221} / {22} = {21}
{2221} / {211} = {111} + {21}
{2221} / {1111} = {111}
{2221} / {221} = {11} + {2}
{2221} / {2111} = {11}
{2221} / {222} = {1}
{2221} / {2211} = {1}
{2221} / {2221} = {0}
{22111} / {0} = {22111}
{22111} / {1} = {21111} + {2211}
{22111} / {2} = {2111}
{22111} / {11} = {11111} + {2111} + {221}
{22111} / {21} = {1111} + {211}
{22111} / {111} = {1111} + {211} + {22}
{22111} / {22} = {111}
{22111} / {211} = {111} + {21}
{22111} / {1111} = {111} + {21}
{22111} / {221} = {11}
{22111} / {2111} = {11} + {2}
{22111} / {11111} = {11}
{22111} / {2211} = {1}
{22111} / {21111} = {1}
{22111} / {22111} = {0}
{211111} / {0} = {211111}
{211111} / {1} = {111111} + {21111}
{211111} / {2} = {11111}
{211111} / {11} = {11111} + {2111}
{211111} / {21} = {1111}
{211111} / {111} = {1111} + {211}
{211111} / {211} = {111}
{211111} / {1111} = {111} + {21}
{211111} / {2111} = {11}
{211111} / {11111} = {11} + {2}
{211111} / {21111} = {1}
{211111} / {111111} = {1}
{211111} / {211111} = {0}
{1111111} / {0} = {1111111}
{1111111} / {1} = {111111}
{1111111} / {11} = {11111}
{1111111} / {111} = {1111}
{1111111} / {1111} = {111}
{1111111} / {11111} = {11}
{1111111} / {111111} = {1}
{1111111} / {1111111} = {0}

{44} / {0} = {44}
{44} / {1} = {43}
{44} / {2} = {42}
{44} / {11} = {33}
{44} / {3} = {41}
{44} / {21} = {32}
{44} / {4} = {4}
{44} / {31} = {31}
{44} / {22} = {22}
{44} / {41} = {3}
{44} / {32} = {21}
{44} / {42} = {2}
{44} / {33} = {11}
{44} / {43} = {1}
{44} / {44} = {0}
{431} / {0} = {431}
{431} / {1} = {331} + {421} + {43}
{431} / {2} = {321} + {33} + {411} + {42}
{431} / {11} = {321} + {33} + {42}
{431} / {3} = {311} + {32} + {41}
{431} / {21} = {221} + {311} +2{32} + {41}
{431} / {111} = {32}
{431} / {4} = {31}
{431} / {31} = {211} + {22} +2{31} + {4}
{431} / {22} = {211} + {22} + {31}
{431} / {211} = {22} + {31}
{431} / {41} = {21} + {3}
{431} / {32} = {111} +2{21} + {3}
{431} / {311} = {21} + {3}
{431} / {221} = {21}
{431} / {42} = {11} + {2}
{431} / {411} = {2}
{431} / {33} = {11} + {2}
{431} / {321} = {11} + {2}
{431} / {43} = {1}
{431} / {421} = {1}
{431} / {331} = {1}
{431} / {431} = {0}
{422} / {0} = {422}

S-Function Division (contd).

{422} / {1} = {322} + {421}
{422} / {2} = {222} + {321} + {42}
{422} / {11} = {321} + {411}
{422} / {3} = {221} + {32}
{422} / {21} = {221} + {311} + {32} + {41}
{422} / {111} = {311}
{422} / {4} = {22}
{422} / {31} = {211} + {22} + {31}
{422} / {22} = {22} + {31} + {4}
{422} / {211} = {211} + {31}
{422} / {41} = {21}
{422} / {32} = {21} + {3}
{422} / {311} = {111} + {21}
{422} / {221} = {21} + {3}
{422} / {42} = {2}
{422} / {411} = {11}
{422} / {321} = {11} + {2}
{422} / {222} = {2}
{422} / {421} = {1}
{422} / {322} = {1}
{422} / {422} = {0}
{4211} / {0} = {4211}
{4211} / {1} = {3211} + {4111} + {421}
{4211} / {2} = {2211} + {3111} + {321} + {411}
{4211} / {11} = {3111} + {321} + {411} + {42}
{4211} / {3} = {2111} + {221} + {311}
{4211} / {21} = {2111} + {221} +2{311} + {32} + {41}
{4211} / {111} = {311} + {32} + {41}
{4211} / {4} = {211}
{4211} / {31} = {1111} +2{211} + {22} + {31}
{4211} / {22} = {211} + {31}
{4211} / {211} = {211} + {22} +2{31} + {4}
{4211} / {1111} = {31}
{4211} / {41} = {111} + {21}
{4211} / {32} = {111} + {21}
{4211} / {311} = {111} +2{21} + {3}
{4211} / {221} = {21} + {3}
{4211} / {2111} = {21} + {3}
{4211} / {42} = {11}
{4211} / {411} = {11} + {2}
{4211} / {321} = {11} + {2}
{4211} / {3111} = {11} + {2}
{4211} / {2211} = {2}
{4211} / {421} = {1}
{4211} / {4111} = {1}
{4211} / {3211} = {1}
{4211} / {4211} = {0}
{41111} / {0} = {41111}
{41111} / {1} = {31111} + {4111}
{41111} / {2} = {21111} + {3111}
{41111} / {11} = {3111} + {411}
{41111} / {3} = {11111} + {2111}
{41111} / {21} = {2111} + {311}
{41111} / {111} = {311} + {41}
{41111} / {4} = {1111}
{41111} / {31} = {1111} + {211}
{41111} / {211} = {211} + {31}
{41111} / {1111} = {31} + {4}
{41111} / {41} = {111}
{41111} / {311} = {111} + {21}
{41111} / {2111} = {21} + {3}
{41111} / {11111} = {3}
{41111} / {411} = {11}
{41111} / {3111} = {11} + {2}
{41111} / {21111} = {2}
{41111} / {4111} = {1}
{41111} / {31111} = {1}
{41111} / {41111} = {0}
{332} / {0} = {332}
{332} / {1} = {322} + {331}
{332} / {2} = {321} + {33}
{332} / {11} = {222} + {321}
{332} / {3} = {32}
{332} / {21} = {221} + {311} + {32}
{332} / {111} = {221}
{332} / {31} = {22} + {31}
{332} / {22} = {211} + {31}
{332} / {211} = {211} + {22}
{332} / {32} = {21} + {3}
{332} / {311} = {21}
{332} / {221} = {111} + {21}
{332} / {33} = {2}
{332} / {321} = {11} + {2}
{332} / {222} = {11}
{332} / {331} = {1}
{332} / {322} = {1}
{332} / {332} = {0}
{3311} / {0} = {3311}
{3311} / {1} = {3211} + {331}
{3311} / {2} = {3111} + {321}
{3311} / {11} = {2211} + {321} + {33}
{3311} / {3} = {311}
{3311} / {21} = {2111} + {221} + {311} + {32}
{3311} / {111} = {221} + {32}
{3311} / {31} = {211} + {31}
{3311} / {22} = {1111} + {211} + {22}
{3311} / {211} = {211} + {22} + {31}
{3311} / {1111} = {22}
{3311} / {32} = {111} + {21}
{3311} / {311} = {21} + {3}
{3311} / {221} = {111} + {21}
{3311} / {2111} = {21}
{3311} / {33} = {11}
{3311} / {321} = {11} + {2}
{3311} / {3111} = {2}
{3311} / {2211} = {11}
{3311} / {331} = {1}
{3311} / {3211} = {1}
{3311} / {3311} = {0}
{3221} / {0} = {3221}
{3221} / {1} = {2221} + {3211} + {322}
{3221} / {2} = {2211} + {222} + {321}
{3221} / {11} = {2211} + {222} + {3111} + {321}
{3221} / {3} = {221}
{3221} / {21} = {2111} +2{221} + {311} + {32}
{3221} / {111} = {2111} + {221} + {311}
{3221} / {31} = {211} + {22}
{3221} / {22} = {211} + {22} + {31}
{3221} / {211} = {1111} +2{211} + {22} + {31}
{3221} / {1111} = {211}
{3221} / {32} = {21}
{3221} / {311} = {111} + {21}
{3221} / {221} = {111} +2{21} + {3}
{3221} / {2111} = {111} + {21}
{3221} / {321} = {11} + {2}
{3221} / {3111} = {11}
{3221} / {222} = {11} + {2}
{3221} / {2211} = {11} + {2}
{3221} / {322} = {1}
{3221} / {3211} = {1}
{3221} / {2221} = {1}
{3221} / {3221} = {0}
{32111} / {0} = {32111}
{32111} / {1} = {22111} + {31111} + {3211}
{32111} / {2} = {21111} + {2211} + {3111}
{32111} / {11} = {21111} + {2211} + {3111} + {321}
{32111} / {3} = {2111}
{32111} / {21} = {11111} +2{2111} + {221} + {311}
{32111} / {111} = {2111} + {221} + {311} + {32}
{32111} / {31} = {1111} + {211}

S-Function Division (contd).

{32111} / {22} = {1111} + {211}
{32111} / {211} = {1111} +2{211} + {22} + {31}
{32111} / {1111} = {211} + {22} + {31}
{32111} / {32} = {111}
{32111} / {311} = {111} + {21}
{32111} / {221} = {111} + {21}
{32111} / {2111} = {111} +2{21} + {3}
{32111} / {11111} = {21}
{32111} / {321} = {11}
{32111} / {3111} = {11} + {2}
{32111} / {2211} = {11} + {2}
{32111} / {21111} = {11} + {2}
{32111} / {3211} = {1}
{32111} / {31111} = {1}
{32111} / {22111} = {1}
{32111} / {32111} = {0}
{311111} / {0} = {311111}
{311111} / {1} = {211111} + {31111}
{311111} / {2} = {111111} + {21111}
{311111} / {11} = {21111} + {3111}
{311111} / {3} = {11111}
{311111} / {21} = {11111} + {2111}
{311111} / {111} = {2111} + {311}
{311111} / {31} = {1111}
{311111} / {211} = {1111} + {211}
{311111} / {1111} = {211} + {31}
{311111} / {311} = {111}
{311111} / {2111} = {111} + {21}
{311111} / {11111} = {21} + {3}
{311111} / {3111} = {11}
{311111} / {21111} = {11} + {2}
{311111} / {111111} = {2}
{311111} / {31111} = {1}
{311111} / {211111} = {1}
{311111} / {311111} = {0}
{2222} / {0} = {2222}
{2222} / {1} = {2221}
{2222} / {2} = {222}
{2222} / {11} = {2211}
{2222} / {21} = {221}
{2222} / {111} = {2111}
{2222} / {22} = {22}
{2222} / {211} = {211}
{2222} / {1111} = {1111}
{2222} / {221} = {21}
{2222} / {2111} = {111}
{2222} / {222} = {2}
{2222} / {2211} = {11}
{2222} / {2221} = {1}
{2222} / {2222} = {0}
{22211} / {0} = {22211}
{22211} / {1} = {22111} + {2221}
{22211} / {2} = {2211}
{22211} / {11} = {21111} + {2211} + {222}
{22211} / {21} = {2111} + {221}
{22211} / {111} = {11111} + {2111} + {221}
{22211} / {22} = {211}
{22211} / {211} = {1111} + {211} + {22}
{22211} / {1111} = {1111} + {211}
{22211} / {221} = {111} + {21}
{22211} / {2111} = {111} + {21}
{22211} / {11111} = {111}
{22211} / {222} = {11}
{22211} / {2211} = {11} + {2}
{22211} / {21111} = {11}
{22211} / {2221} = {1}
{22211} / {22111} = {1}
{22211} / {22211} = {0}
{221111} / {0} = {221111}
{221111} / {1} = {211111} + {22111}
{221111} / {2} = {21111}
{221111} / {11} = {111111} + {21111} + {2211}
{221111} / {21} = {11111} + {2111}
{221111} / {111} = {11111} + {2111} + {221}
{221111} / {22} = {1111}
{221111} / {211} = {1111} + {211}
{221111} / {1111} = {1111} + {211} + {22}
{221111} / {221} = {111}
{221111} / {2111} = {111} + {21}
{221111} / {11111} = {111} + {21}
{221111} / {2211} = {11}
{221111} / {21111} = {11} + {2}
{221111} / {111111} = {11}
{221111} / {22111} = {1}
{221111} / {211111} = {1}
{221111} / {221111} = {0}
{2111111} / {0} = {2111111}
{2111111} / {1} = {1111111} + {211111}
{2111111} / {2} = {111111}
{2111111} / {11} = {111111} + {21111}
{2111111} / {21} = {11111}
{2111111} / {111} = {11111} + {2111}
{2111111} / {211} = {1111}
{2111111} / {1111} = {1111} + {211}
{2111111} / {2111} = {111}
{2111111} / {11111} = {111} + {21}
{2111111} / {21111} = {11}
{2111111} / {111111} = {11} + {2}
{2111111} / {211111} = {1}
{2111111} / {1111111} = {1}
{2111111} / {2111111} = {0}
{11111111} / {0} = {11111111}
{11111111} / {1} = {1111111}
{11111111} / {11} = {111111}
{11111111} / {111} = {11111}
{11111111} / {1111} = {1111}
{11111111} / {11111} = {111}
{11111111} / {111111} = {11}
{11111111} / {1111111} = {1}
{11111111} / {11111111} = {0}

C-1 Branching Rules for the Reduction $U_3 \rightarrow R_3$.

{000} = [0]

{100} = [1]

{200} = [0] + [2]
{210} = [1] + [2]

{300} = [1] + [3]
{310} = [1] + [2] + [3]

{400} = [0] + [2] + [4]
{410} = [1] + [2] + [3] + [4]
{420} = [0] +2[2] + [3] + [4]

{500} = [1] + [3] + [5]
{510} = [1] + [2] + [3] + [4] + [5]
{520} = [1] + [2] +2[3] + [4] + [5]

{600} = [0] + [2] + [4] + [6]
{610} = [1] + [2] + [3] + [4] + [5] + [6]
{620} = [0] +2[2] + [3] +2[4] + [5] + [6]
{630} = [1] + [2] +2[3] +2[4] + [5] + [6]

{700} = [1] + [3] + [5] + [7]
{710} = [1] + [2] + [3] + [4] + [5] + [6] + [7]
{720} = [1] + [2] +2[3] + [4] +2[5] + [6] + [7]
{730} = [1] + [2] +2[3] +2[4] +2[5] + [6] + [7]

{800} = [0] + [2] + [4] + [6] + [8]
{810} = [1] + [2] + [3] + [4] + [5] + [6] + [7] + [8]
{820} = [0] +2[2] + [3] +2[4] + [5] +2[6] + [7] + [8]
{830} = [1] + [2] +2[3] +2[4] +2[5] +2[6] + [7] + [8]
{840} = [0] +2[2] + [3] +3[4] +2[5] +2[6] + [7] + [8]

{900} = [1] + [3] + [5] + [7] + [9]
{910} = [1] + [2] + [3] + [4] + [5] + [6] + [7] + [8] + [9]
{920} = [1] + [2] +2[3] + [4] +2[5] + [6] +2[7] + [8] + [9]
{930} = [1] + [2] +2[3] +2[4] +2[5] +2[6] +2[7] + [8] + [9]
{940} = [1] + [2] +2[3] +2[4] +3[5] +2[6] +2[7] + [8] + [9]

{1000} = [0] + [2] + [4] + [6] + [8] + [10]
{1010} = [1] + [2] + [3] + [4] + [5] + [6] + [7] + [8] + [9] + [10]
{1020} = [0] +2[2] + [3] +2[4] + [5] +2[6] + [7] +2[8] + [9] + [10]
{1030} = [1] + [2] +2[3] +2[4] +2[5] +2[6] +2[7] +2[8] + [9] + [10]
{1040} = [0] +2[2] + [3] +3[4] +2[5] +3[6] +2[7] +2[8] + [9] + [10]
{1050} = [1] + [2] +2[3] +2[4] +3[5] +3[6] +2[7] +2[8] + [9] + [10]

{1100} = [1] + [3] + [5] + [7] + [9] + [11]
{1110} = [1] + [2] + [3] + [4] + [5] + [6] + [7] + [8] + [9] + [10] + [11]
{1120} = [1] + [2] +2[3] + [4] +2[5] + [6] +2[7] + [8] +2[9] + [10] + [11]
{1130} = [1] + [2] +2[3] +2[4] +2[5] +2[6] +2[7] +2[8] +2[9] + [10] + [11]
{1140} = [1] + [2] +2[3] +2[4] +3[5] +2[6] +3[7] +2[8] +2[9] + [10] + [11]
{1150} = [1] + [2] +2[3] +2[4] +3[5] +3[6] +3[7] +2[8] +2[9] + [10] + [11]

{1200} = [0] + [2] + [4] + [6] + [8] + [10] + [12]
{1210} = [1] + [2] + [3] + [4] + [5] + [6] + [7] + [8] + [9] + [10] + [11] + [12]
{1220} = [0] +2[2] + [3] +2[4] + [5] +2[6] + [7] +2[8] + [9] +2[10] + [11] + [12]
{1230} = [1] + [2] +2[3] +2[4] +2[5] +2[6] +2[7] +2[8] +2[9] +2[10] + [11] + [12]
{1240} = [0] +2[2] + [3] +3[4] +2[5] +3[6] +2[7] +3[8] +2[9] +2[10] + [11] + [12]
{1250} = [1] + [2] +2[3] +2[4] +3[5] +3[6] +3[7] +3[8] +2[9] +2[10] + [11] + [12]
{1260} = [0] +2[2] + [3] +3[4] +2[5] +4[6] +3[7] +3[8] +2[9] +2[10] + [11] + [12]

Branching Rules for the Reduction $U_4 \to O_4$.

{0000} = [00]

{1000} = [10]
{1100} = [11]

{2000} = [00] + [20]
{2100} = [10] + [21]
{2110} = [11] + [20]
{2200} = [00] + [20] + [22]

{3000} = [10] + [30]
{3100} = [11] + [20] + [31]
{3110} = [10] + [21] + [30]
{3200} = [10] + [21] + [30] + [32]
{3210} = [11] +2[20] + [22] + [31]
{3300} = [11] + [31] + [33]

{4000} = [00] + [20] + [40]
{4100} = [10] + [21] + [30] + [41]
{4110} = [11] + [20] + [31] + [40]
{4200} = [00] +2[20] + [22] + [31] + [40] + [42]
{4210} = [10] +2[21] +2[30] + [32] + [41]
{4220} = [00] +2[20] + [22] + [31] + [40]
{4300} = [10] + [21] + [30] + [32] + [41] + [43]
{4310} = [11] +2[20] + [22] +2[31] + [33] + [40] + [42]
{4400} = [00] + [20] + [22] + [40] + [42] + [44]

{5000} = [10] + [30] + [50]
{5100} = [11] + [20] + [31] + [40] + [51]
{5110} = [10] + [21] + [30] + [41] + [50]
{5200} = [10] + [21] +2[30] + [32] + [41] + [50] + [52]
{5210} = [11] +2[20] + [22] +2[31] +2[40] + [42] + [51]
{5220} = [10] + [21] +2[30] + [32] + [41] + [50]
{5300} = [11] + [20] +2[31] + [33] + [40] + [42] + [51] + [53]
{5310} = [10] +2[21] +3[30] +2[32] +2[41] + [43] + [50] + [52]
{5320} = [11] +2[20] + [22] +3[31] + [33] +2[40] + [42] + [51]
{5400} = [10] + [21] + [30] + [32] + [41] + [43] + [50] + [52] + [54]
{5410} = [11] +2[20] + [22] +2[31] + [33] +2[40] +2[42] + [44] + [51] + [53]
{5500} = [11] + [31] + [33] + [51] + [53] + [55]

{6000} = [00] + [20] + [40] + [60]
{6100} = [10] + [21] + [30] + [41] + [50] + [61]
{6110} = [11] + [20] + [31] + [40] + [51] + [60]
{6200} = [00] +2[20] + [22] + [31] +2[40] + [42] + [51] + [60] + [62]
{6210} = [10] +2[21] +2[30] + [32] +2[41] +2[50] + [52] + [61]
{6220} = [00] +2[20] + [22] + [31] +2[40] + [42] + [51] + [60]
{6300} = [10] + [21] +2[30] + [32] +2[41] + [43] + [50] + [52] + [61] + [63]
{6310} = [11] +2[20] + [22] +3[31] + [33] +3[40] +2[42] +2[51] + [53] + [60] + [62]
{6320} = [10] +2[21] +3[30] +2[32] +3[41] + [43] +2[50] + [52] + [61]
{6330} = [11] + [20] +2[31] + [33] +2[40] + [42] + [51] + [60]
{6400} = [00] +2[20] + [22] + [31] +2[40] +2[42] + [44] + [51] + [53] + [60] + [62] + [64]
{6410} = [10] +2[21] +3[30] +2[32] +3[41] +2[43] +2[50] +2[52] + [54] + [61]+ [63]
{6420} = [00] +3[20] +2[22] +3[31] + [33] +4[40] +3[42] + [44] +2[51] + [53] + [60] + [62]
{6500} = [10] + [21] + [30] + [32] + [41] + [43] + [50] + [52] + [54] + [61] + [63] + [65]
{6510} = [11] +2[20] + [22] +2[31] + [33] +2[40] +2[42] + [44] +2[51] +2[53] + [55] + [60] + [62]
+ [64]
{6600} = [00] + [20] + [22] + [40] + [42] + [44] + [60] + [62] + [64] + [66]

{7000} = [10] + [30] + [50] + [70]
{7100} = [11] + [20] + [31] + [40] + [51] + [60] + [71]
{7110} = [10] + [21] + [30] + [41] + [50] + [61] + [70]
{7200} = [10] + [21] +2[30] + [32] + [41] +2[50] + [52] + [61] + [70] + [72]
{7210} = [11] +2[20] + [22] +2[31] +2[40] + [42] +2[51] +2[60] + [62] + [71]
{7220} = [10] + [21] +2[30] + [32] + [41] +2[50] + [52] + [61] + [70]
{7300} = [11] + [20] +2[31] + [33] +2[40] + [42] +2[51] + [53] + [60] + [62]+ [71] + [73]
{7310} = [10] +2[21] +3[30] +2[32] +3[41] + [43] +3[50] +2[52] +2[61] + [63] + [70] + [72]
{7320} = [11] +2[20] + [22] +3[31] + [33] +3[40] +2[42] +3[51] + [53] +2[60] + [62] + [71]
{7330} = [10] + [21] +2[30] + [32] +2[41] + [43] +2[50] + [52] + [61] + [70]
{7400} = [10] + [21] +2[30] + [32] +2[41] + [43] +2[50] +2[52] + [54] + [61] + [63] + [70] + [72]
+ [74]

C-2 Branching Rules for the Reduction $U_4 \to O_4$ (contd).

{7410} = [11] +2[20] + [22] +3[31] + [33] +4[40] +3[42] + [44] +3[51] +2[53] +2[60] +2[62] + [64] + [71] + [73]
{7420} = [10] +2[21] +4[30] +3[32] +4[41] +2[43] +4[50] +3[52] + [54] +2[61] + [63] + [70] + [72]
{7430} = [11] +2[20] + [22] +3[31] + [33] +4[40] +3[42] + [44] +3[51] + [53] +2[60] + [62] + [71]
{7500} = [11] + [20] +2[31] + [33] + [40] + [42] +2[51] +2[53] + [55] + [60] + [62] + [64] + [71] + [73] + [75]
{7510} = [10] +2[21] +3[30] +2[32] +3[41] +2[43] +3[50] +3[52] +2[54] +2[61] +2[63] + [65] + [70] + [72] + [74]
{7520} = [11] +2[20] + [22] +4[31] +2[33] +4[40] +3[42] + [44] +4[51] +3[53] + [55] +2[60] +2[62] + [64] + [71] + [73]
{7600} = [10] + [21] + [30] + [32] + [41] + [43] + [50] + [52] + [54] + [61]+ [63] + [65] + [70] + [72] + [74] + [76]
{7610} = [11] +2[20] + [22] +2[31] + [33] +2[40] +2[42] + [44] +2[51] +2[53] + [55] +2[60] +2[62] +2[64] + [66] + [71] + [73] + [75]
{7700} = [11] + [31] + [33] + [51] + [53] + [55] + [71] + [73] + [75] + [77]

{8000} = [00] + [20] + [40] + [60] + [80]
{8100} = [10] + [21] + [30] + [41] + [50] + [61] + [70] + [81]
{8110} = [11] + [20] + [31] + [40] + [51] + [60] + [71] + [80]
{8200} = [00] +2[20] + [22] + [31] +2[40] + [42] + [51] +2[60] + [62] + [71] + [80] + [82]
{8210} = [10] +2[21] +2[30] + [32] +2[41] +2[50] + [52] +2[61] +2[70] + [72] + [81]
{8220} = [00] +2[20] + [22] + [31] +2[40] + [42] + [51] +2[60] + [62] + [71] + [80]
{8300} = [10] + [21] +2[30] + [32] +2[41] + [43] +2[50] + [52] +2[61] + [63] + [70] + [72] + [81] + [83]
{8310} = [11] +2[20] + [22] +3[31] + [33] +3[40] +2[42] +3[51] + [53] +3[60]+2[62] +2[71] + [73] + [80] + [82]
{8320} = [10] +2[21] +3[30] +2[32] +3[41] + [43] +3[50] +2[52] +3[61] + [63] +2[70] + [72] + [81]
{8330} = [11] + [20] +2[31] + [33] +2[40] + [42] +2[51] + [53] +2[60] + [62] + [71] + [80]
{8400} = [00] +2[20] + [22] + [31] +3[40] +2[42] + [44] +2[51] + [53] +2[60] +2[62] + [64] + [71] + [73] + [80] + [82] + [84]
{8410} = [10] +2[21] +3[30] +2[32] +4[41] +2[43] +4[50] +3[52] + [54] +3[61] +2[63] +2[70] +2[72] + [74] + [81] + [83]
{8420} = [00] +3[20] +2[22] +3[31] + [33] +5[40] +4[42] + [44] +4[51] +2[53] +4[60] +3[62] + [64] +2[71] + [73] + [80] + [82]
{8430} = [10] +2[21] +3[30] +2[32] +4[41] +2[43] +4[50] +3[52] + [54] +3[61] + [63] +2[70] + [72] + [81]
{8440} = [00] +2[20] + [22] + [31] +3[40] +2[42] + [44] +2[51] + [53] +2[60] + [62] + [71] + [80]
{8500} = [10] + [21] +2[30] + [32] +2[41] + [43] +2[50] +2[52] + [54] +2[61] +2[63] + [65] + [70] + [72] + [74] + [81] + [83] + [85]
{8510} = [11] +2[20] + [22] +3[31] + [33] +4[40] +3[42] + [44] +4[51] +3[53] + [55] +3[60] +3[62] +2[64] +2[71] +2[73] + [75] + [80] + [82] + [84]
{8520} = [10] +2[21] +4[30] +3[32] +5[41] +3[43] +5[50] +4[52] +2[54] +4[61] +3[63] + [65] +2[70] +2[72] + [74] + [81] + [83]
{8530} = [11] +2[20] + [22] +4[31] +2[33] +5[40] +4[42] + [44] +5[51] +3[53] + [55] +4[60] +3[62] + [64] +2[71] + [73] + [80] + [82]
{8600} = [00] +2[20] + [22] + [31] +2[40] +2[42] + [44] + [51] + [53] +2[60] +2[62] +2[64] + [66] + [71] + [73] + [75] + [80] + [82] + [84] + [86]
{8610} = [10] +2[21] +3[30] +2[32] +3[41] +2[43] +3[50] +3[52] +2[54] +3[61] +3[63] +2[65] +2[70] +2[72] +2[74] + [76] + [81] + [83] + [85]
{8620} = [00] +3[20] +2[22] +3[31] + [33] +5[40] +4[42] +2[44] +4[51] +3[53] + [55] +4[60] +4[62] +3[64] + [66] +2[71] +2[73] + [75] + [80] + [82] + [84]
{8700} = [10] + [21] + [30] + [32] + [41] + [43] + [50] + [52] + [54] + [61] + [63] + [65] + [70] + [72] + [74] + [76] + [81] + [83] + [85] + [87]
{8710} = [11] +2[20] + [22] +2[31] + [33] +2[40] +2[42] + [44] +2[51] +2[53] + [55] +2[60] +2[62] +2[64] + [66] +2[71] +2[73] +2[75] + [77] + [80] + [82] + [84] + [86]
{8800} = [00] + [20] + [22] + [40] + [42] + [44] + [60] + [62] + [64] + [66] + [80] + [82] + [84] + [86] + [88]

{9000} = [10] + [30] + [50] + [70] + [90]
{9100} = [11] + [20] + [31] + [40] + [51] + [60] + [71] + [80] + [91]
{9110} = [10] + [21] + [30] + [41] + [50] + [61] + [70] + [81] + [90]
{9200} = [10] + [21] +2[30] + [32] + [41] +2[50] + [52] + [61] +2[70] + [72] + [81] + [90] + [92]
{9210} = [11] +2[20] + [22] +2[31] +2[40] + [42] +2[51] +2[60] + [62] +2[71] +2[80] + [82] + [91]
{9220} = [10] + [21] +2[30] + [32] + [41] +2[50] + [52] + [61] +2[70] + [72] + [81] + [90]
{9300} = [11] + [20] +2[31] + [33] +2[40] + [42] +2[51] + [53] +2[60] + [62] +2[71] + [73] + [80] + [82] + [91] + [93]
{9310} = [10] +2[21] +3[30] +2[32] +3[41] + [43] +3[50] +2[52] +3[61] + [63] +3[70] +2[72] +2[81] + [83] + [90] + [92]
{9320} = [11] +2[20] + [22] +3[31] + [33] +3[40] +2[42] +3[51] + [53] +3[60] +2[62] +3[71] + [73] +2[80] + [82] + [91]

Branching Rules for the Reduction $U_4 \rightarrow O_4$ (contd).

{9330} = [10] + [21] +2[30] + [32] +2[41] + [43] +2[50] + [52] +2[61] + [63] +2[70] + [72] + [81]
+ [90]
{9400} = [10] + [21] +2[30] + [32] +2[41] + [43] +3[50] +2[52] + [54] +2[61] + [63] +2[70] +2[72]
+ [74] + [81] + [83] + [90] + [92] + [94]
{9410} = [11] +2[20] + [22] +3[31] + [33] +4[40] +3[42] + [44] +4[51] +2[53] +4[60] +3[62] + [64]
+3[71] +2[73] +2[80] +2[82] + [84] + [91] + [93]
{9420} = [10] +2[21] +4[30] +3[32] +4[41] +2[43] +5[50] +4[52] + [54] +4[61] +2[63] +4[70] +3[72]
+ [74] +2[81] + [83] + [90] + [92]
{9430} = [11] +2[20] + [22] +3[31] + [33] +4[40] +3[42] + [44] +4[51] +2[53] +4[60] +3[62] + [64]
+3[71] + [73] +2[80] + [82] + [91]
{9440} = [10] + [21] +2[30] + [32] +2[41] + [43] +3[50] +2[52] + [54] +2[61] + [63] +2[70] + [72]
+ [81] + [90]
{9500} = [11] + [20] +2[31] + [33] +2[40] + [42] +3[51] +2[53] + [55] +2[60] +2[62] + [64] +2[71]
+2[73] + [75] + [80] + [82] + [84] + [91] + [93] + [95]
{9510} = [10] +2[21] +3[30] +2[32] +4[41] +2[43] +5[50] +4[52] +2[54] +4[61] +3[63] + [65] +3[70]
+3[72] +2[74] +2[81] +2[83] + [85] + [90] + [92] + [94]
{9520} = [11] +2[20] + [22] +4[31] +2[33] +5[40] +4[42] + [44] +6[51] +4[53] + [55] +5[60] +4[62]
+2[64] +4[71] +3[73] + [75] +2[80] +2[82] + [84] + [91] + [93]
{9530} = [10] +2[21] +4[30] +3[32] +5[41] +3[43] +6[50] +5[52] +2[54] +5[61] +3[63] + [65] +4[70]
+3[72] + [74] +2[81] + [83] + [90] + [92]
{9540} = [11] +2[20] + [22] +3[31] + [33] +4[40] +3[42] + [44] +5[51] +3[53] + [55] +4[60] +3[62]
+ [64] +3[71] + [73] +2[80] + [82] + [91]
{9600} = [10] + [21] +2[30] + [32] +2[41] + [43] +2[50] +2[52] + [54] +2[61] +2[63] + [65] +2[70]
+2[72] +2[74] + [76] + [81] + [83] + [85] + [90] + [92] + [94] + [96]
{9610} = [11] +2[20] + [22] +3[31] + [33] +4[40] +3[42] + [44] +4[51] +3[53] + [55] +4[60] +4[62]
+3[64] + [66] +3[71] +3[73] +2[75] +2[80] +2[82] +2[84] + [86] + [91] + [93] + [95]
{9620} = [10] +2[21] +4[30] +3[32] +5[41] +3[43] +6[50] +5[52] +3[54] +5[61] +4[63] +2[65] +4[70]
+4[72] +3[74] + [76] +2[81] +2[83] + [85] + [90] + [92] + [94]
{9630} = [11] +2[20] + [22] +4[31] +2[33] +6[40] +5[42] +2[44] +6[51] +4[53] + [55] +6[60] +5[62]
+3[64] + [66] +4[71] +3[73] + [75] +2[80] +2[82] + [84] + [91] + [93]
{9700} = [11] + [20] +2[31] + [33] + [40] + [42] +2[51] +2[53] + [55] + [60] + [62] + [64] +2[71]
+2[73] +2[75] + [77] + [80] + [82] + [84] + [86] + [91] + [93] + [95] + [97]
{9710} = [10] +2[21] +3[30] +2[32] +3[41] +2[43] +3[50] +3[52] +2[54] +3[61] +3[63] +2[65] +3[70]
+3[72] +3[74] +2[76] +2[81] +2[83] +2[85] + [87] + [90] + [92] + [94] + [96]
{9720} = [11] +2[20] + [22] +4[31] +2[33] +4[40] +3[42] + [44] +5[51] +4[53] +2[55] +4[60] +4[62]
+3[64] + [66] +4[71] +4[73] +3[75] + [77] +2[80] +2[82] +2[84] + [86] + [91] + [93] + [95]
{9800} = [10] + [21] + [30] + [32] + [41] + [43] + [50] + [52] + [54] + [61] + [63] + [65] + [70]
+ [72] + [74] + [76] + [81] + [83] + [85] + [87] + [90] + [92] + [94] + [96] + [98]
{9810} = [11] +2[20] + [22] +2[31] + [33] +2[40] +2[42] + [44] +2[51] +2[53] + [55] +2[60] +2[62]
+2[64] + [66] +2[71] +2[73] +2[75] + [77] +2[80] +2[82] +2[84] +2[86] + [88] + [91] + [93]
+ [95] + [97]
{9900} = [11] + [31] + [33] + [51] + [53] + [55] + [71] + [73] + [75] + [77] + [91] + [93] + [95]
+ [97] + [99]

C-3 Branching Rules for the Reduction $U_5 \to R_5$.

{00000} = [00]

{10000} = [10]
{11000} = [11]

{20000} = [00] + [20]
{21000} = [10] + [21]
{21100} = [11] + [21]
{21110} = [11] + [20]
{22000} = [00] + [20] + [22]
{22100} = [10] + [21] + [22]

{30000} = [10] + [30]
{31000} = [11] + [20] + [31]
{31100} = [11] + [21] + [31]
{31110} = [10] + [21] + [30]
{32000} = [10] + [21] + [30] + [32]
{32100} = [11] + [20] + [21] + [22] + [31] + [32]
{32110} = [11] + [20] + [21] + [22] + [31]
{33000} = [11] + [31] + [33]
{33100} = [11] + [21] + [31] + [32] + [33]

{40000} = [00] + [20] + [40]
{41000} = [10] + [21] + [30] + [41]
{41100} = [11] + [21] + [31] + [41]
{41110} = [11] + [20] + [31] + [40]
{42000} = [00] +2[20] + [22] + [31] + [40] + [42]
{42100} = [10] +2[21] + [22] + [30] + [31] + [32] + [41] + [42]
{42110} = [11] +2[21] + [30] + [31] + [32] + [41]
{42200} = [00] +2[20] +2[22] + [31] + [32] + [40] + [42]
{42210} = [10] +2[21] + [22] + [30] + [31] + [32] + [41]
{42220} = [00] +2[20] + [22] + [31] + [40]
{43000} = [10] + [21] + [30] + [32] + [41] + [43]
{43100} = [11] + [20] + [21] + [22] +2[31] + [32] + [33] + [41] + [42] + [43]
{43110} = [11] + [20] + [21] + [22] +2[31] + [32] + [33] + [40] + [42]
{43200} = [10] +2[21] + [22] + [30] + [31] +2[32] + [33] + [41] + [42] + [43]
{43210} = [11] + [20] +2[21] +2[22] + [30] +2[31] +2[32] + [33] + [41] + [42]
{44000} = [00] + [20] + [22] + [40] + [42] + [44]
{44100} = [10] + [21] + [22] + [30] + [32] + [41] + [42] + [43] + [44]
{44200} = [00] +2[20] +2[22] + [31] + [32] + [40] +2[42] + [43] + [44]

{50000} = [10] + [30] + [50]
{51000} = [11] + [20] + [31] + [40] + [51]
{51100} = [11] + [21] + [31] + [41] + [51]
{51110} = [10] + [21] + [30] + [41] + [50]
{52000} = [10] + [21] +2[30] + [32] + [41] + [50] + [52]
{52100} = [11] + [20] + [21] + [22] +2[31] + [32] + [40] + [41] + [42] + [51] + [52]
{52110} = [11] + [20] + [21] + [22] +2[31] + [40] + [41] + [42] + [51]
{52200} = [10] + [21] + [22] +2[30] +2[32] + [41] + [42] + [50] + [52]
{52210} = [11] + [20] + [21] + [22] +2[31] + [32] + [40] + [41] + [42] + [51]
{52220} = [10] + [21] +2[30] + [32] + [41] + [50]
{53000} = [11] + [20] +2[31] + [33] + [40] + [42] + [51] + [53]
{53100} = [11] +2[21] + [30] +2[31] +2[32] + [33] +2[41] + [42] + [43] + [51] + [52] + [53]
{53110} = [10] +2[21] + [22] +2[30] + [31] +2[32] +2[41] + [42] + [43] + [50] + [52]
{53200} = [11] + [20] + [21] + [22] +3[31] +2[32] +2[33] + [40] + [41] +2[42] + [43] + [51] + [52]
+ [53]
{53210} = [11] + [20] +2[21] +2[22] + [30] +3[31] +3[32] + [33] + [40] +2[41] +2[42] + [43] + [51]
+ [52]
{53220} = [11] + [20] + [21] + [22] +3[31] + [32] + [33] + [40] + [41]+ [42] + [51]
{54000} = [10] + [21] + [30] + [32] + [41] + [43] + [50] + [52] + [54]
{54100} = [11] + [20] + [21] + [22] +2[31] + [32] + [33] + [40] + [41] +2[42] + [43] + [44] + [51]
+ [52] + [53] + [54]
{54110} = [11] + [20] + [21] + [22] +2[31] + [32] + [33] + [40] + [41] +2[42] + [43] + [44] + [51]
+ [53]
{54200} = [10] +2[21] + [22] +2[30] + [31] +3[32] + [33] +2[41] +2[42] +2[43] + [44] + [50] +2[52]
+ [53] + [54]
{54210} = [11] + [20] +2[21] +2[22] + [30] +3[31] +3[32] +2[33] + [40] +2[41] +3[42] +2[43] + [44]
+ [51] + [52] + [53]
{55000} = [11] + [31] + [33] + [51] + [53] + [55]
{55100} = [11] + [21] + [31] + [32] + [33] + [41] + [43] + [51] + [52] + [53] + [54] + [55]

Branching Rules for the Reduction $U_5 \to R_5$ (contd).

```
{55200} = [11] + [21] +2[31] + [32] +2[33] + [41] + [42] + [43] + [51] + [52] +2[53] + [54] + [55]

{60000} = [00] + [20] + [40] + [60]
{61000} = [10] + [21] + [30] + [41] + [50] + [61]
{61100} = [11] + [21] + [31] + [41] + [51] + [61]
{61110} = [11] + [20] + [31] + [40] + [51] + [60]
{62000} = [00] +2[20] + [22] + [31] +2[40] + [42] + [51] + [60] + [62]
{62100} = [10] +2[21] + [22] + [30] + [31] + [32] +2[41] + [42] + [50] + [51] + [52] + [61] + [62]
{62110} = [11] +2[21] + [30] + [31] + [32] +2[41] + [50] + [51] + [52] + [61]
{62200} = [00] +2[20] +2[22] + [31] + [32] +2[40] +2[42] + [51] + [52] + [60] + [62]
{62210} = [10] +2[21] + [22] + [30] + [31] + [32] +2[41] + [42] + [50] + [51]+ [52] + [61]
{62220} = [00] +2[20] + [22] + [31] +2[40] + [42] + [51] + [60]
{63000} = [10] + [21] +2[30] + [32] +2[41] + [43] + [50] + [52] + [61] + [63]
{63100} = [11] + [20] + [21] + [22] +3[31] + [32] + [33] + [40] +2[41] +2[42] + [43] +2[51] + [52]
        + [53] + [61] + [62] + [63]
{63110} = [11] + [20] + [21] + [22] +3[31] + [32] + [33] +2[40] + [41] +2[42] +2[51] + [52] + [53]
        + [60] + [62]
{63200} = [10] +2[21] + [22] +2[30] + [31] +3[32] + [33] +3[41] +2[42] +2[43] + [50] + [51] +2[52]
        + [53] + [61] + [62] + [63]
{63210} = [11] + [20] +2[21] +2[22] + [30] +3[31] +3[32] + [33] + [40] +3[41]+3[42] + [43] + [50]
        +2[51] +2[52] + [53] + [61] + [62]
{63220} = [10] +2[21] + [22] +2[30] + [31] +2[32] +3[41] + [42] + [43] + [50] + [51] + [52] + [61]
{63300} = [11] + [21] +2[31] + [32] +2[33] +2[41] + [42] +2[43] + [51] + [52] + [53] + [61] + [63]
{63310} = [11] + [20] + [21] + [22] +3[31] +2[32] +2[33] +2[40] + [41] +3[42] + [43] +2[51] + [52]
        + [53] + [60] + [62]
{63320} = [11] +2[21] + [30] +2[31] +2[32] + [33] +3[41] + [42] + [43] + [50] + [51] + [52] + [61]
{63330} = [11] + [20] +2[31] + [33] +2[40] + [42] + [51] + [60]
{64000} = [00] +2[20] + [22] + [31] +2[40] +2[42] + [44] + [51] + [53] + [60] + [62] + [64]
{64100} = [10] +2[21] + [22] +2[30] + [31] +2[32] +3[41] +2[42] +2[43] + [44] + [50] + [51] +2[52]
        + [53] + [54] + [61] + [62] + [63] + [64]
{64110} = [11] +2[21] + [30] +2[31] +2[32] + [33] +3[41] + [42] +2[43] + [50] + [51] +2[52] + [53]
        + [54] + [61] + [63]
{64200} = [00] +3[20] +3[22] +3[31] +2[32] + [33] +3[40] + [41] +5[42] +2[43] +2[44] +2[51] +2[52]
        +2[53] + [54] + [60] +2[62] + [63] + [64]
{64210} = [10] +3[21] +2[22] +2[30] +3[31] +4[32] +2[33] + [40] +4[41] +4[42] +3[43] + [44] + [50]
        +2[51] +3[52] +2[53] + [54] + [61] + [62] + [63]
{64220} = [00] +3[20] +3[22] +3[31] +2[32] + [33] +3[40] + [41] +4[42] + [43] + [44] +2[51] + [52]
        + [53] + [60] + [62]
{64300} = [10] +2[21] + [22] +2[30] + [31] +3[32] + [33] +3[41] +3[42] +3[43] +2[44] + [50] + [51]
        +2[52] +2[53] + [54] + [61] + [62] + [63] + [64]
{64310} = [11] + [20] +2[21] +2[22] + [30] +4[31] +4[32] +3[33] + [40] +4[41] +4[42] +4[43] + [44]
        + [50] +2[51] +3[52] +2[53] + [54] + [61] + [62] + [63]
{64320} = [10] +3[21] +2[22] +2[30] +3[31] +4[32] +2[33] + [40] +4[41] +4[42] +2[43] + [44] + [50]
        +2[51] +2[52] + [53] + [61] + [62]
{65000} = [10] + [21] + [30] + [32] + [41] + [43] + [50] + [52] + [54]+ [61] + [63] + [65]
{65100} = [11] + [20] + [21] + [22] +2[31] + [32] + [33] + [40] + [41] +2[42] + [43] + [44] +2[51]
        + [52] +2[53] + [54] + [55] + [61] + [62] + [63] + [64] + [65]
{65110} = [11] + [20] + [21] + [22] +2[31] + [32] + [33] + [40] + [41] +2[42] + [43] + [44] +2[51]
        + [52] +2[53] + [54] + [55] + [60] + [62] + [64]
{65200} = [10] +2[21] + [22] +2[30] + [31] +3[32] + [33] +3[41] +2[42] +3[43] + [44] + [50] + [51]
        +3[52] +2[53] +2[54] + [55] + [61] + [62] +2[63] + [64] + [65]
{65210} = [11] + [20] +2[21] +2[22] + [30] +3[31] +3[32] +2[33] + [40] +3[41] +4[42] +3[43] +2[44]
        + [50] +2[51] +3[52] +3[53] +2[54] + [55] + [61] + [62] + [63] + [64]
{65300} = [11] + [20] + [21] + [22] +3[31] +2[32] +2[33] + [40] +2[41] +3[42] +3[43] + [44] +2[51]
        +2[52] +3[53] +2[54] + [55] + [61] + [62] +2[63] + [64] + [65]
{65310} = [11] + [20] +2[21] +2[22] + [30] +4[31] +4[32] +3[33] +2[40] +3[41] +5[42] +4[43] +2[44]
        +3[51] +3[52] +4[53] +2[54] + [55] + [60] +2[62] + [63] + [64]
{66000} = [00] + [20] + [22] + [40] + [42] + [44] + [60] + [62] + [64] + [66]
{66100} = [10] + [21] + [22] + [30] + [32] + [41] + [42] + [43] + [44] + [50] + [52] + [54] + [61]
        + [62] + [63] + [64] + [65] + [66]
{66200} = [00] +2[20] +2[22] + [31] + [32] +2[40] +3[42] + [43] +2[44] + [51] + [52] + [53] + [54]
        + [60] +2[62] + [63] +2[64] + [65] + [66]
{66300} = [10] + [21] + [22] +2[30] +2[32] +2[41] +2[42] +2[43] +2[44] + [50] +2[52] + [53] + [54]
        + [61] + [62] +2[63] +2[64] + [65] + [66]

{70000} = [10] + [30] + [50] + [70]
{71000} = [11] + [20] + [31] + [40] + [51] + [60] + [71]
{71100} = [11] + [21] + [31] + [41] + [51] + [61] + [71]
{71110} = [10] + [21] + [30] + [41] + [50] + [61] + [70]
{72000} = [10] + [21] +2[30] + [32] + [41] +2[50] + [52] + [61] + [70] + [72]
```

C-3 Branching Rules for the Reduction $U_5 \to R_5$ (contd).

{72100} = [11] + [20] + [21] + [22] +2[31] + [32] + [40] + [41] + [42] +2[51] + [52] + [60] + [61]
+ [62] + [71] + [72]

{72110} = [11] + [20] + [21] + [22] +2[31] + [40] + [41] + [42] +2[51] + [60] + [61] + [62] + [71]

{72200} = [10] + [21] + [22] +2[30] +2[32] + [41] + [42] +2[50] +2[52] + [61] + [62] + [70] + [72]

{72210} = [11] + [20] + [21] + [22] +2[31] + [32] + [40] + [41] + [42] +2[51] + [52] + [60] + [61]
+ [62] + [71]

{72220} = [10] + [21] +2[30] + [32] + [41] +2[50] + [52] + [61] + [70]

{73000} = [11] + [20] +2[31] + [33] +2[40] + [42] +2[51] + [53] + [60] + [62] + [71] + [73]

{73100} = [11] +2[21] + [30] +2[31] +2[32] + [33] +3[41] + [42] + [43] + [50] +2[51] +2[52] + [53]
+2[61] + [62] + [63] + [71] + [72] + [73]

{73110} = [10] +2[21] + [22] +2[30] + [31] +2[32] +3[41] + [42] + [43] +2[50] + [51] +2[52] +2[61]
+ [62] + [63] + [70] + [72]

{73200} = [11] + [20] + [21] + [22] +3[31] +2[32] +2[33] +2[40] + [41] +3[42] + [43] +3[51] +2[52]
+2[53] + [60] + [61] +2[62] + [63] + [71] + [72] + [73]

{73210} = [11] + [20] +2[21] +2[22] + [30] +3[31] +3[32] + [33] + [40] +3[41] +3[42] + [43] + [50]
+3[51] +3[52] + [53] + [60] +2[61] +2[62] + [63] + [71] + [72]

{73220} = [11] + [20] + [21] + [22] +3[31] + [32] + [33] +2[40] + [41] +2[42] +3[51] + [52] + [53]
+ [60] + [61] + [62] + [71]

{73300} = [11] + [21] +2[31] + [32] +2[33] +2[41] + [42] +2[43] +2[51] + [52] +2[53] + [61] + [62]
+ [63] + [71] + [73]

{73310} = [10] +2[21] + [22] +2[30] + [31] +3[32] + [33] +3[41] +2[42] +2[43] +2[50] + [51] +3[52]
+ [53] +2[61] + [62] + [63] + [70] + [72]

{73320} = [11] + [20] + [21] + [22] +3[31] + [32] + [33] + [40] +2[41] +2[42] + [43] +3[51] + [52]
+ [53] + [60] + [61] + [62] + [71]

{73330} = [10] + [21] +2[30] + [32] +2[41] + [43] +2[50] + [52] + [61] + [70]

{74000} = [10] + [21] +2[30] + [32] +2[41] + [43] +2[50] +2[52] + [54] + [61] + [63] + [70] + [72]
+ [74]

{74100} = [11] + [20] + [21] + [22] +3[31] + [32] + [33] +2[40] +2[41] +3[42] + [43] + [44] +3[51]
+2[52] +2[53] + [54] + [60] + [61] +2[62] + [63] + [64] + [71] + [72] + [73] + [74]

{74110} = [11] + [20] + [21] + [22] +3[31] + [32] + [33] +2[40] +2[41] +3[42] + [43] + [44] +3[51]
+ [52] +2[53] + [60] + [61] +2[62] + [63] + [64] + [71] + [73]

{74200} = [10] +2[21] + [22] +3[30] + [31] +4[32] + [33] +4[41] +3[42] +3[43] + [44] +3[50] + [51]
+5[52] +2[53] +2[54] +2[61] +2[62] +2[63] + [64] + [70] +2[72] + [73] + [74]

{74210} = [11] + [20] +2[21] +2[22] + [30] +4[31] +4[32] +2[33] +2[40] +4[41] +5[42] +3[43] + [44]
+ [50] +4[51] +4[52] +3[53] + [54] + [60] +2[61] +3[62] +2[63] + [64] + [71] + [72] + [73]

{74220} = [10] +2[21] + [22] +3[30] + [31] +4[32] + [33] +4[41] +2[42] +2[43] +3[50] + [51] +4[52]
+ [53] + [54] +2[61] + [62] + [63] + [70] + [72]

{74300} = [11] + [20] + [21] + [22] +3[31] +2[32] +2[33] +2[40] +2[41] +4[42] +3[43] +2[44] +3[51]
+3[52] +3[53] +2[54] + [60] + [61] +2[62] +2[63] + [64] + [71] + [72] + [73] + [74]

{74310} = [11] + [20] +2[21] +2[22] + [30] +4[31] +4[32] +3[33] +2[40] +4[41] +6[42] +4[43] +2[44]
+ [50] +4[51] +4[52] +4[53] + [54] + [60] +2[61] +3[62] +2[63] + [64] + [71] + [72] + [73]

{74320} = [11] + [20] +2[21] +2[22] + [30] +4[31] +4[32] +2[33] +2[40] +4[41] +5[42] +3[43] + [44]
+ [50] +4[51] +4[52] +2[53] + [54] + [60] +2[61] +2[62] + [63] + [71] + [72]

{74330} = [11] + [20] + [21] + [22] +3[31] + [32] + [33] +2[40] +2[41] +3[42] + [43] + [44] +3[51]
+ [52] + [53] + [60] + [61] + [62] + [71]

{75000} = [11] + [20] +2[31] + [33] + [40] + [42] +2[51] +2[53] + [55] + [60] + [62] + [64] + [71]
+ [73] + [75]

{75100} = [11] +2[21] + [30] +2[31] +2[32] + [33] +3[41] + [42] +2[43] + [50]+2[51] +3[52] +2[53]
+2[54] + [55] +2[61] + [62] +2[63] + [64] + [65] + [71] + [72] + [73] + [74] + [75]

{75110} = [10] +2[21] + [22] +2[30] + [31] +2[32] +3[41] +2[42] +2[43] + [44] +2[50] + [51] +3[52]
+ [53] +2[54] +2[61] + [62] +2[63] + [64] + [65] + [70] + [72] + [74]

{75200} = [11] + [20] + [21] + [22] +4[31] +2[32] +3[33] +2[40] +2[41] +4[42] +2[43] + [44] +4[51]
+3[52] +5[53] +2[54] +2[55] + [60] + [61] +3[62] +2[63] +2[64] + [65] + [71] + [72] +2[73]
+ [74] + [75]

{75210} = [11] + [20] +2[21] +2[22] + [30] +4[31] +4[32] +2[33] +2[40] +4[41] +5[42] +4[43] +2[44]
+ [50] +4[51] +5[52] +4[53] +3[54] + [55] + [60] +2[61] +3[62] +3[63] +2[64] + [65] + [71]
+ [72] + [73] + [74]

{75220} = [11] + [20] + [21] + [22] +4[31] +2[32] +3[33] +2[40] +2[41] +4[42] +2[43] + [44] +4[51]
+2[52] +4[53] + [54] + [55] + [60] + [61] +2[62] + [63] + [64] + [71]+ [73]

{75300} = [11] +2[21] + [30] +3[31] +3[32] +3[33] +4[41] +3[42] +4[43] + [44] + [50] +3[51] +4[52]
+5[53] +3[54] +2[55] +2[61] +2[62] +3[63] +2[64] + [65] + [71] + [72] +2[73] + [74] + [75]

{75310} = [10] +3[21] +2[22] +3[30] +3[31] +6[32] +3[33] + [40] +6[41] +6[42] +6[43] +3[44] +3[50]
+3[51] +7[52] +5[53] +4[54] + [55] +3[61] +3[62] +4[63] +2[64] + [65] + [70] +2[72] + [73]
+ [74]

{75320} = [11] + [20] +2[21] +2[22] + [30] +5[31] +5[32] +4[33] +2[40] +5[41] +6[42] +5[43] +2[44]
+ [50] +5[51] +5[52] +5[53] +2[54] + [55] + [60] +2[61] +3[62] +2[63] + [64] + [71] + [72]
+ [73]

{76000} = [10] + [21] + [30] + [32] + [41] + [43] + [50] + [52] + [54] + [61] + [63] + [65] + [70]
+ [72] + [74] + [76]

{76100} = [11] + [20] + [21] + [22] +2[31] + [32] + [33] + [40] + [41] +2[42] + [43] + [44] +2[51]
+ [52] +2[53] + [54] + [55] + [60] + [61] +2[62] + [63] +2[64] + [65] + [66] + [71] + [72]
+ [73] + [74] + [75] + [76]

Branching Rules for the Reduction $U_5 \to R_5$ (contd).

```
{76110} = [11] + [20] + [21] + [22] +2[31] + [32] + [33] + [40] + [41] +2[42] + [43] + [44] +2[51]
        + [52] +2[53] + [54] + [55] + [60] + [61] +2[62] + [63] +2[64] + [65] + [66] + [71] + [73]
        + [75]
{76200} = [10] +2[21] + [22] +2[30] + [31] +3[32] + [33] +3[41] +2[42] +3[43] + [44] +2[50] + [51]
        +4[52] +2[53] +3[54] + [55] +2[61] +2[62] +3[63] +2[64] +2[65] + [66] + [70] +2[72] + [73]
        +2[74] + [75] + [76]
{76210} = [11] + [20] +2[21] +2[22] + [30] +3[31] +3[32] +2[33] + [40] +3[41] +4[42] +3[43] +2[44]
        + [50] +3[51] +4[52] +4[53] +3[54] +2[55] + [60] +2[61] +3[62] +3[63] +3[64] +2[65] + [66]
        + [71] + [72] + [73] + [74] + [75]
{76300} = [11] + [20] + [21] + [22] +3[31] +2[32] +2[33] +2[40] +2[41] +4[42] +3[43] +2[44] +3[51]
        +3[52] +4[53] +3[54] + [55] + [60] + [61] +3[62] +3[63] +3[64] +2[65] + [66] + [71] + [72]
        +2[73] +2[74] + [75] + [76]
{76310} = [11] + [20] +2[21] +2[22] + [30] +4[31] +4[32] +3[33] +2[40] +4[41] +6[42] +5[43] +3[44]
        + [50] +4[51] +5[52] +6[53] +4[54] +2[55] + [60] +2[61] +4[62] +4[63] +4[64] +2[65] + [66]
        + [71] + [72] +2[73] + [74] + [75]
{77000} = [11] + [31] + [33] + [51] + [53] + [55] + [71] + [73] + [75] + [77]
{77100} = [11] + [21] + [31] + [32] + [33] + [41] + [43] + [51] + [52] + [53] + [54] + [55] + [61]
        + [63] + [65] + [71] + [72] + [73] + [74] + [75] + [76] + [77]
{77200} = [11] + [21] +2[31] + [32] +2[33] + [41] + [42] + [43] +2[51] + [52] +3[53] + [54] +2[55]
        + [61] + [62] + [63] + [64] + [65] + [71] + [72] +2[73] + [74] +2[75] + [76] + [77]
{77300} = [11] + [21] +2[31] + [32] +2[33] +2[41] + [42] +2[43] +2[51] +2[52] +3[53] +2[54] +2[55]
        + [61] + [62] +2[63] + [64] + [65] + [71] + [72] +2[73] +2[74] +2[75] + [76] + [77]

{80000} = [00] + [20] + [40] + [60] + [80]
{81000} = [10] + [21] + [30] + [41] + [50] + [61] + [70] + [81]
{81100} = [11] + [21] + [31] + [41] + [51] + [61] + [71] + [81]
{81110} = [11] + [20] + [31] + [40] + [51] + [60] + [71] + [80]
{82000} = [00] +2[20] + [22] + [31] +2[40] + [42] + [51] +2[60] + [62] + [71] + [80] + [82]
{82100} = [10] +2[21] + [22] + [30] + [31] + [32] +2[41] + [42] + [50] + [51] + [52] +2[61] + [62]
        + [70] + [71] + [72] + [81] + [82]
{82110} = [11] +2[21] + [30] + [31] + [32] +2[41] + [50] + [51] + [52] +2[61] + [70] + [71] + [72]
        + [81]
{82200} = [00] +2[20] +2[22] + [31] + [32] +2[40] +2[42] + [51] + [52] +2[60] +2[62] + [71] + [72]
        + [80] + [82]
{82210} = [10] +2[21] + [22] + [30] + [31] + [32] +2[41] + [42] + [50] + [51] + [52] +2[61] + [62]
        + [70] + [71] + [72] + [81]
{82220} = [00] +2[20] + [22] + [31] +2[40] + [42] + [51] +2[60] + [62] + [71] + [80]
{83000} = [10] + [21] +2[30] + [32] +2[41] + [43] +2[50] + [52] +2[61] + [63] + [70] + [72] + [81]
        + [83]
{83100} = [11] + [20] + [21] + [22] +3[31] + [32] + [33] + [40] +2[41] +2[42] + [43] +3[51] + [52]
        + [53] + [60] +2[61] +2[62] + [63] +2[71] + [72] + [73] + [81] + [82] + [83]
{83110} = [11] + [20] + [21] + [22] +3[31] + [32] + [33] +2[40] + [41] +2[42] +3[51] + [52] + [53]
        +2[60] + [61] +2[62] +2[71] + [72] + [73] + [80] + [82]
{83200} = [10] +2[21] + [22] +2[30] + [31] +3[32] + [33] +3[41] +2[42] +2[43] +2[50] + [51] +3[52]
        + [53] +3[61] +2[62] +2[63] + [70] + [71] +2[72] + [73] + [81] + [82] + [83]
{83210} = [11] + [20] +2[21] +2[22] + [30] +3[31] +3[32] + [33] + [40] +3[41] +3[42] + [43] + [50]
        +3[51] +3[52] + [53] + [60] +3[61] +3[62] + [63] + [70] +2[71] +2[72] + [73] + [81] + [82]
{83220} = [10] +2[21] + [22] +2[30] + [31] +2[32] +3[41] + [42] + [43] +2[50] + [51] +2[52] +3[61]
        + [62] + [63] + [70] + [71] + [72] + [81]
{83300} = [11] + [21] +2[31] + [32] +2[33] +2[41] + [42] +2[43] +2[51] + [52] +2[53] +2[61] + [62]
        +2[63] + [71] + [72] + [73] + [81] + [83]
{83310} = [11] + [20] + [21] + [22] +3[31] +2[32] +2[33] +2[40] + [41] +3[42] + [43] +3[51] +2[52]
        +2[53] +2[60] + [61] +3[62] + [63] +2[71] + [72] + [73] + [80] + [82]
{83320} = [11] +2[21] + [30] +2[31] +2[32] + [33] +3[41] + [42] + [43] + [50] +2[51] +2[52] + [53]
        +3[61] + [62] + [63] + [70] + [71] + [72] + [81]
{83330} = [11] + [20] +2[31] + [33] +2[40] + [42] +2[51] + [53] +2[60] + [62] + [71] + [80]
{84000} = [00] +2[20] + [22] + [31] +3[40] +2[42] + [44] +2[51] + [53] +2[60] +2[62] + [64] + [71]
        + [73] + [80] + [82] + [84]
{84100} = [10] +2[21] + [22] +2[30] + [31] +2[32] +4[41] +2[42] +2[43] + [44] +2[50] +2[51] +3[52]
        + [53] + [54] +3[61] +2[62] +2[63] + [64] + [70] + [71] +2[72] + [73] + [74] + [81] + [82]
        + [83] + [84]
{84110} = [11] +2[21] + [30] +2[31] +2[32] + [33] +4[41] + [42] +2[43] +2[50] +2[51] +3[52] + [53]
        + [54] +3[61] + [62] +2[63] + [70] + [71] +2[72] + [73] + [74] + [81] + [83]
{84200} = [00] +3[20] +3[22] +3[31] +2[32] + [33] +4[40] + [41] +6[42] +2[43] +2[44] +4[51] +3[52]
        +3[53] + [54] +3[60] + [61] +5[62] +2[63] +2[64] +2[71] +2[72] +2[73] + [74] + [80] +2[82]
        + [83] + [84]
{84210} = [10] +3[21] +2[22] +2[30] +3[31] +4[32] +2[33] + [40] +5[41] +5[42] +3[43] + [44] +2[50]
        +4[51] +5[52] +3[53] + [54] + [60] +4[61] +4[62] +3[63] + [64] + [70] +2[71] +3[72] +2[73]
        + [74] + [81] + [82] + [83]
{84220} = [00] +3[20] +3[22] +3[31] +2[32] + [33] +4[40] + [41] +5[42] + [43]+ [44] +4[51] +2[52]
        +2[53] +3[60] + [61] +4[62] + [63] + [64] +2[71] + [72] + [73] + [80]+ [82]
```

Branching Rules for the Reduction $U_5 \to R_5$ (contd).

{84300} = [10] +2[21] + [22] +2[30] + [31] +3[32] + [33] +4[41] +3[42] +4[43] +2[44] +2[50] +2[51]
+4[52] +3[53] +2[54] +3[61] +3[62] +3[63] +2[64] + [70] + [71] +2[72] +2[73] + [74] + [81]
+ [82] + [83] + [84]

{84310} = [11] + [20] +2[21] +2[22] + [30] +4[31] +4[32] +3[33] + [40] +5[41] +5[42] +5[43] + [44]
+2[50] +4[51] +6[52] +4[53] +2[54] + [60] +4[61] +4[62] +4[63] + [64] + [70] +2[71] +3[72]
+2[73] + [74] + [81] + [82] + [83]

{84320} = [10] +3[21] +2[22] +2[30] +3[31] +4[32] +2[33] + [40] +5[41] +5[42] +3[43] + [44] +2[50]
+4[51] +5[52] +3[53] + [54] + [60] +4[61] +4[62] +2[63] + [64] + [70] +2[71] +2[72] + [73]
+ [81] + [82]

{84330} = [11] +2[21] + [30] +2[31] +2[32] + [33] +4[41] + [42] +2[43] +2[50] +2[51] +3[52] + [53]
+ [54] +3[61] + [62] + [63] + [70] + [71] + [72] + [81]

{84400} = [00] +2[20] +2[22] + [31] + [32] +3[40] +4[42] + [43] +3[44] +2[51] +2[52] +2[53] +2[54]
+2[60] +3[62] + [63] +2[64] + [71] + [72] + [73] + [74] + [80] + [82] + [84]

{84410} = [10] +2[21] + [22] +2[30] + [31] +3[32] + [33] +4[41] +3[42] +4[43] +2[44] +2[50] +2[51]
+4[52] +3[53] +2[54] +3[61] +2[62] +3[63] + [64] + [70] + [71] +2[72] + [73] + [74] + [81]
+ [83]

{84420} = [00] +3[20] +3[22] +3[31] +2[32] + [33] +4[40] + [41] +6[42] +2[43] +2[44] +4[51] +3[52]
+3[53] + [54] +3[60] + [61] +4[62] + [63] + [64] +2[71] + [72] + [73] + [80] + [82]

{84430} = [10] +2[21] + [22] +2[30] + [31] +2[32] +4[41] +2[42] +2[43] + [44] +2[50] +2[51] +3[52]
+ [53] + [54] +3[61] + [62] + [63] + [70] + [71] + [72] + [81]

{84440} = [00] +2[20] + [22] + [31] +3[40] +2[42] + [44] +2[51] + [53]+2[60] + [62] + [71] + [80]

{85000} = [10] + [21] +2[30] + [32] +2[41] + [43] +2[50] +2[52] + [54] +2[61]+2[63] + [65] + [70]
+ [72] + [74] + [81] + [83] + [85]

{85100} = [11] + [20] + [21] + [22] +3[31] + [32] + [33] +2[40] +2[41] +3[42] + [43] + [44] +4[51]
+2[52] +3[53] + [54] + [55] + [60] +2[61] +3[62] +2[63] +2[64] + [65] +2[71] + [72] +2[73]
+ [74] + [75] + [81] + [82] + [83] + [84] + [85]

{85110} = [11] + [20] + [21] + [22] +3[31] + [32] + [33] +2[40] +2[41] +3[42] + [43] + [44] +4[51]
+2[52] +3[53] + [54] + [55] +2[60] + [61] +3[62] + [63] +2[64] +2[71] + [72] +2[73] + [74]
+ [75] + [80] + [82] + [84]

{85200} = [10] +2[21] + [22] +3[30] + [31] +4[32] + [33] +5[41] +3[42] +4[43] + [44] +3[50] +2[51]
+6[52] +3[53] +3[54] + [55] +4[61] +3[62] +5[63] +2[64] +2[65] + [70] + [71] +3[72] +2[73]
+2[74] + [75] + [81] + [82] +2[83] + [84] + [85]

{85210} = [11] + [20] +2[21] +2[22] + [30] +4[31] +4[32] +2[33] +2[40] +5[41] +6[42] +4[43] +2[44]
+2[50] +5[51] +6[52] +5[53] +3[54] + [55] + [60] +4[61] +5[62] +4[63] +3[64] + [65] + [70]
+2[71] +3[72] +3[73] +2[74] + [75] + [81] + [82] + [83] + [84]

{85220} = [10] +2[21] + [22] +3[30] + [31] +4[32] + [33] +5[41] +3[42] +4[43] + [44] +3[50] +2[51]
+5[52] +2[53] +2[54] +4[61] +2[62] +4[63] + [64] + [65] + [70] + [71] +2[72] + [73] + [74]
+ [81] + [83]

{85300} = [11] + [20] + [21] + [22] +4[31] +2[32] +3[33] +2[40] +3[41] +5[42] +4[43] +2[44] +5[51]
+4[52] +6[53] +3[54] +2[55] + [60] +3[61] +4[62] +5[63] +3[64] +2[65] +2[71] +2[72] +3[73]
+2[74] + [75] + [81] + [82] +2[83] + [84] + [85]

{85310} = [11] + [20] +2[21] +2[22] + [30] +5[31] +5[32] +4[33] +3[40] +5[41] +8[42] +6[43] +3[44]
+ [50] +7[51] +7[52] +8[53] +4[54] +2[55] +3[60] +3[61] +7[62] +5[63] +4[64] + [65] +3[71]
+3[72] +4[73] +2[74] + [75] + [80] +2[82] + [83] + [84]

{85320} = [11] + [20] +2[21] +2[22] + [30] +5[31] +5[32] +4[33] +2[40] +6[41] +7[42] +6[43] +2[44]
+2[50] +6[51] +7[52] +6[53] +3[54] + [55] + [60] +5[61] +5[62] +5[63] +2[64] + [65] + [70]
+2[71] +3[72] +2[73] + [74] + [81] + [82] + [83]

{85330} = [11] + [20] + [21] + [22] +4[31] +2[32] +3[33] +3[40] +2[41] +5[42] +2[43] + [44] +5[51]
+3[52] +4[53] + [54] + [55] +3[60] + [61] +4[62] + [63] + [64] +2[71] + [72] + [73] + [80]
+ [82]

{85400} = [10] +2[21] + [22] +2[30] + [31] +3[32] + [33] +4[41] +3[42] +4[43] +2[44] +2[50] +2[51]
+5[52] +4[53] +4[54] +2[55] +3[61] +3[62] +4[63] +3[64] +2[65] + [70] + [71] +2[72] +2[73]
+2[74] + [75] + [81] + [82] + [83] + [84] + [85]

{85410} = [11] + [20] +2[21] +2[22] + [30] +4[31] +4[32] +3[33] +2[40] +5[41] +7[42] +6[43] +4[44]
+2[50] +5[51] +7[52] +7[53] +5[54] +2[55] + [60] +4[61] +5[62] +5[63] +4[64] + [65] + [70]
+2[71] +3[72] +3[73] +2[74] + [75] + [81] + [82] + [83] + [84]

{85420} = [10] +3[21] +2[22] +3[30] +3[31] +6[32] +3[33] + [40] +7[41] +7[42] +7[43] +3[44] +3[50]
+5[51] +8[52] +6[53] +4[54] + [55] + [60] +5[61] +5[62] +5[63] +2[64] + [65] + [70] +2[71]
+3[72] +2[73] + [74] + [81] + [82] + [83]

{85430} = [11] + [20] +2[21] +2[22] + [30] +4[31] +4[32] +2[33] +2[40] +5[41] +6[42] +4[43] +2[44]
+2[50] +5[51] +6[52] +4[53] +2[54] + [55] + [60] +4[61] +4[62] +2[63] + [64] + [70] +2[71]
+2[72] + [73] + [81] + [82]

{86000} = [00] +2[20] + [22] + [31] +2[40] +2[42] + [44] + [51] + [53] +2[60] +2[62] +2[64] + [66]
+ [71] + [73] + [75] + [80] + [82] + [84] + [86]

{86100} = [10] +2[21] + [22] +2[30] + [31] +2[32] +3[41] +2[42] +2[43] + [44] +2[50] + [51] +3[52]
+ [53] +2[54] +3[61] +2[62] +3[63] +2[64] +2[65] + [66] + [70] + [71] +2[72] + [73] +2[74]
+ [75] + [76] + [81] + [82] + [83] + [84] + [85] + [86]

{86110} = [11] +2[21] + [30] +2[31] +2[32] + [33] +3[41] + [42] +2[43] + [50] +2[51] +3[52] +2[53]
+2[54] + [55] +3[61] + [62] +3[63] + [64] +2[65] + [70] + [71] +2[72] + [73] +2[74] + [75]
+ [76] + [81] + [83] + [85]

Branching Rules for the Reduction $U_5 \to R_5$ (contd).

{86200} = [00] +3[20] +3[22] +3[31] +2[32] + [33] +4[40] + [41] +6[42] +2[43] +3[44] +4[51] +3[52]
+4[53] +2[54] + [55] +3[60] + [61] +6[62] +3[63] +5[64] +2[65] +2[66] +2[71] +2[72] +3[73]
+2[74] +2[75] + [76] + [80] +2[82] + [83] +2[84] + [85] + [86]

{86210} = [10] +3[21] +2[22] +2[30] +3[31] +4[32] +2[33] + [40] +5[41] +5[42] +4[43] +2[44] +2[50]
+4[51] +6[52] +5[53] +4[54] +2[55] + [60] +4[61] +5[62] +5[63] +4[64] +3[65] + [66] + [70]
+2[71] +3[72] +3[73] +3[74] +2[75] + [76] + [81] + [82] + [83] + [84] + [85]

{86220} = [00] +3[20] +3[22] +3[31] +2[32] + [33] +4[40] + [41] +6[42] +2[43] +3[44] +4[51] +3[52]
+4[53] +2[54] + [55] +3[60] + [61] +5[62] +2[63] +4[64] + [65] + [66] +2[71] + [72] +2[73]
+ [74] + [75] + [80] + [82] + [84]

{86300} = [10] +2[21] + [22] +3[30] + [31] +4[32] + [33] +5[41] +4[42] +5[43] +3[44] +3[50] +2[51]
+6[52] +4[53] +4[54] + [55] +4[61] +4[62] +6[63] +5[64] +3[65] +2[66] + [70] + [71] +3[72]
+3[73] +3[74] +2[75] + [76] + [81] + [82] +2[83] +2[84] + [85] + [86]

{86310} = [11] + [20] +2[21] +2[22] + [30] +5[31] +5[32] +4[33] +2[40] +6[41] +7[42] +7[43] +3[44]
+2[50] +6[51] +8[52] +8[53] +6[54] +3[55] + [60] +5[61] +6[62] +8[63] +5[64] +4[65] + [66]
+ [70] +2[71] +4[72] +4[73] +4[74] +2[75] + [76] + [81] + [82] +2[83] + [84] + [85]

{86320} = [10] +3[21] +2[22] +3[30] +3[31] +6[32] +3[33] + [40] +7[41] +7[42] +7[43] +4[44] +3[50]
+5[51] +8[52] +7[53] +5[54] +2[55] + [60] +5[61] +6[62] +6[63] +5[64] +2[65] + [66] + [70]
+2[71] +3[72] +3[73] +2[74] + [75] + [81] + [82] + [83] + [84]

{86400} = [00] +3[20] +3[22] +3[31] +2[32] + [33] +4[40] + [41] +7[42] +3[43] +4[44] +4[51] +4[52]
+5[53] +4[54] + [55] +3[60] + [61] +6[62] +4[63] +6[64] +3[65] +2[66] +2[71] +2[72] +3[73]
+3[74] +2[75] + [76] + [80] +2[82] + [83] +2[84] + [85] + [86]

{86410} = [10] +3[21] +2[22] +3[30] +3[31] +6[32] +3[33] + [40] +7[41] +7[42] +8[43] +4[44] +3[50]
+5[51] +9[52] +8[53] +7[54] +3[55] + [60] +5[61] +6[62] +8[63] +6[64] +4[65] + [66] + [70]
+2[71] +4[72] +4[73] +4[74] +2[75] + [76] + [81] + [82] +2[83] + [84] + [85]

{86420} = [00] +4[20] +5[22] +6[31] +5[32] +4[33] +6[40] +4[41] +12[42] +7[43] +6[44] + [50] +8[51]
+8[52] +9[53] +6[54] +2[55] +4[60] +3[61] +9[62] +6[63] +6[64] +2[65] + [66] +3[71] +3[72]
+4[73] +2[74] + [75] + [80] +2[82] + [83] + [84]

{87000} = [10] + [21] + [30] + [32] + [41] + [43] + [50] + [52] + [54] + [61] + [63] + [65] + [70]
+ [72] + [74] + [76] + [81] + [83] + [85] + [87]

{87100} = [11] + [20] + [21] + [22] +2[31] + [32] + [33] + [40] + [41] +2[42] + [43] + [44] +2[51]
+ [52] +2[53] + [54] + [55] + [60] + [61] +2[62] + [63] +2[64] + [65] + [66] +2[71] + [72]
+2[73] + [74] +2[75] + [76] + [77] + [81] + [82] + [83] + [84] + [85] + [86] + [87]

{87110} = [11] + [20] + [21] + [22] +2[31] + [32] + [33] + [40] + [41] +2[42] + [43] + [44] +2[51]
+ [52] +2[53] + [54] + [55] + [60] + [61] +2[62] + [63] +2[64] + [65] + [66] +2[71] + [72]
+2[73] + [74] +2[75] + [76] + [77] + [80] + [82] + [84] + [86]

{87200} = [10] +2[21] + [22] +2[30] + [31] +3[32] + [33] +3[41] +2[42] +3[43] + [44] +2[50] + [51]
+4[52] +2[53] +3[54] + [55] +3[61] +2[62] +4[63] +2[64] +3[65] + [66] + [70] + [71] +3[72]
+2[73] +3[74] +2[75] +2[76] + [77] + [81] + [82] +2[83] + [84] +2[85] + [86] + [87]

{87210} = [11] + [20] +2[21] +2[22] + [30] +3[31] +3[32] +2[33] + [40] +3[41] +4[42] +3[43] +2[44]
+ [50] +3[51] +4[52] +4[53] +3[54] +2[55] + [60] +3[61] +4[62] +4[63] +4[64] +3[65] +2[66]
+ [70] +2[71] +3[72] +3[73] +3[74] +3[75] +2[76] + [77] + [81] + [82] + [83] + [84] + [85]
+ [86]

{87300} = [11] + [20] + [21] + [22] +3[31] +2[32] +2[33] +2[40] +2[41] +4[42] +3[43] +2[44] +4[51]
+3[52] +5[53] +3[54] +2[55] + [60] +2[61] +4[62] +4[63] +4[64] +3[65]+ [66] +2[71] +2[72]
+4[73] +3[74] +3[75] +2[76] + [77] + [81] + [82] +2[83] +2[84] +2[85] + [86] + [87]

{87310} = [11] + [20] +2[21] +2[22] + [30] +4[31] +4[32] +3[33] +2[40] +4[41] +6[42] +5[43] +3[44]
+ [50] +5[51] +6[52] +7[53] +5[54] +3[55] +2[60] +3[61] +6[62] +6[63] +6[64] +4[65] +2[66]
+3[71] +3[72] +5[73] +4[74] +4[75] +2[76] + [77] + [80] +2[82] + [83] +2[84] + [85] + [86]

{87400} = [10] +2[21] + [22] +2[30] + [31] +3[32] + [33] +4[41] +3[42] +4[43] +2[44] +2[50] +2[51]
+5[52] +4[53] +4[54] +2[55] +3[61] +3[62] +5[63] +4[64] +3[65] + [66] + [70] + [71] +3[72]
+3[73] +4[74] +3[75] +2[76] + [77] + [81] + [82] +2[83] +2[84] +2[85] + [86] + [87]

{87410} = [11] + [20] +2[21] +2[22] + [30] +4[31] +4[32] +3[33] +2[40] +5[41] +7[42] +6[43] +4[44]
+2[50] +5[51] +7[52] +8[53] +6[54] +3[55] + [60] +4[61] +6[62] +7[63] +7[64] +4[65] +2[66]
+ [70] +2[71] +4[72] +5[73] +5[74] +4[75] +2[76] + [77] + [81] + [82] +2[83] +2[84] + [85]
+ [86]

{88000} = [00] + [20] + [22] + [40] + [42] + [44] + [60] + [62] + [64] + [66] + [80] + [82] + [84]
+ [86] + [88]

{88100} = [10] + [21] + [22] + [30] + [32] + [41] + [42] + [43] + [44] + [50] + [52] + [54] + [61]
+ [62] + [63] + [64] + [65] + [66] + [70] + [72] + [74] + [76] + [81] + [82] + [83] + [84]
+ [85] + [86] + [87] + [88]

{88200} = [00] +2[20] +2[22] + [31] + [32] +2[40] +3[42] + [43] +2[44] + [51] + [52] + [53] + [54]
+2[60] +3[62] + [63] +3[64] + [65] +2[66] + [71] + [72] + [73] + [74] + [75] + [76] + [80]
+2[82] + [83] +2[84] + [85] +2[86] + [87] + [88]

{88300} = [10] + [21] + [22] +2[30] +2[32] +2[41] +2[42] +2[43] +2[44] +2[50] +3[52] + [53] +2[54]
+2[61] +2[62] +3[63] +3[64] +2[65] +2[66] + [70] +2[72] + [73] +2[74] + [75] + [76] + [81]
+ [82] +2[83] +2[84] +2[85] +2[86] + [87] + [88]

{88400} = [00] +2[20] +2[22] + [31] + [32] +3[40] +4[42] + [43] +3[44] +2[51] +2[52] +2[53] +2[54]
+2[60] +4[62] +2[63] +4[64] +2[65] +2[66] + [71] + [72] +2[73] +2[74] + [75] + [76] + [80]
+2[82] + [83] +3[84] +2[85] +2[86] + [87] + [88]

C-4 Branching Rules for the Reduction $U_6 \to O_6$.

{000000} = [000]

{100000} = [100]
{110000} = [110]
{111000} = [111]

{200000} = [000] + [200]
{210000} = [100] + [210]
{211000} = [110] + [211]
{211100} = [111] + [210]
{211110} = [110] + [200]
{220000} = [000] + [200] + [220]
{221000} = [100] + [210] + [221]
{221100} = [110] + [211] + [220]
{222000} = [000] + [200] + [220] + [222]

{300000} = [100] + [300]
{310000} = [110] + [200] + [310]
{311000} = [111] + [210] + [311]
{311100} = [110] + [211] + [310]
{311110} = [100] + [210] + [300]
{320000} = [100] + [210] + [300] + [320]
{321000} = [110] + [200] + [211] + [220] + [310] + [321]
{321100} = [111] +2[210] + [221] + [311] + [320]
{321110} = [110] + [200] + [211] + [220] + [310]
{322000} = [100] + [210] + [221] + [300] + [320] + [322]
{322100} = [110] + [200] + [211] +2[220] + [222] + [310] + [321]
{322110} = [111] +2[210] + [221] + [311]
{330000} = [110] + [310] + [330]
{331000} = [111] + [210] + [311] + [320] + [331]
{331100} = [110] + [211] + [220] + [310] + [321] + [330]
{332000} = [110] + [211] + [310] + [321] + [330] + [332]
{332100} = [111] +2[210] + [221] + [311] +2[320] + [322] + [331]
{333000} = [111] + [311] + [331] + [333]

{400000} = [000] + [200] + [400]
{410000} = [100] + [210] + [300] + [410]
{411000} = [110] + [211] + [310] + [411]
{411100} = [111] + [210] + [311] + [410]
{411110} = [110] + [200] + [310] + [400]
{420000} = [000] +2[200] + [220] + [310] + [400] + [420]
{421000} = [100] +2[210] + [221] + [300] + [311] + [320] + [410] + [421]
{421100} = [110] +2[211] + [220] +2[310] + [321] + [411] + [420]
{421110} = [111] +2[210] + [300] + [311] + [320] + [410]
{422000} = [000] +2[200] +2[220] + [222] + [310] + [321] + [400] + [420] + [422]
{422100} = [100] +2[210] +2[221] + [300] + [311] +2[320] + [322] + [410] + [421]
{422110} = [110] +2[211] + [220] +2[310] + [321] + [411]
{422200} = [000] +2[200] +2[220] + [222] + [310] + [321] + [400] + [420]
{422210} = [100] +2[210] + [221] + [300] + [311] + [320] + [410]
{422220} = [000] +2[200] + [220] + [310] + [400]
{430000} = [100] + [210] + [300] + [320] + [410] + [430]
{431000} = [110] + [200] + [211] + [220] +2[310] + [321] + [330] + [411] + [420] + [431]
{431100} = [111] +2[210] + [221] +2[311] +2[320] + [331] + [410] + [421] + [430]
{431110} = [110] + [200] + [211] + [220] +2[310] + [321] + [330] + [400] + [420]
{432000} = [100] +2[210] + [221] + [300] + [311] +2[320] + [322] + [331] + [410] + [421] + [430] + [432]
{432100} = [110] + [200] +2[211] +3[220] + [222] +3[310] +3[321] +2[330] + [332] + [411] +2[420] + [422] + [431]
{432110} = [111] +3[210] +2[221] + [300] +2[311] +3[320] + [322] + [331] + [410] + [421]
{432200} = [100] +2[210] +2[221] + [300] + [311] +3[320] + [322] + [331] + [410] + [421] + [430]
{432210} = [110] + [200] +2[211] +3[220] + [222] +3[310] +2[321] + [330] + [411] + [420]
{433000} = [110] + [211] + [310] + [321] + [330] + [332] + [411] + [431] + [433]
{433100} = [111] +2[210] + [221] +2[311] +2[320] + [322] +2[331] + [333] + [410] + [421] + [430] + [432]
{433110} = [110] + [200] + [211] +2[220] + [222] +2[310] +2[321] + [330] + [332] + [400] + [420] + [422]
{440000} = [000] + [200] + [220] + [400] + [420] + [440]
{441000} = [100] + [210] + [221] + [300] + [320] + [410] + [421] + [430] + [441]
{441100} = [110] + [211] + [220] + [310] + [321] + [330] + [411] + [420] + [431] + [440]
{442000} = [000] +2[200] +2[220] + [222] + [310] + [321] + [400] +2[420] + [422] + [431] + [440] + [442]
{442100} = [100] +2[210] +2[221] + [300] + [311] +3[320] + [322] + [331] + [410] +2[421] +2[430] + [432] + [441]
{442200} = [000] +2[200] +3[220] + [222] + [310] +2[321] + [330] + [400] +2[420] + [422] + [431] + [440]
{443000} = [100] + [210] + [221] + [300] + [320] + [322] + [410] + [421] + [430] + [432] + [441] + [443]
{443100} = [110] + [200] + [211] +2[220] + [222] +2[310] +2[321] + [330] + [332] + [411] +2[420] + [422] +2[431] + [433] + [440] + [442]
{444000} = [000] + [200] + [220] + [222] + [400] + [420] + [422] + [440] + [442] + [444]

C-5 Branching Rules for the Reduction $U_7 \to R_7$.

{0000000} = [000]

{1000000} = [100]
{1100000} = [110]
{1110000} = [111]

{2000000} = [000] + [200]
{2100000} = [100] + [210]
{2110000} = [110] + [211]
{2111000} = [111] + [211]
{2111100} = [111] + [210]
{2111110} = [110] + [200]
{2200000} = [000] + [200] + [220]
{2210000} = [100] + [210] + [221]
{2211000} = [110] + [211] + [221]
{2211100} = [111] + [211] + [220]
{2220000} = [000] + [200] + [220] + [222]
{2221000} = [100] + [210] + [221] + [222]

{3000000} = [100] + [300]
{3100000} = [110] + [200] + [310]
{3110000} = [111] + [210] + [311]
{3111000} = [111] + [211] + [311]
{3111100} = [110] + [211] + [310]
{3111110} = [100] + [210] + [300]
{3200000} = [100] + [210] + [300] + [320]
{3210000} = [110] + [200] + [211] + [220] + [310] + [321]
{3211000} = [111] + [210] + [211] + [221] + [311] + [321]
{3211100} = [111] + [210] + [211] + [221] + [311] + [320]
{3211110} = [110] + [200] + [211] + [220] + [310]
{3220000} = [100] + [210] + [221] + [300] + [320] + [322]
{3221000} = [110] + [200] + [211] + [220] + [221] + [222] + [310] + [321] + [322]
{3221100} = [111] + [210] + [211] + [220] + [221] + [222] + [311] + [321]
{3221110} = [111] + [210] + [211] + [221] + [311]
{3300000} = [110] + [310] + [330]
{3310000} = [111] + [210] + [311] + [320] + [331]
{3311000} = [111] + [211] + [220] + [311] + [321] + [331]
{3311100} = [110] + [211] + [221] + [310] + [321] + [330]
{3320000} = [110] + [211] + [310] + [321] + [330] + [332]
{3321000} = [111] + [210] + [211] + [221] + [311] + [320] + [321] + [322] + [331] + [332]
{3321100} = [111] + [210] + [211] + [220] + [221] + [222] + [311] + [320] + [321] + [322] + [331]
{3330000} = [111] + [311] + [331] + [333]
{3331000} = [111] + [211] + [311] + [321] + [331] + [332] + [333]

{4000000} = [000] + [200] + [400]
{4100000} = [100] + [210] + [300] + [410]
{4110000} = [110] + [211] + [310] + [411]
{4111000} = [111] + [211] + [311] + [411]
{4111100} = [111] + [210] + [311] + [410]
{4111110} = [110] + [200] + [310] + [400]
{4200000} = [000] +2[200] + [220] + [310] + [400] + [420]
{4210000} = [100] +2[210] + [221] + [300] + [311] + [320] + [410] + [421]
{4211000} = [110] +2[211] + [221] + [310] + [311] + [321] + [411] + [421]
{4211100} = [111] +2[211] + [220] + [310] + [311] + [321] + [411] + [420]
{4211110} = [111] +2[210] + [300] + [311] + [320] + [410]
{4220000} = [000] +2[200] +2[220] + [222] + [310] + [321] + [400] + [420] + [422]
{4221000} = [100] +2[210] +2[221] + [222] + [300] + [311] + [320] + [321] + [322] + [410] + [421]
+ [422]
{4221100} = [110] +2[211] +2[221] + [310] + [311] + [320] + [321] + [322] + [411] + [421]
{4221110} = [111] +2[211] + [220] + [310] + [311] + [321] + [411]
{4222000} = [000] +2[200] +2[220] +2[222] + [310] + [321] + [322] + [400] + [420] + [422]
{4222100} = [100] +2[210] +2[221] + [222] + [300] + [311] + [320] + [321] + [322] + [410] + [421]
{4222110} = [110] +2[211] + [221] + [310] + [311] + [321] + [411]
{4222200} = [000] +2[200] +2[220] + [222] + [310] + [321] + [400] + [420]
{4222210} = [100] +2[210] + [221] + [300] + [311] + [320] + [410]
{4222220} = [000] +2[200] + [220] + [310] + [400]
{4300000} = [100] + [210] + [300] + [320] + [410] + [430]
{4310000} = [110] + [200] + [211] + [220] +2[310] + [321] + [330] + [411] + [420] + [431]
{4311000} = [111] + [210] + [211] + [221] +2[311] + [320] + [321] + [331] + [411] + [421] + [431]
{4311100} = [111] + [210] + [211] + [221] +2[311] + [320] + [321] + [331] + [410] + [421] + [430]

C-5 Branching Rules for the Reduction $U_7 \rightarrow R_7$ (contd).

{4311110} = [110] + [200] + [211] + [220] +2[310] + [321] + [330] + [400] + [420]

{4320000} = [100] +2[210] + [221] + [300] + [311] +2[320] + [322] + [331] + [410] + [421] + [430]
+ [432]

{4321000} = [110] + [200] +2[211] +2[220] + [221] + [222] +2[310] + [311] +3[321] + [322] + [330]
+ [331] + [332] + [411] + [420] + [421] + [422] + [431] + [432]

{4321100} = [111] + [210] +2[211] + [220] +2[221] + [222] + [310] +2[311] + [320] +3[321] + [322]
+ [330] + [331] + [332] + [411] + [420] + [421] + [422] + [431]

{4321110} = [111] +2[210] + [211] +2[221] + [300] +2[311] +2[320] + [321] + [322] + [331] + [410]
+ [421]

{4322000} = [100] +2[210] +2[221] + [222] + [300] + [311] +2[320] + [321] +2[322] + [331] + [332]
+ [410] + [421] + [422] + [430] + [432]

{4322100} = [110] + [200] +2[211] +2[220] +2[221] +2[222] +2[310] + [311] + [320] +3[321] +2[322]
+ [330] + [331] + [332] + [411] + [420] + [421] + [422] + [431]

{4322110} = [111] + [210] +2[211] + [220] +2[221] + [222] + [310] +2[311] + [320] +2[321] + [322]
+ [331] + [411] + [421]

{4322200} = [100] +2[210] +2[221] + [222] + [300] + [311] +2[320] + [321] + [322] + [331] + [410]
+ [421] + [430]

{4322210} = [110] + [200] +2[211] +2[220] + [221] + [222] +2[310] + [311] +2[321] + [330] + [411]
+ [420]

{4330000} = [110] + [211] + [310] + [321] + [330] + [332] + [411] + [431] + [433]

{4331000} = [111] + [210] + [211] + [221] +2[311] + [320] + [321] + [322] +2[331] + [332] + [333]
+ [411] + [421] + [431] + [432] + [433]

{4331100} = [111] + [210] + [211] + [220] + [221] + [222] +2[311] + [320] +2[321] + [322] +2[331]
+ [332] + [333] + [410] + [421] + [422] + [430] + [432]

{4331110} = [110] + [200] + [211] + [220] + [221] + [222] +2[310] +2[321] + [322] + [330] + [332]
+ [400] + [420] + [422]

{4332000} = [110] +2[211] + [221] + [310] + [311] +2[321] + [322] + [330] + [331] +2[332] + [333]
+ [411] + [421] + [431] + [432] + [433]

{4332100} = [111] + [210] +2[211] + [220] +2[221] + [222] + [310] +2[311] + [320] +3[321] +2[322]
+ [330] +2[331] +2[332] + [333] + [411] + [420] + [421] + [422] + [431] + [432]

{4332110} = [111] +2[210] + [211] + [220] +2[221] +2[222] + [300] +2[311] +2[320] +2[321] +2[322]
+ [331] + [332] + [410] + [421] + [422]

{4400000} = [000] + [200] + [220] + [400] + [420] + [440]

{4410000} = [100] + [210] + [221] + [300] + [320] + [410] + [421] + [430] + [441]

{4411000} = [110] + [211] + [221] + [310] + [321] + [330] + [411] + [421] + [431] + [441]

{4411100} = [111] + [211] + [220] + [311] + [321] + [331] + [411] + [420] + [431] + [440]

{4420000} = [000] +2[200] +2[220] + [222] + [310] + [321] + [400] +2[420] + [422] + [431] + [440]
+ [442]

{4421000} = [100] +2[210] +2[221] + [222] + [300] + [311] +2[320] + [321] + [322] + [331] + [410]
+2[421] + [422] + [430] + [431] + [432] + [441] + [442]

{4421100} = [110] +2[211] +2[221] + [310] + [311] + [320] +2[321] + [322] + [330] + [331] + [332]
+ [411] +2[421] + [430] + [431] + [432] + [441]

{4422000} = [000] +2[200] +3[220] +2[222] + [310] +2[321] + [322] + [331] + [400] +2[420] +2[422]
+ [431] + [432] + [440] + [442]

{4422100} = [100] +2[210] +3[221] + [222] + [300] + [311] +2[320] +2[321] +2[322] + [330] + [331]
+ [332] + [410] +2[421] + [422] + [430] + [431] + [432] + [441]

{4422200} = [000] +2[200] +3[220] +2[222] + [310] +2[321] + [322] + [331] + [400] +2[420] + [422]
+ [431] + [440]

{4430000} = [100] + [210] + [221] + [300] + [320] + [322] + [410] + [421] + [430] + [432] + [441]
+ [443]

{4431000} = [110] + [200] + [211] + [220] + [221] + [222] +2[310] +2[321] + [322] + [330] + [332]
+ [411] + [420] + [421] + [422] +2[431] + [432] + [433] + [441] + [442] + [443]

{4431100} = [111] + [210] + [211] + [220] + [221] + [222] +2[311] + [320] +2[321] + [322] +2[331]
+ [332] + [333] + [411] + [420] + [421] + [422] +2[431] + [432] + [433] + [440] + [442]

{4432000} = [100] +2[210] +2[221] + [222] + [300] + [311] +2[320] + [321] +2[322] + [331] + [332]
+ [410] +2[421] + [422] + [430] + [431] +2[432] + [433] + [441] + [442] + [443]

{4432100} = [110] + [200] +2[211] +2[220] +2[221] +2[222] +2[310] + [311] + [320] +4[321] +2[322]
+ [330] +2[331] +2[332] + [333] + [411] + [420] +2[421] +2[422] + [430] +2[431] +2[432]
+ [433] + [441] + [442]

{4440000} = [000] + [200] + [220] + [222] + [400] + [420] + [422] + [440] + [442] + [444]

{4441000} = [100] + [210] + [221] + [222] + [300] + [320] + [322] + [410] + [421] + [422] + [430]
+ [432] + [441] + [442] + [443] + [444]

{4442000} = [000] +2[200] +2[220] +2[222] + [310] + [321] + [322] + [400] +2[420] +2[422] + [431]
+ [432] + [440] +2[442] + [443] + [444]

C-6 Branching Rules for the Reduction $U_8 \rightarrow O_8$.

{00000000} = [0000]

{10000000} = [1000]
{11000000} = [1100]
{11100000} = [1110]
{11110000} = [1111]

{20000000} = [0000] + [2000]
{21000000} = [1000] + [2100]
{21100000} = [1100] + [2110]
{21110000} = [1110] + [2111]
{21111000} = [1111] + [2110]
{21111100} = [1110] + [2100]
{21111110} = [1100] + [2000]
{22000000} = [0000] + [2000] + [2200]
{22100000} = [1000] + [2100] + [2210]
{22110000} = [1100] + [2110] + [2211]
{22111000} = [1110] + [2111] + [2210]
{22111100} = [1111] + [2110] + [2200]
{22200000} = [0000] + [2000] + [2200] + [2220]
{22210000} = [1000] + [2100] + [2210] + [2221]
{22211000} = [1100] + [2110] + [2211] + [2220]
{22220000} = [0000] + [2000] + [2200] + [2220] + [2222]

C-7 Branching Rules for the Reduction $U_9 \to R_9$.

{000000000} = [0000]

{100000000} = [1000]
{110000000} = [1100]
{111000000} = [1110]
{111100000} = [1111]

{200000000} = [0000] + [2000]
{210000000} = [1000] + [2100]
{211000000} = [1100] + [2110]
{211100000} = [1110] + [2111]
{211110000} = [1111] + [2111]
{211111000} = [1111] + [2110]
{211111100} = [1110] + [2100]
{211111110} = [1100] + [2000]
{220000000} = [0000] + [2000] + [2200]
{221000000} = [1000] + [2100] + [2210]
{221100000} = [1100] + [2110] + [2211]
{221110000} = [1110] + [2111] + [2211]
{221111000} = [1111] + [2111] + [2210]
{221111100} = [1111] + [2110] + [2200]
{222000000} = [0000] + [2000] + [2200] + [2220]
{222100000} = [1000] + [2100] + [2210] + [2221]
{222110000} = [1100] + [2110] + [2211] + [2221]
{222111000} = [1110] + [2111] + [2211] + [2220]
{222200000} = [0000] + [2000] + [2200] + [2220] + [2222]
{222210000} = [1000] + [2100] + [2210] + [2221] + [2222]

{300000000} = [1000] + [3000]
{310000000} = [1100] + [2000] + [3100]
{311000000} = [1110] + [2100] + [3110]
{311100000} = [1111] + [2110] + [3111]
{311110000} = [1111] + [2111] + [3111]
{311111000} = [1110] + [2111] + [3110]
{311111100} = [1100] + [2110] + [3100]
{311111110} = [1000] + [2100] + [3000]
{320000000} = [1000] + [2100] + [3000] + [3200]
{321000000} = [1100] + [2000] + [2110] + [2200] + [3100] + [3210]
{321100000} = [1110] + [2100] + [2111] + [2210] + [3110] + [3211]
{321110000} = [1111] + [2110] + [2111] + [2211] + [3111] + [3211]
{321111000} = [1111] + [2110] + [2111] + [2211] + [3111] + [3210]
{321111100} = [1110] + [2100] + [2111] + [2210] + [3110] + [3200]
{321111110} = [1100] + [2000] + [2110] + [2200] + [3100]
{322000000} = [1000] + [2100] + [2210] + [3000] + [3200] + [3220]
{322100000} = [1100] + [2000] + [2110] + [2200] + [2211] + [2220] + [3100]+ [3210] + [3221]
{322110000} = [1110] + [2100] + [2111] + [2210] + [2211] + [2221] + [3110] + [3211] + [3221]
{322111000} = [1111] + [2110] + [2111] + [2210] + [2211] + [2221] + [3111]+ [3211] + [3220]
{322111100} = [1111] + [2110] + [2111] + [2200] + [2211] + [2220] + [3111] + [3210]
{322111110} = [1110] + [2100] + [2111] + [2210] + [3110]
{322200000} = [1000] + [2100] + [2210] + [2221] + [3000] + [3200] + [3220] + [3222]
{322210000} = [1100] + [2000] + [2110] + [2200] + [2211] + [2220] + [2221] + [2222] + [3100]
+ [3210] + [3221] + [3222]
{322211000} = [1110] + [2100] + [2111] + [2210] + [2211] + [2220] + [2221] + [2222] + [3110]
+ [3211] + [3221]
{322211100} = [1111] + [2110] + [2111] + [2210] + [2211] + [2221] + [3111] + [3211]
{322211110} = [1111] + [2110] + [2111] + [2211] + [3111]
{330000000} = [1100] + [3100] + [3300]
{331000000} = [1110] + [2100] + [3110] + [3200] + [3310]
{331100000} = [1111] + [2110] + [2200] + [3111] + [3210] + [3311]
{331110000} = [1111] + [2111] + [2210] + [3111] + [3211] + [3311]
{331111000} = [1110] + [2111] + [2211] + [3110] + [3211] + [3310]
{331111100} = [1100] + [2110] + [2211] + [3100] + [3210] + [3300]
{332000000} = [1100] + [2110] + [3100] + [3210] + [3300] + [3320]
{332100000} = [1110] + [2100] + [2111] + [2210] + [3110] + [3200] + [3211] + [3220] + [3310]
+ [3321]
{332110000} = [1111] + [2110] + [2111] + [2200] + [2211] + [2220] + [3111] + [3210] + [3211]
+ [3221] + [3311] + [3321]
{332111000} = [1111] + [2110] + [2111] + [2210] + [2211] + [2221] + [3111] + [3210] + [3211]
+ [3221] + [3311] + [3320]
{332111100} = [1110] + [2100] + [2111] + [2210] + [2211] + [2221] + [3110] + [3200] + [3211]
+ [3220] + [3310]

{332200000} = [1100] + [2110] + [2211] + [3100] + [3210] + [3221] + [3300] + [3320] + [3322]

{332210000} = [1110] + [2100] + [2111] + [2210] + [2211] + [2221] + [3110] + [3200] + [3211]
+ [3220] + [3221] + [3222] + [3310] + [3321] + [3322]

{332211000} = [1111] + [2110] + [2111] + [2200] + [2210] + [2211] + [2220] + [2221] + [2222]
+ [3111] + [3210] + [3211] + [3220] + [3221] + [3222] + [3311] + [3321]

{332211100} = [1111] + [2110] + [2111] + [2200] + [2210] + [2211] + [2220] + [2221] + [2222]
+ [3111] + [3210] + [3211] + [3221] + [3311]

{333000000} = [1110] + [3110] + [3310] + [3330]

{333100000} = [1111] + [2110] + [3111] + [3210] + [3311] + [3320] + [3331]

{333110000} = [1111] + [2111] + [2210] + [3111] + [3211] + [3220] + [3311] + [3321] + [3331]

{333111000} = [1110] + [2111] + [2211] + [2220] + [3110] + [3211] + [3221] + [3310] + [3321]
+ [3330]

{333200000} = [1110] + [2111] + [3110] + [3211] + [3310] + [3321] + [3330] + [3332]

{333210000} = [1111] + [2110] + [2111] + [2211] + [3111] + [3210] + [3211] + [3221] + [3311]
+ [3320] + [3321] + [3322] + [3331] + [3332]

{333211000} = [1111] + [2110] + [2111] + [2210] + [2211] + [2221] + [3111] + [3210] + [3211]
+ [3220] + [3221] + [3222] + [3311] + [3320] + [3321] + [3322] + [3331]

{333300000} = [1111] + [3111] + [3311] + [3331] + [3333]

{333310000} = [1111] + [2111] + [3111] + [3211] + [3311] + [3321] + [3331] + [3332] + [3333]

{400000000} = [0000] + [2000] + [4000]

{410000000} = [1000] + [2100] + [3000] + [4100]

{411000000} = [1100] + [2110] + [3100] + [4110]

{411100000} = [1110] + [2111] + [3110] + [4111]

{411110000} = [1111] + [2111] + [3111] + [4111]

{411111000} = [1111] + [2110] + [3111] + [4110]

{411111100} = [1110] + [2100] + [3110] + [4100]

{411111110} = [1100] + [2000] + [3100] + [4000]

{420000000} = [0000] +2[2000] + [2200] + [3100] + [4000] + [4200]

{421000000} = [1000] +2[2100] + [2210] + [3000] + [3110] + [3200] + [4100] + [4210]

{421100000} = [1100] +2[2110] + [2211] + [3100] + [3111] + [3210] + [4110] + [4211]

{421110000} = [1110] +2[2111] + [2211] + [3110] + [3111] + [3211] + [4111] + [4211]

{421111000} = [1111] +2[2111] + [2210] + [3110] + [3111] + [3211] + [4111] + [4210]

{421111100} = [1111] +2[2110] + [2200] + [3100] + [3111] + [3210] + [4110] + [4200]

{421111110} = [1110] +2[2100] + [3000] + [3110] + [3200] + [4100]

{422000000} = [0000] +2[2000] +2[2200] + [2220] + [3100] + [3210] + [4000] + [4200] + [4220]

{422100000} = [1000] +2[2100] +2[2210] + [2221] + [3000] + [3110] + [3200] + [3211] + [3220]
+ [4100] + [4210] + [4221]

{422110000} = [1100] +2[2110] +2[2211] + [2221] + [3100] + [3111] + [3210]+ [3211] + [3221]
+ [4110] + [4211] + [4221]

{422111000} = [1110] +2[2111] +2[2211] + [2220] + [3110] + [3111] + [3210] + [3211] + [3221]
+ [4111] + [4211] + [4220]

{422111100} = [1111] +2[2111] +2[2210] + [3110] + [3111] + [3200] + [3211] + [3220] + [4111]
+ [4210]

{422111110} = [1111] +2[2110] + [2200] + [3100] + [3111] + [3210] + [4110]

{422200000} = [0000] +2[2000] +2[2200] +2[2220] + [2222] + [3100] + [3210] + [3221] + [4000]
+ [4200] + [4220] + [4222]

{422210000} = [1000] +2[2100] +2[2210] +2[2221] + [2222] + [3000] + [3110] + [3200] + [3211]
+ [3220] + [3221] + [3222] + [4100] + [4210] + [4221] + [4222]

{422211000} = [1100] +2[2110] +2[2211] +2[2221] + [3100] + [3111] + [3210] + [3211] + [3220]
+ [3221] + [3222] + [4110] + [4211] + [4221]

{422211100} = [1110] +2[2111] +2[2211] + [2220] + [3110] + [3111] + [3210] + [3211] + [3221]
+ [4111] + [4211]

{422211110} = [1111] +2[2111] + [2210] + [3110] + [3111] + [3211] + [4111]

{422220000} = [0000] +2[2000] +2[2200] +2[2220] +2[2222] + [3100] + [3210] + [3221] + [3222]
+ [4000] + [4200] + [4220] + [4222]

{422221000} = [1000] +2[2100] +2[2210] +2[2221] + [2222] + [3000] + [3110] + [3200] + [3211]
+ [3220] + [3221] + [3222] + [4100] + [4210] + [4221]

{422221100} = [1100] +2[2110] +2[2211] + [2221] + [3100] + [3111] + [3210] + [3211] + [3221]
+ [4110] + [4211]

{422221110} = [1110] +2[2111] + [2211] + [3110] + [3111] + [3211] + [4111]

{422222000} = [0000] +2[2000] +2[2200] +2[2220] + [2222] + [3100] + [3210]+ [3221] + [4000]
+ [4200] + [4220]

{422222100} = [1000] +2[2100] +2[2210] + [2221] + [3000] + [3110] + [3200] + [3211] + [3220]
+ [4100] + [4210]

{422222110} = [1100] +2[2110] + [2211] + [3100] + [3111] + [3210] + [4110]

{422222200} = [0000] +2[2000] +2[2200] + [2220] + [3100] + [3210] + [4000] + [4200]

{422222210} = [1000] +2[2100] + [2210] + [3000] + [3110] + [3200] + [4100]

Branching Rules for the Reduction $U_9 \rightarrow R_9$ (contd).

```
{422222220} = [0000] +2[2000] + [2200] + [3100] + [4000]
{430000000} = [1000] + [2100] + [3000] + [3200] + [4100] + [4300]
{431000000} = [1100] + [2000] + [2110] + [2200] +2[3100] + [3210] + [3300] + [4110] + [4200]
            + [4310]
{431100000} = [1110] + [2100] + [2111] + [2210] +2[3110] + [3200] + [3211] + [3310] + [4111]
            + [4210] + [4311]
{431110000} = [1111] + [2110] + [2111] + [2211] +2[3111] + [3210] + [3211] + [3311] + [4111]
            + [4211] + [4311]
{431111000} = [1111] + [2110] + [2111] + [2211] +2[3111] + [3210] + [3211] + [3311] + [4110]
            + [4211] + [4310]
{431111100} = [1110] + [2100] + [2111] + [2210] +2[3110] + [3200] + [3211] + [3310] + [4100]
            + [4210] + [4300]
{431111110} = [1100] + [2000] + [2110] + [2200] +2[3100] + [3210] + [3300] + [4000] + [4200]
{432000000} = [1000] +2[2100] + [2210] + [3000] + [3110] +2[3200] + [3220]+ [3310] + [4100]
            + [4210] + [4300] + [4320]
{432100000} = [1100] + [2000] +2[2110] +2[2200] + [2211] + [2220] +2[3100] + [3111] +3[3210]
            + [3221] + [3300] + [3311] + [3320] + [4110] + [4200] + [4211] + [4220] + [4310]
            + [4321]
{432110000} = [1110] + [2100] +2[2111] +2[2210] + [2211] + [2221] +2[3110] + [3111] + [3200]
            +3[3211] + [3220] + [3221] + [3310] + [3311] + [3321] + [4111] + [4210] + [4211]
            + [4221] + [4311] + [4321]
{432111000} = [1111] + [2110] +2[2111] + [2210] +2[2211] + [2221] + [3110] +2[3111] + [3210]
            +3[3211] + [3220] + [3221] + [3310] + [3311] + [3321] + [4111] + [4210] + [4211]
            + [4221] + [4311] + [4320]
{432111100} = [1111] +2[2110] + [2111] + [2200] +2[2211] + [2220] + [3100] +2[3111] +3[3210]
            + [3211] + [3221] + [3300] + [3311] + [3320] + [4110] + [4200] + [4211] + [4220]
            + [4310]
{432111110} = [1110] +2[2100] + [2111] +2[2210] + [3000] +2[3110] +2[3200] + [3211] + [3220]
            + [3310] + [4100] + [4210]
{432200000} = [1000] +2[2100] +2[2210] + [2221] + [3000] + [3110] +2[3200] + [3211] +2[3220]
            + [3222] + [3310] + [3321] + [4100] + [4210] + [4221] + [4300] + [4320] + [4322]
{432210000} = [1100] + [2000] +2[2110] +2[2200] +2[2211] +2[2220] + [2221] + [2222] +2[3100]
            + [3111] +3[3210] + [3211] +3[3221] + [3222] + [3300] + [3311] + [3320] + [3321]
            + [3322] + [4110] + [4200] + [4211] + [4220] + [4221] + [4222] + [4310]+ [4321]
            + [4322]
{432211000} = [1110] + [2100] +2[2111] +2[2210] +2[2211] + [2220] +2[2221] + [2222]+2[3110]
            + [3111] + [3200] + [3210] +3[3211] + [3220] +3[3221] + [3222] + [3310] + [3311]
            + [3320] + [3321] + [3322] + [4111] + [4210] + [4211] + [4220] + [4221] + [4222]
            + [4311] + [4321]
{432211100} = [1111] + [2110] +2[2111] +2[2210] +2[2211] +2[2221] + [3110] +2[3111] + [3200]
            + [3210] +3[3211] +2[3220] + [3221] + [3222] + [3310] + [3311] + [3321] + [4111]
            + [4210] + [4211] + [4221] + [4311]
{432211110} = [1111] +2[2110] + [2111] + [2200] +2[2211] + [2220] + [3100] +2[3111] +2[3210]
            + [3211] + [3221] + [3311] + [4110] + [4211]
{432220000} = [1000] +2[2100] +2[2210] +2[2221] + [2222] + [3000] + [3110] +2[3200] + [3211]
            +2[3220] + [3221] +2[3222] + [3310] + [3321] + [3322] + [4100] + [4210] + [4221]
            + [4222] + [4300] + [4320] + [4322]
{432221000} = [1100] + [2000] +2[2110] +2[2200] +2[2211] +2[2220] +2[2221] +2[2222] +2[3100]
            + [3111] +3[3210] + [3211] + [3220] +3[3221] +2[3222] + [3300] + [3311] + [3320]
            + [3321] + [3322] + [4110] + [4200] + [4211] + [4220] + [4221] + [4222] + [4310]
            + [4321]
{432221100} = [1110] + [2100] +2[2111] +2[2210] +2[2211] + [2220] +2[2221] + [2222] +2[3110]
            + [3111] + [3200] + [3210] +3[3211] + [3220] +2[3221] + [3222] + [3310] + [3311]
            + [3321] + [4111] + [4210] + [4211] + [4221] + [4311]
{432221110} = [1111] + [2110] +2[2111] + [2210] +2[2211] + [2221] + [3110] +2[3111] + [3210]
            +2[3211] + [3221] + [3311] + [4111] + [4211]
{432222000} = [1000] +2[2100] +2[2210] +2[2221] + [2222] + [3000] + [3110] +2[3200] + [3211]
            +2[3220] + [3221] + [3222] + [3310] + [3321] + [4100] + [4210] + [4221] + [4300]
            + [4320]
{432222100} = [1100] + [2000] +2[2110] +2[2200] +2[2211] +2[2220] + [2221] + [2222] +2[3100]
            + [3111] +3[3210] + [3211] +2[3221] + [3300] + [3311] + [3320] + [4110] + [4200]
            + [4211] + [4220] + [4310]
{432222110} = [1110] + [2100] +2[2111] +2[2210] + [2211] + [2221] +2[3110] + [3111] + [3200]
            +2[3211] + [3220] + [3310] + [4111] + [4210]
{432222200} = [1000] +2[2100] +2[2210] + [2221] + [3000] + [3110] +2[3200] + [3211] + [3220]
            + [3310] + [4100] + [4210] + [4300]
{432222210} = [1100] + [2000] +2[2110] +2[2200] + [2211] + [2220] +2[3100] + [3111] +2[3210]
            + [3300] + [4110] + [4200]
{433000000} = [1100] + [2110] + [3100] + [3210] + [3300] + [3320] + [4110] + [4310] + [4330]
{433100000} = [1110] + [2100] + [2111] + [2210] +2[3110] + [3200] + [3211] + [3220] +2[3310]
            + [3321] + [3330] + [4111] + [4210] + [4311] + [4320] + [4331]
```

Branching Rules for the Reduction $U_9 \rightarrow R_9$ (contd).

{433110000} = [1111] + [2110] + [2111] + [2200] + [2211] + [2220] +2[3111] +2[3210] + [3211]
+ [3221] +2[3311] + [3320] + [3321] + [3331] + [4111] + [4211] + [4220] + [4311]
+ [4321] + [4331]
{433111000} = [1111] + [2110] + [2111] + [2210] + [2211] + [2221] +2[3111] + [3210] +2[3211]
+ [3220] + [3221] +2[3311] + [3320] + [3321] + [3331] + [4110] + [4211] + [4221]
+ [4310] + [4321] + [4330]
{433111100} = [1110] + [2100] + [2111] + [2210] + [2211] + [2221] +2[3110] + [3200] +2[3211]
+ [3220] + [3221] +2[3310] + [3321] + [3330] + [4100] + [4210] + [4221] + [4300]
+ [4320]
{433111110} = [1100] + [2000] + [2110] + [2200] + [2211] + [2220] +2[3100] +2[3210] + [3221]
+ [3300] + [3320] + [4000] + [4200] + [4220]
{433200000} = [1100] +2[2110] + [2211] + [3100] + [3111] +2[3210] + [3221] + [3300] + [3311]
+2[3320] + [3322] + [3331] + [4110] + [4211] + [4310] + [4321] + [4330] + [4332]
{433210000} = [1110] + [2100] +2[2111] +2[2210] + [2211] + [2221] +2[3110] + [3111] + [3200]
+3[3211] +2[3220] + [3221] + [3222] +2[3310] + [3311] +3[3321] + [3322] + [3330]
+ [3331] + [3332] + [4111] + [4210] + [4211] + [4221] + [4311] + [4320] + [4321]
+ [4322] + [4331] + [4332]
{433211000} = [1111] + [2110] +2[2111] + [2200] + [2210] +2[2211] +2[2220] + [2221] + [2222]
+ [3110] +2[3111] +2[3210] +3[3211] + [3220] +3[3221] + [3222] + [3310] +2[3311]
+ [3320] +3[3321] + [3322] + [3330] + [3331] + [3332] + [4111] + [4210] + [4211]
+ [4220] + [4221] + [4222] + [4311] + [4320] + [4321] + [4322] + [4331]
{433211100} = [1111] +2[2110] + [2111] + [2200] + [2210] +2[2211] + [2220] +2[2221] + [2222]
+ [3100] +2[3111] +3[3210] +2[3211] + [3220] +3[3221] + [3222] + [3300] +2[3311]
+2[3320] + [3321] + [3322] + [3331] + [4110] + [4200] + [4211] + [4220] + [4221]
+ [4222] + [4310] + [4321]
{433211110} = [1110] +2[2100] + [2111] +2[2210] + [2211] +2[2221] + [3000]+2[3110] +2[3200]
+2[3211] +2[3220] + [3221] + [3222] + [3310] + [3321] + [4100] + [4210] + [4221]
{433220000} = [1100] +2[2110] +2[2211] + [2221] + [3100] + [3111] +2[3210] + [3211] +2[3221]
+ [3222] + [3300] + [3311] +2[3320] + [3321] +2[3322] + [3331] + [3332] + [4110]
+ [4211] + [4221] + [4310] + [4321] + [4322] + [4330] + [4332]
{433221000} = [1110] + [2100] +2[2111] +2[2210] +2[2211] + [2220] +2[2221] + [2222] +2[3110]
+ [3111] + [3200] + [3210] +3[3211] +2[3220] +3[3221] +2[3222] +2[3310] + [3311]
+ [3320] +3[3321] +2[3322] + [3330] + [3331] + [3332] + [4111] + [4210] + [4211]
+ [4220] + [4221] + [4222] + [4311] + [4320] + [4321] + [4322] + [4331]
{433221100} = [1111] + [2110] +2[2111] + [2200] +2[2210] +2[2211] +2[2220] +2[2221] +2[2222]
+ [3110] +2[3111] + [3200] +2[3210] +3[3211] +2[3220] +3[3221] +2[3222] + [3310]
+2[3311] + [3320] +2[3321] + [3322] + [3331] + [4111] + [4210] + [4211] + [4220]
+ [4221] + [4222] + [4311] + [4321]
{433221110} = [1111] +2[2110] + [2111] + [2200] + [2210] +2[2211] + [2220] +2[2221] + [2222]
+ [3100] +2[3111] +2[3210] +2[3211] + [3220] +2[3221] + [3222] + [3311] + [3321]
+ [4110] + [4211] + [4221]
{433222000} = [1100] +2[2110] +2[2211] +2[2221] + [3100] + [3111] +2[3210] + [3211] + [3220]
+2[3221] + [3222] + [3300] + [3311] +2[3320] + [3321] + [3322] + [3331] + [4110]
+ [4211] + [4221] + [4310] + [4321] + [4330]
{433222100} = [1110] + [2100] +2[2111] +2[2210] +2[2211] + [2220] +2[2221] + [2222] +2[3110]
+ [3111] + [3200] + [3210] +3[3211] +2[3220] +2[3221] + [3222] +2[3310] + [3311]
+2[3321] + [3330] + [4111] + [4210] + [4211] + [4221] + [4311] + [4320]
{433222110} = [1111] + [2110] +2[2111] + [2200] + [2210] +2[2211] +2[2220] + [2221] + [2222]
+ [3110] +2[3111] +2[3210] +2[3211] +2[3221] + [3311] + [3320] + [4111] + [4211]
+ [4220]
{433300000} = [1110] + [2111] + [3110] + [3211] + [3310] + [3321] + [3330] + [3332] + [4111]
+ [4311] + [4331] + [4333]
{433310000} = [1111] + [2110] + [2111] + [2211] +2[3111] + [3210] + [3211] + [3221] +2[3311]
+ [3320] + [3321] + [3322] +2[3331] + [3332] + [3333] + [4111] + [4211] + [4311]
+ [4321] + [4331] + [4332] + [4333]
{433311000} = [1111] + [2110] + [2111] + [2210] + [2211] + [2221] +2[3111] + [3210] +2[3211]
+ [3220] + [3221] + [3222] +2[3311] + [3320] +2[3321] + [3322] +2[3331] + [3332]
+ [3333] + [4110] + [4211] + [4221] + [4310] + [4321] + [4322] + [4330] + [4332]
{433311100} = [1110] + [2100] + [2111] + [2210] + [2211] + [2220] + [2221] + [2222] +2[3110]
+ [3200] +2[3211] + [3220] +2[3221] + [3222] +2[3310] +2[3321] + [3322] + [3330]
+ [3332] + [4100] + [4210] + [4221] + [4222] + [4300] + [4320] + [4322]
{433311110} = [1100] + [2000] + [2110] + [2200] + [2211] + [2220] + [2221] + [2222] +2[3100]
+2[3210] +2[3221] + [3222] + [3300] + [3320] + [3322] + [4000]+ [4200] + [4220]
+ [4222]
{433320000} = [1110] +2[2111] + [2211] + [3110] + [3111] +2[3211] + [3221] + [3310] + [3311]
+2[3321] + [3322] + [3330] + [3331] +2[3332] + [3333] + [4111] + [4211] + [4311]
+ [4321] + [4331] + [4332] + [4333]
{433321000} = [1111] + [2110] +2[2111] + [2210] +2[2211] + [2221] + [3110] +2[3111] + [3210]
+3[3211] + [3220] +2[3221] + [3222] + [3310] +2[3311] + [3320] +3[3321] +2[3322]
+ [3330] +2[3331] +2[3332] + [3333] + [4111] + [4210] + [4211] + [4221] + [4311]
+ [4320] + [4321] + [4322] + [4331] + [4332]

Branching Rules for the Reduction $U_9 \rightarrow R_9$ (contd).

{433321100} = [1111] +2[2110] + [2111] + [2200] + [2210] +2[2211] + [2220] +2[2221] + [2222]
+ [3100] +2[3111] +3[3210] +2[3211] + [3220] +3[3221] +2[3222] + [3300] +2[3311]
+2[3320] +2[3321] +2[3322] + [3331] + [3332] + [4110] + [4200] + [4211] + [4220]
+ [4221] + [4222] + [4310] + [4321] + [4322]
{433321110} = [1110] +2[2100] + [2111] +2[2210] + [2211] + [2220] +2[2221] +2[2222] + [3000]
+2[3110] +2[3200] +2[3211] +2[3220] +2[3221] +2[3222] + [3310] + [3321] + [3322]
+ [4100] + [4210] + [4221] + [4222]
{440000000} = [0000] + [2000] + [2200] + [4000] + [4200] + [4400]
{441000000} = [1000] + [2100] + [2210] + [3000] + [3200] + [4100] + [4210] + [4300] + [4410]
{441100000} = [1100] + [2110] + [2211] + [3100] + [3210] + [3300] + [4110] + [4211] + [4310]
+ [4411]
{441110000} = [1110] + [2111] + [2211] + [3110] + [3211] + [3310] + [4111] + [4211] + [4311]
+ [4411]
{441111000} = [1111] + [2111] + [2210] + [3111] + [3211] + [3311] + [4111] + [4210] + [4311]
+ [4410]
{441111100} = [1111] + [2110] + [2200] + [3111] + [3210] + [3311] + [4110] + [4200] + [4310]
+ [4400]
{442000000} = [0000] +2[2000] +2[2200] + [2220] + [3100] + [3210] + [4000] +2[4200] + [4220]
+ [4310] + [4400] + [4420]
{442100000} = [1000] +2[2100] +2[2210] + [2221] + [3000] + [3110] +2[3200] + [3211] + [3220]
+ [3310] + [4100] +2[4210] + [4221] + [4300] + [4311] + [4320] + [4410] + [4421]
{442110000} = [1100] +2[2110] +2[2211] + [2221] + [3100] + [3111] +2[3210] + [3211] + [3221]
+ [3300] + [3311] + [3320] + [4110] +2[4211] + [4221] + [4310] + [4311] + [4321]
+ [4411] + [4421]
{442111000} = [1110] +2[2111] +2[2211] + [2220] + [3110] + [3111] + [3210] +2[3211] + [3221]
+ [3310] + [3311] + [3321] + [4111] +2[4211] + [4220] + [4310] + [4311] + [4321]
+ [4411] + [4420]
{442111100} = [1111] +2[2111] +2[2210] + [3110] + [3111] + [3200] +2[3211] + [3220] + [3310]
+ [3311] + [3321] + [4111] +2[4210] + [4300] + [4311] + [4320] + [4410]
{442200000} = [0000] +2[2000] +3[2200] +2[2220] + [2222] + [3100] +2[3210] + [3221] + [3311]
+ [4000] +2[4200] +2[4220] + [4222] + [4310] + [4321] + [4400] + [4420] + [4422]
{442210000} = [1000] +2[2100] +3[2210] +2[2221] + [2222] + [3000] + [3110] +2[3200] +2[3211]
+2[3220] + [3221] + [3222] + [3310] + [3311] + [3321] + [4100] +2[4210] +2[4221]
+ [4222] + [4300] + [4311] + [4320] + [4321] + [4322] + [4410] + [4421] + [4422]
{442211000} = [1100] +2[2110] +3[2211] +2[2221] + [3100] + [3111] +2[3210] +2[3211] + [3220]
+2[3221] + [3222] + [3300] + [3310] + [3311] + [3320] + [3321] + [3322] + [4110]
+2[4211] +2[4221] + [4310] + [4311] + [4320] + [4321] + [4322] + [4411] + [4421]
{442211100} = [1110] +2[2111] +3[2211] + [2220] + [3110] + [3111] +2[3210] +2[3211] +2[3221]
+ [3300] + [3310] + [3311] + [3320] + [3321] + [3322] + [4111] +2[4211] + [4220]
+ [4310] + [4311] + [4321] + [4411]
{442220000} = [0000] +2[2000] +3[2200] +3[2220] +2[2222] + [3100] +2[3210] +2[3221] + [3222]
+ [3311] + [3321] + [4000] +2[4200] +2[4220] +2[4222] + [4310] + [4321] + [4322]
+ [4400] + [4420] + [4422]
{442221000} = [1000] +2[2100] +3[2210] +3[2221] + [2222] + [3000] + [3110] +2[3200] +2[3211]
+2[3220] +2[3221] +2[3222] + [3310] + [3311] + [3320] + [3321] + [3322] + [4100]
+2[4210] +2[4221] + [4222] + [4300] + [4311] + [4320] + [4321] + [4322] + [4410]
+ [4421]
{442221100} = [1100] +2[2110] +3[2211] +2[2221] + [3100] + [3111] +2[3210] +2[3211] + [3220]
+2[3221] + [3222] + [3300] + [3310] + [3311] + [3320] + [3321] + [3322] + [4110]
+2[4211] + [4221] + [4310] + [4311] + [4321] + [4411]
{442222000} = [0000] +2[2000] +3[2200] +3[2220] +2[2222] + [3100] +2[3210] +2[3221] + [3222]
+ [3311] + [3321] + [4000] +2[4200] +2[4220] + [4222] + [4310] + [4321] + [4400]
+ [4420]
{442222100} = [1000] +2[2100] +3[2210] +2[2221] + [2222] + [3000] + [3110] +2[3200] +2[3211]
+2[3220] + [3221] + [3222] + [3310] + [3311] + [3321] + [4100] +2[4210] + [4221]
+ [4300] + [4311] + [4320] + [4410]
{442222200} = [0000] +2[2000] +3[2200] +2[2220] + [2222] + [3100] +2[3210] + [3221] + [3311]
+ [4000] +2[4200] + [4220] + [4310] + [4400]
{443000000} = [1000] + [2100] + [2210] + [3000] + [3200] + [3220] + [4100] + [4210] + [4300]
+ [4320] + [4410] + [4430]
{443100000} = [1100] + [2000] + [2110] + [2200] + [2211] + [2220] +2[3100] +2[3210] + [3221]
+ [3300] + [3320] + [4110] + [4200] + [4211] + [4220] +2[4310] + [4321] + [4330]
+ [4411] + [4420] + [4431]
{443110000} = [1110] + [2100] + [2111] + [2210] + [2211] + [2221] +2[3110] + [3200] +2[3211]
+ [3220] + [3221] +2[3310] + [3321] + [3330] + [4111] + [4210] + [4211] + [4221]
+2[4311] + [4320] + [4321] + [4331] + [4411] + [4421] + [4431]
{443111000} = [1111] + [2110] + [2111] + [2210] + [2211] + [2221] +2[3111] + [3210] +2[3211]
+ [3220] + [3221] +2[3311] + [3320] + [3321] + [3331] + [4111] + [4210] + [4211]
+ [4221] +2[4311] + [4320] + [4321] + [4331] + [4410] + [4421] + [4430]
{443111100} = [1111] + [2110] + [2111] + [2200] + [2211] + [2220] +2[3111] +2[3210] + [3211]
+ [3221] +2[3311] + [3320] + [3321] + [3331] + [4110] + [4200] + [4211] + [4220]
+2[4310] + [4321] + [4330] + [4400] + [4420]

C-7 Branching Rules for the Reduction $U_9 \to R_9$ (contd).

{443200000} = [1000] +2[2100] +2[2210] + [2221] + [3000] + [3110] +2[3200] + [3211] +2[3220]
+ [3222] + [3310] + [3321] + [4100] +2[4210] + [4221] + [4300] + [4311] +2[4320]
+ [4322] + [4331] + [4410] + [4421] + [4430] + [4432]

{443210000} = [1100] + [2000] +2[2110] +2[2200] +2[2211] +2[2220] + [2221] + [2222] +2[3100]
+ [3111] +4[3210] + [3211] +3[3221] + [3222] + [3300] +2[3311] +2[3320] + [3321]
+ [3322] + [3331] + [4110] + [4200] +2[4211] +2[4220] + [4221] + [4222] +2[4310]
+ [4311] +3[4321] + [4322] + [4330] + [4331] + [4332] + [4411] + [4420] + [4421]
+ [4422] + [4431] + [4432]

{443211000} = [1110] + [2100] +2[2111] +2[2210] +2[2211] + [2220] +2[2221] + [2222] +2[3110]
+ [3111] + [3200] + [3210] +4[3211] +2[3220] +3[3221] + [3222] +2[3310] +2[3311]
+ [3320] +3[3321] + [3322] + [3330] + [3331] + [3332] + [4111] + [4210] +2[4211]
+ [4220] +2[4221] + [4222] + [4310] +2[4311] + [4320] +3[4321] + [4322] + [4330]
+ [4331] + [4332] + [4411] + [4420] + [4421] + [4422] + [4431]

{443211100} = [1111] + [2110] +2[2111] +2[2210] +2[2211] +2[2221] + [3110] +2[3111] + [3200]
+ [3210] +4[3211] +2[3220] +2[3221] + [3222] +2[3310] +2[3311] + [3320] +3[3321]
+ [3322] + [3330] + [3331] + [3332] + [4111] +2[4210] + [4211]+2[4221] + [4300]
+2[4311] +2[4320] + [4321] + [4322] + [4331] + [4410] + [4421]

{443220000} = [1000] +2[2100] +3[2210] +2[2221] + [2222] + [3000] + [3110] +2[3200] +2[3211]
+3[3220] + [3221] +2[3222] + [3310] + [3311] +2[3321] + [3322] + [3331] + [4100]
+2[4210] +2[4221] + [4222] + [4300] + [4311] +2[4320] + [4321] +2[4322] + [4331]
+ [4332] + [4410] + [4421] + [4422] + [4430] + [4432]

{443221000} = [1100] + [2000] +2[2110] +2[2200] +3[2211] +3[2220] +2[2221] +2[2222] +2[3100]
+ [3111] +4[3210] +2[3211] + [3220] +5[3221] +2[3222] + [3300] + [3310] +2[3311]
+2[3320] +3[3321] +2[3322] + [3330] + [3331] + [3332] + [4110] + [4200] +2[4211]
+2[4220] +2[4221] +2[4222] +2[4310] + [4311] + [4320] +3[4321] +2[4322] + [4330]
+ [4331] + [4332] + [4411] + [4420] + [4421] + [4422] + [4431]

{443221100} = [1110] + [2100] +2[2111] +2[2210] +3[2211] + [2220] +3[2221] + [2222] +2[3110]
+ [3111] + [3200] +2[3210] +4[3211] +2[3220] +4[3221] +2[3222] + [3300] +2[3310]
+2[3311] +2[3320] +3[3321] +2[3322] + [3330] + [3331] + [3332] + [4111] + [4210]
+2[4211] + [4220] +2[4221] + [4222] + [4310] +2[4311] + [4320]+2[4321] + [4322]
+ [4331] + [4411] + [4421]

{443222000} = [1000] +2[2100] +3[2210] +3[2221] + [2222] + [3000] + [3110] +2[3200] +2[3211]
+3[3220] +2[3221] +2[3222] + [3310] + [3311] + [3320] +2[3321] + [3322] + [3331]
+ [4100] +2[4210] +2[4221] + [4222] + [4300] + [4311] +2[4320] + [4321] + [4322]
+ [4331] + [4410] + [4421] + [4430]

{443222100} = [1100] + [2000] +2[2110] +2[2200] +3[2211] +3[2220] +2[2221] +2[2222] +2[3100]
+ [3111] +4[3210] +2[3211] + [3220] +4[3221] +2[3222] + [3300] + [3310] +2[3311]
+2[3320] +2[3321] + [3322] + [3331] + [4110] + [4200] +2[4211] +2[4220] + [4221]
+ [4222] +2[4310] + [4311] +2[4321] + [4330] + [4411] + [4420]

{443300000} = [1100] + [2110] + [2211] + [3100] + [3210] + [3221] + [3300] + [3320] + [3322]
+ [4110] + [4211] + [4310] + [4321] + [4330] + [4332] + [4411] + [4431] + [4433]

{443310000} = [1110] + [2100] + [2111] + [2210] + [2211] + [2221] +2[3110] + [3200] +2[3211]
+ [3220] + [3221] + [3222] +2[3310] +2[3321] + [3322] + [3330] + [3332] + [4111]
+ [4210] + [4211] + [4221] +2[4311] + [4320] + [4321] + [4322] +2[4331] + [4332]
+ [4333] + [4411] + [4421] + [4431] + [4432] + [4433]

{443311000} = [1111] + [2110] + [2111] + [2200] + [2210] + [2211] + [2220] + [2221] + [2222]
+2[3111] +2[3210] +2[3211] + [3220] +2[3221] + [3222] +3[3311] + [3320] +2[3321]
+ [3322] +2[3331] + [3332] + [3333] + [4111] + [4210] + [4211] + [4220] + [4221]
+ [4222] +2[4311] + [4320] +2[4321] + [4322] +2[4331] + [4332] + [4333] + [4410]
+ [4421] + [4422] + [4430] + [4432]

{443311100} = [1111] + [2110] + [2111] + [2200] + [2210] + [2211] + [2220] + [2221] + [2222]
+2[3111] +2[3210] +2[3211] + [3220] +2[3221] + [3222] +3[3311] + [3320] +2[3321]
+ [3322] +2[3331] + [3332] + [3333] + [4110] + [4200] + [4211] + [4220] + [4221]
+ [4222] +2[4310] +2[4321] + [4322] + [4330] + [4332] + [4400] + [4420] + [4422]

{443320000} = [1100] +2[2110] +2[2211] + [2221] + [3100] + [3111] +2[3210] + [3211] +2[3221]
+ [3222] + [3300] + [3311] +2[3320] + [3321] +2[3322] + [3331] + [3332] + [4110]
+2[4211] + [4221] + [4310] + [4311] +2[4321] + [4322] + [4330] + [4331] +2[4332]
+ [4333] + [4411] + [4421] + [4431] + [4432] + [4433]

{443321000} = [1110] + [2100] +2[2111] +2[2210] +2[2211] + [2220] +2[2221] + [2222] +2[3110]
+ [3111] + [3200] + [3210] +4[3211] +2[3220] +3[3221] +2[3222] +2[3310] +2[3311]
+ [3320] +4[3321] +2[3322] + [3330] +2[3331] +2[3332] + [3333] + [4111] + [4210]
+2[4211] + [4220] +2[4221] + [4222] + [4310] +2[4311] + [4320] +3[4321] +2[4322]
+ [4330] +2[4331] +2[4332] + [4333] + [4411] + [4420] + [4421] + [4422] + [4431]
+ [4432]

{443321100} = [1111] + [2110] +2[2111] + [2200] +2[2210] +2[2211] +2[2220] +2[2221] +2[2222]
+ [3110] +2[3111] + [3200] +2[3210] +4[3211] +2[3220] +4[3221] +2[3222] +2[3310]
+3[3311] + [3320] +4[3321] +2[3322] + [3330] +2[3331] +2[3332] + [3333] + [4111]
+2[4210] + [4211] + [4220] +2[4221] +2[4222] + [4300] +2[4311] +2[4320] +2[4321]
+2[4322] + [4331] + [4332] + [4410] + [4421] + [4422]

{444000000} = [0000] + [2000] + [2200] + [2220] + [4000] + [4200] + [4220] + [4400] + [4420]
+ [4440]

C-7 Branching Rules for the Reduction $U_9 \to R_9$ (contd).

{444100000} = [1000] + [2100] + [2210] + [2221] + [3000] + [3200] + [3220] + [4100] + [4210]
+ [4221] + [4300] + [4320] + [4410] + [4421] + [4430] + [4441]

{444110000} = [1100] + [2110] + [2211] + [2221] + [3100] + [3210] + [3221]+ [3300] + [3320]
+ [4110] + [4211] + [4221] + [4310] + [4321] + [4330] + [4411] + [4421] + [4431]
+ [4441]

{444111000} = [1110] + [2111] + [2211] + [2220] + [3110] + [3211] + [3221] + [3310] + [3321]
+ [3330] + [4111] + [4211] + [4220] + [4311] + [4321] + [4331] + [4411] + [4420]
+ [4431] + [4440]

{444200000} = [0000] +2[2000] +2[2200] +2[2220] + [2222] + [3100] + [3210] + [3221] + [4000]
+2[4200] +2[4220] + [4222] + [4310] + [4321] + [4400] +2[4420] + [4422] + [4431]
+ [4440] + [4442]

{444210000} = [1000] +2[2100] +2[2210] +2[2221] + [2222] + [3000] + [3110] +2[3200] + [3211]
+2[3220] + [3221] + [3222] + [3310] + [3321] + [4100] +2[4210] +2[4221] + [4222]
+ [4300] + [4311] +2[4320] + [4321] + [4322] + [4331] + [4410] +2[4421] + [4422]
+ [4430] + [4431] + [4432] + [4441] + [4442]

{444211000} = [1100] +2[2110] +2[2211] +2[2221] + [3100] + [3111] +2[3210] + [3211] + [3220]
+2[3221] + [3222] + [3300] + [3311] +2[3320] + [3321] + [3322] + [3331] + [4110]
+2[4211] +2[4221] + [4310] + [4311] + [4320] +2[4321] + [4322] + [4330] + [4331]
+ [4332] + [4411] +2[4421] + [4430] + [4431] + [4432] + [4441]

{444220000} = [0000] +2[2000] +3[2200] +3[2220] +2[2222] + [3100] +2[3210] +2[3221] + [3222]
+ [3311] + [3321] + [4000] +2[4200] +3[4220] +2[4222] + [4310] +2[4321] + [4322]
+ [4331] + [4400] +2[4420] +2[4422] + [4431] + [4432] + [4440] + [4442]

{444221000} = [1000] +2[2100] +3[2210] +3[2221] + [2222] + [3000] + [3110] +2[3200] +2[3211]
+3[3220] +2[3221] +2[3222] + [3310] + [3311] + [3320] +2[3321] + [3322] + [3331]
+ [4100] +2[4210] +3[4221] + [4222] + [4300] + [4311] +2[4320] +2[4321] +2[4322]
+ [4330] + [4331] + [4332] + [4410] +2[4421] + [4422] + [4430] + [4431] + [4432]
+ [4441]

{444222000} = [0000] +2[2000] +3[2200] +4[2220] +2[2222] + [3100] +2[3210] +3[3221] + [3222]
+ [3311] +2[3321] + [3330] + [4000] +2[4200] +3[4220] +2[4222] + [4310] +2[4321]
+ [4322] + [4331] + [4400] +2[4420] + [4422] + [4431] + [4440]

{444300000} = [1000] + [2100] + [2210] + [2221] + [3000] + [3200] + [3220] + [3222] + [4100]
+ [4210] + [4221] + [4300] + [4320] + [4322] + [4410] + [4421] + [4430] + [4432]
+ [4441] + [4443]

{444310000} = [1100] + [2000] + [2110] + [2200] + [2211] + [2220] + [2221] + [2222] +2[3100]
+2[3210] +2[3221] + [3222] + [3300] + [3320] + [3322] + [4110] + [4200] + [4211]
+ [4220] + [4221] + [4222] +2[4310] +2[4321] + [4322] + [4330] + [4332] + [4411]
+ [4420] + [4421] + [4422] +2[4431] + [4432] + [4433] + [4441] + [4442] + [4443]

{444311000} = [1110] + [2100] + [2111] + [2210] + [2211] + [2220] + [2221] + [2222] +2[3110]
+ [3200] +2[3211] + [3220] +2[3221] + [3222] +2[3310] +2[3321] + [3322] + [3330]
+ [3332] + [4111] + [4210] + [4211] + [4220] + [4221] + [4222] +2[4311] + [4320]
+2[4321] + [4322] +2[4331] + [4332] + [4333] + [4411] + [4420] + [4421] + [4422]
+2[4431] + [4432] + [4433] + [4440] + [4442]

{444320000} = [1000] +2[2100] +2[2210] +2[2221] + [2222] + [3000] + [3110] +2[3200] + [3211]
+2[3220] + [3221] +2[3222] + [3310] + [3321] + [3322] + [4100] +2[4210] +2[4221]
+ [4222] + [4300] + [4311] +2[4320] + [4321] +2[4322] + [4331] + [4332] + [4410]
+2[4421] + [4422] + [4430] + [4431] +2[4432] + [4433] + [4441] + [4442] + [4443]

{444321000} = [1100] + [2000] +2[2110] +2[2200] +2[2211] +2[2220] +2[2221] +2[2222] +2[3100]
+ [3111] +4[3210] + [3211] + [3220] +4[3221] +2[3222] + [3300]+2[3311] +2[3320]
+2[3321] +2[3322] + [3331] + [3332] + [4110] + [4200] +2[4211] +2[4220] +2[4221]
+2[4222] +2[4310] + [4311] + [4320] +4[4321] +2[4322] + [4330] +2[4331] +2[4332]
+ [4333] + [4411] + [4420] +2[4421] +2[4422] + [4430] +2[4431] +2[4432] + [4433]
+ [4441] + [4442]

{444400000} = [0000] + [2000] + [2200] + [2220] + [2222] + [4000] + [4200] + [4220] + [4222]
+ [4400] + [4420] + [4422] + [4440] + [4442] + [4444]

{444410000} = [1000] + [2100] + [2210] + [2221] + [2222] + [3000] + [3200] + [3220] + [3222]
+ [4100] + [4210] + [4221] + [4222] + [4300] + [4320] + [4322] + [4410] + [4421]
+ [4422] + [4430] + [4432] + [4441] + [4442] + [4443] + [4444]

{444420000} = [0000] +2[2000] +2[2200] +2[2220] +2[2222] + [3100] + [3210]+ [3221] + [3222]
+ [4000] +2[4200] +2[4220] +2[4222] + [4310] + [4321] + [4322] + [4400] +2[4420]
+2[4422] + [4431] + [4432] + [4440] +2[4442] + [4443] + [4444]

C-8 Branching Rules for the Reduction $U_{11} \to R_{11}$.

{00000000000} = [00000]

{10000000000} = [10000]
{11000000000} = [11000]
{11100000000} = [11100]
{11110000000} = [11110]
{11111000000} = [11111]

{20000000000} = [00000] + [20000]
{21000000000} = [10000] + [21000]
{21100000000} = [11000] + [21100]
{21110000000} = [11100] + [21110]
{21111000000} = [11110] + [21111]
{21111100000} = [11111] + [21111]
{21111110000} = [11111] + [21110]
{21111111000} = [11110] + [21100]
{21111111100} = [11100] + [21000]
{21111111110} = [11000] + [20000]
{22000000000} = [00000] + [20000] + [22000]
{22100000000} = [10000] + [21000] + [22100]
{22110000000} = [11000] + [21100] + [22110]
{22111000000} = [11100] + [21110] + [22111]
{22111100000} = [11110] + [21111] + [22111]
{22111110000} = [11111] + [21111] + [22110]
{22111111000} = [11111] + [21110] + [22100]
{22111111100} = [11110] + [21100] + [22000]
{22200000000} = [00000] + [20000] + [22000] + [22200]
{22210000000} = [10000] + [21000] + [22100] + [22210]
{22211000000} = [11000] + [21100] + [22110] + [22211]
{22211100000} = [11100] + [21110] + [22111] + [22211]
{22211110000} = [11110] + [21111] + [22111] + [22210]
{22211111000} = [11111] + [21111] + [22110] + [22200]
{22220000000} = [00000] + [20000] + [22000] + [22200] + [22220]
{22221000000} = [10000] + [21000] + [22100] + [22210] + [22221]
{22221100000} = [11000] + [21100] + [22110] + [22211] + [22221]
{22221110000} = [11100] + [21110] + [22111] + [22211] + [22220]
{22222000000} = [00000] + [20000] + [22000] + [22200] + [22220] + [22222]
{22222100000} = [10000] + [21000] + [22100] + [22210] + [22221] + [22222]

C-9 Branching Rules for the Reduction $U_{13} \to R_{13}$.

{0000000000000} = [000000]

{1000000000000} = [100000]
{1100000000000} = [110000]
{1110000000000} = [111000]
{1111000000000} = [111100]
{1111100000000} = [111110]
{1111110000000} = [111111]

{2000000000000} = [000000] + [200000]
{2100000000000} = [100000] + [210000]
{2110000000000} = [110000] + [211000]
{2111000000000} = [111000] + [211100]
{2111100000000} = [111100] + [211110]
{2111110000000} = [111110] + [211111]
{2111111000000} = [111111] + [211111]
{2111111100000} = [111111] + [211110]
{2111111110000} = [111110] + [211100]
{2111111111000} = [111100] + [211000]
{2111111111100} = [111000] + [210000]
{2111111111110} = [110000] + [200000]
{2200000000000} = [000000] + [200000] + [220000]
{2210000000000} = [100000] + [210000] + [221000]
{2211000000000} = [110000] + [211000] + [221100]
{2211100000000} = [111000] + [211100] + [221110]
{2211110000000} = [111100] + [211110] + [221111]
{2211111000000} = [111110] + [211111] + [221111]
{2211111100000} = [111111] + [211111] + [221110]
{2211111110000} = [111111] + [211110] + [221100]
{2211111111000} = [111110] + [211100] + [221000]
{2211111111100} = [111100] + [211000] + [220000]
{2220000000000} = [000000] + [200000] + [220000] + [222000]
{2221000000000} = [100000] + [210000] + [221000] + [222100]
{2221100000000} = [110000] + [211000] + [221100] + [222110]
{2221110000000} = [111000] + [211100] + [221110] + [222111]
{2221111000000} = [111100] + [211110] + [221111] + [222111]
{2221111100000} = [111110] + [211111] + [221111] + [222110]
{2221111110000} = [111111] + [211111] + [221110] + [222100]
{2221111111000} = [111111] + [211110] + [221100] + [222000]
{2222000000000} = [000000] + [200000] + [220000] + [222000] + [222200]
{2222100000000} = [100000] + [210000] + [221000] + [222100] + [222210]
{2222110000000} = [110000] + [211000] + [221100] + [222110] + [222211]
{2222111000000} = [111000] + [211100] + [221110] + [222111] + [222211]
{2222111100000} = [111100] + [211110] + [221111] + [222111] + [222210]
{2222111110000} = [111110] + [211111] + [221111] + [222110] + [222200]
{2222200000000} = [000000] + [200000] + [220000] + [222000] + [222200] + [222220]
{2222210000000} = [100000] + [210000] + [221000] + [222100] + [222210] + [222221]
{2222211000000} = [110000] + [211000] + [221100] + [222110] + [222211] + [222221]
{2222211100000} = [111000] + [211100] + [221110] + [222111] + [222211] + [222220]
{2222220000000} = [000000] + [200000] + [220000] + [222000] + [222200] + [222220] + [222222]
{2222221000000} = [100000] + [210000] + [221000] + [222100] + [222210] + [222221] + [222222]

C-10 Branching Rules for the Reduction $U_4 \rightarrow Sp_4$.

$\{0000\} = \langle 00\rangle$

$\{1000\} = \langle 10\rangle$
$\{1100\} = \langle 00\rangle + \langle 11\rangle$

$\{2000\} = \langle 20\rangle$
$\{2100\} = \langle 10\rangle + \langle 21\rangle$
$\{2110\} = \langle 11\rangle + \langle 20\rangle$
$\{2200\} = \langle 00\rangle + \langle 11\rangle + \langle 22\rangle$

$\{3000\} = \langle 30\rangle$
$\{3100\} = \langle 20\rangle + \langle 31\rangle$
$\{3110\} = \langle 21\rangle + \langle 30\rangle$
$\{3200\} = \langle 10\rangle + \langle 21\rangle + \langle 32\rangle$
$\{3210\} = \langle 11\rangle + \langle 20\rangle + \langle 22\rangle + \langle 31\rangle$
$\{3300\} = \langle 00\rangle + \langle 11\rangle + \langle 22\rangle + \langle 33\rangle$

$\{4000\} = \langle 40\rangle$
$\{4100\} = \langle 30\rangle + \langle 41\rangle$
$\{4110\} = \langle 31\rangle + \langle 40\rangle$
$\{4200\} = \langle 20\rangle + \langle 31\rangle + \langle 42\rangle$
$\{4210\} = \langle 21\rangle + \langle 30\rangle + \langle 32\rangle + \langle 41\rangle$
$\{4220\} = \langle 22\rangle + \langle 31\rangle + \langle 40\rangle$
$\{4300\} = \langle 10\rangle + \langle 21\rangle + \langle 32\rangle + \langle 43\rangle$
$\{4310\} = \langle 11\rangle + \langle 20\rangle + \langle 22\rangle + \langle 31\rangle + \langle 33\rangle + \langle 42\rangle$
$\{4400\} = \langle 00\rangle + \langle 11\rangle + \langle 22\rangle + \langle 33\rangle + \langle 44\rangle$

C-11 Branching Rules for the Reduction $U_6 \rightarrow Sp_6$.

{000000} = <000>

{100000} = <100>
{110000} = <000> + <110>
{111000} = <100> + <111>

{200000} = <200>
{210000} = <100> + <210>
{211000} = <110> + <200> + <211>
{211100} = <100> + <111> + <210>
{211110} = <110> + <200>
{220000} = <000> + <110> + <220>
{221000} = <100> + <111> + <210> + <221>
{221100} = <000> +2<110> + <211> + <220>
{222000} = <200> + <211> + <222>

{300000} = <300>
{310000} = <200> + <310>
{311000} = <210> + <300> + <311>
{311100} = <200> + <211> + <310>
{311110} = <210> + <300>
{320000} = <100> + <210> + <320>
{321000} = <110> + <200> + <211> + <220> + <310> + <321>
{321100} = <100> + <111> +2<210> + <221> + <311> + <320>
{321110} = <110> + <200> + <211> + <220> + <310>
{322000} = <210> + <221> + <300> + <311> + <322>
{322100} = <110> + <200> +2<211> + <220> + <222> + <310> + <321>
{322110} = <111> +2<210> + <221> + <300> + <311>
{330000} = <000> + <110> + <220> + <330>
{331000} = <100> + <111> + <210> + <221> + <320> + <331>
{331100} = <000> +2<110> + <211> +2<220> + <321> + <330>
{332000} = <200> + <211> + <222> + <310> + <321> + <332>
{332100} = <100> + <111> +2<210> +2<221> + <311> + <320> + <322> + <331>
{333000} = <300> + <311> + <322> + <333>

{400000} = <400>
{410000} = <300> + <410>
{411000} = <310> + <400> + <411>
{411100} = <300> + <311> + <410>
{411110} = <310> + <400>
{420000} = <200> + <310> + <420>
{421000} = <210> + <300> + <311> + <320> + <410> + <421>
{421100} = <200> + <211> +2<310> + <321> + <411> + <420>
{421110} = <210> + <300> + <311> + <320> + <410>
{422000} = <220> + <310> + <321> + <400> + <411> + <422>
{422100} = <210> + <221> + <300> +2<311> + <320> + <322> + <410> + <421>
{422110} = <211> + <220> +2<310> + <321> + <400> + <411>
{422200} = <200> + <211> + <222> + <310> + <321> + <420>
{422210} = <210> + <221> + <300> + <311> + <320> + <410>
{422220} = <220> + <310> + <400>
{430000} = <100> + <210> + <320> + <430>
{431000} = <110> + <200> + <211> + <220> + <310> + <321> + <330> + <420> + <431>
{431100} = <100> + <111> +2<210> + <221> + <311> +2<320> + <331> + <421> + <430>
{431110} = <110> + <200> + <211> + <220> + <310> + <321> + <330> + <420>
{432000} = <210> + <221> + <300> + <311> + <320> + <322> + <331> + <410> + <421> + <432>
{432100} = <110> + <200> +2<211> +2<220> + <222> +2<310> +3<321> + <330> + <332> + <411> + <420>
+ <422> + <431>
{432110} = <111> +2<210> +2<221> + <300> +2<311> +2<320> + <322> + <331> + <410> + <421>
{432200} = <100> + <111> +2<210> +2<221> + <311> +2<320> + <322> + <331> + <421> + <430>
{432210} = <110> + <200> +2<211> +2<220> + <222> +2<310> +2<321> + <330> + <411> + <420>
{433000} = <310> + <321> + <332> + <400> + <411> + <422> + <433>
{433100} = <210> + <221> + <300> +2<311> + <320> +2<322> + <331> + <333> + <410> + <421> + <432>
{433110} = <211> + <220> + <222> +2<310> +2<321> + <332> + <400> + <411> + <422>
{440000} = <000> + <110> + <220> + <330> + <440>
{441000} = <100> + <111> + <210> + <221> + <320> + <331> + <430> + <441>
{441100} = <000> +2<110> + <211> +2<220> + <321> +2<330> + <431> + <440>
{442000} = <200> + <211> + <222> + <310> + <321> + <332> + <420> + <431> + <442>
{442100} = <100> + <111> +2<210> +2<221> + <311> +2<320> + <322> +2<331> + <421> + <430> + <432>
+ <441>
{442200} = <000> +2<110> + <211> +3<220> +2<321> +2<330> + <422> + <431> + <440>
{443000} = <300> + <311> + <322> + <333> + <410> + <421> + <432> + <443>
{443100} = <200> + <211> + <222> +2<310> +2<321> +2<332> + <411> + <420> + <422> + <431> + <433>
+ <442>
{444000} = <400> + <411> + <422> + <433> + <444>

C-12 Branching Rules for the Reduction $U_8 \to Sp_8$.

{00000000} = <0000>

{10000000} = <1000>
{11000000} = <0000> + <1100>
{11100000} = <1000> + <1110>
{11110000} = <0000> + <1100> + <1111>

{20000000} = <2000>
{21000000} = <1000> + <2100>
{21100000} = <1100> + <2000> + <2110>
{21110000} = <1000> + <1110> + <2100> + <2111>
{21111000} = <1100> + <1111> + <2000> + <2110>
{21111100} = <1000> + <1110> + <2100>
{21111110} = <1100> + <2000>
{22000000} = <0000> + <1100> + <2200>
{22100000} = <1000> + <1110> + <2100> + <2210>
{22110000} = <0000> +2<1100> + <1111> + <2110> + <2200> + <2211>
{22111000} = <1000> +2<1110> + <2100> + <2111> + <2210>
{22111100} = <0000> +2<1100> + <1111> + <2110> + <2200>
{22200000} = <2000> + <2110> + <2220>
{22210000} = <1000> + <1110> + <2100> + <2111> + <2210> + <2221>
{22211000} = <1100> + <1111> + <2000> +2<2110> + <2211> + <2220>
{22220000} = <0000> + <1100> + <1111> + <2200> + <2211> + <2222>

C-13 Branching Rules for the Reduction $U_{10} \to Sp_{10}$.

{0000000000} = <00000>

{1000000000} = <10000>
{1100000000} = <00000> + <11000>
{1110000000} = <10000> + <11100>
{1111000000} = <00000> + <11000> + <11110>
{1111100000} = <10000> + <11100> + <11111>

{2000000000} = <20000>
{2100000000} = <10000> + <21000>
{2110000000} = <11000> + <20000> + <21100>
{2111000000} = <10000> + <11100> + <21000> + <21110>
{2111100000} = <11000> + <11110> + <20000> + <21100> + <21111>
{2111110000} = <10000> + <11100> + <11111> + <21000> + <21110>
{2111111000} = <11000> + <11110> + <20000> + <21100>
{2111111100} = <10000> + <11100> + <21000>
{2111111110} = <11000> + <20000>
{2200000000} = <00000> + <11000> + <22000>
{2210000000} = <10000> + <11100> + <21000> + <22100>
{2211000000} = <00000> +2<11000> + <11110> + <21100> + <22000> + <22110>
{2211100000} = <10000> +2<11100> + <11111> + <21000> + <21110> + <22100> + <22111>
{2211110000} = <00000> +2<11000> +2<11110> + <21100> + <21111> + <22000> + <22110>
{2211111000} = <10000> +2<11100> + <11111> + <21000> + <21110> + <22100>
{2211111100} = <00000> +2<11000> + <11110> + <21100> + <22000>
{2220000000} = <20000> + <21100> + <22200>
{2221000000} = <10000> + <11100> + <21000> + <21110> + <22100> + <22210>
{2221100000} = <11000> + <11110> + <20000> +2<21100> + <21111> + <22110> + <22200> + <22211>
{2221110000} = <10000> +2<11100> + <11111> + <21000> +2<21110> + <22100> + <22111> + <22210>
{2221111000} = <11000> +2<11110> + <20000> +2<21100> + <21111> + <22110> + <22200>
{2222000000} = <00000> + <11000> + <11110> + <22000> + <22110> + <22220>
{2222100000} = <10000> + <11100> + <11111> + <21000> + <21110> + <22100> + <22111> + <22210> + <22221>
{2222110000} = <00000> +2<11000> +2<11110> + <21100> + <21111> + <22000> +2<22110> + <22211> + <22220>
{2222200000} = <20000> + <21100> + <21111> + <22200> + <22211> + <22222>

Branching Rules for the Reduction $U_{12} \to Sp_{12}$.

```
{000000000000} = <000000>

{100000000000} = <100000>
{110000000000} = <000000> + <110000>
{111000000000} = <100000> + <111000>
{111100000000} = <000000> + <110000> + <111100>
{111110000000} = <100000> + <111000> + <111110>
{111111000000} = <000000> + <110000> + <111100> + <111111>

{200000000000} = <200000>
{210000000000} = <100000> + <210000>
{211000000000} = <110000> + <200000> + <211000>
{211100000000} = <100000> + <111000> + <210000> + <211100>
{211110000000} = <110000> + <111100> + <200000> + <211000> + <211110>
{211111000000} = <100000> + <111000> + <111110> + <210000> + <211100> + <211111>
{211111100000} = <110000> + <111100> + <111111> + <200000> + <211000> + <211110>
{211111110000} = <100000> + <111000> + <111110> + <210000> + <211100>
{211111111000} = <110000> + <111100> + <200000> + <211000>
{211111111100} = <100000> + <111000> + <210000>
{211111111110} = <110000> + <200000>
{220000000000} = <000000> + <110000> + <220000>
{221000000000} = <100000> + <111000> + <210000> + <221000>
{221100000000} = <000000> +2<110000> + <111100> + <211000> + <220000> + <221100>
{221110000000} = <100000> +2<111000> + <111110> + <210000> + <211100> + <221000> + <221110>
{221111000000} = <000000> +2<110000> +2<111100> + <111111> + <211000> + <211110> + <220000>
               + <221100> + <221111>
{221111100000} = <100000> +2<111000> +2<111110> + <210000> + <211100> + <211111> + <221000>
               + <221110>
{221111110000} = <000000> +2<110000> +2<111100> + <111111> + <211000> + <211110> + <220000>
               + <221100>
{221111111000} = <100000> +2<111000> + <111110> + <210000> + <211100> + <221000>
{221111111100} = <000000> +2<110000> + <111100> + <211000> + <220000>
{222000000000} = <200000> + <211000> + <222000>
{222100000000} = <100000> + <111000> + <210000> + <211100> + <221000> + <222100>
{222110000000} = <110000> + <111100> + <200000> +2<211000> + <211110> + <221100> + <222000>
               + <222110>
{222111000000} = <100000> +2<111000> + <111110> + <210000> +2<211100> + <211111> + <221000>
               + <221110> + <222100> + <222111>
{222111100000} = <110000> +2<111100> + <111111> + <200000> +2<211000> +2<211110> + <221100>
               + <221111> + <222000> + <222110>
{222111110000} = <100000> +2<111000> +2<111110> + <210000> +2<211100> + <211111> + <221000>
               + <221110> + <222100>
{222111111000} = <110000> +2<111100> + <111111> + <200000> +2<211000> + <211110> + <221100>
               + <222000>
{222200000000} = <000000> + <110000> + <111100> + <220000> + <221100> + <222200>
{222210000000} = <100000> + <111000> + <111110> + <210000> + <211100> + <221000> + <221110>
               + <222100> + <222210>
{222211000000} = <000000> +2<110000> +2<111100> + <111111> + <211000> + <211110> + <220000>
               +2<221100> + <221111> + <222110> + <222200> + <222211>
{222211100000} = <100000> +2<111000> +2<111110> + <210000> +2<211100> + <211111> + <221000>
               +2<221110> + <222100> + <222111> + <222210>
{222211110000} = <000000> +2<110000> +3<111100> + <111111> + <211000> +2<211110> + <220000>
               +2<221100> + <221111> + <222110> + <222200>
{222220000000} = <200000> + <211000> + <211110> + <222000> + <222110> + <222220>
{222221000000} = <100000> + <111000> + <111110> + <210000> + <211100> + <211111> + <221000>
               + <221110> + <222100> + <222111> + <222210> + <222221>
{222221100000} = <110000> + <111100> + <111111> + <200000> +2<211000> +2<211110> + <221100>
               + <221111> + <222000> +2<222110> + <222211> + <222220>
{222222000000} = <000000> + <110000> + <111100> + <111111> + <220000> + <221100> + <221111>
               + <222200> + <222211> + <222222>
```

C-15 Branching Rules for the Reduction $U_{14} \to Sp_{14}$.

{00000000000000} = <0000000>

{10000000000000} = <1000000>
{11000000000000} = <0000000> + <1100000>
{11100000000000} = <1000000> + <1110000>
{11110000000000} = <0000000> + <1100000> + <1111000>
{11111000000000} = <1000000> + <1110000> + <1111100>
{11111100000000} = <0000000> + <1100000> + <1111000> + <1111110>
{11111110000000} = <1000000> + <1110000> + <1111100> + <1111111>

{20000000000000} = <2000000>
{21000000000000} = <1000000> + <2100000>
{21100000000000} = <1100000> + <2000000> + <2110000>
{21110000000000} = <1000000> + <1110000> + <2100000> + <2111000>
{21111000000000} = <1100000> + <1111000> + <2000000> + <2110000> + <2111100>
{21111100000000} = <1000000> + <1110000> + <1111100> + <2100000> + <2111000> + <2111110>
{21111110000000} = <1100000> + <1111000> + <1111110> + <2000000> + <2110000> + <2111100> + <2111111>
{21111111000000} = <1000000> + <1110000> + <1111100> + <1111111> + <2100000> + <2111000> + <2111110>
{21111111100000} = <1100000> + <1111000> + <1111110> + <2000000> + <2110000> + <2111100>
{21111111110000} = <1000000> + <1110000> + <1111100> + <2100000> + <2111000>
{21111111111000} = <1100000> + <1111000> + <2000000> + <2110000>
{21111111111100} = <1000000> + <1110000> + <2100000>
{21111111111110} = <1100000> + <2000000>
{22000000000000} = <0000000> + <1100000> + <2200000>
{22100000000000} = <1000000> + <1110000> + <2100000> + <2210000>
{22110000000000} = <0000000> +2<1100000> + <1111000> + <2110000> + <2200000> + <2211000>
{22111000000000} = <1000000> +2<1110000> + <1111100> + <2100000> + <2111000> + <2210000> + <2211100>
{22111100000000} = <0000000> +2<1100000> +2<1111000> + <1111110> + <2110000> + <2111100> + <2200000>
+ <2211000> + <2211110>
{22111110000000} = <1000000> +2<1110000> +2<1111100> + <1111111> + <2100000> + <2111000> + <2111110>
+ <2210000> + <2211100> + <2211111>
{22111111000000} = <0000000> +2<1100000> +2<1111000> +2<1111110> + <2110000> + <2111100> + <2111111>
+ <2200000> + <2211000> + <2211110>
{22111111100000} = <1000000> +2<1110000> +2<1111100> + <1111111> + <2100000> + <2111000> + <2111110>
+ <2210000> + <2211100>
{22111111110000} = <0000000> +2<1100000> +2<1111000> + <1111110> + <2110000> + <2111100> + <2200000>
+ <2211000>
{22111111111000} = <1000000> +2<1110000> + <1111100> + <2100000> + <2111000> + <2210000>
{22111111111100} = <0000000> +2<1100000> + <1111000> + <2110000> + <2200000>
{22200000000000} = <2000000> + <2110000> + <2220000>
{22210000000000} = <1000000> + <1110000> + <2100000> + <2111000> + <2210000> + <2221000>
{22211000000000} = <1100000> + <1111000> + <2000000> +2<2110000> + <2111100> + <2211000> + <2220000>
+ <2221100>
{22211100000000} = <1000000> +2<1110000> + <1111100> + <2100000> +2<2111000> + <2111110> + <2210000>
+ <2211100> + <2221000> + <2221110>
{22211110000000} = <1100000> +2<1111000> + <1111110> + <2000000> +2<2110000> +2<2111100> + <2111111>
+ <2211000> + <2211110> + <2220000> + <2221100> + <2221111>
{22211111000000} = <1000000> +2<1110000> +2<1111100> + <1111111> + <2100000> +2<2111000> +2<2111110>
+ <2210000> + <2211100> + <2211111> + <2221000> + <2221110>
{22211111100000} = <1100000> +2<1111000> +2<1111110> + <2000000> +2<2110000> +2<2111100> + <2111111>
+ <2211000> + <2211110> + <2220000> + <2221100>
{22211111110000} = <1000000> +2<1110000> +2<1111100> + <1111111> + <2100000> +2<2111000> + <2111110>
+ <2210000> + <2211100> + <2221000>
{22211111111000} = <1100000> +2<1111000> + <1111110> + <2000000> +2<2110000> + <2111100> + <2211000>
+ <2220000>
{22220000000000} = <0000000> + <1100000> + <1111000> + <2200000> + <2211000> + <2222000>
{22221000000000} = <1000000> + <1110000> + <1111100> + <2100000> + <2111000> + <2210000> + <2211100>
+ <2221000> + <2222100>
{22221100000000} = <0000000> +2<1100000> +2<1111000> + <1111110> + <2110000> + <2111100> + <2200000>
+2<2211000> + <2211110> + <2221100> + <2222000> + <2222110>
{22221110000000} = <1000000> +2<1110000> +2<1111100> + <1111111> + <2100000> +2<2111000> + <2111110>
+ <2210000> +2<2211100> + <2211111> + <2221000> + <2221110> + <2222100> + <2222111>
{22221111000000} = <0000000> +2<1100000> +3<1111000> +2<1111110> + <2110000> +2<2111100> + <2111111>
+ <2200000> +2<2211000> +2<2211110> + <2221100> + <2221111> + <2222000> + <2222110>
{22221111100000} = <1000000> +2<1110000> +3<1111100> + <1111111> + <2100000> +2<2111000> +2<2111110>
+ <2210000> +2<2211100> + <2211111> + <2221000> + <2221110> + <2222100>
{22221111110000} = <0000000> +2<1100000> +3<1111000> +2<1111110> + <2110000> +2<2111100> + <2111111>
+ <2200000> +2<2211000> + <2211110> + <2221100> + <2222000>
{22222000000000} = <2000000> + <2110000> + <2111100> + <2220000> + <2221100> + <2222200>
{22222100000000} = <1000000> + <1110000> + <1111100> + <2100000> + <2111000> + <2111110> + <2210000>
+ <2211100> + <2221000> + <2221110> + <2222100> + <2222210>

C-15 Branching Rules for the Reduction $U_{14} \rightarrow Sp_{14}$ (contd).

{22222110000000} = <1100000> + <1111000> + <1111110> + <2000000> +2<2110000> +2<2111100> + <2111111>
+ <2211000> + <2211110> + <2220000> +2<2221100> + <2221111> + <2222110> + <2222200>
+ <2222211>

{22222111000000} = <1000000> +2<1110000> +2<1111100> + <1111111> + <2100000> +2<2111000> +2<2111110>
+ <2210000> +2<2211100> + <2211111> + <2221000> +2<2221110> + <2222100> + <2222111>
+ <2222210>

{22222111100000} = <1100000> +2<1111000> +2<1111110> + <2000000> +2<2110000> +3<2111100> + <2111111>
+ <2211000> +2<2211110> + <2220000> +2<2221100> + <2221111> + <2222110> + <2222200>

{22222200000000} = <0000000> + <1100000> + <1111000> + <1111110> + <2200000> + <2211000> + <2211110>
+ <2222000> + <2222110> + <2222220>

{22222210000000} = <1000000> + <1110000> + <1111100> + <1111111> + <2100000> + <2111000> + <2111110>
+ <2210000> + <2211100> + <2211111> + <2221000> + <2221110> + <2222100> + <2222111>
+ <2222210> + <2222221>

{22222211000000} = <0000000> +2<1100000> +2<1111000> +2<1111110> + <2110000> + <2111100> + <2111111>
+ <2200000> +2<2211000> +2<2211110> + <2221100> + <2221111> + <2222000> +2<2222110>
+ <2222211> + <2222220>

{22222220000000} = <2000000> + <2110000> + <2111100> + <2111111> + <2220000> + <2221100> + <2221111>
+ <2222200> + <2222211> + <2222222>

C-16 Branching Rules for the Reduction $R_5 \to R_3$.

[00] = [0]

[10] = [2]
[11] = [1] + [3]

[20] = [2] + [4]
[21] = [1] + [2] + [3] + [4] + [5]
[22] = [0] + [2] + [3] + [4] + [6]

[30] = [0] + [3] + [4] + [6]
[31] = [1] + [2] +2[3] + [4] +2[5] + [6] + [7]
[32] = [1] +2[2] + [3] +2[4] +2[5] + [6] + [7] + [8]
[33] = [1] +2[3] + [4] + [5] + [6] + [7] + [9]

[40] = [2] + [4] + [5] + [6] + [8]
[41] = [1] + [2] +2[3] +2[4] +2[5] +2[6] +2[7] + [8] + [9]
[42] = [0] + [1] +2[2] +2[3] +3[4] +2[5] +3[6] +2[7] +2[8] + [9] + [10]
[43] = [1] +2[2] +2[3] +2[4] +3[5] +2[6] +2[7] +2[8] + [9] + [10] + [11]
[44] = [0] + [2] + [3] +2[4] + [5] +2[6] + [7] + [8] + [9] + [10] + [12]

[50] = [2] + [4] + [5] + [6] + [7] + [8] + [10]
[51] = [1] + [2] +2[3] +2[4] +3[5] +2[6] +3[7] +2[8] +2[9] + [10] + [11]
[52] = [0] + [1] +2[2] +3[3] +3[4] +3[5] +4[6] +3[7] +3[8] +3[9] +2[10] + [11] + [12]
[53] =2[1] +2[2] +3[3] +3[4] +4[5] +3[6] +4[7] +3[8] +3[9] +2[10] +2[11] + [12] + [13]
[54] = [1] +2[2] +2[3] +3[4] +3[5] +3[6] +3[7] +3[8] +2[9] +2[10] +2[11] + [12] + [13] + [14]
[55] = [1] +2[3] + [4] +2[5] +2[6] +2[7] + [8] +2[9] + [10] + [11] + [12] + [13] + [15]

[60] = [0] + [3] + [4] +2[6] + [7] + [8] + [9] + [10] + [12]
[61] = [1] + [2] +2[3] +2[4] +3[5] +3[6] +3[7] +3[8] +3[9] +2[10] +2[11] + [12] + [13]
[62] = [1] +3[2] +2[3] +4[4] +4[5] +4[6] +4[7] +5[8] +3[9] +4[10] +3[11] +2[12] + [13] + [14]
[63] = [0] +2[1] +2[2] +4[3] +4[4] +4[5] +5[6] +5[7] +4[8] +5[9] +4[10] +3[11] +3[12] +2[13] + [14] + [15]
[64] = [0] + [1] +3[2] +3[3] +4[4] +4[5] +5[6] +4[7] +5[8] +4[9] +4[10] +3[11] +3[12] +2[13] +2[14] + [15] + [16]
[65] = [1] +2[2] +2[3] +3[4] +4[5] +3[6] +4[7] +4[8] +3[9] +3[10] +3[11] +2[12] +2[13] +2[14] + [15] + [16] + [17]
[66] = [0] + [2] + [3] +2[4] + [5] +3[6] +2[7] +2[8] +2[9] +2[10] + [11] +2[12] + [13] + [14] + [15] + [16] + [18]

[70] = [2] + [4] + [5] + [6] + [7] +2[8] + [9] + [10] + [11] + [12] + [14]
[71] = [1] + [2] +2[3] +2[4] +3[5] +3[6] +4[7] +3[8] +4[9] +3[10] +3[11] +2[12] +2[13] + [14] + [15]
[72] = [0] + [1] +2[2] +3[3] +4[4] +4[5] +5[6] +5[7] +5[8] +5[9] +5[10] +4[11] +4[12] +3[13] +2[14] + [15] + [16]
[73] =2[1] +3[2] +4[3] +4[4] +6[5] +5[6] +6[7] +6[8] +6[9] +5[10] +6[11] +4[12] +4[13] +3[14] +2[15] + [16] + [17]
[74] = [0] +2[1] +3[2] +4[3] +5[4] +5[5] +6[6] +6[7] +6[8] +6[9] +6[10] +5[11] +5[12] +4[13] +3[14] +3[15] +2[16] + [17] + [18]
[75] =2[1] +2[2] +4[3] +4[4] +5[5] +5[6] +6[7] +5[8] +6[9] +5[10] +5[11] +4[12] +4[13] +3[14] +3[15] +2[16] +2[17] + [18] + [19]
[76] = [1] +2[2] +2[3] +3[4] +4[5] +4[6] +4[7] +5[8] +4[9] +4[10] +4[11] +3[12] +3[13] +3[14] +2[15] +2[16] +2[17] + [18] + [19] + [20]
[77] = [1] +2[3] + [4] +2[5] +2[6] +3[7] +2[8] +3[9] +2[10] +2[11] +2[12] +2[13] + [14] +2[15] + [16] + [17] + [18] + [19] + [21]

[80] = [2] + [4] + [5] + [6] + [7] +2[8] + [9] +2[10] + [11] + [12] + [13] + [14] + [16]
[81] = [1] + [2] +2[3] +2[4] +3[5] +3[6] +4[7] +4[8] +4[9] +4[10] +4[11] +3[12] +3[13] +2[14] +2[15] + [16] + [17]
[82] = [0] + [1] +2[2] +3[3] +4[4] +4[5] +6[6] +5[7] +6[8] +6[9] +6[10] +5[11] +6[12] +4[13] +4[14] +3[15] +2[16] + [17] + [18]
[83] =2[1] +3[2] +4[3] +5[4] +6[5] +6[6] +7[7] +7[8] +7[9] +7[10] +7[11] +6[12] +6[13] +5[14] +4[15] +3[16] +2[17] + [18] + [19]
[84] = [0] +2[1] +4[2] +4[3] +6[4] +6[5] +7[6] +7[7] +8[8] +7[9] +8[10] +7[11] +7[12] +6[13] +6[14] +4[15] +4[16] +3[17] +2[18] + [19] + [20]
[85] = [0] +2[1] +3[2] +5[3] +5[4] +6[5] +7[6] +7[7] +7[8] +8[9] +7[10] +7[11] +7[12] +6[13] +5[14] +5[15] +4[16] +3[17] +3[18] +2[19] + [20] + [21]
[86] = [0] + [1] +3[2] +3[3] +5[4] +5[5] +6[6] +6[7] +7[8] +6[9] +7[10] +6[11] +6[12] +5[13] +5[14] +4[15] +4[16] +3[17] +3[18] +2[19] +2[20] + [21] + [22]
[87] = [1] +2[2] +2[3] +3[4] +4[5] +4[6] +5[7] +5[8] +5[9] +5[10] +5[11] +4[12] +4[13] +4[14] +3[15] +3[16] +3[17] +2[18] +2[19] +2[20] + [21] + [22] + [23]
[88] = [0] + [2] + [3] +2[4] + [5] +3[6] +2[7] +3[8] +3[9] +3[10] +2[11] +3[12] +2[13] +2[14] +2[15] +2[16] + [17] +2[18] + [19] + [20] + [21] + [22] + [24]

 Branching Rules for the Reduction $R_7 \to R_3$.

```
[000] = [0]

[100] = [3]
[110] = [1] + [3] + [5]
[111] = [0] + [2] + [3] + [4] + [6]

[200] = [2] + [4] + [6]
[210] = [1] +2[2] + [3] +2[4] +2[5] + [6] + [7] + [8]
[211] =2[1] +2[2] +3[3] +3[4] +3[5] +2[6] +2[7] + [8] + [9]
[220] = [0] +3[2] + [3] +3[4] +2[5] +2[6] + [7] +2[8] + [10]
[221] =3[1] +3[2] +5[3] +4[4] +5[5] +4[6] +4[7] +2[8] +2[9] + [10] + [11]
[222] =2[0] + [1] +2[2] +3[3] +4[4] +2[5] +4[6] +2[7] +2[8] +2[9] + [10] + [12]

[300] = [1] + [3] + [4] + [5] + [6] + [7] + [9]
[310] =2[1] +2[2] +4[3] +3[4] +4[5] +3[6] +4[7] +2[8] +2[9] + [10] + [11]
[311] =2[0] +2[1] +5[2] +5[3] +7[4] +6[5] +7[6] +5[7] +5[8] +3[9] +3[10] + [11] + [12]
[320] = [0] +3[1] +4[2] +6[3] +6[4] +7[5] +6[6] +6[7] +5[8] +4[9] +3[10] +2[11] + [12] + [13]
[321] =2[0] +6[1] +10[2] +12[3] +14[4] +15[5] +14[6] +13[7] +12[8] +9[9] +7[10] +5[11] +3[12] +2[13]
      + [14]
[322] = [0] +5[1] +7[2] +10[3] +10[4] +12[5] +11[6] +11[7] +9[8] +8[9] +6[10] +5[11] +3[12] +2[13]
      + [14] + [15]
[330] =4[1] +2[2] +7[3] +5[4] +7[5] +6[6] +7[7] +4[8] +6[9] +3[10] +3[11] +2[12] +2[13] + [15]
[331] =3[0] +5[1] +11[2] +11[3] +16[4] +14[5] +17[6] +14[7] +14[8] +11[9] +10[10] +7[11] +6[12]
      +3[13] +3[14] + [15] + [16]
[332] = [0] +7[1] +9[2] +13[3] +14[4] +17[5] +15[6] +16[7] +14[8] +13[9] +10[10] +9[11] +6[12]
      +5[13] +3[14] +2[15] + [16] + [17]
[333] =3[0] +2[1] +5[2] +7[3] +8[4] +7[5] +10[6] +8[7] +8[8] +7[9] +7[10] +4[11] +5[12] +3[13]
      +2[14] +2[15] + [16] + [18]

[400] = [0] + [2] + [3] +2[4] + [5] +2[6] + [7] +2[8] + [9] + [10] + [12]
[410] = [0] +2[1] +4[2] +5[3] +6[4] +6[5] +7[6] +6[7] +6[8] +5[9] +4[10] +3[11] +2[12] + [13] + [14]
[411] = [0] +5[1] +6[2] +10[3] +10[4] +12[5] +12[6] +12[7] +10[8] +10[9] +7[10] +6[11] +4[12] +3[13]
      + [14] + [15]
[420] =3[0] +4[1] +9[2] +9[3] +13[4] +12[5] +15[6] +12[7] +13[8] +10[9] +10[10] +7[11] +6[12] +3[13]
      +3[14] + [15] + [16]
[421] =3[0] +12[1] +17[2] +24[3] +27[4] +31[5] +30[6] +31[7] +28[8] +26[9] +21[10] +18[11] +13[12]
      +10[13] +6[14] +4[15] +2[16] + [17]
[422] =4[0] +8[1] +16[2] +18[3] +24[4] +24[5] +27[6] +25[7] +25[8] +21[9] +20[10] +15[11] +13[12]
      +9[13] +7[14] +4[15] +3[16] + [17] + [18]
[430] =3[0] +6[1] +10[2] +14[3] +16[4] +17[5] +19[6] +18[7] +17[8] +16[9] +14[10] +11[11] +10[12]
      +7[13] +5[14] +4[15] +2[16] + [17] + [18]
[431] =5[0] +16[1] +25[2] +34[3] +39[4] +44[5] +45[6] +46[7] +43[8] +40[9] +35[10] +30[11] +24[12]
      +19[13] +14[14] +10[15] +6[16] +4[17] +2[18] + [19]
[432] =5[0] +17[1] +28[2] +35[3] +43[4] +48[5] +49[6] +50[7] +49[8] +44[9] +40[10] +35[11] +28[12]
      +23[13] +18[14] +12[15] +9[16] +6[17] +3[18] +2[19] + [20]
[433] =3[0] +11[1] +15[2] +22[3] +25[4] +29[5] +30[6] +31[7] +29[8] +29[9] +25[10] +23[11] +19[12]
      +16[13] +12[14] +10[15] +7[16] +5[17] +3[18] +2[19] + [20] + [21]
[440] =4[0] +3[1] +10[2] +9[3] +15[4] +13[5] +17[6] +14[7] +17[8] +13[9] +14[10] +10[11] +11[12]
      +7[13] +7[14] +4[15] +4[16] +2[17] +2[18] + [20]
[441] =3[0] +15[1] +19[2] +30[3] +32[4] +39[5] +39[6] +42[7] +38[8] +39[9] +33[10] +31[11] +25[12]
      +22[13] +16[14] +14[15] +9[16] +7[17] +4[18] +3[19] + [20] + [21]
[442] =7[0] +14[1] +27[2] +32[3] +42[4] +44[5] +50[6] +48[7] +50[8] +45[9] +44[10] +37[11] +34[12]
      +27[13] +23[14] +17[15] +14[16] +9[17] +7[18] +4[19] +3[20] + [21] + [22]
[443] =3[0] +13[1] +19[2] +28[3] +31[4] +37[5] +38[6] +41[7] +39[8] +39[9] +35[10] +33[11] +28[12]
      +25[13] +20[14] +17[15] +12[16] +10[17] +7[18] +5[19] +3[20] +2[21] + [22] + [23]
[444] =4[0] +5[1] +9[2] +12[3] +16[4] +15[5] +20[6] +18[7] +19[8] +19[9] +18[10] +15[11] +16[12]
      +12[13] +11[14] +9[15] +8[16] +5[17] +5[18] +3[19] +2[20] +2[21] + [22] + [24]
```

Branching Rules for the Reduction $R_9 \to R_3$.

[0000] = [0]

[1000] = [4]

[1100] = [1] + [3] + [5] + [7]

[1110] = [1] +2[3] + [4] + [5] + [6] + [7] + [9]

[1111] = [0] +2[2] + [3] +2[4] + [5] +2[6] + [7] + [8] + [10]

[2000] = [2] + [4] + [6] + [8]

[2100] = [1] +2[2] +2[3] +2[4] +3[5] +2[6] +2[7] +2[8] + [9] + [10] + [11]

[2110] =3[1] +3[2] +5[3] +5[4] +6[5] +5[6] +5[7] +4[8] +4[9] +2[10] +2[11] + [12] + [13]

[2111] = [0] +3[1] +6[2] +6[3] +8[4] +8[5] +8[6] +7[7] +7[8] +5[9] +4[10] +3[11] +2[12] + [13]
+ [14]

[2200] =2[0] +4[2] +2[3] +5[4] +3[5] +5[6] +3[7] +4[8] +2[9] +3[10] + [11] +2[12] + [14]

[2210] =3[0] +4[1] +8[2] +9[3] +12[4] +11[5] +13[6] +11[7] +11[8] +9[9] +8[10] +5[11] +5[12] +3[13]
+2[14] + [15] + [16]

[2211] = [0] +9[1] +10[2] +17[3] +16[4] +21[5] +18[6] +20[7] +16[8] +16[9] +12[10] +11[11] +7[12]
+6[13] +3[14] +3[15] + [16] + [17]

[2220] =4[0] +3[1] +8[2] +9[3] +13[4] +10[5] +15[6] +11[7] +12[8] +10[9] +10[10] +6[11] +7[12]
+4[13] +3[14] +2[15] +2[16] + [18]

[2221] =2[0] +10[1] +14[2] +20[3] +22[4] +25[5] +25[6] +26[7] +23[8] +22[9] +18[10] +16[11] +12[12]
+10[13] +7[14] +5[15] +3[16] +2[17] + [18] + [19]

[2222] =3[0] +4[1] +11[2] +9[3] +15[4] +15[5] +16[6] +14[7] +17[8] +12[9] +13[10] +10[11] +9[12]
+6[13] +6[14] +3[15] +3[16] +2[17] + [18] + [20]

[3000] = [0] + [2] + [3] + [4] + [5] +2[6] + [7] + [8] + [9] + [10] + [12]

[3100] =3[1] +3[2] +6[3] +5[4] +7[5] +6[6] +7[7] +5[8] +6[9] +4[10] +4[11] +2[12] +2[13] + [14]
+ [15]

[3110] = [0] +7[1] +8[2] +13[3] +13[4] +17[5] +15[6] +17[7] +14[8] +14[9] +11[10] +10[11] +7[12]
+6[13] +3[14] +3[15] + [16] + [17]

[3111] =5[0] +7[1] +15[2] +17[3] +23[4] +22[5] +26[6] +23[7] +24[8] +20[9] +19[10] +14[11] +13[12]
+8[13] +7[14] +4[15] +3[16] + [17] + [18]

[3200] =2[0] +5[1] +9[2] +11[3] +14[4] +14[5] +16[6] +15[7] +15[8] +13[9] +12[10] +10[11] +8[12]
+6[13] +5[14] +3[15] +2[16] + [17] + [18]

[3210] =5[0] +17[1] +28[2] +35[3] +43[4] +48[5] +49[6] +50[7] +49[8] +44[9] +40[10] +35[11] +28[12]
+23[13] +18[14] +12[15] +9[16] +6[17] +3[18] +2[19] + [20]

[3211] =9[0] +28[1] +44[2] +59[3] +70[4] +78[5] +82[6] +83[7] +80[8] +75[9] +67[10] +59[11] +49[12]
+40[13] +31[14] +23[15] +16[16] +11[17] +7[18] +4[19] +2[20] + [21]

[3220] =8[0] +18[1] +33[2] +40[3] +51[4] +55[5] +60[6] +59[7] +60[8] +54[9] +52[10] +44[11] +39[12]
+31[13] +26[14] +19[15] +15[16] +10[17] +7[18] +4[19] +3[20] + [21] + [22]

[3221] =12[0] +41[1] +63[2] +87[3] +101[4] +116[5] +121[6] +126[7] +121[8] +117[9] +106[10] +96[11]
+81[12] +69[13] +55[14] +44[15] +32[16] +24[17] +16[18] +11[19] +6[20] +4[21] +2[22] + [23]

[3222] =10[0] +25[1] +41[2] +55[3] +67[4] +73[5] +81[6] +81[7] +81[8] +78[9] +72[10] +64[11] +58[12]
+48[13] +40[14] +32[15] +25[16] +18[17] +14[18] +9[19] +6[20] +4[21] +2[22] + [23] + [24]

[3300] =9[1] +7[2] +17[3] +14[4] +20[5] +18[6] +22[7] +17[8] +21[9] +15[10] +16[11] +12[12] +12[13]
+7[14] +8[15] +4[16] +4[17] +2[18] +2[19] + [21]

[3310] =5[0] +25[1] +34[2] +51[3] +56[4] +68[5] +68[6] +74[7] +69[8] +70[9] +61[10] +58[11] +48[12]
+43[13] +33[14] +28[15] +20[16] +16[17] +10[18] +8[19] +4[20] +3[21] + [22] + [23]

[3311] =18[0] +34[1] +66[2] +79[3] +104[4] +108[5] +125[6] +120[7] +126[8] +115[9] +113[10] +96[11]
+90[12] +72[13] +63[14] +48[15] +40[16] +27[17] +22[18] +13[19] +10[20] +5[21] +4[22] + [23]
+ [24]

[3320] =9[0] +38[1] +53[2] +78[3] +88[4] +106[5] +108[6] +117[7] +112[8] +113[9] +102[10] +97[11]
+83[12] +75[13] +60[14] +52[15] +39[16] +32[17] +22[18] +17[19] +11[20] +8[21] +4[22] +3[23]
+ [24] + [25]

[3321] =26[0] +73[1] +122[2] +159[3] +196[4] +220[5] +239[6] +246[7] +249[8] +239[9] +228[10]
+208[11] +187[12] +162[13] +139[14] +113[15] +92[16] +71[17] +54[18] +39[19] +28[20] +18[21]
+12[22] +7[23] +4[24] +2[25] + [26]

[3322] =14[0] +57[1] +82[2] +119[3] +136[4] +163[5] +169[6] +183[7] +178[8] +180[9] +166[10]
+159[11] +139[12] +127[13] +105[14] +92[15] +72[16] +60[17] +44[18] +35[19] +24[20] +18[21]
+11[22] +8[23] +4[24] +3[25] + [26] + [27]

[3330] =8[0] +29[1] +37[2] +61[3] +66[4] +79[5] +84[6] +91[7] +86[8] +92[9] +82[10] +79[11] +71[12]
+65[13] +52[14] +49[15] +37[16] +31[17] +24[18] +19[19] +12[20] +11[21] +6[22] +4[23] +3[24]
+2[25] + [27]

[3331] =24[0] +56[1] +101[2] +127[3] +163[4] +178[5] +201[6] +204[7] +213[8] +204[9] +201[10]
+183[11] +172[12] +149[13] +134[14] +110[15] +95[16] +74[17] +61[18] +45[19] +35[20] +24[21]
+18[22] +11[23] +8[24] +4[25] +3[26] + [27] + [28]

[3332] =16[0] +57[1] +91[2] +122[3] +148[4] +173[5] +184[6] +197[7] +200[8] +198[9] +190[10]
+182[11] +164[12] +150[13] +132[14] +113[15] +95[16] +80[17] +62[18] +50[19] +38[20] +28[21]
+20[22] +15[23] +9[24] +6[25] +4[26] +2[27] + [28] + [29]

[3333] =13[0] +22[1] +43[2] +54[3] +70[4] +74[5] +90[6] +88[7] +94[8] +91[9] +93[10] +83[11] +83[12]
+71[13] +66[14] +57[15] +50[16] +39[17] +36[18] +26[19] +22[20] +16[21] +13[22] +8[23] +7[24]
+4[25] +3[26] +2[27] + [28] + [30]

Branching Rules for the Reduction $R_9 \rightarrow R_3$ (contd).

[4000] = [0] +2[2] + [3] +3[4] +2[5] +3[6] +2[7] +3[8] +2[9] +3[10] + [11] +2[12] + [13] + [14]
+ [16]

[4100] = [0] +5[1] +7[2] +10[3] +12[4] +13[5] +14[6] +15[7] +14[8] +14[9] +12[10] +11[11] +9[12]
+8[13] +6[14] +4[15] +3[16] +2[17] + [18] + [19]

[4110] =3[0] +13[1] +17[2] +27[3] +29[4] +35[5] +36[6] +39[7] +36[8] +37[9] +32[10] +30[11] +25[12]
+22[13] +16[14] +14[15] +9[16] +7[17] +4[18] +3[19] + [20] + [21]

[4111] =7[0] +17[1] +30[2] +38[3] +48[4] +52[5] +58[6] +58[7] +59[8] +55[9] +53[10] +46[11] +41[12]
+34[13] +28[14] +21[15] +17[16] +11[17] +8[18] +5[19] +3[20] + [21] + [22]

[4200] =7[0] +10[1] +22[2] +24[3] +34[4] +34[5] +41[6] +38[7] +42[8] +36[9] +38[10] +31[11] +30[12]
+23[13] +21[14] +15[15] +13[16] +8[17] +7[18] +3[19] +3[20] + [21] + [22]

[4210] =15[0] +39[1] +67[2] +87[3] +108[4] +120[5] +132[6] +134[7] +137[8] +131[9] +125[10] +113[11]
+102[12] +86[13] +74[14] +59[15] +47[16] +35[17] +26[18] +17[19] +12[20] +7[21] +4[22] +2[23]
+ [24]

[4211] =20[0] +71[1] +107[2] +150[3] +176[4] +206[5] +217[6] +231[7] +227[8] +226[9] +209[10]
+196[11] +172[12] +153[13] +126[14] +106[15] +82[16] +65[17] +46[18] +34[19] +22[20] +15[21]
+8[22] +5[23] +2[24] + [25]

[4220] =21[0] +47[1] +87[2] +107[3] +139[4] +150[5] +171[6] +171[7] +180[8] +170[9] +167[10]
+150[11] +141[12] +119[13] +107[14] +85[15] +72[16] +54[17] +44[18] +30[19] +23[20] +14[21]
+10[22] +5[23] +4[24] + [25] + [26]

[4221] =34[0] +109[1] +171[2] +235[3] +281[4] +326[5] +350[6] +370[7] +371[8] +368[9] +349[10]
+328[11] +295[12] +264[13] +225[14] +192[15] +155[16] +125[17] +95[18] +72[19] +51[20]
+36[21] +23[22] +15[23] +8[24] +5[25] +2[26] + [27]

[4222] =28[0] +68[1] +120[2] +153[3] +195[4] +215[5] +242[6] +248[7] +258[8] +249[9] +246[10]
+225[11] +211[12] +185[13] +165[14] +137[15] +118[16] +92[17] +75[18] +56[19] +43[20] +29[21]
+22[22] +13[23] +9[24] +5[25] +3[26] + [27] + [28]

[4300] =7[0] +21[1] +33[2] +45[3] +54[4] +63[5] +67[6] +71[7] +71[8] +70[9] +67[10] +62[11] +56[12]
+50[13] +42[14] +36[15] +29[16] +23[17] +17[18] +13[19] +9[20] +6[21] +4[22] +2[23] + [24]
+ [25]

[4310] =24[0] +77[1] +123[2] +167[3] +201[4] +233[5] +251[6] +266[7] +268[8] +266[9] +254[10]
+240[11] +217[12] +195[13] +169[14] +144[15] +118[16] +96[17] +74[18] +57[19] +41[20] +29[21]
+19[22] +13[23] +7[24] +4[25] +2[26] + [27]

[4311] =46[0] +131[1] +216[2] +288[3] +354[4] +403[5] +442[6] +463[7] +473[8] +466[9] +451[10]
+422[11] +389[12] +347[13] +305[14] +259[15] +217[16] +175[17] +139[18] +106[19] +79[20]
+56[21] +39[22] +25[23] +16[24] +9[25] +5[26] +2[27] + [28]

[4320] =41[0] +127[1] +207[2] +277[3] +340[4] +391[5] +426[6] +451[7] +462[8] +458[9] +444[10]
+422[11] +388[12] +352[13] +312[14] +268[15] +227[16] +188[17] +150[18] +118[19] +90[20]
+66[21] +47[22] +33[23] +21[24] +13[25] +8[26] +4[27] +2[28] + [29]

[4321] =93[0] +276[1] +449[2] +606[3] +742[4] +852[5] +935[6] +988[7] +1013[8] +1010[9] +981[10]
+932[11] +865[12] +785[13] +698[14] +606[15] +515[16] +428[17] +347[18] +274[19] +211[20]
+158[21] +114[22] +80[23] +54[24] +34[25] +21[26] +12[27] +6[28] +3[29] + [30]

[4322] =70[0] +206[1] +333[2] +454[3] +554[4] +638[5] +703[6] +746[7] +765[8] +769[9] +750[10]
+716[11] +670[12] +613[13] +548[14] +483[15] +414[16] +348[17] +287[18] +231[19] +180[20]
+139[21] +103[22] +74[23] +52[24] +35[25] +22[26] +14[27] +8[28] +4[29] +2[30] + [31]

[4330] =33[0] +112[1] +173[2] +242[3] +290[4] +341[5] +369[6] +399[7] +405[8] +412[9] +399[10]
+387[11] +358[12] +334[13] +296[14] +265[15] +226[16] +194[17] +158[18] +131[19] +101[20]
+80[21] +59[22] +44[23] +30[24] +22[25] +13[26] +9[27] +5[28] +3[29] + [30] + [31]

[4331] =89[0] +253[1] +419[2] +561[3] +694[4] +797[5] +884[6] +936[7] +971[8] +972[9] +960[10]
+918[11] +867[12] +796[13] +722[14] +638[15] +556[16] +471[17] +395[18] +320[19] +258[20]
+199[21] +153[22] +112[23] +81[24] +56[25] +38[26] +24[27] +15[28] +8[29] +5[30] +2[31]
+ [32]

[4332] =81[0] +251[1] +401[2] +550[3] +671[4] +782[5] +859[6] +922[7] +951[8] +965[9] +947[10]
+919[11] +866[12] +807[13] +731[14] +656[15] +572[16] +495[17] +415[18] +345[19] +278[20]
+222[21] +170[22] +130[23] +95[24] +69[25] +47[26] +32[27] +20[28] +13[29] +7[30] +4[31]
+2[32] + [33]

[4333] =41[0] +119[1] +197[2] +265[3] +330[4] +379[5] +423[6] +452[7] +472[8] +478[9] +476[10]
+461[11] +441[12] +412[13] +380[14] +341[15] +305[16] +264[17] +227[18] +191[19] +157[20]
+127[21] +102[22] +78[23] +60[24] +44[25] +32[26] +22[27] +16[28] +10[29] +6[30] +4[31]
+2[32] + [33] + [34]

[4400] =12[0] +16[1] +39[2] +41[3] +62[4] +60[5] +76[6] +71[7] +82[8] +74[9] +79[10] +68[11] +70[12]
+58[13] +57[14] +44[15] +43[16] +31[17] +29[18] +20[19] +18[20] +11[21] +10[22] +5[23] +5[24]
+2[25] +2[26] + [28]

[4410] =33[0] +78[1] +137[2] +178[3] +226[4] +251[5] +287[6] +296[7] +311[8] +307[9] +306[10]
+286[11] +275[12] +246[13] +225[14] +195[15] +171[16] +140[17] +120[18] +93[19] +75[20]
+56[21] +43[22] +29[23] +22[24] +13[25] +9[26] +5[27] +3[28] + [29] + [30]

[4411] =41[0] +155[1] +230[2] +331[3] +389[4] +465[5] +497[6] +544[7] +547[8] +563[9] +540[10]
+530[11] +486[12] +458[13] +403[14] +365[15] +308[16] +268[17] +216[18] +181[19] +138[20]
+111[21] +80[22] +61[23] +40[24] +30[25] +17[26] +12[27] +6[28] +4[29] + [30] + [31]

[4420] =61[0] +146[1] +259[2] +333[3] +426[4] +477[5] +543[6] +564[7] +598[8] +590[9] +595[10]
+561[11] +542[12] +491[13] +455[14] +398[15] +355[16] +297[17] +256[18] +204[19] +170[20]
+129[21] +103[22] +73[23] +56[24] +37[25] +27[26] +16[27] +11[28] +5[29] +4[30] + [31] + [32]

[4421] =112[0] +348[1] +557[2] +764[3] +932[4] +1087[5] +1195[6] +1284[7] +1325[8] +1346[9]
+1323[10] +1285[11] +1212[12] +1131[13] +1026[14] +922[15] +805[16] +697[17] +585[18]
+487[19] +392[20] +313[21] +240[22] +183[23] +133[24] +96[25] +65[26] +44[27] +27[28] +17[29]
+9[30] +5[31] +2[32] + [33]
[4422] =100[0] +261[1] +453[2] +589[3] +749[4] +849[5] +959[6] +1010[7] +1069[8] +1066[9] +1074[10]
+1029[11] +992[12] +915[13] +852[14] +755[15] +679[16] +581[17] +503[18] +414[19] +347[20]
+272[21] +220[22] +165[23] +127[24] +90[25] +67[26] +43[27] +31[28] +18[29] +12[30] +6[31]
+4[32] + [33] + [34]
[4430] =64[0] +172[1] +293[2] +387[3] +488[4] +556[5] +627[6] +664[7] +700[8] +704[9] +707[10]
+681[11] +657[12] +610[13] +567[14] +507[15] +456[16] +393[17] +341[18] +284[19] +237[20]
+189[21] +153[22] +116[23] +90[24] +65[25] +48[26] +32[27] +23[28] +14[29] +9[30] +5[31]
+3[32] + [33] + [34]
[4431] =136[0] +433[1] +690[2] +953[3] +1161[4] +1363[5] +1502[6] +1626[7] +1684[8] +1727[9]
+1708[10] +1677[11] +1595[12] +1508[13] +1384[14] +1263[15] +1119[16] +987[17] +844[18]
+720[19] +592[20] +487[21] +385[22] +305[23] +230[24] +174[25] +125[26] +90[27] +60[28]
+41[29] +25[30] +16[31] +8[32] +5[33] +2[34] + [35]
[4432] =149[0] +435[1] +716[2] +969[3] +1202[4] +1395[5] +1557[6] +1672[7] +1754[8] +1788[9]
+1789[10] +1750[11] +1685[12] +1588[13] +1477[14] +1346[15] +1210[16] +1067[17] +928[18]
+790[19] +665[20] +547[21] +443[22] +351[23] +274[24] +207[25] +155[26] +112[27] +79[28]
+54[29] +36[30] +22[31] +14[32] +8[33] +4[34] +2[35] + [36]
[4433] =73[0] +245[1] +381[2] +536[3] +647[4] +771[5] +846[6] +928[7] +961[8] +998[9] +991[10]
+987[11] +944[12] +908[13] +839[14] +782[15] +700[16] +632[17] +548[18] +480[19] +403[20]
+342[21] +277[22] +228[23] +177[24] +142[25] +105[26] +81[27] +57[28] +42[29] +28[30] +20[31]
+12[32] +8[33] +4[34] +3[35] + [36] + [37]
[4440] =47[0] +105[1] +186[2] +244[3] +314[4] +349[5] +408[6] +425[7] +456[8] +461[9] +470[10]
+450[11] +449[12] +415[13] +394[14] +357[15] +329[16] +284[17] +258[18] +215[19] +186[20]
+153[21] +128[22] +98[23] +83[24] +60[25] +47[26] +34[27] +26[28] +16[29] +13[30] +7[31]
+5[32] +3[33] +2[34] + [36]
[4441] =90[0] +284[1] +453[2] +627[3] +767[4] +904[5] +1001[6] +1090[7] +1137[8] +1174[9] +1173[10]
+1162[11] +1118[12] +1070[13] +995[14] +922[15] +831[16] +746[17] +651[18] +567[19] +479[20]
+404[21] +330[22] +269[23] +211[24] +167[25] +125[26] +95[27] +68[28] +49[29] +33[30] +23[31]
+14[32] +9[33] +5[34] +3[35] + [36] + [37]
[4442] =112[0] +310[1] +528[2] +698[3] +884[4] +1019[5] +1151[6] +1234[7] +1315[8] +1336[9]
+1360[10] +1335[11] +1304[12] +1237[13] +1174[14] +1075[15] +989[16] +882[17] +785[18]
+678[19] +589[20] +490[21] +413[22] +334[23] +271[24] +211[25] +167[26]+123[27] +94[28]
+67[29] +48[30] +32[31] +23[32] +13[33] +9[34] +5[35] +3[36] + [37] + [38]
[4443] =76[0] +229[1] +369[2] +512[3] +630[4] +738[5] +829[6] +901[7] +947[8] +985[9] +991[10]
+986[11] +963[12] +926[13] +873[14] +818[15] +748[16] +678[17] +606[18] +534[19] +460[20]
+397[21] +332[22] +276[23] +226[24] +181[25] +142[26] +112[27] +84[28] +63[29] +46[30]
+33[31] +22[32] +16[33] +10[34] +6[35] +4[36] +2[37] + [38] + [39]
[4444] =36[0] +84[1] +151[2] +194[3] +253[4] +287[5] +332[6] +352[7] +383[8] +388[9] +404[10]
+394[11] +396[12] +376[13] +364[14] +336[15] +319[16] +285[17] +262[18] +229[19] +206[20]
+174[21] +153[22] +125[23] +107[24] +86[25] +71[26] +54[27] +45[28] +32[29] +26[30] +18[31]
+14[32] +9[33] +7[34] +4[35] +3[36] +2[37] + [38] + [40]

C-19 Branching Rules for the Reduction $R_{11} \rightarrow R_3$.

[00000] = [0]

[10000] = [5]
[11000] = [1] + [3] + [5] + [7] + [9]
[11100] = [0] + [2] + [3] +2[4] + [5] +2[6] + [7] + [8] + [9] + [10] + [12]
[11110] = [0] +3[2] + [3] +3[4] +2[5] +3[6] +2[7] +3[8] + [9] +2[10] + [11] + [12] + [14]
[11111] =2[1] + [2] +4[3] +2[4] +4[5] +3[6] +4[7] +2[8] +3[9] +2[10] +2[11] + [12] + [13] + [15]

[20000] = [2] + [4] + [6] + [8] + [10]
[21000] = [1] +2[2] +2[3] +3[4] +3[5] +3[6] +3[7] +3[8] +2[9] +2[10] +2[11] + [12] + [13] + [14]
[21100] =4[1] +4[2] +7[3] +7[4] +9[5] +8[6] +9[7] +8[8] +8[9] +6[10] +6[11] +4[12] +4[13] +2[14] +2[15] + [16] + [17]
[21110] = [0] +6[1] +9[2] +12[3] +14[4] +16[5] +16[6] +17[7] +16[8] +15[9] +13[10] +12[11] +9[12] +8[13] +6[14] +4[15] +3[16] +2[17] + [18] + [19]
[21111] =3[0] +7[1] +13[2] +15[3] +20[4] +21[5] +23[6] +22[7] +23[8] +20[9] +19[10] +16[11] +14[12] +11[13] +9[14] +6[15] +5[16] +3[17] +2[18] + [19] + [20]
[22000] =3[0] +5[2] +3[3] +7[4] +4[5] +8[6] +5[7] +7[8] +5[9] +6[10] +3[11] +5[12] +2[13] +3[14] + [15] +2[16] + [18]
[22100] =2[0] +9[1] +11[2] +18[3] +19[4] +23[5] +23[6] +25[7] +23[8] +24[9] +20[10] +19[11] +16[12] +14[13] +10[14] +9[15] +6[16] +5[17] +3[18] +2[19] + [20] + [21]
[22110] =2[0] +19[1] +23[2] +38[3] +39[4] +50[5] +48[6] +55[7] +49[8] +52[9] +44[10] +44[11] +35[12] +33[13] +25[14] +22[15] +15[16] +13[17] +8[18] +7[19] +3[20] +3[21] + [22] + [23]
[22111] =11[0] +20[1] +39[2] +47[3] +62[4] +64[5] +75[6] +72[7] +76[8] +70[9] +69[10] +59[11] +56[12] +45[13] +40[14] +31[15] +26[16] +18[17] +15[18] +9[19] +7[20] +4[21] +3[22] + [23] + [24]
[22200] =7[0] +7[1] +18[2] +19[3] +29[4] +26[5] +35[6] +30[7] +35[8] +30[9] +32[10] +25[11] +27[12] +20[13] +19[14] +14[15] +13[16] +8[17] +8[18] +4[19] +4[20] +2[21] +2[22] + [24]
[22210] =9[0] +26[1] +46[2] +57[3] +72[4] +81[5] +88[6] +91[7] +94[8] +89[9] +87[10] +80[11] +72[12] +63[13] +56[14] +45[15] +38[16] +30[17] +23[18] +17[19] +13[20] +8[21] +6[22] +4[23] +2[24] + [25] + [26]
[22211] =12[0] +44[1] +64[2] +93[3] +107[4] +127[5] +134[6] +144[7] +141[8] +143[9] +133[10] +127[11] +113[12] +103[13] +86[14] +76[15] +60[16] +50[17] +38[18] +30[19] +21[20] +16[21] +10[22] +7[23] +4[24] +3[25] + [26] + [27]
[22220] =12[0] +19[1] +44[2] +46[3] +68[4] +69[5] +82[6] +79[7] +89[8] +79[9] +84[10] +73[11] +72[12] +60[13] +58[14] +44[15] +42[16] +31[17] +27[18] +19[19] +17[20] +10[21] +9[22] +5[23] +4[24] +2[25] +2[26] + [28]
[22221] =12[0] +49[1] +71[2] +102[3] +119[4] +143[5] +149[6] +163[7] +162[8] +165[9] +155[10] +151[11] +135[12] +126[13] +108[14] +96[15] +79[16] +68[17] +52[18] +43[19] +32[20] +25[21] +17[22] +13[23] +8[24] +6[25] +3[26] +2[27] + [28] + [29]
[22222] =13[0] +22[1] +43[2] +54[3] +70[4] +74[5] +90[6] +88[7] +94[8] +91[9] +93[10] +83[11] +83[12] +71[13] +66[14] +57[15] +50[16] +39[17] +36[18] +26[19] +22[20] +16[21] +13[22] +8[23] +7[24] +4[25] +3[26] +2[27] + [28] + [30]

Branching Rules for the Reduction $R_{13} \to R_3$.

[000000] = [0]

[100000] = [6]
[110000] = [1] + [3] + [5] + [7] + [9] + [11]
[111000] = [1] +2[3] + [4] +2[5] +2[6] +2[7] + [8] +2[9] + [10] + [11] + [12] + [13] + [15]
[111100] =2[0] +3[2] +2[3] +4[4] +2[5] +5[6] +3[7] +4[8] +3[9] +4[10] +2[11] +3[12] + [13] +2[14]
+ [15] + [16] + [18]
[111110] =2[0] + [1] +4[2] +3[3] +6[4] +5[5] +7[6] +5[7] +7[8] +5[9] +6[10] +4[11] +5[12] +3[13]
+3[14] +2[15] +2[16] + [17] + [18] + [20]
[111111] =4[1] +2[2] +7[3] +5[4] +8[5] +7[6] +9[7] +6[8] +9[9] +6[10] +7[11] +5[12] +5[13] +3[14]
+4[15] +2[16] +2[17] + [18] + [19] + [21]

[200000] = [2] + [4] + [6] + [8] + [10] + [12]
[210000] = [1] +2[2] +2[3] +3[4] +4[5] +3[6] +4[7] +4[8] +3[9] +3[10] +3[11] +2[12] +2[13] +2[14]
+ [15] + [16] + [17]
[211000] =5[1] +5[2] +9[3] +9[4] +12[5] +11[6] +13[7] +12[8] +13[9] +11[10] +11[11] +9[12] +9[13]
+6[14] +6[15] +4[16] +4[17] +2[18] +2[19] + [20] + [21]
[211100] =3[0] +8[1] +14[2] +18[3] +23[4] +25[5] +28[6] +29[7] +30[8] +29[9] +28[10] +26[11] +24[12]
+21[13] +19[14] +15[15] +13[16] +10[17] +8[18] +6[19] +4[20] +3[21] +2[22] + [23] + [24]
[211110] =5[0] +13[1] +25[2] +29[3] +39[4] +43[5] +48[6] +49[7] +53[8] +49[9] +50[10] +46[11]
+43[12] +38[13] +35[14] +28[15] +25[16] +20[17] +16[18] +12[19] +10[20] +6[21] +5[22]
+3[23] +2[24] + [25] + [26]
[211111] =5[0] +19[1] +29[2] +41[3] +48[4] +57[5] +61[6] +66[7] +66[8] +67[9] +64[10] +62[11]
+56[12] +52[13] +45[14] +40[15] +33[16] +28[17] +22[18] +18[19] +13[20] +10[21] +7[22]
+5[23] +3[24] +2[25] + [26] + [27]
[220000] =3[0] +7[2] +3[3] +9[4] +6[5] +10[6] +7[7] +11[8] +7[9] +10[10] +7[11] +8[12] +5[13] +7[14]
+3[15] +5[16] +2[17] +3[18] + [19] +2[20] + [22]
[221000] =5[0] +10[1] +20[2] +24[3] +32[4] +34[5] +40[6] +40[7] +43[8] +41[9] +42[10] +38[11]
+37[12] +32[13] +30[14] +25[15] +22[16] +17[17] +15[18] +11[19] +9[20] +6[21] +5[22] +3[23]
+2[24] + [25] + [26]
[221100] =5[0] +34[1] +43[2] +70[3] +75[4] +97[5] +97[6] +113[7] +108[8] +117[9] +107[10] +110[11]
+97[12] +96[13] +81[14] +77[15] +62[16] +57[17] +44[18] +39[19] +28[20] +24[21] +16[22]
+14[23] +8[24] +7[25] +3[26] +3[27] + [28] + [29]
[221110] =13[0] +57[1] +83[2] +120[3] +140[4] +171[5] +180[6] +201[7] +202[8] +210[9] +202[10]
+202[11] +185[12] +178[13] +158[14] +145[15] +124[16] +111[17] +90[18] +78[19] +61[20]
+50[21] +37[22] +30[23] +20[24] +16[25] +10[26] +7[27] +4[28] +3[29] + [30] + [31]
[221111] =30[0] +64[1] +120[2] +149[3] +197[4] +216[5] +251[6] +258[7] +279[8] +273[9] +280[10]
+263[11] +259[12] +234[13] +222[14] +193[15] +177[16] +148[17] +131[18] +105[19] +90[20]
+69[21] +57[22] +41[23] +33[24] +22[25] +17[26] +10[27] +8[28] +4[29] +3[30] + [31] + [32]
[222000] =12[0] +13[1] +32[2] +36[3] +53[4] +50[5] +68[6] +62[7] +72[8] +67[9] +73[10] +62[11]
+68[12] +56[13] +56[14] +47[15] +45[16] +34[17] +34[18] +24[19] +22[20] +16[21] +14[22]
+8[23] +9[24] +4[25] +4[26] +2[27] +2[28] + [30]
[222100] =21[0] +62[1] +97[2] +138[3] +166[4] +194[5] +216[6] +233[7] +241[8] +250[9] +246[10]
+241[11] +232[12] +219[13] +200[14] +185[15] +163[16] +144[17] +125[18] +106[19] +87[20]
+73[21] +57[22] +45[23] +35[24] +26[25] +18[26] +14[27] +9[28] +6[29] +4[30] +2[31] + [32]
+ [33]
[222110] =35[0] +127[1] +193[2] +276[3] +329[4] +395[5] +430[6] +474[7] +487[8] +508[9] +500[10]
+500[11] +474[12] +457[13] +419[14] +391[15] +346[16] +313[17] +268[18] +235[19] +194[20]
+165[21] +131[22] +108[23] +82[24] +65[25] +47[26] +36[27] +24[28] +18[29] +11[30] +8[31]
+4[32] +3[33] + [34] + [35]
[222111] =62[0] +164[1] +276[2] +372[3] +466[4] +534[5] +608[6] +647[7] +685[8] +700[9] +707[10]
+689[11] +675[12] +635[13] +597[14] +548[15] +499[16] +440[17] +392[18] +334[19] +286[20]
+238[21] +197[22] +156[23] +127[24] +96[25] +74[26] +55[27] +40[28] +27[29] +20[30] +12[31]
+8[32] +5[33] +3[34] + [35] + [36]
[222200] =33[0] +58[1] +121[2] +141[3] +198[4] +210[5] +254[6] +256[7] +288[8] +279[9] +296[10]
+277[11] +284[12] +257[13] +254[14] +222[15] +214[16] +181[17] +169[18] +138[19] +126[20]
+99[21] +88[22] +66[23] +58[24] +41[25] +35[26] +23[27] +20[28] +12[29] +10[30] +5[31]
+5[32] +2[33] +2[34] + [36]
[222210] =62[0] +171[1] +298[2] +387[3] +495[4] +570[5] +645[6] +692[7] +742[8] +751[9] +771[10]
+758[11] +742[12] +706[13] +676[14] +618[15] +574[16] +514[17] +460[18] +400[19] +351[20]
+292[21] +250[22] +204[23] +166[24] +131[25] +106[26] +78[27] +61[28] +44[29] +32[30]
+22[31] +16[32] +9[33] +7[34] +4[35] +2[36] + [37] + [38]
[222211] =78[0] +275[1] +422[2] +601[3] +723[4] +869[5] +953[6] +1056[7] +1096[8] +1151[9] +1147[10]
+1157[11] +1114[12] +1086[13] +1014[14] +959[15] +869[16] +799[17] +703[18] +629[19]
+537[20] +468[21] +387[22] +328[23] +262[24] +216[25] +166[26] +133[27] +98[28] +76[29]
+53[30] +40[31] +26[32] +19[33] +11[34] +8[35] +4[36] +3[37] + [38] + [39]
[222220] =56[0] +131[1] +245[2] +304[3] +404[4] +455[5] +527[6] +556[7] +611[8] +609[9] +640[10]
+623[11] +623[12] +589[13] +576[14] +524[15] +499[16] +445[17] +408[18] +354[19] +321[20]
+266[21] +236[22] +192[23] +163[24] +129[25] +109[26] +80[27] +67[28] +48[29] +38[30]
+26[31] +21[32] +12[33] +10[34] +6[35] +4[36] +2[37] +2[38] + [40]
[222221] =84[0] +271[1] +429[2] +598[3] +734[4] +872[5] +971[6] +1067[7] +1123[8] +1175[9] +1186[10]
+1194[11] +1166[12] +1136[13] +1076[14] +1019[15] +940[16] +866[17] +776[18] +697[19]
+609[20] +533[21] +452[22] +385[23] +318[24] +264[25] +211[26] +170[27] +131[28] +103[29]
+76[30] +58[31] +41[32] +30[33] +20[34] +14[35] +9[36] +6[37] +3[38] +2[39] + [40] + [41]
[222222] =59[0] +130[1] +233[2] +307[3] +396[4] +444[5] +526[6] +554[7] +601[8] +617[9] +641[10]
+626[11] +637[12] +603[13] +588[14] +550[15] +520[16] +467[17] +439[18] +383[19] +345[20]
+299[21] +263[22] +217[23] +191[24] +152[25] +128[26] +102[27] +83[28] +61[29] +52[30]
+36[31] +28[32] +20[33] +15[34] +9[35] +8[36] +4[37] +3[38] +2[39] + [40] + [42]

 Branching Rules for the Reduction $R_{15} \rightarrow R_3$.

```
[0000000] = [0]

[1000000] = [7]
[1100000] = [1] + [3] + [5] + [7] + [9] + [11] + [13]
[1110000] = [0] + [2] + [3] +2[4] + [5] +3[6] +2[7] +2[8] +2[9] +2[10] + [11] +2[12] + [13] + [14]
          + [15] + [16] + [18]
[1111000] =2[0] +4[2] +2[3] +5[4] +3[5] +6[6] +4[7] +6[8] +4[9] +6[10] +4[11] +5[12] +3[13] +4[14]
          +2[15] +3[16] + [17] +2[18] + [19] + [20] + [22]
[1111100] =4[1] +3[2] +7[3] +6[4] +10[5] +8[6] +11[7] +9[8] +11[9] +9[10] +10[11] +8[12] +9[13]
          +6[14] +7[15] +5[16] +5[17] +3[18] +3[19] +2[20] +2[21] + [22] + [23] + [25]
[1111110] =6[1] +4[2] +12[3] +9[4] +14[5] +13[6] +17[7] +13[8] +18[9] +14[10] +16[11] +13[12]
          +14[13] +10[14] +12[15] +8[16] +8[17] +6[18] +6[19] +3[20] +4[21] +2[22] +2[23] + [24]
          + [25] + [27]
[1111111] =4[0] +3[1] +10[2] +9[3] +16[4] +14[5] +19[6] +17[7] +21[8] +18[9] +21[10] +17[11] +19[12]
          +15[13] +16[14] +12[15] +13[16] +9[17] +9[18] +6[19] +6[20] +4[21] +4[22] +2[23] +2[24]
          + [25] + [26] + [28]

[2000000] = [2] + [4] + [6] + [8] + [10] + [12] + [14]
[2100000] = [1] +2[2] +2[3] +3[4] +4[5] +4[6] +4[7] +5[8] +4[9] +4[10] +4[11] +3[12] +3[13] +3[14]
          +2[15] +2[16] +2[17] + [18] + [19] + [20]
[2110000] =6[1] +6[2] +11[3] +11[4] +15[5] +14[6] +17[7] +16[8] +18[9] +16[10] +17[11] +15[12]
          +15[13] +12[14] +12[15] +9[16] +9[17] +6[18] +6[19] +4[20] +4[21] +2[22] +2[23] + [24]
          + [25]
[2111000] =3[0] +12[1] +19[2] +26[3] +32[4] +38[5] +41[6] +45[7] +47[8] +48[9] +47[10] +47[11]
          +44[12] +42[13] +39[14] +35[15] +31[16] +28[17] +23[18] +20[19] +16[20] +13[21] +10[22]
          +8[23] +6[24] +4[25] +3[26] +2[27] + [28] + [29]
[2111100] =9[0] +22[1] +41[2] +50[3] +67[4] +75[5] +86[6] +90[7] +99[8] +97[9] +101[10] +97[11]
          +96[12] +89[13] +86[14] +76[15] +71[16] +62[17] +55[18] +46[19] +41[20] +32[21] +27[22]
          +21[23] +17[24] +12[25] +10[26] +6[27] +5[28] +3[29] +2[30] + [31] + [32]
[2111110] =13[0] +36[1] +62[2] +82[3] +104[4] +119[5] +136[6] +145[7] +155[8] +158[9] +161[10]
          +158[11] +155[12] +147[13] +140[14] +128[15] +118[16] +105[17] +94[18] +81[19] +70[20]
          +58[21] +49[22] +39[23] +32[24] +24[25] +19[26] +14[27] +10[28] +7[29] +5[30] +3[31]
          +2[32] + [33] + [34]
[2111111] =13[0] +48[1] +75[2] +105[3] +127[4] +153[5] +167[6] +185[7] +193[8] +201[9] +201[10]
          +203[11] +194[12] +189[13] +177[14] +166[15] +150[16] +138[17] +120[18] +107[19] +91[20]
          +78[21] +64[22] +54[23] +42[24] +34[25] +26[26] +20[27] +14[28] +11[29] +7[30] +5[31]
          +3[32] +2[33] + [34] + [35]
[2200000] =4[0] +8[2] +4[3] +11[4] +7[5] +13[6] +9[7] +14[8] +10[9] +14[10] +10[11] +13[12] +9[13]
          +11[14] +7[15] +9[16] +5[17] +7[18] +3[19] +5[20] +2[21] +3[22] + [23] +2[24] + [26]
[2210000] =4[0] +17[1] +24[2] +37[3] +42[4] +52[5] +56[6] +63[7] +63[8] +68[9] +66[10] +67[11]
          +63[12] +62[13] +56[14] +54[15] +47[16] +43[17] +37[18] +33[19] +26[20] +23[21] +18[22]
          +15[23] +11[24] +9[25] +6[26] +5[27] +3[28] +2[29] + [30] + [31]
[2211000] =9[0] +55[1] +72[2] +115[3] +127[4] +164[5] +170[6] +199[7] +197[8] +217[9] +208[10]
          +219[11] +203[12] +206[13] +186[14] +183[15] +160[16] +153[17] +130[18] +121[19] +99[20]
          +90[21] +71[22] +63[23] +47[24] +41[25] +29[26] +25[27] +16[28] +14[29] +8[30] +7[31]
          +3[32] +3[33] + [34] + [35]
[2211100] =41[0] +99[1] +177[2] +229[3] +297[4] +338[5] +392[6] +417[7] +454[8] +462[9] +481[10]
          +473[11] +475[12] +453[13] +441[14] +409[15] +387[16] +349[17] +322[18] +282[19] +254[20]
          +216[21] +189[22] +156[23] +133[24] +106[25] +88[26] +67[27] +54[28] +39[29] +31[30]
          +21[31] +16[32] +10[33] +7[34] +4[35] +3[36] + [37] + [38]
[2211110] =69[0] +160[1] +291[2] +374[3] +490[4] +553[5] +648[6] +687[7] +752[8] +766[9] +802[10]
          +788[11] +798[12] +762[13] +747[14] +695[15] +664[16] +600[17] +560[18] +493[19] +448[20]
          +384[21] +341[22] +283[23] +246[24] +198[25] +167[26] +130[27] +107[28] +79[29] +64[30]
          +45[31] +35[32] +23[33] +18[34] +10[35] +8[36] +4[37] +3[38] + [39] + [40]
[2211111] =63[0] +228[1] +347[2] +499[3] +601[4] +729[5] +803[6] +899[7] +941[8] +1001[9] +1008[10]
          +1032[11] +1008[12] +1000[13] +949[14] +916[15] +847[16] +797[17] +718[18] +659[19]
          +579[20] +519[21] +443[22] +388[23] +322[24] +275[25] +221[26] +184[27] +143[28] +116[29]
          +86[30] +68[31] +48[32] +37[33] +24[34] +18[35] +11[36] +8[37] +4[38] +3[39] + [40] + [41]
[2220000] =18[0] +22[1] +52[2] +59[3] +87[4] +86[5] +114[6] +109[7] +128[8] +123[9] +136[10]
          +123[11] +135[12] +119[13] +122[14] +108[15] +108[16] +90[17] +90[18] +73[19] +69[20]
          +56[21] +52[22] +38[23] +37[24] +26[25] +23[26] +16[27] +15[28] +8[29] +9[30] +4[31]
          +4[32] +2[33] +2[34] + [36]
[2221000] =41[0] +117[1] +195[2] +264[3] +333[4] +389[5] +442[6] +482[7] +517[8] +538[9] +554[10]
          +556[11] +554[12] +540[13] +522[14] +495[15] +467[16] +431[17] +396[18] +357[19] +320[20]
          +281[21] +246[22] +210[23] +179[24] +149[25] +123[26] +99[27] +80[28] +62[29] +48[30]
          +36[31] +27[32] +19[33] +14[34] +9[35] +6[36] +4[37] +2[38] + [39] + [40]
[2221100] =86[0] +298[1] +462[2] +658[3] +800[4] +966[5] +1074[6] +1200[7] +1266[8] +1346[9]
          +1368[10] +1402[11] +1382[12] +1375[13] +1319[14] +1278[15] +1196[16] +1131[17] +1033[18]
          +955[19] +851[20] +769[21] +669[22] +591[23] +501[24] +433[25] +357[26] +301[27] +241[28]
          +198[29] +153[30] +123[31] +91[32] +71[33] +50[34] +38[35] +25[36] +19[37] +11[38] +8[39]
          +4[40] +3[41] + [42] + [43]
```

[2221110] =165[0] +503[1] +811[2] +1131[3] +1401[4] +1663[5] +1883[6] +2079[7] +2219[8] +2347[9]
+2411[10] +2454[11] +2452[12] +2425[13] +2352[14] +2274[15] +2153[16] +2027[17] +1881[18]
+1732[19] +1566[20] +1415[21] +1251[22] +1101[23] +955[24] +822[25] +692[26] +585[27]
+480[28] +393[29] +316[30] +252[31] +194[32] +152[33] +113[34] +84[35] +61[36] +44[37]
+29[38] +21[39] +13[40] +8[41] +5[42] +3[43] + [44] + [45]

[2221111] =227[0] +639[1] +1076[2] +1455[3] +1842[4] +2158[5] +2470[6] +2702[7] +2920[8] +3055[9]
+3174[10] +3211[11] +3232[12] +3180[13] +3117[14] +2991[15] +2862[16] +2683[17] +2510[18]
+2302[19] +2106[20] +1889[21] +1693[22] +1484[23] +1301[24] +1115[25] +956[26] +799[27]
+670[28] +545[29] +446[30] +353[31] +281[32] +215[33] +167[34] +123[35] +92[36] +65[37]
+47[38] +31[39] +22[40] +13[41] +9[42] +5[43] +3[44] + [45] + [46]

[2222000] =70[0] +140[1] +279[2] +337[3] +464[4] +512[5] +614[6] +642[7] +725[8] +727[9] +782[10]
+766[11] +793[12] +756[13] +764[14] +708[15] +698[16] +635[17] +609[18] +540[19] +511[20]
+440[21] +407[22] +344[23] +311[24] +255[25] +228[26] +180[27] +158[28] +122[29] +104[30]
+76[31] +66[32] +45[33] +38[34] +25[35] +21[36] +12[37] +11[38] +5[39] +5[40] +2[41]
+2[42] + [44]

[2222100] =159[0] +518[1] +833[2] +1158[3] +1434[4] +1719[5] +1932[6] +2151[7] +2301[8] +2437[9]
+2512[10] +2579[11] +2572[12] +2566[13] +2503[14] +2430[15] +2314[16] +2203[17] +2048[18]
+1908[19] +1741[20] +1584[21] +1415[22] +1266[23] +1102[24] +965[25] +826[26] +705[27]
+588[28] +494[29] +400[30] +328[31] +260[32] +206[33] +158[34] +124[35] +90[36] +68[37]
+49[38] +35[39] +23[40] +17[41] +10[42] +7[43] +4[44] +2[45] + [46] + [47]

[2222110] =302[0] +1003[1] +1582[2] +2235[3] +2747[4] +3310[5] +3715[6] +4155[7] +4431[8] +4729[9]
+4862[10] +5013[11] +5009[12] +5022[13] +4894[14] +4789[15] +4564[16] +4369[17] +4075[18]
+3821[19] +3492[20] +3210[21] +2873[22] +2590[23] +2271[24] +2008[25] +1722[26] +1493[27]
+1252[28] +1064[29] +870[30] +724[31] +577[32] +470[33] +363[34] +289[35] +216[36]
+168[37] +120[38] +91[39] +62[40] +46[41] +29[42] +21[43] +12[44] +9[45] +4[46] +3[47]
+ [48] + [49]

[2222111] =462[0] +1301[1] +2202[2] +2970[3] +3777[4] +4435[5] +5091[6] +5588[7] +6073[8] +6377[9]
+6667[10] +6787[11] +6877[12] +6816[13] +6741[14] +6521[15] +6303[16] +5974[17] +5650[18]
+5245[19] +4868[20] +4425[21] +4027[22] +3593[23] +3203[24] +2799[25] +2452[26] +2095[27]
+1798[28] +1506[29] +1264[30] +1033[31] +851[32] +676[33] +544[34] +422[35] +330[36]
+247[37] +190[38] +136[39] +101[40] +70[41] +50[42] +32[43] +23[44] +13[45] +9[46] +5[47]
+3[48] + [49] + [50]

[2222200] =197[0] +486[1] +868[2] +1134[3] +1478[4] +1701[5] +1991[6] +2152[7] +2372[8] +2463[9]
+2609[10] +2628[11] +2696[12] +2648[13] +2648[14] +2544[15] +2485[16] +2337[17] +2238[18]
+2061[19] +1936[20] +1748[21] +1611[22] +1426[23] +1290[24] +1118[25] +995[26] +843[27]
+736[28] +609[29] +524[30] +422[31] +356[32] +279[33] +231[34] +176[35] +143[36] +104[37]
+84[38] +58[39] +46[40] +30[41] +24[42] +14[43] +11[44] +6[45] +5[46] +2[47] +2[48] + [50]

[2222210] =439[0] +1268[1] +2107[2] +2886[3] +3641[4] +4302[5] +4931[6] +5443[7] +5900[8] +6245[9]
+6520[10] +6675[11] +6777[12] +6760[13] +6688[14] +6527[15] +6321[16] +6035[17] +5732[18]
+5368[19] +4997[20] +4596[21] +4200[22] +3786[23] +3401[24] +3010[25] +2649[26] +2303[27]
+1991[28] +1693[29] +1437[30] +1198[31] +994[32] +812[33] +660[34] +524[35] +418[36]
+324[37] +250[38] +189[39] +143[40] +103[41] +76[42] +53[43] +37[44] +25[45] +17[46]
+10[47] +7[48] +4[49] +2[50] + [51] + [52]

[2222211] =606[0] +1889[1] +3058[2] +4257[3] +5304[4] +6343[5] +7200[6] +8020[7] +8641[8] +9206[9]
+9565[10] +9863[11] +9962[12] +10004[13] +9865[14] +9683[15] +9347[16] +8987[17] +8504[18]
+8020[19] +7447[20] +6895[21] +6284[22] +5714[23] +5113[24] +4566[25] +4010[26] +3517[27]
+3030[28] +2609[29] +2203[30] +1861[31] +1539[32] +1275[33] +1030[34] +836[35] +659[36]
+523[37] +401[38] +311[39] +231[40] +175[41] +125[42] +92[43] +63[44] +45[45] +29[46]
+20[47] +12[48] +8[49] +4[50] +3[51] + [52] + [53]

[2222220] =356[0] +918[1] +1594[2] +2130[3] +2745[4] +3187[5] +3721[6] +4059[7] +4456[8] +4685[9]
+4948[10] +5032[11] +5175[12] +5135[13] +5139[14] +5001[15] +4900[16] +4663[17] +4491[18]
+4196[19] +3956[20] +3638[21] +3373[22] +3036[23] +2776[24] +2454[25] +2198[26] +1914[27]
+1687[28] +1434[29] +1248[30] +1040[31] +885[32] +725[33] +608[34] +481[35] +400[36]
+309[37] +249[38] +189[39] +150[40] +107[41] +86[42] +59[43] +45[44] +30[45] +23[46]
+13[47] +11[48] +6[49] +4[50] +2[51] +2[52] + [54]

[2222221] =563[0] +1762[1] +2852[2] +3975[3] +4954[4] +5934[5] +6745[6] +7529[7] +8124[8] +8677[9]
+9038[10] +9347[11] +9468[12] +9542[13] +9444[14] +9308[15] +9022[16] +8716[17] +8290[18]
+7862[19] +7341[20] +6839[21] +6276[22] +5747[23] +5181[24] +4665[25] +4133[26] +3658[27]
+3183[28] +2769[29] +2366[30] +2023[31] +1694[32] +1422[33] +1168[34] +962[35] +772[36]
+624[37] +489[38] +387[39] +295[40] +228[41] +169[42] +128[43] +91[44] +67[45] +46[46]
+33[47] +21[48] +15[49] +9[50] +6[51] +3[52] +2[53] + [54] + [55]

[2222222] =308[0] +853[1] +1463[2] +1957[3] +2514[4] +2955[5] +3411[6] +3757[7] +4124[8] +4344[9]
+4589[10] +4707[11] +4820[12] +4822[13] +4835[14] +4723[15] +4638[16] +4457[17] +4283[18]
+4040[19] +3827[20] +3539[21] +3293[22] +3003[23] +2743[24] +2457[25] +2216[26] +1944[27]
+1725[28] +1494[29] +1298[30] +1102[31] +949[32] +786[33] +663[34] +543[35] +448[36]
+357[37] +293[38] +225[39] +181[40] +138[41] +107[42] +78[43] +62[44] +42[45] +32[46]
+22[47] +16[48] +10[49] +8[50] +4[51] +3[52] +2[53] + [54] + [56]

C-22 Branching Rules for the Reduction $R_{17} \rightarrow R_3$.

[00000000] = [0]

[10000000] = [8]

[11000000] = [1] + [3] + [5] + [7] + [9] + [11] + [13] + [15]

[11100000] = [1] +2[3] + [4] +2[5] +2[6] +3[7] +2[8] +3[9] +2[10] +2[11] +2[12] +2[13] + [14] +2[15] + [16] + [17] + [18] + [19] + [21]

[11110000] =2[0] +5[2] +2[3] +6[4] +4[5] +7[6] +5[7] +8[8] +5[9] +8[10] +6[11] +7[12] +5[13] +7[14] +4[15] +5[16] +3[17] +4[18] +2[19] +3[20] + [21] +2[22] + [23] + [24] + [26]

[11111000] =3[0] +2[1] +7[2] +7[3] +11[4] +10[5] +15[6] +13[7] +16[8] +14[9] +17[10] +14[11] +16[12] +13[13] +14[14] +12[15] +12[16] +9[17] +10[18] +7[19] +7[20] +5[21] +5[22] +3[23] +3[24] +2[25] +2[26] + [27] + [28] + [30]

[11111100] =9[1] +7[2] +18[3] +15[4] +23[5] +21[6] +28[7] +24[8] +31[9] +26[10] +30[11] +26[12] +29[13] +23[14] +26[15] +20[16] +21[17] +17[18] +17[19] +12[20] +13[21] +9[22] +9[23] +6[24] +6[25] +3[26] +4[27] +2[28] +2[29] + [30] + [31] + [33]

[11111110] =12[1] +13[2] +23[3] +23[4] +34[5] +31[6] +40[7] +38[8] +43[9] +40[10] +45[11] +39[12] +42[13] +37[14] +38[15] +32[16] +33[17] +26[18] +26[19] +21[20] +20[21] +15[22] +15[23] +10[24] +10[25] +7[26] +6[27] +4[28] +4[29] +2[30] +2[31] + [32] + [33] + [35]

[11111111] =8[0] +8[1] +21[2] +22[3] +35[4] +33[5] +45[6] +42[7] +51[8] +48[9] +54[10] +48[11] +54[12] +47[13] +49[14] +43[15] +44[16] +36[17] +37[18] +29[19] +29[20] +23[21] +22[22] +16[23] +16[24] +11[25] +10[26] +7[27] +7[28] +4[29] +4[30] +2[31] +2[32] + [33] + [34] + [36]

C-23 Branching Rules for the Reduction $R_{19} \rightarrow R_3$.

[000000000] = [0]

[100000000] = [9]

[110000000] = [1] + [3] + [5] + [7] + [9] + [11] + [13] + [15] + [17]

[111000000] = [0] + [2] + [3] +2[4] + [5] +3[6] +2[7] +3[8] +3[9] +3[10] +2[11] +3[12] +2[13] +2[14] +2[15] +2[16] + [17] +2[18] + [19] + [20] + [21] + [22] + [24]

[111100000] =3[0] +5[2] +3[3] +7[4] +4[5] +9[6] +6[7] +9[8] +7[9] +10[10] +7[11] +10[12] +7[13] +9[14] +7[15] +8[16] +5[17] +7[18] +4[19] +5[20] +3[21] +4[22] +2[23] +3[24] + [25] +2[26] + [27] + [28] + [30]

[111110000] =6[1] +6[2] +12[3] +11[4] +17[5] +16[6] +21[7] +19[8] +23[9] +21[10] +24[11] +21[12] +23[13] +20[14] +22[15] +18[16] +19[17] +16[18] +16[19] +13[20] +13[21] +10[22] +10[23] +7[24] +7[25] +5[26] +5[27] +3[28] +3[29] +2[30] +2[31] + [32]+ [33] + [35]

[111111000] =13[1] +11[2] +26[3] +23[4] +34[5] +33[6] +43[7] +38[8] +49[9] +43[10] +50[11] +45[12] +50[13] +43[14] +48[15] +40[16] +42[17] +36[18] +37[19] +29[20] +31[21] +24[22] +24[23] +19[24] +18[25] +13[26] +14[27] +9[28] +9[29] +6[30] +6[31] +3[32] +4[33] +2[34] +2[35] + [36] + [37] + [39]

[111111100] =10[0] +12[1] +29[2] +33[3] +48[4] +49[5] +65[6] +64[7] +75[8] +74[9] +83[10] +78[11] +86[12] +79[13] +83[14] +77[15] +78[16] +69[17] +71[18] +61[19] +60[20] +52[21] +50[22] +41[23] +40[24] +32[25] +30[26] +24[27] +22[28] +16[29] +16[30] +11[31] +10[32] +7[33] +6[34] +4[35] +4[36] +2[37] +2[38] + [39] + [40] + [42]

[111111110] =13[0] +17[1] +42[2] +45[3] +69[4] +70[5] +91[6] +90[7] +108[8] +105[9] +118[10] +112[11] +123[12] +114[13] +121[14] +110[15] +114[16] +102[17] +103[18] +90[19] +90[20] +77[21] +75[22] +63[23] +61[24] +49[25] +47[26] +37[27] +35[28] +27[29] +25[30] +18[31] +17[32] +12[33] +11[34] +7[35] +7[36] +4[37] +4[38] +2[39] +2[40] + [41] + [42] + [44]

[111111111] =5[0] +32[1] +37[2] +66[3] +70[4] +94[5] +98[6] +118[7] +117[8] +136[9] +131[10] +143[11] +138[12] +146[13] +135[14] +142[15] +129[16] +130[17] +118[18] +116[19] +102[20] +101[21] +86[22] +82[23] +70[24] +66[25] +53[26] +51[27] +40[28] +37[29] +29[30] +26[31] +19[32] +18[33] +12[34] +11[35] +8[36] +7[37] +4[38] +4[39] +2[40] +2[41] + [42] + [43] + [45]

C-24 Branching Rules for the Reduction $R_{21} \to R_3$.

[0000000000] = [0]

[1000000000] = [10]

[1100000000] = [1] + [3] + [5] + [7] + [9] + [11] + [13] + [15] + [17] + [19]

[1110000000] = [1] +2[3] + [4] +2[5] +2[6] +3[7] +2[8] +4[9] +3[10] +3[11] +3[12] +3[13] +2[14] +3[15] +2[16] +2[17] +2[18] +2[19] + [20] +2[21] + [22] + [23] + [24] + [25] + [27]

[1111000000] =3[0] +6[2] +3[3] +8[4] +5[5] +10[6] +7[7] +11[8] +8[9] +12[10] +9[11] +12[12] +9[13] +12[14] +9[15] +11[16] +8[17] +10[18] +7[19] +8[20] +5[21] +7[22] +4[23] +5[24] +3[25] +4[26] +2[27] +3[28] + [29] +2[30] + [31] + [32] + [34]

[1111100000] =4[0] +4[1] +11[2] +11[3] +18[4] +18[5] +24[6] +23[7] +29[8] +27[9] +32[10] +29[11] +33[12] +30[13] +32[14] +29[15] +31[16] +27[17] +28[18] +24[19] +25[20] +21[21] +21[22] +17[23] +17[24] +14[25] +13[26] +10[27] +10[28] +7[29] +7[30] +5[31] +5[32] +3[33] +3[34] +2[35] +2[36] + [37] + [38] + [40]

[1111110000] = [0] +18[1] +16[2] +36[3] +33[4] +49[5] +48[6] +62[7] +57[8] +72[9] +66[10] +76[11] +71[12] +79[13] +70[14] +79[15] +69[16] +73[17] +65[18] +67[19] +57[20] +60[21] +50[22] +50[23] +42[24] +42[25] +33[26] +34[27] +26[28] +25[29] +20[30] +19[31] +13[32] +14[33] +9[34] +9[35] +6[36] +6[37] +3[38] +4[39] +2[40] +2[41] + [42] + [43] + [45]

[1111111000] =4[0] +30[1] +37[2] +62[3] +68[4] +91[5] +95[6] +116[7] +117[8] +134[9] +133[10] +146[11] +142[12] +152[13] +145[14] +151[15] +142[16] +145[17] +134[18] +135[19] +122[20] +121[21] +108[22] +105[23] +92[24] +89[25] +76[26] +72[27] +61[28] +57[29] +47[30] +44[31] +35[32] +32[33] +25[34] +23[35] +17[36] +16[37] +11[38] +10[39] +7[40] +6[41] +4[42] +4[43] +2[44] +2[45] + [46] + [47] + [49]

[1111111100] =22[0] +33[1] +75[2] +85[3] +126[4] +133[5] +169[6] +173[7] +205[8] +205[9] +231[10] +226[11] +247[12] +238[13] +252[14] +239[15] +249[16] +232[17] +236[18] +217[19] +218[20] +197[21] +195[22] +173[23] +169[24] +148[25] +142[26] +122[27] +117[28] +98[29] +92[30] +76[31] +71[32] +57[33] +53[34] +41[35] +38[36] +29[37] +26[38] +19[39] +18[40] +12[41] +11[42] +7[43] +7[44] +4[45] +4[46] +2[47] +2[48] + [49] + [50] + [52]

[1111111110] =28[0] +48[1] +99[2] +122[3] +169[4] +184[5] +232[6] +242[7] +278[8] +287[9] +317[10] +315[11] +341[12] +333[13] +348[14] +338[15] +345[16] +326[17] +332[18] +309[19] +305[20] +283[21] +276[22] +249[23] +242[24] +215[25] +204[26] +181[27] +169[28] +145[29] +137[30] +115[31] +105[32] +88[33] +80[34] +64[35] +59[36] +46[37] +41[38] +32[39] +28[40] +20[41] +19[42] +13[43] +11[44] +8[45] +7[46] +4[47] +4[48] +2[49] +2[50] + [51] + [52] + [54]

[1111111111] =13[0] +75[1] +99[2] +159[3] +180[4] +236[5] +252[6] +301[7] +311[8] +352[9] +356[10] +388[11] +384[12] +408[13] +396[14] +412[15] +394[16] +401[17] +378[18] +379[19] +352[20] +347[21] +318[22] +309[23] +278[24] +268[25] +237[26] +225[27] +197[28] +184[29] +158[30] +147[31] +123[32] +113[33] +93[34] +85[35] +68[36] +62[37] +48[38] +43[39] +33[40] +29[41] +21[42] +19[43] +13[44] +12[45] +8[46] +7[47] +4[48] +4[49] +2[50] +2[51] + [52] + [53] + [55]

C-25 Branching Rules for the Reduction $Sp_4 \to R_3$.

<00> = [0]

<10> = {3}
<11> = [2]

<20> = [1] + [3]
<21> = {1} + {5} + {7}
<22> = [2] + [4]

<30> = {3} + {5} + {9}
<31> = [1] + [2] + [3] + [4] + [5]
<32> = {3} + {5} + {7} + {9} + {11}
<33> = [0] + [3] + [4] + [6]

<40> = [0] + [2] + [3] + [4] + [6]
<41> = {1} + {3} + {5} +2{7} + {9} + {11} + {13}
<42> = [1] + [2] +2[3] + [4] +2[5] + [6] + [7]
<43> = {3} + {5} + {7} +2{9} + {11} + {13} + {15}
<44> = [2] + [4] + [5] + [6] + [8]

C-26 Branching Rules for the Reduction $Sp_6 \to R_3$.

<000> = [0]

<100> = {5}
<110> = [2] + [4]
<111> = {3} + {9}

<200> = [1] + [3] + [5]
<210> = {1} + {3} + {5} +2{7} + {9} + {11} + {13}
<211> = [1] + [2] +2[3] + [4] + [5] + [6] + [7]
<220> = [0] +2[2] + [3] +2[4] + [5] +2[6] + [8]
<221> = {1} + {3} +2{5} +2{7} +2{9} +2{11} + {13} + {15} + {17}
<222> = [1] +2[3] + [4] + [5] + [6] + [7] + [9]

C-27 Branching Rules for the Reduction $Sp_8 \to R_3$.

<0000> = [0]

<1000> = {7}
<1100> = [2] + [4] + [6]
<1110> = {3} + {5} + {9} + {11} + {15}
<1111> = [2] + [4] + [5] + [8]

<2000> = [1] + [3] + [5] + [7]
<2100> = {1} + {3} +2{5} +2{7} +2{9} +2{11} +2{13} + {15} + {17} + {19}
<2110> =2[1] +2[2] +4[3] +3[4] +4[5] +3[6] +3[7] +2[8] +2[9] + [10] + [11]
<2111> = {1} +2{3} +2{5} +3{7} +3{9} +3{11} +3{13} +2{15} +2{17} + {19} + {21} + {23}
<2200> =2[0] +3[2] +2[3] +4[4] +2[5] +4[6] +2[7] +3[8] + [9] +2[10] + [12]
<2210> =2{1} +4{3} +5{5} +7{7} +7{9} +7{11} +7{13} +6{15} +5{17} +4{19} +3{21} +2{23} + {25} + {27}
<2211> =2[0] +2[1] +5[2] +5[3] +7[4] +6[5] +7[6] +6[7] +6[8] +4[9] +4[10] +2[11] +2[12] + [13]
+ [14]
<2220> =4[1] +2[2] +7[3] +5[4] +7[5] +6[6] +7[7] +4[8] +6[9] +3[10] +3[11] +2[12] +2[13] + [15]
<2221> =3{1} +3{3} +6{5} +7{7} +7{9} +8{11} +8{13} +7{15} +6{17} +6{19} +4{21} +3{23} +3{25} + {27}
+ {29} + {31}
<2222> = [0] + [1] +3[2] +2[3] +5[4] +3[5] +5[6] +3[7] +4[8] +3[9] +3[10] +2[11] +2[12] + [13]
+ [14] + [16]

C-28 Branching Rules for the Reduction $Sp_{10} \to R_3$.

<00000> = [0]

<10000> = {9}
<11000> = [2] + [4] + [6] + [8]
<11100> = {3} + {5} + {7} + {9} + {11} + {13} + {15} + {17} + {21}
<11110> = [0] + [2] + [3] +2[4] + [5] +2[6] + [7] + [8] + [9] + [10] + [12]
<11111> = {1} + {5} + {7} + {9} + {11} + {13} + {15} + {17} + {19} + {25}

<20000> = [1] + [3] + [5] + [7] + [9]
<21000> = {1} + {3} +2{5} +3{7} +2{9} +3{11} +3{13} +2{15} +2{17} +2{19} + {21} + {23} + {25}
<21100> =3[1] +3[2] +6[3] +5[4] +7[5] +6[6] +7[7] +5[8] +5[9] +4[10] +4[11] +2[12] +2[13] + [14]
+ [15]
<21110> =2{1} +5{3} +6{5} +7{7} +9{9} +9{11} +9{13} +9{15} +8{17} +7{19} +6{21} +5{23} +3{25} +3{27}
+2{29} + {31} + {33}
<21111> =3[1] +4[2] +5[3] +6[4] +8[5] +6[6] +7[7] +7[8] +6[9] +5[10] +5[11] +3[12] +3[13] +2[14]
+ [15] + [16] + [17]
<22000> =2[0] +5[2] +2[3] +6[4] +4[5] +6[6] +4[7] +6[8] +3[9] +5[10] +2[11] +3[12] + [13] +2[14]
+ [16]
<22100> =4{1} +7{3} +11{5} +13{7} +15{9} +16{11} +17{13} +16{15} +15{17} +14{19} +12{21} +10{23}
+8{25} +6{27} +5{29} +3{31} +2{33} + {35} + {37}
<22110> =5[0] +7[1] +16[2] +17[3] +24[4] +23[5] +28[6] +25[7] +27[8] +23[9] +23[10] +18[11] +17[12]
+12[13] +11[14] +7[15] +6[16] +3[17] +3[18] + [19] + [20]
<22111> =5{1} +9{3} +13{5} +17{7} +19{9} +21{11} +22{13} +21{15} +21{17} +19{19} +17{21} +15{23}
+13{25} +10{27} +8{29} +6{31} +4{33} +3{35} +2{37} + {39} + {41}
<22200> =9[1] +7[2] +17[3] +14[4] +20[5] +18[6] +22[7] +17[8] +21[9] +15[10] +16[11] +12[12] +12[13]
+7[14] +8[15] +4[16] +4[17] +2[18] +2[19] + [21]
<22210> =8{1} +18{3} +25{5} +31{7} +37{9} +40{11} +41{13} +43{15} +41{17} +38{19} +36{21} +31{23}
+27{25} +23{27} +18{29} +14{31} +11{33} +8{35} +5{37} +4{39} +2{41} + {43} + {45}
<22211> =2[0] +13[1] +18[2] +26[3] +29[4] +36[5] +35[6] +39[7] +37[8] +37[9] +33[10] +32[11] +26[12]
+24[13] +19[14] +16[15] +12[16] +10[17] +6[18] +5[19] +3[20] +2[21] + [22] + [23]
<22220> =7[0] +7[1] +18[2] +19[3] +29[4] +26[5] +35[6] +30[7] +35[8] +30[9] +32[10] +25[11] +27[12]
+20[13] +19[14] +14[15] +13[16] +8[17] +8[18] +4[19] +4[20] +2[21] +2[22] + [24]
<22221> =9{1} +14{3} +21{5} +28{7} +31{9} +35{11} +38{13} +37{15} +38{17} +37{19} +33{21} +31{23}
+28{25} +23{27} +20{29} +17{31} +13{33} +10{35} +8{37} +5{39} +4{41} +3{43} + {45} + {47}
+ {49}
<22222> =7[1] +7[2] +12[3] +12[4] +18[5] +15[6] +19[7] +17[8] +18[9] +16[10] +17[11] +13[12] +14[13]
+10[14] +10[15] +7[16] +7[17] +4[18] +4[19] +3[20] +2[21] + [22] + [23] + [25]

C-29 Branching Rules for the Reduction $Sp_{12} \to R_3$.

<000000> = [0]

<100000> = {11}

<110000> = [2] + [4] + [6] + [8] + [10]

<111000> = {3} + {5} + {7} +2{9} + {11} + {13} +2{15} + {17} + {19} + {21} + {23} + {27}

<111100> = [0] +2[2] + [3] +3[4] +2[5] +3[6] +2[7] +3[8] +2[9] +2[10] + [11] +2[12] + [13] + [14] + [16]

<111110> = {1} + {3} +2{5} +3{7} +2{9} +3{11} +3{13} +3{15} +3{17} +3{19} +2{21} +2{23} +2{25} + {27} + {29} + {31} + {35}

<111111> = [0] + [2] +2[3] +2[4] + [5] +3[6] +2[7] +2[8] +2[9] +2[10] + [11] +2[12] + [13] + [14] + [15] + [18]

<200000> = [1] + [3] + [5] + [7] + [9] + [11]

<210000> = {1} + {3} +2{5} +3{7} +3{9} +3{11} +4{13} +3{15} +3{17} +3{19} +2{21} +2{23} +2{25} + {27} + {29} + {31}

<211000> =4[1] +4[2] +8[3] +7[4] +10[5] +9[6] +11[7] +9[8] +10[9] +8[10] +8[11] +6[12] +6[13] +4[14] +4[15] +2[16] +2[17] + [18] + [19]

<211100> =4{1} +8{3} +11{5} +14{7} +16{9} +18{11} +19{13} +19{15} +19{17} +18{19} +17{21} +15{23} +13{25} +11{27} +9{29} +7{31} +6{33} +4{35} +3{37} +2{39} + {41} + {43}

<211110> = [0] +8[1] +11[2] +16[3] +18[4] +23[5] +22[6] +25[7] +24[8] +24[9] +22[10] +22[11] +18[12] +17[13] +14[14] +12[15] +9[16] +8[17] +5[18] +4[19] +3[20] +2[21] + [22] + [23]

<211111> =3{1} +7{3} +10{5} +12{7} +15{9} +16{11} +17{13} +18{15} +18{17} +17{19} +16{21} +15{23} +13{25} +12{27} +10{29} +8{31} +7{33} +5{35} +4{37} +3{39} +2{41} + {43} + {45} + {47}

<220000> =3[0] +6[2] +3[3] +8[4] +5[5] +9[6] +6[7] +9[8] +6[9] +8[10] +5[11] +7[12] +3[13] +5[14] +2[15] +3[16] + [17] +2[18] + [20]

<221000> =6{1} +12{3} +17{5} +22{7} +25{9} +29{11} +30{13} +31{15} +31{17} +30{19} +28{21} +26{23} +23{25} +20{27} +17{29} +14{31} +11{33} +9{35} +6{37} +5{39} +3{41} +2{43} + {45} + {47}

<221100> =10[0] +16[1] +34[2] +38[3] +54[4] +55[5] +67[6] +64[7] +71[8] +65[9] +68[10] +60[11] +59[12] +49[13] +47[14] +37[15] +34[16] +25[17] +22[18] +15[19] +13[20] +8[21] +7[22] +3[23] +3[24] + [25] + [26]

<221110> =17{1} +33{3} +48{5} +61{7} +73{9} +82{11} +88{13} +92{15} +92{17} +92{19} +88{21} +83{23} +76{25} +69{27} +60{29} +53{31} +44{33} +36{35} +29{37} +23{39} +17{41} +13{43} +9{45} +6{47} +4{49} +3{51} + {53} + {55}

<221111> =8[0] +17[1] +32[2] +39[3] +52[4] +56[5] +65[6] +66[7] +71[8] +68[9] +70[10] +64[11] +62[12] +55[13] +51[14] +43[15] +39[16] +31[17] +27[18] +21[19] +17[20] +12[21] +10[22] +6[23] +5[24] +3[25] +2[26] + [27] + [28]

<222000> = [0] +17[1] +15[2] +33[3] +30[4] +42[5] +40[6] +49[7] +42[8] +51[9] +42[10] +45[11] +38[12] +39[13] +29[14] +31[15] +22[16] +21[17] +15[18] +14[19] +8[20] +9[21] +4[22] +4[23] +2[24] +2[25] + [27]

<222100> =24{1} +47{3} +69{5} +88{7} +105{9} +118{11} +128{13} +134{15} +137{17} +136{19} +132{21} +126{23} +117{25} +107{27} +96{29} +84{31} +72{33} +61{35} +50{37} +40{39} +32{41} +24{43} +18{45} +13{47} +9{49} +6{51} +4{53} +2{55} + {57} + {59}

<222110> =13[0] +57[1] +83[2] +120[3] +140[4] +171[5] +180[6] +201[7] +202[8] +210[9] +202[10] +202[11] +185[12] +178[13] +158[14] +145[15] +124[16] +111[17] +90[18] +78[19] +61[20] +50[21] +37[22] +30[23] +20[24] +16[25] +10[26] +7[27] +4[28] +3[29] + [30] + [31]

<222111> =26{1} +56{3} +79{5} +101{7} +123{9} +138{11} +149{13} +160{15} +163{17} +162{19} +162{21} +154{23} +145{25} +136{27} +123{29} +109{31} +97{33} +83{35} +69{37} +59{39} +47{41} +37{43} +30{45} +22{47} +16{49} +12{51} +8{53} +5{55} +4{57} +2{59} + {61} + {63}

<222200> =19[0] +28[1] +64[2] +71[3] +103[4] +105[5] +130[6] +126[7] +145[8] +135[9] +145[10] +130[11] +135[12] +117[13] +116[14] +97[15] +93[16] +75[17] +70[18] +53[19] +49[20] +35[21] +31[22] +21[23] +19[24] +11[25] +10[26] +5[27] +5[28] +2[29] +2[30] + [32]

<222210> =47{1} +89{3} +132{5} +172{7} +202{9} +231{11} +254{13} +267{15} +277{17} +281{19} +275{21} +268{23} +256{25} +238{27} +219{29} +199{31} +175{33} +154{35} +133{37} +111{39} +92{41} +76{43} +59{45} +47{47} +36{49} +26{51} +19{53} +14{55} +9{57} +6{59} +4{61} +2{63} + {65} + {67}

<222211> =24[0] +54[1] +99[2] +127[3] +164[4] +183[5] +213[6] +222[7] +238[8] +239[9] +245[10] +234[11] +232[12] +214[13] +204[14] +183[15] +169[16] +145[17] +131[18] +109[19] +94[20] +76[21] +64[22] +49[23] +40[24] +29[25] +23[26] +16[27] +12[28] +7[29] +6[30] +3[31] +2[32] + [33] + [34]

<222220> =9[0] +53[1] +69[2] +110[3] +121[4] +156[5] +161[6] +186[7] +184[8] +200[9] +190[10] +198[11] +182[12] +182[13] +162[14] +158[15] +135[16] +128[17] +106[18] +97[19] +78[20] +70[21] +53[22] +47[23] +34[24] +29[25] +20[26] +17[27] +10[28] +9[29] +5[30] +4[31] +2[32] +2[33] + [35]

<222221> =35{1} +67{3} +99{5} +127{7} +153{9} +175{11} +192{13} +205{15} +213{17} +217{19} +217{21} +213{23} +204{25} +193{27} +181{29} +166{31} +150{33} +134{35} +117{37} +101{39} +87{41} +72{43} +59{45} +48{47} +38{49} +30{51} +23{53} +17{55} +12{57} +9{59} +6{61} +4{63} +3{65} + {67} + {69} + {71}

<222222> =12[0] +19[1] +38[2] +46[3] +64[4] +66[5] +85[6] +83[7] +93[8] +92[9] +98[10] +91[11] +96[12] +86[13] +85[14] +77[15] +74[16] +62[17] +60[18] +49[19] +45[20] +37[21] +33[22] +24[23] +23[24] +16[25] +14[26] +10[27] +8[28] +5[29] +5[30] +3[31] +2[32] + [33] + [34] + [36]

Branching Rules for the Reduction $Sp_{14} \to R_3$.

<0000000> = [0]

<1000000> = {13}
<1100000> = [2] + [4] + [6] + [8] + [10] + [12]
<1110000> = {3} + {5} + {7} +2{9} +2{11} + {13} +2{15} +2{17} + {19} +2{21} + {23} + {25} + {27}
+ {29} + {33}
<1111000> = [0] +3[2] + [3] +4[4] +3[5] +4[6] +3[7] +5[8] +3[9] +4[10] +3[11] +3[12] +2[13] +3[14]
+ [15] +2[16] + [17] + [18] + [20]
<1111100> = {1} +2{3} +4{5} +4{7} +5{9} +5{11} +6{13} +6{15} +6{17} +6{19} +6{21} +5{23} +5{25}
+4{27} +4{29} +3{31} +2{33} +2{35} +2{37} + {39} + {41} + {45}
<1111110> =2[0] + [1] +3[2] +4[3] +6[4] +4[5] +8[6] +6[7] +7[8] +7[9] +7[10] +5[11] +7[12] +5[13]
+5[14] +4[15] +4[16] +2[17] +3[18] +2[19] + [20] + [21] + [22] + [24]
<1111111> =2{1} + {3} +2{5} +4{7} +3{9} +4{11} +5{13} +4{15} +5{17} +5{19} +4{21} +4{23} +5{25}
+3{27} +3{29} +3{31} +2{33} +2{35} +2{37} + {39} + {41} + {43} + {49}

<2000000> = [1] + [3] + [5] + [7] + [9] + [11] + [13]
<2100000> = {1} + {3} +2{5} +3{7} +3{9} +4{11} +4{13} +4{15} +4{17} +4{19} +3{21} +3{23} +3{25}
+2{27} +2{29} +2{31} + {33} + {35} + {37}
<2110000> =5[1] +5[2] +10[3] +9[4] +13[5] +12[6] +15[7] +13[8] +15[9] +13[10] +14[11] +11[12]
+11[13] +9[14] +9[15] +6[16] +6[17] +4[18] +4[19] +2[20] +2[21] + [22] + [23]
<2111000> =6{1} +12{3} +17{5} +22{7} +26{9} +29{11} +32{13} +33{15} +34{17} +34{19} +33{21} +32{23}
+29{25} +27{27} +24{29} +21{31} +18{33} +15{35} +12{37} +10{39} +8{41} +6{43} +4{45}
+3{47} +2{49} + {51} + {53}
<2111100> =3[0] +16[1] +22[2] +33[3] +38[4] +48[5] +49[6] +56[7] +56[8] +59[9] +56[10] +57[11]
+52[12] +51[13] +45[14] +42[15] +36[16] +33[17] +26[18] +23[19] +18[20] +15[21] +11[22]
+9[23] +6[24] +5[25] +3[26] +2[27] + [28] + [29]
<2111110> =11{1} +22{3} +32{5} +41{7} +49{9} +56{11} +61{13} +65{15} +67{17} +68{19} +67{21} +65{23}
+62{25} +58{27} +53{29} +48{31} +43{33} +37{35} +32{37} +27{39} +22{41} +18{43} +14{45}
+11{47} +8{49} +6{51} +4{53} +3{55} +2{57} + {59} + {61}
<2111111> =3[0] +12[1] +18[2] +26[3] +30[4] +37[5] +40[6] +45[7] +45[8] +47[9] +47[10] +47[11]
+44[12] +43[13] +39[14] +37[15] +32[16] +29[17] +25[18] +22[19] +18[20] +15[21] +12[22]
+10[23] +7[24] +6[25] +4[26] +3[27] +2[28] + [29] + [30] + [31]
<2200000> =4[0] +7[2] +4[3] +10[4] +6[5] +12[6] +8[7] +12[8] +9[9] +12[10] +8[11] +11[12] +7[13]
+9[14] +5[15] +7[16] +3[17] +5[18] +2[19] +3[20] + [21] +2[22] + [24]
<2210000> =9{1} +17{3} +25{5} +32{7} +39{9} +43{11} +48{13} +50{15} +52{17} +52{19} +51{21} +49{23}
+47{25} +43{27} +39{29} +35{31} +31{33} +26{35} +22{37} +18{39} +15{41} +11{43} +9{45}
+6{47} +5{49} +3{51} +2{53} + {55} + {57}
<2211000> =17[0] +29[1] +61[2] +71[3] +99[4] +105[5] +128[6] +128[7] +144[8] +138[9] +148[10]
+137[11] +140[12] +126[13] +124[14] +108[15] +103[16] +86[17] +80[18] +64[19] +58[20]
+44[21] +39[22] +28[23] +24[24] +16[25] +14[26] +8[27] +7[28] +3[29] +3[30] + [31] + [32]
<2211100> =40{1} +80{3} +117{5} +151{7} +182{9} +208{11} +229{13} +245{15} +256{17} +260{19}
+262{21} +256{23} +248{25} +235{27} +221{29} +202{31} +185{33} +164{35} +145{37} +125{39}
+107{41} +89{43} +74{45} +59{47} +47{49} +36{51} +28{53} +20{55} +15{57} +10{59} +7{61}
+4{63} +3{65} + {67} + {69}
<2211110> =29[0] +67[1] +123[2] +156[3] +205[4] +230[5] +268[6] +282[7] +308[8] +310[9] +323[10]
+314[11] +315[12] +296[13] +288[14] +263[15] +248[16] +220[17] +202[18] +173[19] +155[20]
+129[21] +112[22] +90[23] +76[24] +58[25] +48[26] +35[27] +28[28] +19[29] +15[30] +9[31]
+7[32] +4[33] +3[34] + [35] + [36]
<2211111> =34{1} +67{3} +98{5} +127{7} +154{9} +176{11} +195{13} +209{15} +220{17} +225{19} +227{21}
+225{23} +219{25} +210{27} +199{29} +185{31} +171{33} +154{35} +138{37} +121{39} +106{41}
+90{43} +76{45} +63{47} +52{49} +41{51} +33{53} +25{55} +19{57} +14{59} +10{61} +7{63}
+5{65} +3{67} +2{69} + {71} + {73}
<2220000> =2[0] +28[1] +28[2] +56[3] +54[4] +75[5] +74[6] +91[7] +83[8] +99[9] +88[10] +96[11]
+86[12] +90[13] +75[14] +79[15] +64[16] +63[17] +51[18] +49[19] +36[20] +36[21] +25[22]
+23[23] +16[24] +15[25] +8[26] +9[27] +4[28] +4[29] +2[30] +2[31] + [33]
<2221000> =53{1} +101{3} +150{5} +196{7} +233{9} +269{11} +298{13} +318{15} +334{17} +344{19}
+344{21} +341{23} +333{25} +317{27} +300{29} +280{31} +255{33} +231{35} +207{37} +180{39}
+156{41} +134{43} +111{45} +92{47} +75{49} +59{51} +46{53} +36{55} +26{57} +19{59} +14{61}
+9{63} +6{65} +4{67} +2{69} + {71} + {73}
<2221100> =43[0] +162[1] +244[2] +353[3] +422[4] +513[5] +562[6] +630[7] +655[8] +697[9] +697[10]
+713[11] +691[12] +684[13] +644[14] +619[15] +567[16] +531[17] +473[18] +432[19] +374[20]
+333[21] +280[22] +243[23] +198[24] +167[25] +131[26] +108[27] +81[28] +65[29] +46[30]
+36[31] +24[32] +18[33] +11[34] +8[35] +4[36] +3[37] + [38] + [39]
<2221110> =128{1} +260{3} +380{5} +491{7} +599{9} +687{11} +761{13} +827{15} +870{17} +897{19}
+915{21} +912{23} +895{25} +872{27} +832{29} +784{31} +733{33} +674{35} +610{37} +550{39}
+486{41} +423{43} +368{45} +312{47} +261{49} +218{51} +177{53} +141{55} +113{57} +87{59}
+65{61} +50{63} +36{65} +25{67} +18{69} +12{71} +7{73} +5{75} +3{77} + {79} + {81}
<2221111> =41[0] +145[1] +230[2] +319[3] +392[4] +473[5] +522[6] +582[7] +617[8] +648[9] +660[10]
+675[11] +660[12] +654[13] +628[14] +600[15] +561[16] +528[17] +477[18] +438[19] +390[20]
+346[21] +300[22] +263[23] +219[24] +187[25] +154[26] +126[27] +100[28] +82[29] +61[30]
+48[31] +36[32] +26[33] +18[34] +14[35] +8[36] +6[37] +4[38] +2[39] + [40] + [41]

<2222000> =45[0] +80[1] +166[2] +196[3] +275[4] +295[5] +360[6] +368[7] +418[8] +412[9] +445[10]
+425[11] +443[12] +412[13] +415[14] +376[15] +370[16] +326[17] +312[18] +268[19] +251[20]
+209[21] +192[22] +154[23] +139[24] +108[25] +95[26] +70[27] +62[28] +43[29] +37[30]
+24[31] +21[32] +12[33] +11[34] +5[35] +5[36] +2[37] +2[38] + [40]

<2222100> =166{1} +325{3} +482{5} +627{7} +756{9} +874{11} +976{13} +1053{15} +1116{17} +1159{19}
+1180{21} +1184{23} +1172{25} +1140{27} +1099{29} +1044{31} +978{33} +907{35} +833{37}
+751{39} +673{41} +595{43} +519{45} +447{47} +382{49} +319{51} +266{53} +217{55} +174{57}
+138{59} +109{61} +82{63} +62{65} +46{67} +33{69} +23{71} +16{73} +10{75} +7{77} +4{79}
+2{81} + {83} + {85}

<2222110> =130[0] +335[1] +585[2] +772[3] +994[4] +1147[5] +1327[6] +1434[7] +1562[8] +1617[9]
+1689[10] +1692[11] +1711[12] +1666[13] +1639[14] +1556[15] +1493[16] +1384[17] +1298[18]
+1175[19] +1078[20] +953[21] +855[22] +738[23] +648[24] +545[25] +468[26] +383[27]
+321[28] +255[29] +209[30] +160[31] +128[32] +94[33] +73[34] +51[35] +39[36] +25[37]
+19[38] +11[39] +8[40] +4[41] +3[42] + [43] + [44]

<2222111> =177{1} +350{3} +516{5} +672{7} +816{9} +943{11} +1053{13} +1144{15} +1216{17} +1266{19}
+1296{21} +1307{23} +1300{25} +1274{27} +1235{29} +1181{31} +1118{33} +1045{35} +966{37}
+883{39} +800{41} +714{43} +632{45} +553{47} +478{49} +408{51} +345{53} +287{55} +237{57}
+192{59} +154{61} +121{63} +95{65} +72{67} +54{69} +40{71} +29{73} +20{75} +14{77} +9{79}
+6{81} +4{83} +2{85} + {87} + {89}

<2222200> =57[0] +235[1] +337[2] +508[3] +595[4] +741[5] +807[6] +922[7] +956[8] +1040[9] +1043[10]
+1088[11] +1064[12] +1078[13] +1023[14] +1014[15] +941[16] +907[17] +825[18] +778[19]
+690[20] +640[21] +553[22] +501[23] +424[24] +378[25] +309[26] +271[27] +216[28] +184[29]
+143[30] +120[31] +88[32] +74[33] +51[34] +42[35] +28[36] +23[37] +13[38] +11[39] +6[40]
+5[41] +2[42] +2[43] + [45]

<2222210> =269{1} +533{3} +785{5} +1025{7} +1244{9} +1441{11} +1613{13} +1755{15} +1869{17}
+1953{19} +2006{21} +2030{23} +2025{25} +1995{27} +1941{29} +1867{31} +1776{33} +1669{35}
+1554{37} +1431{39} +1303{41} +1176{43} +1048{45} +925{47} +808{49} +698{51} +596{53}
+503{55} +420{57} +346{59} +282{61} +227{63} +179{65} +140{67} +108{69} +81{71} +61{73}
+44{75} +31{77} +22{79} +15{81} +10{83} +6{85} +4{87} +2{89} + {91} + {93}

<2222211> =103[0] +342[1] +541[2] +764[3] +938[4] +1129[5] +1268[6] +1417[7] +1509[8] +1609[9]
+1654[10] +1703[11] +1699[12] +1701[13] +1656[14] +1618[15] +1539[16] +1470[17] +1369[18]
+1281[19] +1167[20] +1070[21] +956[22] +859[23] +750[24] +661[25] +565[26] +488[27]
+407[28] +344[29] +280[30] +232[31] +183[32] +148[33] +114[34] +90[35] +66[36] +51[37]
+36[38] +27[39] +18[40] +13[41] +8[42] +6[43] +3[44] +2[45] + [46] + [47]

<2222220> =104[0] +235[1] +429[2] +556[3] +732[4] +829[5] +985[6] +1052[7] +1165[8] +1205[9]
+1280[10] +1277[11] +1321[12] +1286[13] +1288[14] +1232[15] +1206[16] +1123[17] +1082[18]
+987[19] +927[20] +833[21] +768[22] +671[23] +612[24] +523[25] +465[26] +391[27] +342[28]
+277[29] +241[30] +190[31] +160[32] +124[33] +103[34] +75[35] +63[36] +44[37] +35[38]
+24[39] +19[40] +11[41] +10[42] +5[43] +4[44] +2[45] +2[46] + [48]

<2222221> =183{1} +353{3} +527{5} +690{7} +834{9} +971{11} +1091{13} +1188{15} +1272{17} +1336{19}
+1374{21} +1399{23} +1408{25} +1390{27} +1363{29} +1323{31} +1264{33} +1201{35} +1130{37}
+1047{39} +966{41} +882{43} +794{45} +711{47} +632{49} +551{51} +479{53} +414{55} +350{57}
+295{59} +247{61} +201{63} +165{65} +133{67} +104{69} +82{71} +64{73} +48{75} +36{77}
+27{79} +19{81} +13{83} +10{85} +6{87} +4{89} +3{91} + {93} + {95} + {97}

<2222222> =28[0] +118[1] +172[2] +256[3] +304[4] +377[5] +414[6] +476[7] +498[8] +544[9] +551[10]
+580[11] +574[12] +587[13] +568[14] +566[15] +536[16] +524[17] +486[18] +467[19] +423[20]
+399[21] +356[22] +329[23] +288[24] +262[25] +224[26] +200[27] +168[28] +148[29] +120[30]
+105[31] +83[32] +71[33] +55[34] +46[35] +34[36] +29[37] +20[38] +17[39] +11[40] +9[41]
+6[42] +5[43] +3[44] +2[45] + [46] + [47] + [49]

C-31 Branching Rules for the Reduction $Sp_{16} \rightarrow R_3$.

<00000000> = [0]

<10000000> = {15}

<11000000> = [2] + [4] + [6] + [8] + [10] + [12] + [14]

<11100000> = {3} + {5} + {7} +2{9} +2{11} +2{13} +2{15} +2{17} +2{19} +2{21} +2{23} + {25} +2{27}
+ {29} + {31} + {33} + {35} + {39}

<11110000> =2[0] +3[2] +2[3] +5[4] +3[5] +6[6] +4[7] +6[8] +5[9] +6[10] +4[11] +6[12] +4[13] +4[14]
+3[15] +4[16] +2[17] +3[18] + [19] +2[20] + [21] + [22] + [24]

<11111000> =2{1} +3{3} +5{5} +7{7} +7{9} +9{11} +9{13} +10{15} +10{17} +11{19} +10{21} +10{23}
+10{25} +9{27} +8{29} +8{31} +6{33} +6{35} +5{37} +4{39} +3{41} +3{43} +2{45} +2{47}
+ {49} + {51} + {55}

<11111100> =3[0] +2[1] +7[2] +7[3] +11[4] +10[5] +15[6] +13[7] +16[8] +14[9] +17[10] +14[11] +16[12]
+13[13] +14[14] +12[15] +12[16] +9[17] +10[18] +7[19] +7[20] +5[21] +5[22] +3[23] +3[24]
+2[25] +2[26] + [27] + [28] + [30]

<11111110> =2{1} +6{3} +7{5} +9{7} +12{9} +13{11} +14{13} +16{15} +16{17} +16{19} +17{21} +17{23}
+15{25} +16{27} +14{29} +13{31} +12{33} +11{35} +9{37} +9{39} +7{41} +6{43} +5{45} +4{47}
+3{49} +3{51} +2{53} + {55} + {57} + {59} + {63}

<11111111> = [0] +2[1] +5[2] +4[3] +8[4] +8[5] +9[6] +9[7] +12[8] +10[9] +11[10] +11[11] +11[12]
+10[13] +11[14] +8[15] +9[16] +8[17] +7[18] +6[19] +6[20] +4[21] +4[22] +3[23] +3[24]
+2[25] +2[26] + [27] + [28] + [29] + [32]

C-32 Branching Rules for the Reduction $Sp_{18} \rightarrow R_3$.

<000000000> = [0]

<100000000> = {17}

<110000000> = [2] + [4] + [6] + [8] + [10] + [12] + [14] + [16]

<111000000> = {3} + {5} + {7} +2{9} +2{11} +2{13} +3{15} +2{17} +2{19} +3{21} +2{23} +2{25} +2{27}
+2{29} + {31} +2{33} + {35} + {37} + {39} + {41} + {45}

<111100000> =2[0] +4[2] +2[3] +6[4] +4[5] +7[6] +5[7] +8[8] +6[9] +8[10] +6[11] +8[12] +6[13] +7[14]
+5[15] +6[16] +4[17] +5[18] +3[19] +4[20] +2[21] +3[22] + [23] +2[24] + [25] + [26]
+ [28]

<111110000> =3{1} +4{3} +7{5} +9{7} +11{9} +12{11} +14{13} +14{15} +16{17} +16{19} +16{21} +16{23}
+17{25} +15{27} +15{29} +14{31} +13{33} +12{35} +11{37} +9{39} +9{41} +7{43} +6{45}
+5{47} +5{49} +3{51} +3{53} +2{55} +2{57} + {59} + {61} + {65}

<111111000> =5[0] +4[1] +11[2] +12[3] +19[4] +17[5] +26[6] +23[7] +28[8] +27[9] +31[10] +27[11]
+32[12] +27[13] +29[14] +26[15] +27[16] +22[17] +24[18] +19[19] +19[20] +16[21] +15[22]
+11[23] +12[24] +8[25] +8[26] +6[27] +5[28] +3[29] +4[30] +2[31] +2[32] + [33] + [34]
+ [36]

<111111100> =5{1} +11{3} +16{5} +21{7} +25{9} +29{11} +33{13} +35{15} +38{17} +39{19} +41{21}
+41{23} +41{25} +40{27} +40{29} +37{31} +36{33} +33{35} +32{37} +28{39} +26{41} +23{43}
+21{45} +18{47} +16{49} +13{51} +12{53} +9{55} +8{57} +6{59} +5{61} +4{63} +3{65} +2{67}
+2{69} + {71} + {73} + {77}

<111111110> =4[0] +6[1] +16[2] +16[3] +25[4] +27[5] +32[6] +33[7] +40[8] +37[9] +43[10] +41[11]
+43[12] +41[13] +43[14] +38[15] +40[16] +36[17] +35[18] +31[19] +31[20] +25[21] +25[22]
+21[23] +19[24] +16[25] +15[26] +11[27] +11[28] +8[29] +7[30] +5[31] +5[32] +3[33]
+3[34] +2[35] + [36] + [37] + [38] + [40]

<111111111> =3{1} +8{3} +10{5} +12{7} +18{9} +19{11} +20{13} +24{15} +25{17} +25{19} +28{21} +27{23}
+27{25} +28{27} +26{29} +25{31} +26{33} +23{35} +21{37} +21{39} +19{41} +16{43} +16{45}
+13{47} +12{49} +11{51} +9{53} +7{55} +7{57} +5{59} +4{61} +4{63} +3{65} +2{67} +2{69}
+ {71} + {73} + {75} + {81}

D-1 Reduction of the Kronecker Products for R_4.

[00] x [00] = [00]

[10] x [00] = [10]
[10] x [10] = [00] + [1-1] + [11] + [20]
[11] x [00] = [11]
[11] x [10] = [10] + [21]
[11] x [11] = [00] + [11] + [22]
[11] x [1-1] = [20]
[1-1] x [00] = [1-1]
[1-1] x [10] = [10] + [2-1]
[1-1] x [11] = [20]
[1-1] x [1-1] = [00] + [1-1] + [2-2]

[20] x [00] = [20]
[20] x [10] = [10] + [2-1] + [21] + [30]
[20] x [11] = [1-1] + [20] + [31]
[20] x [1-1] = [11] + [20] + [3-1]
[20] x [20] = [00] + [1-1] + [11] + [2-2] + [20] + [22] + [3-1] + [31] + [40]
[21] x [00] = [21]
[21] x [10] = [11] + [20] + [22] + [31]
[21] x [11] = [10] + [21] + [32]
[21] x [1-1] = [21] + [30]
[21] x [20] = [10] + [2-1] + [21] + [30] + [32] + [41]
[21] x [21] = [00] + [1-1] + [11] + [20] + [22] + [31] + [33] + [42]
[21] x [2-1] = [20] + [3-1] + [31] + [40]
[2-1] x [00] = [2-1]
[2-1] x [10] = [1-1] + [2-2] + [20] + [3-1]
[2-1] x [11] = [2-1] + [30]
[2-1] x [1-1] = [10] + [2-1] + [3-2]
[2-1] x [20] = [10] + [2-1] + [21] + [3-2] + [30] + [4-1]
[2-1] x [21] = [20] + [3-1] + [31] + [40]
[2-1] x [2-1] = [00] + [1-1] + [11] + [2-2] + [20] + [3-3] + [3-1] + [4-2]
[22] x [00] = [22]
[22] x [10] = [21] + [32]
[22] x [11] = [11] + [22] + [33]
[22] x [1-1] = [31]
[22] x [20] = [20] + [31] + [42]
[22] x [21] = [10] + [21] + [32] + [43]
[22] x [2-1] = [30] + [41]
[22] x [22] = [00] + [11] + [22] + [33] + [44]
[22] x [2-2] = [40]
[2-2] x [00] = [2-2]
[2-2] x [10] = [2-1] + [3-2]
[2-2] x [11] = [3-1]
[2-2] x [1-1] = [1-1] + [2-2] + [3-3]
[2-2] x [20] = [20] + [3-1] + [4-2]
[2-2] x [21] = [30] + [4-1]
[2-2] x [2-1] = [10] + [2-1] + [3-2] + [4-3]
[2-2] x [22] = [40]
[2-2] x [2-2] = [00] + [1-1] + [2-2] + [3-3] + [4-4]

[30] x [00] = [30]
[30] x [10] = [20] + [3-1] + [31] + [40]
[30] x [11] = [2-1] + [30] + [41]
[30] x [1-1] = [21] + [30] + [4-1]
[30] x [20] = [10] + [2-1] + [21] + [3-2] + [30] + [32] + [4-1] + [41] + [50]
[30] x [21] = [1-1] + [2-2] + [20] + [3-1] + [31] + [40] + [42] + [51]
[30] x [2-1] = [11] + [20] + [22] + [3-1] + [31] + [4-2] + [40] + [5-1]
[30] x [22] = [2-1] + [30] + [41] + [52]
[30] x [2-2] = [21] + [30] + [4-1] + [5-2]
[30] x [30] = [00] + [1-1] + [11] + [2-2] + [20] + [22] + [3-3] + [3-1] + [31] + [33] + [4-2] + [40]
+ [42] + [5-1] + [51] + [60]
[31] x [00] = [31]
[31] x [10] = [21] + [30] + [32] + [41]
[31] x [11] = [20] + [31] + [42]
[31] x [1-1] = [22] + [31] + [40]
[31] x [20] = [11] + [20] + [22] + [3-1] + [31] + [33] + [40] + [42] + [51]
[31] x [21] = [10] + [2-1] + [21] + [30] + [32] + [41] + [43] + [52]
[31] x [2-1] = [21] + [30] + [32] + [4-1] + [41] + [50]
[31] x [22] = [1-1] + [20] + [31] + [42] + [53]

Reduction of the Kronecker Products for R_4 (contd).

[31] x [2-2] = [31] + [40] + [5-1]
[31] x [30] = [10] + [2-1] + [21] + [3-2] + [30] + [32] + [4-1] + [41] + [43] + [50] + [52] + [61]
[31] x [31] = [00] + [1-1] + [11] + [2-2] + [20] + [22] + [3-1] + [31] + [33] + [40] + [42] + [44]
+ [51] + [53] + [62]
[31] x [3-1] = [20] + [3-1] + [31] + [4-2] + [40] + [42] + [5-1] + [51] + [60]
[3-1] x [00] = [3-1]
[3-1] x [10] = [2-1] + [3-2] + [30] + [4-1]
[3-1] x [11] = [2-2] + [3-1] + [40]
[3-1] x [1-1] = [20] + [3-1] + [4-2]
[3-1] x [20] = [1-1] + [2-2] + [20] + [3-3] + [3-1] + [31] + [4-2] + [40] + [5-1]
[3-1] x [21] = [2-1] + [3-2] + [30] + [4-1] + [41] + [50]
[3-1] x [2-1] = [10] + [2-1] + [21] + [3-2] + [30] + [4-3] + [4-1] + [5-2]
[3-1] x [22] = [3-1] + [40] + [51]
[3-1] x [2-2] = [11] + [20] + [3-1] + [4-2] + [5-3]
[3-1] x [30] = [10] + [2-1] + [21] + [3-2] + [30] + [32] + [4-3] + [4-1] + [41] + [5-2] + [50]
+ [6-1]
[3-1] x [31] = [20] + [3-1] + [31] + [4-2] + [40] + [42] + [5-1] + [51] + [60]
[3-1] x [3-1] = [00] + [1-1] + [11] + [2-2] + [20] + [22] + [3-3] + [3-1] + [31] + [4-4] + [4-2]
+ [40] + [5-3] + [5-1] + [6-2]
[32] x [00] = [32]
[32] x [10] = [22] + [31] + [33] + [42]
[32] x [11] = [21] + [32] + [43]
[32] x [1-1] = [32] + [41]
[32] x [20] = [21] + [30] + [32] + [41] + [43] + [52]
[32] x [21] = [11] + [20] + [22] + [31] + [33] + [42] + [44] + [53]
[32] x [2-1] = [31] + [40] + [42] + [51]
[32] x [22] = [10] + [21] + [32] + [43] + [54]
[32] x [2-2] = [41] + [50]
[32] x [30] = [20] + [3-1] + [31] + [40] + [42] + [51] + [53] + [62]
[32] x [31] = [10] + [2-1] + [21] + [30] + [32] + [41] + [43] + [52] + [54] + [63]
[32] x [3-1] = [30] + [4-1] + [41] + [50] + [52] + [61]
[32] x [32] = [00] + [1-1] + [11] + [20] + [22] + [31] + [33] + [42] + [44] + [53] + [55] + [64]
[32] x [3-2] = [40] + [5-1] + [51] + [60]
[3-2] x [00] = [3-2]
[3-2] x [10] = [2-2] + [3-3] + [3-1] + [4-2]
[3-2] x [11] = [3-2] + [4-1]
[3-2] x [1-1] = [2-1] + [3-2] + [4-3]
[3-2] x [20] = [2-1] + [3-2] + [30] + [4-3] + [4-1] + [5-2]
[3-2] x [21] = [3-1] + [4-2] + [40] + [5-1]
[3-2] x [2-1] = [1-1] + [2-2] + [20] + [3-3] + [3-1] + [4-4] + [4-2] + [5-3]
[3-2] x [22] = [4-1] + [50]
[3-2] x [2-2] = [10] + [2-1] + [3-2] + [4-3] + [5-4]
[3-2] x [30] = [20] + [3-1] + [31] + [4-2] + [40] + [5-3] + [5-1] + [6-2]
[3-2] x [31] = [30] + [4-1] + [41] + [5-2] + [50] + [6-1]
[3-2] x [3-1] = [10] + [2-1] + [21] + [3-2] + [30] + [4-3] + [4-1] + [5-4] + [5-2] + [6-3]
[3-2] x [32] = [40] + [5-1] + [51] + [60]
[3-2] x [3-2] = [00] + [1-1] + [11] + [2-2] + [20] + [3-3] + [3-1] + [4-4] + [4-2] + [5-5] + [5-3]
+ [6-4]
[33] x [00] = [33]
[33] x [10] = [32] + [43]
[33] x [11] = [22] + [33] + [44]
[33] x [1-1] = [42]
[33] x [20] = [31] + [42] + [53]
[33] x [21] = [21] + [32] + [43] + [54]
[33] x [2-1] = [41] + [52]
[33] x [22] = [11] + [22] + [33] + [44] + [55]
[33] x [2-2] = [51]
[33] x [30] = [30] + [41] + [52] + [63]
[33] x [31] = [20] + [31] + [42] + [53] + [64]
[33] x [3-1] = [40] + [51] + [62]
[33] x [32] = [10] + [21] + [32] + [43] + [54] + [65]
[33] x [3-2] = [50] + [61]
[33] x [33] = [00] + [11] + [22] + [33] + [44] + [55] + [66]
[33] x [3-3] = [60]
[3-3] x [00] = [3-3]
[3-3] x [10] = [3-2] ± [4-3]
[3-3] x [11] = [4-2]
[3-3] x [1-1] = [2-2] + [3-3] + [4-4]
[3-3] x [20] = [3-1] + [4-2] + [5-3]
[3-3] x [21] = [4-1] + [5-2]

Reduction of the Kronecker Products for R_4 (contd).

[3-3] x [2-1] = [2-1] + [3-2] + [4-3] + [5-4]
[3-3] x [22] = [5-1]
[3-3] x [2-2] = [1-1] + [2-2] + [3-3] + [4-4] + [5-5]
[3-3] x [30] = [30] + [4-1] + [5-2] + [6-3]
[3-3] x [31] = [40] + [5-1] + [6-2]
[3-3] x [3-1] = [20] + [3-1] + [4-2] + [5-3] + [6-4]
[3-3] x [32] = [50] + [6-1]
[3-3] x [3-2] = [10] + [2-1] + [3-2] + [4-3] + [5-4] + [6-5]
[3-3] x [33] = [60]
[3-3] x [3-3] = [00] + [1-1] + [2-2] + [3-3] + [4-4] + [5-5] + [6-6]

[40] x [00] = [40]
[40] x [10] = [30] + [4-1] + [41] + [50]
[40] x [11] = [3-1] + [40] + [51]
[40] x [1-1] = [31] + [40] + [5-1]
[40] x [20] = [20] + [3-1] + [31] + [4-2] + [40] + [42] + [5-1] + [51] + [60]
[40] x [21] = [2-1] + [3-2] + [30] + [4-1] + [41] + [50] + [52] + [61]
[40] x [2-1] = [21] + [30] + [32] + [4-1] + [41] + [5-2] + [50] + [6-1]
[40] x [22] = [2-2] + [3-1] + [40] + [51] + [62]
[40] x [2-2] = [22] + [31] + [40] + [5-1] + [6-2]
[40] x [30] = [10] + [2-1] + [21] + [3-2] + [30] + [32] + [4-3] + [4-1] + [41] + [43] + [5-2] + [50]
+ [52] + [6-1] + [61] + [70]
[40] x [31] = [1-1] + [2-2] + [20] + [3-3] + [3-1] + [31] + [4-2] + [40] + [42] + [5-1] + [51]
+ [53] + [60] + [62] + [71]
[40] x [3-1] = [11] + [20] + [22] + [3-1] + [31] + [33] + [4-2] + [40] + [42] + [5-3] + [5-1] + [51]
+ [6-2] + [60] + [7-1]
[40] x [32] = [2-1] + [3-2] + [30] + [4-1] + [41] + [50] + [52] + [61] + [63] + [72]
[40] x [3-2] = [21] + [30] + [32] + [4-1] + [41] + [5-2] + [50] + [6-3] + [6-1] + [7-2]
[40] x [33] = [3-1] + [40] + [51] + [62] + [73]
[40] x [3-3] = [31] + [40] + [5-1] + [6-2] + [7-3]
[40] x [40] = [00] + [1-1] + [11] + [2-2] + [20] + [22] + [3-3] + [3-1] + [31] + [33] + [4-4]
+ [4-2] + [40] + [42] + [44] + [5-3] + [5-1] + [51] + [53] + [6-2] + [60] + [62] + [7-1]
+ [71] + [80]
[41] x [00] = [41]
[41] x [10] = [31] + [40] + [42] + [51]
[41] x [11] = [30] + [41] + [52]
[41] x [1-1] = [32] + [41] + [50]
[41] x [20] = [21] + [30] + [32] + [4-1] + [41] + [43] + [50] + [52] + [61]
[41] x [21] = [20] + [3-1] + [31] + [40] + [42] + [51] + [53] + [62]
[41] x [2-1] = [22] + [31] + [33] + [40] + [42] + [5-1] + [51] + [60]
[41] x [22] = [2-1] + [30] + [41] + [52] + [63]
[41] x [2-2] = [32] + [41] + [50] + [6-1]
[41] x [30] = [11] + [20] + [22] + [3-1] + [31] + [33] + [4-2] + [40] + [42] + [44] + [5-1] + [51]
+ [53] + [60] + [62] + [71]
[41] x [31] = [10] + [2-1] + [21] + [3-2] + [30] + [32] + [4-1] + [41] + [43] + [50] + [52] + [54]
+ [61] + [63] + [72]
[41] x [3-1] = [21] + [30] + [32] + [4-1] + [41] + [43] + [5-2] + [50] + [52] + [6-1] + [61] + [70]
[41] x [32] = [1-1] + [2-2] + [20] + [3-1] + [31] + [40] + [42] + [51] + [53] + [62] + [64] + [73]
[41] x [3-2] = [31] + [40] + [42] + [5-1] + [51] + [6-2] + [60] + [7-1]
[41] x [33] = [2-1] + [30] + [41] + [52] + [63] + [74]
[41] x [3-3] = [41] + [50] + [6-1] + [7-2]
[41] x [40] = [10] + [2-1] + [21] + [3-2] + [30] + [32] + [4-3] + [4-1] + [41] + [43] + [5-2] + [50]
+ [52] + [54] + [6-1] + [61] + [63] + [70] + [72] + [81]
[41] x [41] = [00] + [1-1] + [11] + [2-2] + [20] + [22] + [3-3] + [3-1] + [31] + [33] + [4-2] + [40]
+ [42] + [44] + [5-1] + [51] + [53] + [55] + [60] + [62] + [64] + [71] + [73] + [82]
[41] x [4-1] = [20] + [3-1] + [31] + [4-2] + [40] + [42] + [5-3] + [5-1] + [51] + [53] + [6-2]
+ [60] + [62] + [7-1] + [71] + [80]
[4-1] x [00] = [4-1]
[4-1] x [10] = [3-1] + [4-2] + [40] + [5-1]
[4-1] x [11] = [3-2] + [4-1] + [50]
[4-1] x [1-1] = [30] + [4-1] + [5-2]
[4-1] x [20] = [2-1] + [3-2] + [30] + [4-3] + [4-1] + [41] + [5-2] + [50] + [6-1]
[4-1] x [21] = [2-2] + [3-3] + [3-1] + [4-2] + [40] + [5-1] + [51] + [60]
[4-1] x [2-1] = [20] + [3-1] + [31] + [4-2] + [40] + [5-3] + [5-1] + [6-2]
[4-1] x [22] = [3-2] + [4-1] + [50] + [61]
[4-1] x [2-2] = [21] + [30] + [4-1] + [5-2] + [6-3]
[4-1] x [30] = [1-1] + [2-2] + [20] + [3-3] + [3-1] + [31] + [4-4] + [4-2] + [40] + [42] + [5-3]
+ [5-1] + [51] + [6-2] + [60] + [7-1]
[4-1] x [31] = [2-1] + [3-2] + [30] + [4-3] + [4-1] + [41] + [5-2] + [50] + [52] + [6-1] + [61]
+ [70]

Reduction of the Kronecker Products for R_4 (contd).

[4-1] x [3-1] = [10] + [2-1] + [21] + [3-2] + [30] + [32] + [4-3] + [4-1] + [41] + [5-4] + [5-2]
+ [50] + [6-3] + [6-1] + [7-2]
[4-1] x [32] = [3-1] + [4-2] + [40] + [5-1] + [51] + [60] + [62] + [71]
[4-1] x [3-2] = [11] + [20] + [22] + [3-1] + [31] + [4-2] + [40] + [5-3] + [5-1] + [6-4] + [6-2]
+ [7-3]
[4-1] x [33] = [4-1] + [50] + [61] + [72]
[4-1] x [3-3] = [21] + [30] + [4-1] + [5-2] + [6-3] + [7-4]
[4-1] x [40] = [10] + [2-1] + [21] + [3-2] + [30] + [32] + [4-3] + [4-1] + [41] + [43] + [5-4]
+ [5-2] + [50] + [52] + [6-3] + [6-1] + [61] + [7-2] + [70] + [8-1]
[4-1] x [41] = [20] + [3-1] + [31] + [4-2] + [40] + [42] + [5-3] + [5-1] + [51] + [53] + [6-2]
+ [60] + [62] + [7-1] + [71] + [80]
[4-1] x [4-1] = [00] + [1-1] + [11] + [2-2] + [20] + [22] + [3-3] + [3-1] + [31] + [33] + [4-4]
+ [4-2] + [40] + [42] + [5-5] + [5-3] + [5-1] + [51] + [6-4] + [6-2] + [60] + [7-3]
+ [7-1] + [8-2]
[42] x [00] = [42]
[42] x [10] = [32] + [41] + [43] + [52]
[42] x [11] = [31] + [42] + [53]
[42] x [1-1] = [33] + [42] + [51]
[42] x [20] = [22] + [31] + [33] + [40] + [42] + [44] + [51] + [53] + [62]
[42] x [21] = [21] + [30] + [32] + [41] + [43] + [52] + [54] + [63]
[42] x [2-1] = [32] + [41] + [43] + [50] + [52] + [61]
[42] x [22] = [20] + [31] + [42] + [53] + [64]
[42] x [2-2] = [42] + [51] + [60]
[42] x [30] = [21] + [30] + [32] + [4-1] + [41] + [43] + [50] + [52] + [54] + [61] + [63] + [72]
[42] x [31] = [11] + [20] + [22] + [3-1] + [31] + [33] + [40] + [42] + [44] + [51] + [53] + [55]
+ [62] + [64] + [73]
[42] x [3-1] = [31] + [40] + [42] + [5-1] + [51] + [53] + [60] + [62] + [71]
[42] x [32] = [10] + [2-1] + [21] + [30] + [32] + [41] + [43] + [52] + [54]+ [63] + [65] + [74]
[42] x [3-2] = [41] + [50] + [52] + [6-1] + [61] + [70]
[42] x [33] = [1-1] + [20] + [31] + [42] + [53] + [64] + [75]
[42] x [3-3] = [51] + [60] + [7-1]
[42] x [40] = [20] + [3-1] + [31] + [4-2] + [40] + [42] + [5-1] + [51] + [53] + [60] + [62] + [64]
+ [71] + [73] + [82]
[42] x [41] = [10] + [2-1] + [21] + [3-2] + [30] + [32] + [4-1] + [41] + [43] + [50] + [52] + [54]
+ [61] + [63] + [65] + [72] + [74] + [83]
[42] x [4-1] = [30] + [4-1] + [41] + [5-2] + [50] + [52] + [6-1] + [61] + [63] + [70] + [72] + [81]
[42] x [42] = [00] + [1-1] + [11] + [2-2] + [20] + [22] + [3-1] + [31] + [33]+ [40] + [42] + [44]
+ [51] + [53] + [55] + [62] + [64] + [66] + [73] + [75] + [84]
[42] x [4-2] = [40] + [5-1] + [51] + [6-2] + [60] + [62] + [7-1] + [71] + [80]
[4-2] x [00] = [4-2]
[4-2] x [10] = [3-2] + [4-3] + [4-1] + [5-2]
[4-2] x [11] = [3-3] + [4-2] + [5-1]
[4-2] x [1-1] = [3-1] + [4-2] + [5-3]
[4-2] x [20] = [2-2] + [3-3] + [3-1] + [4-4] + [4-2] + [40] + [5-3] + [5-1] + [6-2]
[4-2] x [21] = [3-2] + [4-3] + [4-1] + [5-2] + [50] + [6-1]
[4-2] x [2-1] = [2-1] + [3-2] + [30] + [4-3] + [4-1] + [5-4] + [5-2] + [6-3]
[4-2] x [22] = [4-2] + [5-1] + [60]
[4-2] x [2-2] = [20] + [3-1] + [4-2] + [5-3] + [6-4]
[4-2] x [30] = [2-1] + [3-2] + [30] + [4-3] + [4-1] + [41] + [5-4] + [5-2] + [50] + [6-3] + [6-1]
+ [7-2]
[4-2] x [31] = [3-1] + [4-2] + [40] + [5-3] + [5-1] + [51] + [6-2] + [60] + [7-1]
[4-2] x [3-1] = [1-1] + [2-2] + [20] + [3-3] + [3-1] + [31] + [4-4] + [4-2] + [40] + [5-5] + [5-3]
+ [5-1] + [6-4] + [6-2] + [7-3]
[4-2] x [32] = [4-1] + [5-2] + [50] + [6-1] + [61] + [70]
[4-2] x [3-2] = [10] + [2-1] + [21] + [3-2] + [30] + [4-3] + [4-1] + [5-4] + [5-2] + [6-5] + [6-3]
+ [7-4]
[4-2] x [33] = [5-1] + [60] + [71]
[4-2] x [3-3] = [11] + [20] + [3-1] + [4-2] + [5-3] + [6-4] + [7-5]
[4-2] x [40] = [20] + [3-1] + [31] + [4-2] + [40] + [42] + [5-3] + [5-1] + [51] + [6-4] + [6-2]
+ [60] + [7-3] + [7-1] + [8-2]
[4-2] x [41] = [30] + [4-1] + [41] + [5-2] + [50] + [52] + [6-3] + [6-1] + [61] + [7-2] + [70]
+ [8-1]
[4-2] x [4-1] = [10] + [2-1] + [21] + [3-2] + [30] + [32] + [4-3] + [4-1] + [41] + [5-4] + [5-2]
+ [50] + [6-5] + [6-3] + [6-1] + [7-4] + [7-2] + [8-3]
[4-2] x [42] = [40] + [5-1] + [51] + [6-2] + [60] + [62] + [7-1] + [71]+ [80]
[4-2] x [4-2] = [00] + [1-1] + [11] + [2-2] + [20] + [22] + [3-3] + [3-1] + [31] + [4-4] + [4-2]
+ [40] + [5-5] + [5-3] + [5-1] + [6-6] + [6-4] + [6-2] + [7-5] + [7-3] + [8-4]
[43] x [00] = [43]
[43] x [10] = [33] + [42] + [44] + [53]
[43] x [11] = [32] + [43] + [54]

Reduction of the Kronecker Products for R_4 (contd).

[43] x [1-1] = [43] + [52]
[43] x [20] = [32] + [41] + [43] + [52] + [54] + [63]
[43] x [21] = [22] + [31] + [33] + [42] + [44] + [53] + [55] + [64]
[43] x [2-1] = [42] + [51] + [53] + [62]
[43] x [22] = [21] + [32] + [43] + [54] + [65]
[43] x [2-2] = [52] + [61]
[43] x [30] = [31] + [40] + [42] + [51] + [53] + [62] + [64] + [73]
[43] x [31] = [21] + [30] + [32] + [41] + [43] + [52] + [54] + [63] + [65] + [74]
[43] x [3-1] = [41] + [50] + [52] + [61] + [63] + [72]
[43] x [32] = [11] + [20] + [22] + [31] + [33] + [42] + [44] + [53] + [55] + [64] + [66] + [75]
[43] x [3-2] = [51] + [60] + [62] + [71]
[43] x [33] = [10] + [21] + [32] + [43] + [54] + [65] + [76]
[43] x [3-3] = [61] + [70]
[43] x [40] = [30] + [4-1] + [41] + [50] + [52] + [61] + [63] + [72] + [74] + [83]
[43] x [41] = [20] + [3-1] + [31] + [40] + [42] + [51] + [53] + [62] + [64]+ [73] + [75] + [84]
[43] x [4-1] = [40] + [5-1] + [51] + [60] + [62] + [71] + [73] + [82]
[43] x [42] = [10] + [2-1] + [21] + [30] + [32] + [41] + [43] + [52] + [54] + [63] + [65] + [74]
+ [76] + [85]
[43] x [4-2] = [50] + [6-1] + [61] + [70] + [72] + [81]
[43] x [43] = [00] + [1-1] + [11] + [20] + [22] + [31] + [33] + [42] + [44] + [53] + [55] + [64]
+ [66] + [75] + [77] + [86]
[43] x [4-3] = [60] + [7-1] + [71] + [80]
[4-3] x [00] = [4-3]
[4-3] x [10] = [3-3] + [4-4] + [4-2] + [5-3]
[4-3] x [11] = [4-3] + [5-2]
[4-3] x [1-1] = [3-2] + [4-3] + [5-4]
[4-3] x [20] = [3-2] + [4-3] + [4-1] + [5-4] + [5-2] + [6-3]
[4-3] x [21] = [4-2] + [5-3] + [5-1] + [6-2]
[4-3] x [2-1] = [2-2] + [3-3] + [3-1] + [4-4] + [4-2] + [5-5] + [5-3] + [6-4]
[4-3] x [22] = [5-2] + [6-1]
[4-3] x [2-2] = [2-1] + [3-2] + [4-3] + [5-4] + [6-5]
[4-3] x [30] = [3-1] + [4-2] + [40] + [5-3] + [5-1] + [6-4] + [6-2] + [7-3]
[4-3] x [31] = [4-1] + [5-2] + [50] + [6-3] + [6-1] + [7-2]
[4-3] x [3-1] = [2-1] + [3-2] + [30] + [4-3] + [4-1] + [5-4] + [5-2] + [6-5] + [6-3] + [7-4]
[4-3] x [32] = [5-1] + [6-2] + [60] + [7-1]
[4-3] x [3-2] = [1-1] + [2-2] + [20] + [3-3] + [3-1] + [4-4] + [4-2] + [5-5] + [5-3] + [6-6] + [6-4]
+ [7-5]
[4-3] x [33] = [6-1] + [70]
[4-3] x [3-3] = [10] + [2-1] + [3-2] + [4-3] + [5-4] + [6-5] + [7-6]
[4-3] x [40] = [30] + [4-1] + [41] + [5-2] + [50] + [6-3] + [6-1] + [7-4] + [7-2] + [8-3]
[4-3] x [41] = [40] + [5-1] + [51] + [6-2] + [60] + [7-3] + [7-1] + [8-2]
[4-3] x [4-1] = [20] + [3-1] + [31] + [4-2] + [40] + [5-3] + [5-1] + [6-4] + [6-2] + [7-5] + [7-3]
+ [8-4]
[4-3] x [42] = [50] + [6-1] + [61] + [7-2] + [70] + [8-1]
[4-3] x [4-2] = [10] + [2-1] + [21] + [3-2] + [30] + [4-3] + [4-1] + [5-4] + [5-2] + [6-5] + [6-3]
+ [7-6] + [7-4] + [8-5]
[4-3] x [43] = [60] + [7-1] + [71] + [80]
[4-3] x [4-3] = [00] + [1-1] + [11] + [2-2] + [20] + [3-3] + [3-1] + [4-4] + [4-2] + [5-5] + [5-3]
+ [6-6] + [6-4] + [7-7] + [7-5] + [8-6]
[44] x [00] = [44]
[44] x [10] = [43] + [54]
[44] x [11] = [33] + [44] + [55]
[44] x [1-1] = [53]
[44] x [20] = [42] + [53] + [64]
[44] x [21] = [32] + [43] + [54] + [65]
[44] x [2-1] = [52] + [63]
[44] x [22] = [22] + [33] + [44] + [55] + [66]
[44] x [2-2] = [62]
[44] x [30] = [41] + [52] + [63] + [74]
[44] x [31] = [31] + [42] + [53] + [64] + [75]
[44] x [3-1] = [51] + [62] + [73]
[44] x [32] = [21] + [32] + [43] + [54] + [65] + [76]
[44] x [3-2] = [61] + [72]
[44] x [33] = [11] + [22] + [33] + [44] + [55] + [66] + [77]
[44] x [3-3] = [71]
[44] x [40] = [40] + [51] + [62] + [73] + [84]
[44] x [41] = [30] + [41] + [52] + [63] + [74] + [85]
[44] x [4-1] = [50] + [61] + [72] + [83]
[44] x [42] = [20] + [31] + [42] + [53] + [64] + [75] + [86]
[44] x [4-2] = [60] + [71] + [82]

Reduction of the Kronecker Products for R_4 (contd).

[44] x [43] = [10] + [21] + [32] + [43] + [54] + [65] + [76] + [87]
[44] x [4-3] = [70] + [81]
[44] x [44] = [00] + [11] + [22] + [33] + [44] + [55] + [66] + [77] + [88]
[44] x [4-4] = [80]
[4-4] x [00] = [4-4]
[4-4] x [10] = [4-3] + [5-4]
[4-4] x [11] = [5-3]
[4-4] x [1-1] = [3-3] + [4-4] + [5-5]
[4-4] x [20] = [4-2] + [5-3] + [6-4]
[4-4] x [21] = [5-2] + [6-3]
[4-4] x [2-1] = [3-2] + [4-3] + [5-4] + [6-5]
[4-4] x [22] = [6-2]
[4-4] x [2-2] = [2-2] + [3-3] + [4-4] + [5-5] + [6-6]
[4-4] x [30] = [4-1] + [5-2] + [6-3] + [7-4]
[4-4] x [31] = [5-1] + [6-2] + [7-3]
[4-4] x [3-1] = [3-1] + [4-2] + [5-3] + [6-4] + [7-5]
[4-4] x [32] = [6-1] + [7-2]
[4-4] x [3-2] = [2-1] + [3-2] + [4-3] + [5-4] + [6-5] + [7-6]
[4-4] x [33] = [7-1]
[4-4] x [3-3] = [1-1] + [2-2] + [3-3] + [4-4] + [5-5] + [6-6] + [7-7]
[4-4] x [40] = [40] + [5-1] + [6-2] + [7-3] + [8-4]
[4-4] x [41] = [50] + [6-1] + [7-2] + [8-3]
[4-4] x [4-1] = [30] + [4-1] + [5-2] + [6-3] + [7-4] + [8-5]
[4-4] x [42] = [60] + [7-1] + [8-2]
[4-4] x [4-2] = [20] + [3-1] + [4-2] + [5-3] + [6-4] + [7-5] + [8-6]
[4-4] x [43] = [70] + [8-1]
[4-4] x [4-3] = [10] + [2-1] + [3-2] + [4-3] + [5-4] + [6-5] + [7-6] + [8-7]
[4-4] x [44] = [80]
[4-4] x [4-4] = [00] + [1-1] + [2-2] + [3-3] + [4-4] + [5-5] + [6-6] + [7-7] + [8-8]

D-2 Reduction of the Kronecker Products for R_5.

```
[00] x [00] = [00]

[10] x [00] = [10]
[10] x [10] = [00] + [11] + [20]
[11] x [00] = [11]
[11] x [10] = [10] + [11] + [21]
[11] x [11] = [00] + [10] + [11] + [20] + [21] + [22]

[20] x [00] = [20]
[20] x [10] = [10] + [21] + [30]
[20] x [11] = [11] + [20] + [21] + [31]
[20] x [20] = [00] + [11] + [20] + [22] + [31] + [40]
[21] x [00] = [21]
[21] x [10] = [11] + [20] + [21] + [22] + [31]
[21] x [11] = [10] + [11] + [20] +2[21] + [22] + [30] + [31] + [32]
[21] x [20] = [10] + [11] +2[21] + [22] + [30] + [31] + [32] + [41]
[21] x [21] = [00] + [10] +2[11] +2[20] +3[21] +2[22] + [30] +3[31] +2[32] + [33] + [40] + [41]
            + [42]
[22] x [00] = [22]
[22] x [10] = [21] + [22] + [32]
[22] x [11] = [11] + [21] + [22] + [31] + [32] + [33]
[22] x [20] = [20] + [21] + [22] + [31] + [32] + [42]
[22] x [21] = [10] + [11] + [20] +2[21] + [22] + [30] +2[31] +2[32] + [33] + [41] + [42] + [43]
[22] x [22] = [00] + [10] + [11] + [20] + [21] + [22] + [30] + [31] + [32] + [33] + [40] + [41]
            + [42] + [43] + [44]

[30] x [00] = [30]
[30] x [10] = [20] + [31] + [40]
[30] x [11] = [21] + [30] + [31] + [41]
[30] x [20] = [10] + [21] + [30] + [32] + [41] + [50]
[30] x [21] = [11] + [20] + [21] + [22] +2[31] + [32] + [40] + [41] + [42] + [51]
[30] x [22] = [21] + [22] + [30] + [31] + [32] + [41] + [42] + [52]
[30] x [30] = [00] + [11] + [20] + [22] + [31] + [33] + [40] + [42] + [51] + [60]
[31] x [00] = [31]
[31] x [10] = [21] + [30] + [31] + [32] + [41]
[31] x [11] = [20] + [21] + [22] + [30] +2[31] + [32] + [40] + [41] + [42]
[31] x [20] = [11] + [20] + [21] + [22] +2[31] + [32] + [33] + [40] + [41] + [42] + [51]
[31] x [21] = [10] + [11] + [20] +3[21] +2[22] +2[30] +3[31] +3[32] + [33] + [40] +3[41] +2[42]
            + [43] + [50] + [51] + [52]
[31] x [22] = [11] + [20] +2[21] + [22] + [30] +3[31] +2[32] + [33] + [40] +2[41] +2[42] + [43]
            + [51] + [52] + [53]
[31] x [30] = [10] + [11] +2[21] + [22] + [30] + [31] +2[32] + [33] +2[41] + [42] + [43] + [50]
            + [51] + [52] + [61]
[31] x [31] = [00] + [10] +2[11] +2[20] +3[21] +3[22] + [30] +4[31] +4[32] +2[33] +2[40] +3[41]
            +4[42] +2[43] + [44] + [50] +3[51] +2[52] + [53] + [60] + [61] + [62]
[32] x [00] = [32]
[32] x [10] = [22] + [31] + [32] + [33] + [42]
[32] x [11] = [21] + [22] + [31] +2[32] + [33] + [41] + [42] + [43]
[32] x [20] = [21] + [22] + [30] + [31] +2[32] + [33] + [41] + [42] + [43] + [52]
[32] x [21] = [11] + [20] +2[21] +2[22] + [30] +3[31] +3[32] +2[33] + [40] +2[41] +3[42] +2[43]
            + [44] + [51] + [52] + [53]
[32] x [22] = [10] + [11] + [20] +2[21] + [22] + [30] +2[31] +2[32] + [33] + [40] +2[41] +2[42]
            +2[43] + [44] + [50] + [51] + [52] + [53] + [54]
[32] x [30] = [20] + [21] + [22] +2[31] +2[32] + [33] + [40] + [41] +2[42] + [43] + [51] + [52]
            + [53] + [62]
[32] x [31] = [10] + [11] + [20] +3[21] +2[22] +2[30] +4[31] +4[32] +2[33] + [40] +4[41] +4[42]
            +3[43] + [44] + [50] +2[51] +3[52] +2[53] + [54] + [61] + [62] + [63]
[32] x [32] = [00] + [10] +2[11] +2[20] +3[21] +2[22] +2[30] +4[31] +3[32] +2[33] +2[40] +4[41]
            +4[42] +3[43] +2[44] + [50] +3[51] +3[52] +3[53] +2[54] + [55] + [60] + [61] + [62]
            + [63] + [64]
[33] x [00] = [33]
[33] x [10] = [32] + [33] + [43]
[33] x [11] = [22] + [32] + [33] + [42] + [43] + [44]
[33] x [20] = [31] + [32] + [33] + [42] + [43] + [53]
[33] x [21] = [21] + [22] + [31] +2[32] + [33] + [41] +2[42] +2[43] + [44] + [52] + [53] + [54]
[33] x [22] = [11] + [21] + [22] + [31] + [32] + [33] + [41] + [42] + [43] + [44] + [51] + [52]
            + [53] + [54] + [55]
[33] x [30] = [30] + [31] + [32] + [33] + [41] + [42] + [43] + [52] + [53] + [63]
[33] x [31] = [20] + [21] + [22] + [30] +2[31] +2[32] + [33] + [40] +2[41] +3[42] +2[43] + [44]
            + [51] +2[52] +2[53] + [54] + [62] + [63] + [64]
```

[33] x [32] = [10] + [11] + [20] +2[21] + [22] + [30] +2[31] +2[32] + [33] + [40] +2[41] +2[42]
+2[43] + [44] + [50] +2[51] +2[52] +2[53] +2[54] + [55] + [61] + [62] + [63] + [64]
+ [65]
[33] x [33] = [00] + [10] + [11] + [20] + [21] + [22] + [30] + [31] + [32] + [33] + [40] + [41]
+ [42] + [43] + [44] + [50] + [51] + [52] + [53] + [54] + [55] + [60] + [61] + [62]
+ [63] + [64] + [65] + [66]

[40] x [00] = [40]
[40] x [10] = [30] + [41] + [50]
[40] x [11] = [31] + [40] + [41] + [51]
[40] x [20] = [20] + [31] + [40] + [42] + [51] + [60]
[40] x [21] = [21] + [30] + [31] + [32] +2[41] + [42] + [50] + [51] + [52] + [61]
[40] x [22] = [22] + [31] + [32] + [40] + [41] + [42] + [51] + [52] + [62]
[40] x [30] = [10] + [21] + [30] + [32] + [41] + [43] + [50] + [52] + [61] + [70]
[40] x [31] = [11] + [20] + [21] + [22] +2[31] + [32] + [33] + [40] + [41] +2[42] + [43] +2[51]
+ [52] + [53] + [60] + [61] + [62] + [71]
[40] x [32] = [21] + [22] + [30] + [31] +2[32] + [33] +2[41] +2[42] + [43] + [50] + [51] +2[52]
+ [53] + [61] + [62] + [63] + [72]
[40] x [33] = [31] + [32] + [33] + [40] + [41] + [42] + [43] + [51] + [52] + [53] + [62] + [63]
+ [73]
[40] x [40] = [00] + [11] + [20] + [22] + [31] + [33] + [40] + [42] + [44] + [51] + [53] + [60]
+ [62] + [71] + [80]
[41] x [00] = [41]
[41] x [10] = [31] + [40] + [41] + [42] + [51]
[41] x [11] = [30] + [31] + [32] + [40] +2[41] + [42] + [50] + [51] + [52]
[41] x [20] = [21] + [30] + [31] + [32] +2[41] + [42] + [43] + [50] + [51] + [52] + [61]
[41] x [21] = [20] + [21] + [22] + [30] +3[31] +2[32] + [33] +2[40] +3[41] +3[42] + [43] + [50]
+3[51] +2[52] + [53] + [60] + [61] + [62]
[41] x [22] = [21] + [22] + [30] +2[31] +2[32] + [33] + [40] +3[41] +2[42] + [43] + [50] +2[51]
+2[52] + [53] + [61] + [62] + [63]
[41] x [30] = [11] + [20] + [21] + [22] +2[31] + [32] + [33] + [40] + [41] +2[42] + [43] + [44]
+2[51] + [52] + [53] + [60] + [61] + [62] + [71]
[41] x [31] = [10] + [11] + [20] +3[21] +2[22] +2[30] +3[31] +4[32] +2[33] + [40] +4[41] +4[42]
+3[43] + [44] +2[50] +3[51] +4[52] +2[53] + [54] + [60] +3[61] +2[62] + [63] + [70]
+ [71] + [72]
[41] x [32] = [11] + [20] +2[21] +2[22] + [30] +4[31] +4[32] +2[33] +2[40] +4[41] +5[42] +3[43]
+ [44] + [50] +4[51] +4[52] +3[53] + [54] + [60] +2[61] +3[62] +2[63] + [64] + [71]
+ [72] + [73]
[41] x [33] = [21] + [22] + [30] +2[31] +2[32] + [33] + [40] +3[41] +3[42] +2[43] + [44] + [50]
+2[51] +3[52] +2[53] + [54] + [61] +2[62] +2[63] + [64] + [72] + [73] + [74]
[41] x [40] = [10] + [11] +2[21] + [22] + [30] + [31] +2[32] + [33] +2[41] + [42] +2[43] + [44]
+ [50] + [51] +2[52] + [53] + [54] +2[61] + [62] + [63] + [70] + [71] + [72] + [81]
[41] x [41] = [00] + [10] +2[11] +2[20] +3[21] +3[22] + [30] +4[31] +4[32] +3[33] +2[40] +3[41]
+5[42] +4[43] +2[44] + [50] +4[51] +4[52] +4[53] +2[54] + [55] +2[60] +3[61] +4[62]
+2[63] + [64] + [70] +3[71] +2[72] + [73] + [80] + [81] + [82]
[42] x [00] = [42]
[42] x [10] = [32] + [41] + [42] + [43] + [52]
[42] x [11] = [31] + [32] + [33] + [41] +2[42] + [43] + [51] + [52] + [53]
[42] x [20] = [22] + [31] + [32] + [33] + [40] + [41] +2[42] + [43] + [44] + [51] + [52] + [53]
+ [62]
[42] x [21] = [21] + [22] + [30] +2[31] +3[32] +2[33] + [40] +3[41] +3[42] +3[43] + [44] + [50]
+2[51] +3[52] +2[53] + [54] + [61] + [62] + [63]
[42] x [22] = [20] + [21] + [22] + [30] +2[31] +2[32] + [33] + [40] +2[41] +3[42] +2[43] + [44]
+ [50] +2[51] +2[52] +2[53] + [54] + [60] + [61] + [62] + [63] + [64]
[42] x [30] = [21] + [22] + [30] + [31] +2[32] + [33] +2[41] +2[42] +2[43] + [44] + [50] + [51]
+2[52] + [53] + [54] + [61] + [62] + [63] + [72]
[42] x [31] = [11] + [20] +2[21] +2[22] + [30] +4[31] +4[32] +3[33] +2[40] +4[41] +5[42] +4[43]
+2[44] + [50] +4[51] +4[52] +4[53] +2[54] + [55] + [60] +2[61] +3[62] +2[63] + [64]
+ [71] + [72] + [73]
[42] x [32] = [10] + [11] + [20] +3[21] +2[22] +2[30] +4[31] +4[32] +2[33] +2[40] +5[41] +5[42]
+4[43] +2[44] +2[50] +4[51] +5[52] +4[53] +3[54] + [55] + [60]+3[61] +3[62] +3[63]
+2[64] + [65] + [70] + [71] + [72] + [73] + [74]
[42] x [33] = [11] + [20] +2[21] + [22] + [30] +3[31] +2[32] + [33] + [40] +3[41] +3[42] +2[43]
+ [44] + [50] +3[51] +3[52] +3[53] +2[54] + [55] + [60] +2[61] +2[62] +2[63] +2[64]
+ [65] + [71] + [72] + [73] + [74] + [75]
[42] x [40] = [20] + [21] + [22] +2[31] +2[32] + [33] + [40] + [41] +3[42] +2[43] + [44] +2[51]
+2[52] +2[53] + [54] + [60] + [61] +2[62] + [63] + [64] + [71] + [72] + [73] + [82]
[42] x [41] = [10] + [11] + [20] +3[21] +2[22] +2[30] +4[31] +5[32] +3[33] + [40] +5[41] +6[42]
+5[43] +2[44] +2[50] +4[51] +6[52] +5[53] +3[54] + [55] + [60] +4[61] +4[62] +4[63]
+2[64] + [65] + [70] +2[71] +3[72] +2[73] + [74] + [81] + [82] + [83]

D-2 Reduction of the Kronecker Products for R_5 (contd).

[42] x [42] = [00] + [10] +2[11] +2[20] +3[21] +3[22] +2[30] +5[31] +5[32] +3[33] +3[40] +6[41]
+7[42] +5[43] +3[44] +2[50] +6[51] +7[52] +6[53] +4[54] +2[55] +2[60] +4[61] +6[62]
+5[63] +4[64] +2[65] + [66] + [70] +3[71] +3[72] +3[73] +2[74] + [75] + [80] + [81]
+ [82] + [83] + [84]
[43] x [00] = [43]
[43] x [10] = [33] + [42] + [43] + [44] + [53]
[43] x [11] = [32] + [33] + [42] +2[43] + [44] + [52] + [53] + [54]
[43] x [20] = [32] + [33] + [41] + [42] +2[43] + [44] + [52] + [53] + [54] + [63]
[43] x [21] = [22] + [31] +2[32] +2[33] + [41] +3[42] +3[43] +2[44] + [51] +2[52] +3[53] +2[54]
+ [55] + [62] + [63] + [64]
[43] x [22] = [21] + [22] + [31] +2[32] + [33] + [41] +2[42] +2[43] + [44] + [51] +2[52] +2[53]
+2[54] + [55] + [61] + [62] + [63] + [64] + [65]
[43] x [30] = [31] + [32] + [33] + [40] + [41] +2[42] +2[43] + [44] + [51] + [52] +2[53] + [54]
+ [62] + [63] + [64] + [73]
[43] x [31] = [21] + [22] + [30] +2[31] +3[32] +2[33] + [40] +3[41] +4[42] +4[43] +2[44] + [50]
+2[51] +4[52] +4[53] +3[54] + [55] + [61] +2[62] +3[63] +2[64] + [65] + [72] + [73]
+ [74]
[43] x [32] = [11] + [20] +2[21] +2[22] + [30] +3[31] +3[32] +2[33] + [40] +3[41] +4[42] +3[43]
+2[44] + [50] +3[51] +4[52] +4[53] +3[54] +2[55] + [60] +2[61] +3[62] +3[63] +3[64]
+2[65] + [66] + [71] + [72] + [73] + [74] + [75]
[43] x [33] = [10] + [11] + [20] +2[21] + [22] + [30] +2[31] +2[32] + [33] + [40] +2[41] +2[42]
+2[43] + [44] + [50] +2[51] +2[52] +2[53] +2[54] + [55] + [60] +2[61] +2[62] +2[63]
+2[64] +2[65] + [66] + [70] + [71] + [72] + [73] + [74] + [75] + [76]
[43] x [40] = [30] + [31] + [32] + [33] +2[41] +2[42] +2[43] + [44] + [50] + [51] +2[52] +2[53]
+ [54] + [61] + [62] +2[63] + [64] + [72] + [73] + [74] + [83]
[43] x [41] = [20] + [21] + [22] + [30] +3[31] +3[32] +2[33] +2[40] +4[41] +5[42] +4[43] +2[44]
+ [50] +4[51] +5[52] +5[53] +3[54] + [55] + [60] +2[61] +4[62] +4[63] +3[64] + [65]
+ [71] +2[72] +3[73] +2[74] + [75] + [82] + [83] + [84]
[43] x [42] = [10] + [11] + [20] +3[21] +2[22] +2[30] +4[31] +4[32] +2[33] +2[40] +5[41] +5[42]
+4[43] +2[44] +2[50] +5[51] +6[52] +5[53] +4[54] +2[55] + [60] +4[61] +5[62] +5[63]
+4[64] +3[65] + [66] + [70] +2[71] +3[72] +3[73] +3[74] +2[75] + [76] + [81] + [82]
+ [83] + [84] + [85]
[43] x [43] = [00] + [10] +2[11] +2[20] +3[21] +2[22] +2[30] +4[31] +3[32] +2[33] +2[40] +4[41]
+4[42] +3[43] +2[44] +2[50] +4[51] +4[52] +4[53] +3[54] +2[55] +2[60] +4[61] +4[62]
+4[63] +4[64] +3[65] +2[66] + [70] +3[71] +3[72] +3[73] +3[74] +3[75] +2[76] + [77]
+ [80] + [81] + [82] + [83] + [84] + [85] + [86]
[44] x [00] = [44]
[44] x [10] = [43] + [44] + [54]
[44] x [11] = [33] + [43] + [44] + [53] + [54] + [55]
[44] x [20] = [42] + [43] + [44] + [53] + [54] + [64]
[44] x [21] = [32] + [33] + [42] +2[43] + [44] + [52] +2[53] +2[54] + [55] + [63] + [64] + [65]
[44] x [22] = [22] + [32] + [33] + [42] + [43] + [44] + [52] + [53] + [54] + [55] + [62] + [63]
+ [64] + [65] + [66]
[44] x [30] = [41] + [42] + [43] + [44] + [52] + [53] + [54] + [63] + [64] + [74]
[44] x [31] = [31] + [32] + [33] + [41] +2[42] +2[43] + [44] + [51] +2[52] +3[53] +2[54] + [55]
+ [62] +2[63] +2[64] + [65] + [73] + [74] + [75]
[44] x [32] = [21] + [22] + [31] +2[32] + [33] + [41] +2[42] +2[43] + [44] + [51] +2[52] +2[53]
+2[54] + [55] + [61] +2[62] +2[63] +2[64] +2[65] + [66] + [72] + [73] + [74] + [75]
+ [76]
[44] x [33] = [11] + [21] + [22] + [31] + [32] + [33] + [41] + [42] + [43] + [44] + [51] + [52]
+ [53] + [54] + [55] + [61] + [62] + [63] + [64] + [65] + [66] + [71] + [72] + [73]
+ [74] + [75] + [76] + [77]
[44] x [40] = [40] + [41] + [42] + [43] + [44] + [51] + [52] + [53] + [54] + [62] + [63] + [64]
+ [73] + [74] + [84]
[44] x [41] = [30] + [31] + [32] + [33] + [40] +2[41] +2[42] +2[43] + [44] + [50] +2[51] +3[52]
+3[53] +2[54] + [55] + [61] +2[62] +3[63] +2[64] + [65] + [72] +2[73] +2[74] + [75]
+ [83] + [84] + [85]
[44] x [42] = [20] + [21] + [22] + [30] +2[31] +2[32] + [33] + [40] +2[41] +3[42] +2[43] + [44]
+ [50] +2[51] +3[52] +3[53] +2[54] + [55] + [60] +2[61] +3[62] +3[63] +3[64] +2[65]
+ [66] + [71] +2[72] +2[73] +2[74] +2[75] + [76] + [82] + [83] + [84] + [85] + [86]
[44] x [43] = [10] + [11] + [20] +2[21] + [22] + [30] +2[31] +2[32] + [33] + [40] +2[41] +2[42]
+2[43] + [44] + [50] +2[51] +2[52] +2[53] +2[54] + [55] + [60] +2[61] +2[62] +2[63]
+2[64] +2[65] + [66] + [70] +2[71] +2[72] +2[73] +2[74] +2[75] +2[76] + [77] + [81]
+ [82] + [83] + [84] + [85] + [86] + [87]
[44] x [44] = [00] + [10] + [11] + [20] + [21] + [22] + [30] + [31] + [32] + [33] + [40] + [41]
+ [42] + [43] + [44] + [50] + [51] + [52] + [53] + [54] + [55] + [60] + [61] + [62]
+ [63] + [64] + [65] + [66] + [70] + [71] + [72] + [73] + [74] + [75] + [76] + [77]
+ [80] + [81] + [82] + [83] + [84] + [85] + [86] + [87] + [88]

D-3 Reduction of the Kronecker Products for R_6.

[000] x [000] = [000]

[100] x [000] = [100]
[100] x [100] = [000] + [110] + [200]
[110] x [000] = [110]
[110] x [100] = [100] + [11-1] + [111] + [210]
[110] x [110] = [000] +2[110] + [200] + [21-1] + [211] + [220]
[110] x [11-1] = [100] + [11-1] + [210] + [22-1]
[111] x [000] = [111]
[111] x [100] = [110] + [211]
[111] x [110] = [100] + [111] + [210] + [221]
[111] x [111] = [200] + [211] + [222]
[111] x [11-1] = [000] + [110] + [220]
[11-1] x [000] = [11-1]
[11-1] x [100] = [110] + [21-1]
[11-1] x [11-1] = [200] + [21-1] + [22-2]

[200] x [000] = [200]
[200] x [100] = [100] + [210] + [300]
[200] x [110] = [110] + [200] + [21-1] + [211] + [310]
[200] x [111] = [11-1] + [210] + [311]
[200] x [11-1] = [111] + [210] + [31-1]
[200] x [200] = [000] + [110] + [200] + [220] + [310] + [400]
[210] x [000] = [210]
[210] x [100] = [110] + [200] + [21-1] + [211] + [220] + [310]
[210] x [110] = [100] + [11-1] + [111] +3[210] + [22-1] + [221] + [300] + [31-1] + [311] + [320]
[210] x [111] = [110] + [200] + [21-1] + [211] + [220] + [310] + [321]
[210] x [11-1] = [110] + [200] + [21-1] + [211] + [220] + [310] + [32-1]
[210] x [200] = [100] + [11-1] + [111] +2[210] + [22-1] + [221] + [300] + [31-1] + [311] + [320]
+ [410]
[210] x [210] = [000] +3[110] +2[200] +3[21-1] +3[211] + [22-2] +3[220] + [222] +4[310] +2[32-1]
+2[321] + [330] + [400] + [41-1] + [411] + [420]
[210] x [21-1] = [100] + [11-1] + [111] +3[210] +2[22-1] + [221] + [300] +2[31-1] + [311] + [32-2]
+2[320] + [33-1] + [410] + [42-1]
[211] x [000] = [211]
[211] x [100] = [111] + [210] + [221] + [311]
[211] x [110] = [110] + [200] +2[211] + [220] + [222] + [310] + [321]
[211] x [111] = [210] + [221] + [300] + [311] + [322]
[211] x [11-1] = [100] + [111] + [210] + [221] + [320]
[211] x [200] = [110] + [21-1] + [211] + [220] + [310] + [321] + [411]
[211] x [210] = [100] + [11-1] + [111] +3[210] + [22-1] +2[221] + [300] + [31-1] +2[311] +2[320]
+ [322] + [331] + [410] + [421]
[211] x [211] = [200] + [21-1] + [211] + [220] + [222] +2[310] +2[321] + [332] + [400] + [411]
+ [422]
[211] x [21-1] = [000] +2[110] + [200] + [21-1] + [211] +2[220] + [310] + [32-1] + [321] + [330]
+ [420]
[21-1] x [000] = [21-1]
[21-1] x [100] = [11-1] + [210] + [22-1] + [31-1]
[21-1] x [110] = [110] + [200] +2[21-1] + [22-2] + [220] + [310] + [32-1]
[21-1] x [111] = [100] + [11-1] + [210] + [22-1] + [320]
[21-1] x [11-1] = [210] + [22-1] + [300] + [31-1] + [32-2]
[21-1] x [200] = [110] + [21-1] + [211] + [220] + [310] + [32-1] + [41-1]
[21-1] x [21-1] = [200] + [21-1] + [211] + [22-2] + [220] +2[310] +2[32-1] + [33-2] + [400] + [41-1]
+ [42-2]
[220] x [000] = [220]
[220] x [100] = [210] + [22-1] + [221] + [320]
[220] x [110] = [110] + [21-1] + [211] +2[220] + [310] + [32-1] + [321] + [330]
[220] x [111] = [111] + [210] + [221] + [31-1] + [320] + [331]
[220] x [11-1] = [11-1] + [210] + [22-1] + [311] + [320] + [33-1]
[220] x [200] = [200] + [21-1] + [211] + [22-2] + [220] + [222] + [310] + [32-1] + [321] + [420]
[220] x [210] = [100] + [11-1] + [111] +3[210] +2[22-1] +2[221] + [300] +2[31-1] +2[311] + [32-2]
+3[320] + [322] + [33-1] + [331] + [410] + [42-1] + [421] + [430]
[220] x [211] = [110] + [200] + [21-1] +2[211] + [220] + [222] +2[310] + [32-1] +2[321] + [330]
+ [332] + [41-1] + [420] + [431]
[220] x [21-1] = [110] + [200] +2[21-1] + [211] + [22-2] + [220] +2[310] +2[32-1] + [321] + [33-2]
+ [330] + [411] + [420] + [43-1]
[220] x [220] = [000] +2[110] + [200] + [21-1] + [211] +3[220] +2[310] +2[32-1] +2[321] +2[330]
+ [400] + [41-1] + [411] + [42-2] + [420] + [422] + [43-1] + [431] + [440]
[220] x [22-1] = [100] + [11-1] +2[210] +2[22-1] + [300] + [31-1] + [311] + [32-2] +2[320] +2[33-1]
+ [410] + [42-1] + [421] + [43-2] + [430] + [44-1]

Reduction of the Kronecker Products for R_6 (contd).

```
[220] x [22-2] = [200] + [21-1] + [22-2] + [310] + [32-1] + [33-2] + [420] + [43-1] + [44-2]
[221] x [000] = [221]
[221] x [100] = [211] + [220] + [222] + [321]
[221] x [110] = [111] + [210] +2[221] + [311] + [320] + [322] + [331]
[221] x [111] = [211] + [222] + [310] + [321] + [332]
[221] x [11-1] = [110] + [211] + [220] + [321] + [330]
[221] x [200] = [210] + [22-1] + [221] + [311] + [320] + [322] + [421]
[221] x [210] = [110] + [200] + [21-1] +2[211] +2[220] + [222] +2[310] + [32-1] +3[321] + [330]
              + [332] + [411] + [420] + [422] + [431]
[221] x [211] = [210] + [221] + [300] + [31-1] +2[311] + [320] +2[322] + [331] + [333] + [410]
              + [421] + [432]
[221] x [21-1] = [100] + [11-1] + [111] +2[210] + [22-1] + [221] + [311] +2[320] + [33-1] + [331]
               + [421] + [430]
[221] x [220] = [100] + [111] +2[210] +2[221] + [300] + [31-1] + [311] +2[320] + [322] +2[331]
              + [410] + [42-1] + [421] + [430] + [432] + [441]
[221] x [221] = [200] + [211] + [222] +2[310] +2[321] +2[332] + [400] + [41-1] + [411] + [420]
              + [422] + [431] + [433] + [442]
[221] x [22-1] = [000] +2[110] + [200] + [21-1] + [211] +2[220] + [310] + [32-1] + [321] +2[330]
               + [420] + [43-1] + [431] + [440]
[221] x [22-2] = [100] + [11-1] + [210] + [22-1] + [320] + [33-1] + [430] + [44-1]
[22-1] x [000] = [22-1]
[22-1] x [100] = [21-1] + [22-2] + [220] + [32-1]
[22-1] x [110] = [11-1] + [210] +2[22-1] + [31-1] + [32-2] + [320] + [33-1]
[22-1] x [111] = [110] + [21-1] + [220] + [32-1] + [330]
[22-1] x [11-1] = [21-1] + [22-2] + [310] + [32-1] + [33-2]
[22-1] x [200] = [210] + [22-1] + [221] + [31-1] + [32-2] + [320] + [42-1]
[22-1] x [210] = [110] + [200] +2[21-1] + [211] + [22-2] +2[220] +2[310] +3[32-1] + [321] + [33-2]
               + [330] + [41-1] + [42-2] + [420] + [43-1]
[22-1] x [211] = [100] + [11-1] + [111] +2[210] + [22-1] + [221] + [31-1] +2[320] + [33-1] + [331]
               + [42-1] + [430]
[22-1] x [21-1] = [210] + [22-1] + [300] +2[31-1] + [311] +2[32-2] + [320] + [33-3] + [33-1] + [410]
                + [42-1] + [43-2]
[22-1] x [22-1] = [200] + [21-1] + [22-2] +2[310] +2[32-1] +2[33-2] + [400] + [41-1] + [411]
                + [42-2] + [420] + [43-3] + [43-1] + [44-2]
[22-1] x [22-2] = [300] + [31-1] + [32-2] + [33-3] + [410] + [42-1] + [43-2] + [44-3]
[222] x [000] = [222]
[222] x [100] = [221] + [322]
[222] x [110] = [211] + [222] + [321] + [332]
[222] x [111] = [311] + [322] + [333]
[222] x [11-1] = [111] + [221] + [331]
[222] x [200] = [220] + [321] + [422]
[222] x [210] = [210] + [221] + [311] + [320] + [322] + [331] + [421] + [432]
[222] x [211] = [310] + [321] + [332] + [411] + [422] + [433]
[222] x [21-1] = [110] + [211] + [220] + [321] + [330] + [431]
[222] x [220] = [200] + [211] + [222] + [310] + [321] + [332] + [420] + [431] + [442]
[222] x [221] = [300] + [311] + [322] + [333] + [410] + [421] + [432] + [443]
[222] x [22-1] = [100] + [111] + [210] + [221] + [320] + [331] + [430] + [441]
[222] x [222] = [400] + [411] + [422] + [433] + [444]
[222] x [22-2] = [000] + [110] + [220] + [330] + [440]
[22-2] x [000] = [22-2]
[22-2] x [100] = [22-1] + [32-2]
[22-2] x [110] = [21-1] + [22-2] + [32-1] + [33-2]
[22-2] x [111] = [11-1] + [22-1] + [33-1]
[22-2] x [11-1] = [31-1] + [32-2] + [33-3]
[22-2] x [200] = [220] + [32-1] + [42-2]
[22-2] x [210] = [210] + [22-1] + [31-1] + [32-2] + [320] + [33-1] + [42-1] + [43-2]
[22-2] x [211] = [110] + [21-1] + [220] + [32-1] + [330] + [43-1]
[22-2] x [21-1] = [310] + [32-1] + [33-2] + [41-1] + [42-2] + [43-3]
[22-2] x [22-2] = [400] + [41-1] + [42-2] + [43-3] + [44-4]
```

D-4 Reduction of the Kronecker Products for R_7.

```
[000] x [000] = [000]

[100] x [000] = [100]
[100] x [100] = [000] + [110] + [200]
[110] x [000] = [110]
[110] x [100] = [100] + [111] + [210]
[110] x [110] = [000] + [110] + [111] + [200] + [211] + [220]
[111] x [000] = [111]
[111] x [100] = [110] + [111] + [211]
[111] x [110] = [100] + [110] + [111] + [210] + [211] + [221]
[111] x [111] = [000] + [100] + [110] + [111] + [200] + [210] + [211] + [220] + [221] + [222]

[200] x [000] = [200]
[200] x [100] = [100] + [210] + [300]
[200] x [110] = [110] + [200] + [211] + [310]
[200] x [111] = [111] + [210] + [211] + [311]
[200] x [200] = [000] + [110] + [200] + [220] + [310] + [400]
[210] x [000] = [210]
[210] x [100] = [110] + [200] + [211] + [220] + [310]
[210] x [110] = [100] + [111] +2[210] + [211] + [221] + [300] + [311] + [320]
[210] x [111] = [110] + [111] + [200] + [210] +2[211] + [220] + [221] + [310] + [311] + [321]
[210] x [200] = [100] + [111] +2[210] + [221] + [300] + [311] + [320] + [410]
[210] x [210] = [000] +2[110] + [111] +2[200] +3[211] +2[220] + [221] + [222] +3[310] + [311]
               +2[321] + [330] + [400] + [411] + [420]
[211] x [000] = [211]
[211] x [100] = [111] + [210] + [211] + [221] + [311]
[211] x [110] = [110] + [111] + [200] + [210] +2[211] + [220] + [221] + [222] + [310] + [311]
               + [321]
[211] x [111] = [100] + [110] + [111] + [200] +2[210] +2[211] + [220] +2[221] + [222] + [300]
               + [310] + [311] + [320] + [321] + [322]
[211] x [200] = [110] + [111] +2[211] + [220] + [221] + [310] + [311] + [321] + [411]
[211] x [210] = [100] + [110] +2[111] +3[210] +3[211] + [220] +3[221] + [222] + [300] + [310]
               +3[311] +2[320] +2[321] + [322] + [331] + [410] + [411] + [421]
[211] x [211] = [000] + [100] +2[110] +2[111] +2[200] +3[210] +4[211] +3[220] +4[221] +2[222]
               + [300] +3[310] +3[311] +2[320] +4[321] +2[322] + [330] + [331] + [332] + [400]
               + [410] + [411] + [420] + [421] + [422]
[220] x [000] = [220]
[220] x [100] = [210] + [221] + [320]
[220] x [110] = [110] + [211] + [220] + [221] + [310] + [321] + [330]
[220] x [111] = [111] + [210] + [211] + [220] + [221] + [311] + [320] + [321] + [331]
[220] x [200] = [200] + [211] + [220] + [222] + [310] + [321] + [420]
[220] x [210] = [100] + [111] +2[210] + [211] +2[221] + [222] + [300] +2[311] +2[320] + [321]
               + [322] + [331] + [410] + [421] + [430]
[220] x [211] = [110] + [111] + [200] + [210] +3[211] + [220] +2[221] + [222] +2[310] +2[311]
               + [320] +3[321] + [322] + [330] + [331] + [332] + [411] + [420] + [421] + [431]
[220] x [220] = [000] + [110] + [111] + [200] + [211] +2[220] + [221] + [222] + [310] + [311]
               +2[321] + [322] + [330] + [331] + [400] + [411] + [420] + [422] + [431] + [440]
[221] x [000] = [221]
[221] x [100] = [211] + [220] + [221] + [222] + [321]
[221] x [110] = [111] + [210] + [211] + [220] +2[221] + [222] + [311] + [320] + [321] + [322]
               + [331]
[221] x [111] = [110] + [111] + [210] +2[211] + [220] +2[221] + [222] + [310] + [311] + [320]
               +2[321] + [322] + [330] + [331] + [332]
[221] x [200] = [210] + [211] +2[221] + [222] + [311] + [320] + [321] + [322] + [421]
[221] x [210] = [110] + [111] + [200] + [210] +3[211] +2[220] +3[221] +2[222] +2[310] +2[311]
               + [320] +4[321] +2[322] + [330] + [331] + [332] + [411] + [420] + [421] + [422]
               + [431]
[221] x [211] = [100] + [110] +2[111] + [200] +3[210] +4[211] +2[220] +4[221] +2[222] + [300]
               +2[310] +4[311] +3[320] +5[321] +3[322] + [330] +3[331] +2[332] + [333] + [410]
               + [411] + [420] +2[421] + [422] + [430] + [431] + [432]
[221] x [220] = [100] + [110] + [111] +2[210] +2[211] + [220] +3[221] + [222] + [300] + [310]
               +2[311] +2[320] +3[321] +2[322] + [330] +2[331] + [332] + [410] + [411] +2[421]
               + [422] + [430] + [431] + [432] + [441]
[221] x [221] = [000] + [100] +2[110] +2[111] +2[200] +3[210] +4[211] +3[220] +4[221] +2[222]
               + [300] +3[310] +4[311] +3[320] +6[321] +3[322] +2[330] +4[331] +3[332] + [333]
               + [400] + [410] +2[411] +2[420] +3[421] +2[422] + [430] +3[431] +2[432] + [433]
               + [440] + [441] + [442]
[222] x [000] = [222]
[222] x [100] = [221] + [222] + [322]
[222] x [110] = [211] + [221] + [222] + [321] + [322] + [332]
```

Reduction of the Kronecker Products for R_7 (contd).

```
[222] x [111] = [111] + [211] + [221] + [222] + [311] + [321] + [322] + [331] + [332] + [333]
[222] x [200] = [220] + [221] + [222] + [321] + [322] + [422]
[222] x [210] = [210] + [211] + [220] +2[221] + [222] + [311] + [320] +2[321] +2[322] + [331]
              + [332] + [421] + [422] + [432]
[222] x [211] = [110] + [111] + [210] +2[211] + [220] +2[221] + [222] + [310] +2[311] + [320]
              +3[321] +2[322] + [330] +2[331] +2[332] + [333] + [411] + [421] + [422] + [431]
              + [432] + [433]
[222] x [220] = [200] + [210] + [211] + [220] + [221] + [222] + [310] + [311] + [320] +2[321]
              + [322] + [331] + [332] + [420] + [421] + [422] + [431] + [432] + [442]
[222] x [221] = [100] + [110] + [111] + [200] +2[210] +2[211] + [220] +2[221] + [222] + [300]
              +2[310] +2[311] +2[320] +3[321] +2[322] + [330] +2[331] +2[332] + [333] + [410]
              + [411] + [420] +2[421] + [422] + [430] +2[431] +2[432] + [433] + [441] + [442]
              + [443]
[222] x [222] = [000] + [100] + [110] + [111] + [200] + [210] + [211] + [220] + [221] + [222]
              + [300] + [310] + [311] + [320] + [321] + [322] + [330] + [331] + [332] + [333]
              + [400] + [410] + [411] + [420] + [421] + [422] + [430] + [431] + [432] + [433]
              + [440] + [441] + [442] + [443] + [444]
```

D-5 Reduction of the Kronecker Products for O_8.

```
[0000] x [0000] = [0000]

[1000] x [0000] = [1000]
[1000] x [1000] = [0000] + [1100] + [2000]
[1100] x [0000] = [1100]
[1100] x [1000] = [1000] + [1110] + [2100]
[1100] x [1100] = [0000] + [1100] + [1111] + [2000] + [2110] + [2200]
[1110] x [0000] = [1110]
[1110] x [1000] = [1100] + [1111] + [2110]
[1110] x [1100] = [1000] +2[1110] + [2100] + [2111] + [2210]
[1110] x [1110] = [0000] +2[1100] + [1111] + [2000] +2[2110] + [2200] + [2211] + [2220]
[1111] x [0000] = [1111]
[1111] x [1000] =2[1110] + [2111]
[1111] x [1100] =2[1100] + [1111] +2[2110] + [2211]
[1111] x [1110] =2[1000] +2[1110] +2[2100] + [2111] +2[2210] + [2221]
[1111] x [1111] =2[0000] +2[1100] + [1111] +2[2000] +2[2110] +2[2200] + [2211] +2[2220] + [2222]

[2000] x [0000] = [2000]
[2000] x [1000] = [1000] + [2100] + [3000]
[2000] x [1100] = [1100] + [2000] + [2110] + [3100]
[2000] x [1110] = [1110] + [2100] + [2111] + [3110]
[2000] x [1111] = [1111] +2[2110] + [3111]
[2000] x [2000] = [0000] + [1100] + [2000] + [2200] + [3100] + [4000]
[2100] x [0000] = [2100]
[2100] x [1000] = [1100] + [2000] + [2110] + [2200] + [3100]
[2100] x [1100] = [1000] + [1110] +2[2100] + [2111] + [2210] + [3000] + [3110] + [3200]
[2100] x [1110] = [1100] + [1111] + [2000] +3[2110] + [2200] + [2211] + [3100] + [3111] + [3210]
[2100] x [1111] =2[1110] +2[2100] +2[2111] +2[2210] +2[3110] + [3211]
[2100] x [2000] = [1000] + [1110] +2[2100] + [2210] + [3000] + [3110] + [3200] + [4100]
[2100] x [2100] = [0000] +2[1100] + [1111] +2[2000] +3[2110] +2[2200] + [2211] + [2220] +3[3100]
                + [3111] +2[3210] + [3300] + [4000] + [4110] + [4200]
[2110] x [0000] = [2110]
[2110] x [1000] = [1110] + [2100] + [2111] + [2210] + [3110]
[2110] x [1100] = [1100] + [1111] + [2000] +3[2110] + [2200] + [2211] + [2220] + [3100] + [3111]
                + [3210]
[2110] x [1110] = [1000] +2[1110] +3[2100] +2[2111] +3[2210] + [2221] + [3000] +2[3110] + [3200]
                + [3211] + [3220]
[2110] x [1111] =2[1100] + [1111] +2[2000] +4[2110] +2[2200] +2[2211] +2[2220] +2[3100] + [3111]
                +2[3210] + [3221]
[2110] x [2000] = [1100] + [1111] +2[2110] + [2200] + [2211] + [3100] + [3111] + [3210] + [4110]
[2110] x [2100] = [1000] +3[1110] +3[2100] +3[2111] +4[2210] + [2221] + [3000] +4[3110] +2[3200]
                +2[3211] + [3220] + [3310] + [4100] + [4111] + [4210]
[2110] x [2110] = [0000] +3[1100] +2[1111] +2[2000] +7[2110] +4[2200] +4[2211] +3[2220] + [2222]
                +4[3100] +3[3111] +6[3210] +2[3221] + [3300] + [3311] + [3320] + [4000] +2[4110]
                + [4200] + [4211] + [4220]
[2111] x [0000] = [2111]
[2111] x [1000] = [1111] +2[2110] + [2211] + [3111]
[2111] x [1100] =2[1110] +2[2100] +2[2111] +2[2210] + [2221] +2[3110] + [3211]
[2111] x [1110] =2[1100] + [1111] +2[2000] +4[2110] +2[2200] +2[2211] +2[2220] + [2222] +2[3100]
                + [3111] +2[3210] + [3221]
[2111] x [1111] =2[1000] +2[1110] +4[2100] +2[2111] +4[2210] +2[2221] +2[3000] +2[3110] +2[3200]
                + [3211] +2[3220] + [3222]
[2111] x [2000] =2[1110] +2[2111] +2[2210] +2[3110] + [3211] + [4111]
[2111] x [2100] =2[1100] +2[1111] +6[2110] +2[2200] +3[2211] +2[2220] +2[3100] +3[3111] +4[3210]
                + [3221] + [3311] +2[4110] + [4211]
[2111] x [2110] =2[1000] +4[1110] +6[2100] +4[2111] +8[2210] +3[2221] +2[3000] +6[3110] +4[3200]
                +4[3211] +4[3220] + [3222] +2[3310] + [3321] +2[4100] + [4111] +2[4210] + [4221]
[2111] x [2111] =2[0000] +4[1100] +2[1111] +4[2000] +8[2110] +6[2200] +5[2211] +6[2220] + [2222]
                +6[3100] +3[3111] +8[3210] +4[3221] +2[3300] + [3311] +2[3320] + [3322] +2[4000]
                +2[4110] +2[4200] + [4211] +2[4220] + [4222]
[2200] x [0000] = [2200]
[2200] x [1000] = [2100] + [2210] + [3200]
[2200] x [1100] = [1100] + [2110] + [2200] + [2211] + [3100] + [3210] + [3300]
[2200] x [1110] = [1110] + [2100] + [2111] +2[2210] + [3110] + [3200] + [3211] + [3310]
[2200] x [1111] = [1111] +2[2110] +2[2200] + [2211] + [3111] +2[3210] + [3311]
[2200] x [2000] = [2000] + [2110] + [2200] + [2220] + [3100] + [3210] + [4200]
[2200] x [2100] = [1000] + [1110] +2[2100] + [2111] +2[2210] + [2221] + [3000] +2[3110] +2[3200]
                + [3211] + [3220] + [3310] + [4100] + [4210] + [4300]
[2200] x [2110] = [1100] + [1111] + [2000] +4[2110] + [2200] +2[2211] +2[2220] +2[3100] +2[3111]
                +4[3210] + [3221] + [3300] + [3311] + [3320] + [4110] + [4200] + [4211] + [4310]
```

Reduction of the Kronecker Products for O_8 (contd).

[2200] x [2111] =2[1110] +2[2100] +3[2111] +4[2210] + [2221] +4[3110] +2[3200] +3[3211] +2[3220] +2[3310] + [3321] + [4111] +2[4210] + [4311]

[2200] x [2200] = [0000] + [1100] + [1111] + [2000] + [2110] +2[2200] + [2211] + [2220] + [2222] + [3100] + [3111] +2[3210] + [3221] + [3300] + [3311] + [4000] + [4110] + [4200] + [4220] + [4310] + [4400]

[2210] x [0000] = [2210]

[2210] x [1000] = [2110] + [2200] + [2211] + [2220] + [3210]

[2210] x [1100] = [1110] + [2100] + [2111] +3[2210] + [2221] + [3110] + [3200] + [3211] + [3220] + [3310]

[2210] x [1110] = [1100] + [1111] +3[2110] +2[2200] +2[2211] +2[2220] + [3100] + [3111] +3[3210] + [3221] + [3300] + [3311] + [3320]

[2210] x [1111] =2[1110] +2[2100] +2[2111] +4[2210] + [2221] +2[3110] +2[3200] +2[3211] +2[3220] +2[3310] + [3321]

[2210] x [2000] = [2100] + [2111] +2[2210] + [2221] + [3110] + [3200] + [3211] + [3220] + [4210]

[2210] x [2100] = [1100] + [1111] + [2000] +4[2110] +2[2200] +3[2211] +3[2220] + [2222] +2[3100] +2[3111] +5[3210] +2[3221] + [3300] + [3311] + [3320] + [4110] + [4200] + [4211] + [4220] + [4310]

[2210] x [2110] = [1000] +3[1110] +4[2100] +4[2111] +7[2210] +3[2221] + [3000] +6[3110] +4[3200] +5[3211] +5[3220] + [3222] +4[3310] +2[3321] + [3330] + [4100] + [4111] +3[4210] + [4221] + [4300] + [4311] + [4320]

[2210] x [2111] =2[1100] +2[1111] +2[2000] +8[2110] +4[2200] +5[2211] +4[2220] + [2222] +4[3100] +4[3111] +10[3210] +4[3221] +2[3300] +3[3311] +4[3320] + [3322] + [3331] +2[4110] +2[4200] +2[4211] +2[4220] +2[4310] + [4321]

[2210] x [2200] = [1000] +2[1110] +2[2100] +2[2111] +4[2210] +2[2221] + [3000] +3[3110] +2[3200] +3[3211] +3[3220] + [3222] +3[3310] + [3321] + [4100] + [4111] +2[4210] + [4221] + [4300] + [4311] + [4320] + [4410]

[2210] x [2210] = [0000] +3[1100] +2[1111] +2[2000] +7[2110] +4[2200] +5[2211] +5[2220] + [2222] +4[3100] +4[3111] +10[3210] +5[3221] +3[3300] +4[3311] +5[3320] + [3322] + [3331] + [4000] +3[4110] +2[4200] +3[4211] +3[4220] + [4222] +4[4310] +2[4321] + [4330] + [4400] + [4411] + [4420]

[2211] x [0000] = [2211]

[2211] x [1000] = [2111] +2[2210] + [2221] + [3211]

[2211] x [1100] = [1111] +2[2110] +2[2200] +2[2211] +2[2220] + [2222] + [3111] +2[3210] + [3221] + [3311]

[2211] x [1110] =2[1110] +2[2100] +2[2111] +4[2210] +2[2221] +2[3110] +2[3200] +2[3211] +2[3220] + [3222] +2[3310] + [3321]

[2211] x [1111] =2[1100] + [1111] +4[2110] +2[2200] +3[2211] +2[2220] + [2222] +2[3100] + [3111] +4[3210] +2[3221] +2[3300] + [3311] +2[3320] + [3322]

[2211] x [2000] =2[2110] +2[2211] +2[2220] + [3111] +2[3210] + [3221] + [4211]

[2211] x [2100] =2[1110] +2[2100] +3[2111] +6[2210] +3[2221] +4[3110] +2[3200] +4[3211] +4[3220] + [3222] +2[3310] + [3321] + [4111] +2[4210] + [4221] + [4311]

[2211] x [2110] =2[1100] +2[1111] +2[2000] +8[2110] +4[2200] +5[2211] +6[2220] + [2222] +4[3100] +4[3111] +10[3210] +5[3221] +2[3300] +3[3311] +4[3320] + [3322] + [3331] +2[4110] +2[4200] +2[4211] +2[4220] + [4222] +2[4310] + [4321]

[2211] x [2111] =2[1000] +4[1110] +6[2100] +5[2111] +10[2210] +4[2221] +2[3000] +8[3110] +6[3200] +7[3211] +8[3220] +2[3222] +6[3310] +4[3321] +2[3330] + [3332] +2[4100] + [4111] +4[4210] +2[4221] +2[4300] + [4311] +2[4320] + [4322]

[2211] x [2200] =2[1100] + [1111] +4[2110] +2[2200] +4[2211] +2[2220] + [2222] +2[3100] +2[3111] +6[3210] +3[3221] +2[3300] +2[3311] +2[3320] + [3322] +2[4110] +2[4211] +2[4220] +2[4310] + [4321] + [4411]

[2211] x [2210] =2[1000] +4[1110] +6[2100] +5[2111] +10[2210] +4[2221] +2[3000] +8[3110] +6[3200] +10[3211] +8[3220] +2[3222] +8[3310] +5[3321] +2[3330] + [3332] +2[4100] +2[4111] +6[4210] +3[4221] +2[4300] +3[4311] +4[4320] + [4322] + [4331] +2[4410] + [4421]

[2211] x [2211] =2[0000] +4[1100] +3[1111] +4[2000] +10[2110] +8[2200] +7[2211] +6[2220] +2[2222] +6[3100] +6[3111] +16[3210] +7[3221] +4[3300] +7[3311] +8[3320] +3[3322] +3[3331] + [3333] +2[4000] +4[4110] +4[4200] +4[4211] +6[4220] + [4222] +6[4310] +4[4321] +2[4330] + [4332] +2[4400] + [4411] +2[4420] + [4422]

[2220] x [0000] = [2220]

[2220] x [1000] = [2210] + [2221] + [3220]

[2220] x [1100] = [2110] + [2211] +2[2220] + [3210] + [3221] + [3320]

[2220] x [1110] = [1110] + [2111] +2[2210] + [2221] + [3110] + [3211] +2[3220] + [3310] + [3321] + [3330]

[2220] x [1111] = [1111] +2[2110] + [2211] +2[2220] + [3111] +2[3210] + [3221] + [3311] +2[3320] + [3331]

[2220] x [2000] = [2200] + [2211] + [2220] + [2222] + [3210] + [3221] + [4220]

[2220] x [2100] = [2100] + [2111] +3[2210] +2[2221] + [3110] + [3200] +2[3211] +3[3220] + [3222] + [3310] + [3321] + [4210] + [4221] + [4320]

[2220] x [2110] = [1100] + [1111] +3[2110] +2[2200] +3[2211] +2[2220] + [2222] + [3100] +2[3111] +5[3210] +3[3221] + [3300] +2[3311] +3[3320] + [3322] + [3331] + [4110] + [4211] +2[4220] + [4310] + [4321] + [4330]

Reduction of the Kronecker Products for O_8 (contd).

[2220] x [2111] =2[1110] +2[2100] +3[2111] +4[2210] +2[2221] +4[3110] +2[3200] +4[3211] +4[3220] + [3222] +4[3310] +3[3321] +2[3330] + [3332] + [4111] +2[4210] + [4221] + [4311] +2[4320] + [4331]

[2220] x [2200] = [2000] +2[2110] + [2200] + [2211] +3[2220] + [3100] + [3111] +3[3210] +2[3221] + [3311] +2[3320] + [4200] + [4211] + [4220] + [4222] + [4310] + [4321] + [4420]

[2220] x [2210] = [1000] +2[1110] +3[2100] +2[2111] +5[2210] +2[2221] + [3000] +4[3110] +3[3200] +4[3211] +5[3220] + [3222] +4[3310] +3[3321] +2[3330] + [4100] + [4111] +3[4210] +2[4221] + [4300] +2[4311] +3[4320] + [4322] + [4331] + [4410] + [4421] + [4430]

[2220] x [2211] =2[1100] + [1111] +2[2000] +6[2110] +2[2200] +3[2211] +4[2220] +4[3100] +3[3111] +8[3210] +4[3221] +2[3300] +3[3311] +6[3320] + [3322] +2[3331] +2[4110] +2[4200] +3[4211] +2[4220] + [4222] +4[4310] +3[4321] +2[4330] + [4332] + [4411] +2[4420] + [4431]

[2220] x [2220] = [0000] +2[1100] + [1111] + [2000] +2[2110] +3[2200] +2[2211] + [2220] + [2222] +2[3100] + [3111] +4[3210] +2[3221] +2[3300] +2[3311] +2[3320] + [3322] + [3331] + [4000] +2[4110] + [4200] + [4211] +3[4220] +2[4310] +2[4321] +2[4330] + [4400] + [4411] + [4420] + [4422] + [4431] + [4440]

[2221] x [0000] = [2221]

[2221] x [1000] = [2211] +2[2220] + [2222] + [3221]

[2221] x [1100] = [2111] +2[2210] +2[2221] + [3211] +2[3220] + [3222] + [3321]

[2221] x [1110] = [1111] +2[2110] +2[2211] +2[2220] + [2222] + [3111] +2[3210] +2[3221] + [3311] +2[3320] + [3322] + [3331]

[2221] x [1111] =2[1110] +2[2111] +2[2210] +2[2221] +2[3110] +2[3211] +2[3220] + [3222] +2[3310] +2[3321] +2[3330] + [3332]

[2221] x [2000] =2[2210] +2[2221] + [3211] +2[3220] + [3222] + [4221]

[2221] x [2100] =2[2110] +2[2200] +3[2211] +4[2220] + [2222] + [3111] +4[3210] +4[3221] + [3311] +2[3320] + [3322] + [4211] +2[4220] + [4222] + [4321]

[2221] x [2110] =2[1110] +2[2100] +3[2111] +6[2210] +3[2221] +4[3110] +2[3200] +5[3211] +6[3220] +2[3222] +4[3310] +4[3321] +2[3330] + [3332] + [4111] +2[4210] +2[4221] + [4311] +2[4320] + [4322] + [4331]

[2221] x [2111] =2[1100] +2[1111] +6[2110] +2[2200] +4[2211] +4[2220] + [2222] +2[3100] +4[3111] +8[3210] +5[3221] +2[3300] +4[3311] +6[3320] +2[3322] +3[3331] + [3333] +2[4110] +2[4211] +2[4220] + [4222] +2[4310] +2[4321] +2[4330] + [4332]

[2221] x [2200] =2[2100] + [2111] +4[2210] +2[2221] +2[3110] +2[3200] +3[3211] +4[3220] + [3222] +2[3310] +2[3321] +2[4210] +2[4221] + [4311] +2[4320] + [4322] + [4421]

[2221] x [2210] =2[1100] + [1111] +2[2000] +6[2110] +4[2200] +4[2211] +4[2220] + [2222] +4[3100] +3[3111] +10[3210] +5[3221] +2[3300] +4[3311] +6[3320] +2[3322] +2[3331] +2[4110] +2[4200] +3[4211] +4[4220] + [4222] +4[4310] +4[4321] +2[4330] + [4332] + [4411] +2[4420] + [4422] + [4431]

[2221] x [2211] =2[1000] +4[1110] +6[2100] +4[2111] +8[2210] +3[2221] +2[3000] +8[3110] +6[3200] +7[3211] +8[3220] +2[3222] +8[3310] +6[3321] +4[3330] +2[3332] +2[4100] +2[4111] +6[4210] +3[4221] +2[4300] +4[4311] +6[4320] +2[4322] +3[4331] + [4333] +2[4410] +2[4421] +2[4430] + [4432]

[2221] x [2220] =2[1000] +2[1110] +4[2100] +2[2111] +4[2210] +2[2221] +2[3000] +4[3110] +4[3200] +3[3211] +4[3220] + [3222] +4[3310] +3[3321] +2[3330] + [3332] +2[4100] + [4111] +4[4210] +2[4221] +2[4300] +2[4311] +4[4320] + [4322] +2[4331] +2[4410] +2[4421] +2[4430] + [4432] + [4441]

[2221] x [2221] =2[0000] +4[1100] +2[1111] +4[2000] +6[2110] +4[2200] +3[2211] +4[2220] + [2222] +6[3100] +3[3111] +8[3210] +4[3221] +4[3300] +3[3311] +6[3320] +2[3322] +3[3331] + [3333] +2[4000] +4[4110] +4[4200] +3[4211] +4[4220] + [4222] +6[4310] +4[4321] +4[4330] +2[4332] +2[4400] +2[4411] +4[4420] + [4422] +3[4431] + [4433] +2[4440] + [4442]

[2222] x [0000] = [2222]

[2222] x [1000] = [2221] + [3222]

[2222] x [1100] = [2211] + [2222] + [3221] + [3322]

[2222] x [1110] = [2111] + [2221] + [3211] + [3222] + [3321] + [3332]

[2222] x [1111] = [1111] + [2211] + [2222] + [3111] + [3221] + [3311] + [3322] + [3331] + [3333]

[2222] x [2000] =2[2220] + [3221] + [4222]

[2222] x [2100] =2[2210] + [2221] + [3211] +2[3220] + [3222] + [3321] + [4221] + [4322]

[2222] x [2110] =2[2110] + [2211] +2[2220] + [3111] +2[3210] +2[3221] + [3311] +2[3320] + [3322] + [3331] + [4211] + [4222] + [4321] + [4332]

[2222] x [2111] =2[1110] + [2111] +2[2210] + [2221] +2[3110] +2[3211] +2[3220] + [3222] +2[3310] +2[3321] +2[3330] + [3332] + [4111] + [4221] + [4311] + [4322] + [4331] + [4333]

[2222] x [2200] =2[2200] + [2211] + [2222] +2[3210] + [3221] + [3311] + [3322] +2[4220] + [4321] + [4422]

[2222] x [2210] =2[2100] + [2111] +2[2210] + [2221] +2[3110] +2[3200] +2[3211] +2[3220] + [3222] +2[3310] +2[3321] + [3332] +2[4210] + [4221] + [4311] +2[4320] + [4322] + [4331] + [4421] + [4432]

[2222] x [2211] =2[1100] + [1111] +2[2110] +2[2200] +2[2211] + [2222] +2[3100] + [3111] +4[3210] +2[3221] +2[3300] +2[3311] +2[3320] +2[3322] + [3331] + [3333] +2[4110] + [4211] +2[4220] +2[4310] +2[4321] +2[4330] + [4332] + [4411] + [4422] + [4431] + [4433]

[2222] x [2220] =2[2000] +2[2110] +2[2220] +2[3100] + [3111] +2[3210] + [3221] +2[3320] + [3331] +2[4200] + [4211] + [4222] +2[4310] + [4321] + [4332] +2[4420] + [4431] + [4442]

[2222] x [2221] =2[1000] +2[1110] +2[2100] + [2111] +2[2210] + [2221] +2[3000] +2[3110] +2[3200] + [3211] +2[3220] + [3222] +2[3310] + [3321] +2[3330] + [3332] +2[4100] + [4111] +2[4210] + [4221] +2[4300] + [4311] +2[4320] + [4322] + [4331] + [4333] +2[4410] + [4421] +2[4430] + [4432] + [4441] + [4443]

[2222] x [2222] =2[0000] +2[1100] + [1111] +2[2200] + [2211] + [2222] +2[3300] + [3311] + [3322] + [3333] +2[4000] +2[4110] +2[4220] +2[4330] +2[4400] + [4411] + [4422] + [4433] +2[4440] + [4444]

D-6 Reduction of the Kronecker Products for R_9.

```
[0000] x [0000] = [0000]

[1000] x [0000] = [1000]
[1000] x [1000] = [0000] + [1100] + [2000]
[1100] x [0000] = [1100]
[1100] x [1000] = [1000] + [1110] + [2100]
[1100] x [1100] = [0000] + [1100] + [1111] + [2000] + [2110] + [2200]
[1110] x [0000] = [1110]
[1110] x [1000] = [1100] + [1111] + [2110]
[1110] x [1100] = [1000] + [1110] + [1111] + [2100] + [2111] + [2210]
[1110] x [1110] = [0000] + [1100] + [1110] + [1111] + [2000] + [2110] + [2111] + [2200] + [2211]
                + [2220]
[1111] x [0000] = [1111]
[1111] x [1000] = [1110] + [1111] + [2111]
[1111] x [1100] = [1100] + [1110] + [1111] + [2110] + [2111] + [2211]
[1111] x [1110] = [1000] + [1100] + [1110] + [1111] + [2100] + [2110] + [2111] + [2210] + [2211]
                + [2221]
[1111] x [1111] = [0000] + [1000] + [1100] + [1110] + [1111] + [2000] + [2100] + [2110] + [2111]
                + [2200] + [2210] + [2211] + [2220] + [2221] + [2222]

[2000] x [0000] = [2000]
[2000] x [1000] = [1000] + [2100] + [3000]
[2000] x [1100] = [1100] + [2000] + [2110] + [3100]
[2000] x [1110] = [1110] + [2100] + [2111] + [3110]
[2000] x [1111] = [1111] + [2110] + [2111] + [3111]
[2000] x [2000] = [0000] + [1100] + [2000] + [2200] + [3100] + [4000]
[2100] x [0000] = [2100]
[2100] x [1000] = [1100] + [2000] + [2110] + [2200] + [3100]
[2100] x [1100] = [1000] + [1110] +2[2100] + [2111] + [2210] + [3000] + [3110] + [3200]
[2100] x [1110] = [1100] + [1111] + [2000] +2[2110] + [2111] + [2200] + [2211] + [3100] + [3111]
                + [3210]
[2100] x [1111] = [1110] + [1111] + [2100] + [2110] +2[2111] + [2210] + [2211] + [3110] + [3111]
                + [3211]
[2100] x [2000] = [1000] + [1110] +2[2100] + [2210] + [3000] + [3110] + [3200] + [4100]
[2100] x [2100] = [0000] +2[1100] + [1111] +2[2000] +3[2110] +2[2200] + [2211] + [2220] +3[3100]
                + [3111] +2[3210] + [3300] + [4000] + [4110] + [4200]
[2110] x [0000] = [2110]
[2110] x [1000] = [1110] + [2100] + [2111] + [2210] + [3110]
[2110] x [1100] = [1100] + [1111] + [2000] +2[2110] + [2111] + [2200] + [2211] + [2220] + [3100]
                + [3111] + [3210]
[2110] x [1110] = [1000] + [1110] + [1111] +2[2100] + [2110] +2[2111] +2[2210] + [2211] + [2221]
                + [3000] + [3110] + [3111] + [3200] + [3211] + [3220]
[2110] x [1111] = [1100] + [1110] + [1111] + [2000] + [2100] +2[2110] +2[2111] + [2200] + [2210]
               +2[2211] + [2220] + [2221] + [3100] + [3110] + [3111] + [3210] + [3211] + [3221]
[2110] x [2000] = [1100] + [1111] +2[2110] + [2200] + [2211] + [3100] + [3111] + [3210] + [4110]
[2110] x [2100] = [1000] +2[1110] + [1111] +3[2100] +3[2111] +3[2210] + [2211] + [2221] + [3000]
               +3[3110] + [3111] +2[3200] +2[3211] + [3220] + [3310] + [4100] + [4111] + [4210]
[2110] x [2110] = [0000] +2[1100] + [1110] +2[1111] +2[2000] +4[2110] +3[2111] +3[2200] + [2210]
               +4[2211] +2[2220] + [2221] + [2222] +3[3100] + [3110] +3[3111] +4[3210] +2[3211]
               +2[3221] + [3300] + [3311] + [3320] + [4000] + [4110] + [4111] + [4200] + [4211]
                + [4220]
[2111] x [0000] = [2111]
[2111] x [1000] = [1111] + [2110] + [2111] + [2211] + [3111]
[2111] x [1100] = [1110] + [1111] + [2100] + [2110] +2[2111] + [2210] + [2211] + [2221] + [3110]
                + [3111] + [3211]
[2111] x [1110] = [1100] + [1110] + [1111] + [2000] + [2100] +2[2110] +2[2111] + [2200] + [2210]
               +2[2211] + [2220] + [2221] + [2222] + [3100] + [3110] + [3111] + [3210] + [3211]
                + [3221]
[2111] x [1111] = [1000] + [1100] + [1110] + [1111] + [2000] +2[2100] +2[2110] +2[2111] + [2200]
               +2[2210] +2[2211] + [2220] +2[2221] + [2222] + [3000] + [3100] + [3110] + [3111]
                + [3200] + [3210] + [3211] + [3220] + [3221] + [3222]
[2111] x [2000] = [1110] + [1111] +2[2111] + [2210] + [2211] + [3110] + [3111] + [3211] + [4111]
[2111] x [2100] = [1100] + [1110] +2[1111] +3[2110] +3[2111] + [2200] + [2210] +3[2211] + [2220]
                + [2221] + [3100] + [3110] +3[3111] +2[3210] +2[3211] + [3221] + [3311] + [4110]
                + [4111] + [4211]
[2111] x [2110] = [1000] + [1100] +2[1110] +2[1111] +3[2100] +3[2110] +4[2111] + [2200] +4[2210]
               +4[2211] + [2220] +3[2221] + [2222] + [3000] + [3100] +3[3110] +3[3111] +2[3200]
               +2[3210] +4[3211] +2[3220] +2[3221] + [3222] + [3310] + [3311] + [3321] + [4100]
                + [4110] + [4111] + [4210] + [4211] + [4221]
[2111] x [2111] = [0000] + [1000] +2[1100] +2[1110] +2[1111] +2[2000] +3[2100] +4[2110] +4[2111]
               +3[2200] +4[2210] +5[2211] +3[2220] +4[2221] +2[2222] + [3000] +3[3100] +3[3110]
               +3[3111] +2[3200] +4[3210] +4[3211] +2[3220] +4[3221] +2[3222] + [3300] + [3310]
                + [3311] + [3320] + [3321] + [3322] + [4000] + [4100] + [4110] + [4111] + [4200]
                + [4210] + [4211] + [4220] + [4221] + [4222]
```

Reduction of the Kronecker Products for R_9 (contd).

[2200] x [0000] = [2200]
[2200] x [1000] = [2100] + [2210] + [3200]
[2200] x [1100] = [1100] + [2110] + [2200] + [2211] + [3100] + [3210] + [3300]
[2200] x [1110] = [1110] + [2100] + [2111] + [2210] + [2211] + [3110] + [3200] + [3211] + [3310]
[2200] x [1111] = [1111] + [2110] + [2111] + [2200] + [2210] + [2211] + [3111] + [3210] + [3211]
+ [3311]
[2200] x [2000] = [2000] + [2110] + [2200] + [2220] + [3100] + [3210] + [4200]
[2200] x [2100] = [1000] + [1110] +2[2100] + [2111] +2[2210] + [2221] + [3000] +2[3110] +2[3200]
+ [3211] + [3220] + [3310] + [4100] + [4210] + [4300]
[2200] x [2110] = [1100] + [1111] + [2000] +3[2110] + [2111] + [2200] +2[2211] + [2220] + [2221]
+2[3100] +2[3111] +3[3210] + [3211] + [3221] + [3300] + [3311] + [3320] + [4110]
+ [4200] + [4211] + [4310]
[2200] x [2111] = [1110] + [1111] + [2100] + [2110] +3[2111] +2[2210] +2[2211] + [2220] + [2221]
+2[3110] +2[3111] + [3200] + [3210] +3[3211] + [3220] + [3221] + [3310] + [3311]
+ [3321] + [4111] + [4210] + [4211] + [4311]
[2200] x [2200] = [0000] + [1100] + [1111] + [2000] + [2110] +2[2200] + [2211] + [2220] + [2222]
+ [3100] + [3111] +2[3210] + [3221] + [3300] + [3311] + [4000]+ [4110] + [4200]
+ [4220] + [4310] + [4400]
[2210] x [0000] = [2210]
[2210] x [1000] = [2110] + [2200] + [2211] + [2220] + [3210]
[2210] x [1100] = [1110] + [2100] + [2111] +2[2210] + [2211] + [2221] + [3110] + [3200] + [3211]
+ [3220] + [3310]
[2210] x [1110] = [1100] + [1111] +2[2110] + [2111] + [2200] + [2210] +2[2211] + [2220] + [2221]
+ [3100] + [3111] +2[3210] + [3211] + [3221] + [3300] + [3311] + [3320]
[2210] x [1111] = [1110] + [1111] + [2100] + [2110] +2[2111] + [2200] +2[2210] +2[2211] + [2220]
+ [2221] + [3110] + [3111] + [3200] + [3210] +2[3211] + [3220]+ [3221] + [3310]
+ [3311] + [3321]
[2210] x [2000] = [2100] + [2111] +2[2210] + [2221] + [3110] + [3200] + [3211] + [3220] + [4210]
[2210] x [2100] = [1100] + [1111] + [2000] +3[2110] + [2111] +2[2200] +3[2211] +2[2220] + [2221]
+ [2222] +2[3100] +2[3111] +4[3210] + [3211] +2[3221] + [3300] + [3311] + [3320]
+ [4110] + [4200] + [4211] + [4220] + [4310]
[2210] x [2110] = [1000] +2[1110] + [1111] +3[2100] + [2110] +4[2111] +4[2210] +3[2211] + [2220]
+3[2221] + [2222] + [3000] +4[3110] +2[3111] +3[3200] + [3210] +5[3211] +3[3220]
+2[3221] + [3222] +3[3310] + [3311] +2[3321] + [3330] + [4100] + [4111] +2[4210]
+ [4211] + [4221] + [4300] + [4311] + [4320]
[2210] x [2111] = [1100] + [1110] +2[1111] + [2000] + [2100] +4[2110] +4[2111] +2[2200] +3[2210]
+5[2211] +2[2220] +3[2221] + [2222] +2[3100] +2[3110] +4[3111] + [3200] +5[3210]
+5[3211] +2[3220] +4[3221] + [3222] + [3300] + [3310] +3[3311] +2[3320] +2[3321]
+ [3322] + [3331] + [4110] + [4111] + [4200] + [4210] +2[4211] + [4220] + [4221]
+ [4310] + [4311] + [4321]
[2210] x [2200] = [1000] + [1110] + [1111] +2[2100] +2[2111] +3[2210] + [2211] +2[2221] + [2222]
+ [3000] +2[3110] + [3111] +2[3200] +3[3211] +2[3220] + [3221] + [3222] +2[3310]
+ [3311] + [3321] + [4100] + [4111] +2[4210] + [4221] + [4300] + [4311] + [4320]
+ [4410]
[2210] x [2210] = [0000] +2[1100] + [1110] +2[1111] +2[2000] +4[2110] +3[2111] +3[2200] + [2210]
+5[2211] +3[2220] +3[2221] +2[2222] +3[3100] + [3110] +4[3111] +6[3210] +4[3211]
+ [3220] +5[3221] +2[3222] +2[3300] + [3310] +4[3311] +3[3320] +2[3321] + [3322]
+ [3331] + [4000] +2[4110] + [4111] +2[4200] +3[4211] +2[4220] + [4221] + [4222]
+3[4310] + [4311] +2[4321] + [4330] + [4400] + [4411] + [4420]
[2211] x [0000] = [2211]
[2211] x [1000] = [2111] + [2210] + [2211] + [2221] + [3211]
[2211] x [1100] = [1111] + [2110] + [2111] + [2200] + [2210] +2[2211] + [2220] + [2221] + [2222]
+ [3111] + [3210] + [3211] + [3221] + [3311]
[2211] x [1110] = [1110] + [1111] + [2100] + [2110] +2[2111] + [2200] +2[2210] +2[2211] + [2220]
+2[2221] + [2222] + [3110] + [3111] + [3200] + [3210] +2[3211] + [3220] + [3221]
+ [3222] + [3310] + [3311] + [3321]
[2211] x [1111] = [1100] + [1110] + [1111] + [2100] +2[2110] +2[2111] + [2200] +2[2210] +3[2211]
+ [2220] +2[2221] + [2222] + [3100] + [3110] + [3111] + [3200] +2[3210] +2[3211]
+ [3220] +2[3221] + [3222] + [3300] + [3310] + [3311] + [3320] + [3321] + [3322]
[2211] x [2000] = [2110] + [2111] +2[2211] + [2220] + [2221] + [3111] + [3210] + [3211] + [3221]
+ [4211]
[2211] x [2100] = [1110] + [1111] + [2100] + [2110] +3[2111] +3[2210] +3[2211] + [2220] +3[2221]
+ [2222] +2[3110] +2[3111] + [3200] + [3210] +4[3211] +2[3220] +2[3221] + [3222]
+ [3310] + [3311] + [3321] + [4111] + [4210] + [4211] + [4221] + [4311]
[2211] x [2110] = [1100] + [1110] +2[1111] + [2000] + [2100] +4[2110] +4[2111] +2[2200] +3[2210]
+5[2211] +3[2220] +4[2221] +2[2222] +2[3100] +2[3110] +4[3111] + [3200] +5[3210]
+5[3211] +2[3220] +5[3221] +2[3222] + [3300] + [3310] +3[3311] +2[3320] +2[3321]
+ [3322] + [3331] + [4110] + [4111] + [4200] + [4210] +2[4211] + [4220] + [4221]
+ [4222] + [4310] + [4311] + [4321]
[2211] x [2111] = [1000] + [1100] +2[1110] +2[1111] + [2000] +3[2100] +4[2110] +5[2111] +2[2200]
+5[2210] +6[2211] +3[2220] +5[2221] +2[2222] + [3000] +2[3100] +4[3110] +4[3111]
+3[3200] +5[3210] +7[3211] +4[3220] +6[3221] +3[3222] + [3300] +3[3310] +3[3311]
+2[3320] +4[3321] +2[3322] + [3330] + [3331] + [3332] + [4100] + [4110] + [4111]
+ [4200] +2[4210] +2[4211] + [4220] +2[4221] + [4222] + [4300] + [4310] + [4311]

```
                + [4320] + [4321] + [4322]
[2211] x [2200] = [1100] + [1110] + [1111] +2[2110] +2[2111] + [2200] + [2210] +4[2211] + [2220]
                +2[2221] + [2222] + [3100] + [3110] +2[3111] +3[3210] +3[3211] + [3220] +3[3221]
                + [3222] + [3300] + [3310] +2[3311] + [3320] + [3321] + [3322] + [4110] + [4111]
                +2[4211] + [4220] + [4221] + [4310] + [4311] + [4321] + [4411]
[2211] x [2210] = [1000] + [1100] +2[1110] +2[1111] +3[2100] +3[2110] +5[2111] + [2200] +5[2210]
                +6[2211] +2[2220] +5[2221] +2[2222] + [3000] + [3100] +4[3110] +4[3111] +3[3200]
                +4[3210] +8[3211] +4[3220] +6[3221] +3[3222] + [3300] +4[3310] +4[3311] +2[3320]
                +5[3321] +2[3322] + [3330] + [3331] + [3332] + [4100] + [4110] +2[4111] +3[4210]
                +3[4211] + [4220] +3[4221] + [4222] + [4300] + [4310] +3[4311] +2[4320] +2[4321]
                + [4322] + [4331] + [4410] + [4411] + [4421]
[2211] x [2211] = [0000] + [1000] +2[1100] +2[1110] +3[1111] +2[2000] +3[2100] +5[2110] +6[2111]
                +4[2200] +6[2210] +8[2211] +4[2220] +6[2221] +3[2222] + [3000] +3[3100] +4[3110]
                +6[3111] +3[3200] +8[3210] +10[3211] +5[3220] +9[3221] +4[3222] +2[3300] +4[3310]
                +7[3311] +4[3320] +7[3321] +4[3322] + [3330] +3[3331] +2[3332] + [3333] + [4000]
                + [4100] +2[4110] +2[4111] +2[4200] +3[4210] +4[4211] +3[4220] +4[4221] +2[4222]
                + [4300] +3[4310] +3[4311] +2[4320] +4[4321] +2[4322] + [4330] + [4331] + [4332]
                + [4400] + [4410] + [4411] + [4420] + [4421] + [4422]
[2220] x [0000] = [2220]
[2220] x [1000] = [2210] + [2221] + [3220]
[2220] x [1100] = [2110] + [2211] + [2220] + [2221] + [3210] + [3221] + [3320]
[2220] x [1110] = [1110] + [2111] + [2210] + [2211] + [2220] + [2221] + [3110] + [3211] + [3220]
                + [3221] + [3310] + [3321] + [3330]
[2220] x [1111] = [1111] + [2110] + [2111] + [2210] + [2211] + [2220] + [2221] + [3111] + [3210]
                + [3211] + [3220] + [3221] + [3311] + [3320] + [3321] + [3331]
[2220] x [2000] = [2200] + [2211] + [2220] + [2222] + [3210] + [3221] + [4220]
[2220] x [2100] = [2100] + [2111] +2[2210] + [2211] +2[2221] + [2222] + [3110] + [3200] +2[3211]
                +2[3220] + [3221] + [3222] + [3310] + [3321] + [4210] + [4221] + [4320]
[2220] x [2110] = [1100] + [1111] +2[2110] + [2111] + [2200] + [2210] +3[2211] + [2220] +2[2221]
                + [2222] + [3100] +2[3111] +3[3210] +2[3211] + [3220] +3[3221] + [3222] + [3300]
                +2[3311] +2[3320] + [3321] + [3322] + [3331] + [4110] + [4211] + [4220] + [4221]
                + [4310] + [4321] + [4330]
[2220] x [2111] = [1110] + [1111] + [2100] + [2110] +3[2111] + [2200] +2[2210] +3[2211] + [2220]
                +2[2221] + [2222] +2[3110] +2[3111] + [3200] +2[3210] +4[3211] +2[3220] +3[3221]
                + [3222] +2[3310] +2[3311] + [3320] +3[3321] + [3322] + [3330] + [3331] + [3332]
                + [4111] + [4210] + [4211] + [4220] + [4221] + [4311] + [4320] + [4321] + [4331]
[2220] x [2200] = [2000] + [2110] + [2111] + [2200] + [2211] +2[2220] + [2221] + [2222] + [3100]
                + [3111] +2[3210] + [3211] +2[3221] + [3222] + [3311] + [3320] + [3321] + [4200]
                + [4211] + [4220] + [4222] + [4310] + [4321] + [4420]
[2220] x [2210] = [1000] + [1110] + [1111] +2[2100] + [2110] +2[2111] +3[2210] +2[2211] + [2220]
                +3[2221] + [2222] + [3000] +2[3110] +2[3111] +2[3200] + [3210] +4[3211] +3[3220]
                +3[3221] +2[3222] +2[3310] +2[3311] + [3320] +3[3321] + [3322] + [3330] + [3331]
                + [4100] + [4111] +2[4210] + [4211] +2[4221] + [4222] + [4300] +2[4311] +2[4320]
                + [4321] + [4322] + [4331] + [4410] + [4421] + [4430]
[2220] x [2211] = [1100] + [1110] + [1111] + [2000] + [2100] +3[2110] +3[2111] + [2200] +2[2210]
                +4[2211] +2[2220] +3[2221] + [2222] +2[3100] +2[3110] +3[3111] + [3200] +4[3210]
                +5[3211] +2[3220] +5[3221] +2[3222] + [3300] +2[3310] +3[3311] +3[3320] +4[3321]
                +2[3322] + [3330] +2[3331] + [3332] + [4110] + [4111] + [4200] + [4210] +3[4211]
                + [4220] +2[4221] + [4222] +2[4310] +2[4311] + [4320] +3[4321] + [4322] + [4330]
                + [4331] + [4332] + [4411] + [4420] + [4421] + [4431]
[2220] x [2220] = [0000] + [1100] + [1110] + [1111] + [2000] + [2110] + [2111] +2[2200] + [2210]
                +2[2211] +2[2220] + [2221] + [2222] + [3100] + [3110] + [3111] +2[3210] +2[3211]
                + [3220] +3[3221] + [3222] + [3300] + [3310] +2[3311] + [3320] +2[3321] + [3322]
                + [3330] + [3331] + [4000] + [4110] + [4111] + [4200]+ [4211] +2[4220] + [4221]
                + [4222] + [4310] + [4311] +2[4321] + [4322] + [4330] + [4331] + [4400] + [4411]
                + [4420] + [4422] + [4431] + [4440]
[2221] x [0000] = [2221]
[2221] x [1000] = [2211] + [2220] + [2221] + [2222] + [3221]
[2221] x [1100] = [2111] + [2210] + [2211] + [2220] +2[2221] + [2222] + [3211] + [3220] + [3221]
                + [3222] + [3321]
[2221] x [1110] = [1111] + [2110] + [2111] + [2210] +2[2211] + [2220] +2[2221] + [2222] + [3111]
                + [3210] + [3211] + [3220] +2[3221] + [3222] + [3311] + [3320] + [3321] + [3322]
                + [3331]
[2221] x [1111] = [1110] + [1111] + [2110] +2[2111] + [2210] +2[2211] + [2220] +2[2221] + [2222]
                + [3110] + [3111] + [3210] +2[3211] + [3220] +2[3221] + [3222] + [3310] + [3311]
                + [3320] +2[3321] + [3322] + [3330] + [3331] + [3332]
[2221] x [2000] = [2210] + [2211] +2[2221] + [2222] + [3211] + [3220] + [3221] + [3222] + [4221]
[2221] x [2100] = [2110] + [2111] + [2200] + [2210] +3[2211] +2[2220] +3[2221] +2[2222] + [3111]
                +2[3210] +2[3211] + [3220] +4[3221] +2[3222] + [3311] + [3320] + [3321] + [3322]
                + [4211] + [4220] + [4221] + [4222] + [4321]
```

[2221] x [2110] = [1110] + [1111] + [2100] + [2110] +3[2111] + [2200] +3[2210] +4[2211] +2[2220]
+4[2221] +2[2222] +2[3110] +2[3111] + [3200] +2[3210] +5[3211] +3[3220] +5[3221]
+3[3222] +2[3310] +2[3311] + [3320] +4[3321] +2[3322] + [3330] + [3331] + [3332]
+ [4111] + [4210] + [4211] + [4220] +2[4221] + [4222] + [4311] + [4320] + [4321]
+ [4322] + [4331]

[2221] x [2111] = [1100] + [1110] +2[1111] + [2100] +3[2110] +4[2111] + [2200] +3[2210] +5[2211]
+2[2220] +4[2221] +2[2222] + [3100] +2[3110] +4[3111] + [3200] +4[3210] +6[3211]
+3[3220] +6[3221] +3[3222] + [3300] +2[3310] +4[3311] +3[3320] +5[3321] +3[3322]
+ [3330] +3[3331] +2[3332] + [3333] + [4110] + [4111] + [4210] +2[4211] + [4220]
+2[4221] + [4222] + [4310] + [4311] + [4320] +2[4321] + [4322] + [4330] + [4331]
+ [4332]

[2221] x [2200] = [2100] + [2110] + [2111] +2[2210] +2[2211] + [2220] +3[2221] + [2222] + [3110]
+ [3111] + [3200] + [3210] +3[3211] +2[3220] +3[3221] +2[3222] + [3310] + [3311]
+ [3320] +2[3321] + [3322] + [4210] + [4211] +2[4221] + [4222] + [4311] + [4320]
+ [4321] + [4322] + [4421]

[2221] x [2210] = [1100] + [1110] + [1111] + [2000] + [2100] +3[2110] +3[2111] +2[2200] +3[2210]
+5[2211] +3[2220] +4[2221] +2[2222] +2[3100] +2[3110] +3[3111] + [3200] +5[3210]
+6[3211] +3[3220] +7[3221] +3[3222] + [3300] +2[3310] +4[3311] +3[3320] +5[3321]
+3[3322] + [3330] +2[3331] + [3332] + [4110] + [4111] + [4200] + [4210] +3[4211]
+2[4220] +3[4221] +2[4222] +2[4310] +2[4311] + [4320] +4[4321] +2[4322] + [4330]
+ [4331] + [4332] + [4411] + [4420] + [4421] + [4422] + [4431]

[2221] x [2211] = [1000] + [1100] +2[1110] +2[1111] + [2000] +3[2100] +4[2110] +5[2111] +2[2200]
+5[2210] +6[2211] +3[2220] +5[2221] +2[2222] + [3000] +2[3100] +4[3110] +5[3111]
+3[3200] +6[3210] +9[3211] +5[3220] +8[3221] +4[3222] + [3300] +4[3310] +6[3311]
+4[3320] +8[3321] +4[3322] +2[3330] +4[3331] +3[3332] + [3333] + [4100] + [4110]
+2[4111] + [4200] +3[4210] +4[4211] +2[4220] +4[4221] +2[4222] + [4300] +2[4310]
+4[4311] +3[4320] +5[4321] +3[4322] + [4330] +3[4331] +2[4332] + [4333] + [4410]
+ [4411] + [4420] +2[4421] + [4422] + [4430] + [4431] + [4432]

[2221] x [2220] = [1000] + [1100] + [1110] + [1111] +2[2100] +2[2110] +2[2111] + [2200] +3[2210]
+3[2211] + [2220] +3[2221] + [2222] + [3000] + [3100] +2[3110] +2[3111] +2[3200]
+3[3210] +4[3211] +3[3220] +4[3221] +2[3222] + [3300] +2[3310] +3[3311] +2[3320]
+4[3321] +2[3322] + [3330] +2[3331] + [3332] + [4100] + [4110] + [4111] +2[4210]
+2[4211] + [4220] +3[4221] + [4222] + [4300] + [4310] +2[4311] +2[4320] +3[4321]
+2[4322] + [4330] +2[4331] + [4332] + [4410] + [4411] +2[4421] + [4422] + [4430]
+ [4431] + [4432] + [4441]

[2221] x [2221] = [0000] + [1000] +2[1100] +2[1110] +2[1111] +2[2000] +3[2100] +4[2110] +4[2111]
+3[2200] +4[2210] +5[2211] +3[2220] +4[2221] +2[2222] + [3000] +3[3100] +4[3110]
+4[3111] +3[3200] +6[3210] +7[3211] +4[3220] +7[3221] +3[3222] +2[3300] +4[3310]
+5[3311] +4[3320] +7[3321] +4[3322] +2[3330] +4[3331] +3[3332] + [3333] + [4000]
+ [4100] +2[4110] +2[4111] +2[4200] +3[4210] +4[4211] +3[4220] +4[4221] +2[4222]
+ [4300] +3[4310] +4[4311] +3[4320] +6[4321] +3[4322] +2[4330] +4[4331] +3[4332]
+ [4333] + [4400] + [4410] +2[4411] +2[4420] +3[4421] +2[4422] + [4430] +3[4431]
+2[4432] + [4433] + [4440] + [4441] + [4442]

[2222] x [0000] = [2222]

[2222] x [1000] = [2221] + [2222] + [3222]

[2222] x [1100] = [2211] + [2221] + [2222] + [3221] + [3222] + [3322]

[2222] x [1110] = [2111] + [2211] + [2221] + [2222] + [3211] + [3221] + [3222]+ [3321] + [3322]
+ [3332]

[2222] x [1111] = [1111] + [2111] + [2211] + [2221] + [2222] + [3111] + [3211] + [3221] + [3222]
+ [3311] + [3321] + [3322] + [3331] + [3332] + [3333]

[2222] x [2000] = [2220] + [2221] + [2222] + [3221] + [3222] + [4222]

[2222] x [2100] = [2210] + [2211] + [2220] +2[2221] + [2222] + [3211] + [3220] +2[3221] +2[3222]
+ [3321] + [3322] + [4221] + [4222] + [4322]

[2222] x [2110] = [2110] + [2111] + [2210] +2[2211] + [2220] +2[2221] + [2222] + [3111] + [3210]
+2[3211] + [3220] +3[3221] +2[3222] + [3311] + [3320] +2[3321] +2[3322] + [3331]
+ [3332] + [4211] + [4221] + [4222] + [4321] + [4322] + [4332]

[2222] x [2111] = [1110] + [1111] + [2110] +2[2111] + [2210] +2[2211] + [2220] +2[2221] + [2222]
+ [3110] +2[3111] + [3210] +3[3211] + [3220] +3[3221] +2[3222] + [3310] +2[3311]
+ [3320] +3[3321] +2[3322] + [3330] +2[3331] +2[3332] + [3333] + [4111] + [4211]
+ [4221] + [4222] + [4311] + [4321] + [4322] + [4331] + [4332] + [4333]

[2222] x [2200] = [2200] + [2210] + [2211] + [2220] + [2221] + [2222] + [3210] + [3211] + [3220]
+2[3221] + [3222] + [3311] + [3321] + [3322] + [4220] + [4221] + [4222] + [4321]
+ [4322] + [4422]

[2222] x [2210] = [2100] + [2110] + [2111] + [2200] +2[2210] +2[2211] + [2220] +2[2221] + [2222]
+ [3110] + [3111] + [3200] +2[3210] +3[3211] +2[3220] +3[3221] +2[3222] + [3310]
+2[3311] + [3320] +3[3321] +2[3322] + [3331] + [3332] + [4210] + [4211] + [4220]
+2[4221] + [4222] + [4311] + [4320] +2[4321] +2[4322] + [4331] + [4332] + [4421]
+ [4422] + [4432]

[2222] x [2211] = [1100] + [1110] + [1111] + [2100] +2[2110] +2[2111] + [2200] +2[2210] +3[2211]
+ [2220] +2[2221] + [2222] + [3100] +2[3110] +2[3111] + [3200] +3[3210] +4[3211]
+2[3220] +4[3221] +2[3222] + [3300] +2[3310] +3[3311] +2[3320] +4[3321] +3[3322]
+ [3330] +2[3331] +2[3332] + [3333] + [4110] + [4111] + [4210] +2[4211] + [4220]
+2[4221] + [4222] + [4310] +2[4311] + [4320] +3[4321] +2[4322] + [4330] +2[4331]
+2[4332] + [4333] + [4411] + [4421] + [4422] + [4431] + [4432] + [4433]

D-6 Reduction of the Kronecker Products for R_9 (contd).

[2222] x [2220] = [2000] + [2100] + [2110] + [2111] + [2200] + [2210] + [2211] + [2220] + [2221]
+ [2222] + [3100] + [3110] + [3111] + [3200] +2[3210] +2[3211] + [3220] +2[3221]
+ [3222] + [3310] + [3311] + [3320] +2[3321] + [3322] + [3331] + [3332] + [4200]
+ [4210] + [4211] + [4220] + [4221] + [4222] + [4310] + [4311] + [4320] +2[4321]
+ [4322] + [4331] + [4332] + [4420] + [4421] + [4422] + [4431] + [4432] + [4442]

[2222] x [2221] = [1000] + [1100] + [1110] + [1111] + [2000] +2[2100] +2[2110] +2[2111] + [2200]
+2[2210] +2[2211] + [2220] +2[2221] + [2222] + [3000] +2[3100] +2[3110] +2[3111]
+2[3200] +3[3210] +3[3211] +2[3220] +3[3221] +2[3222] + [3300] +2[3310] +2[3311]
+2[3320] +3[3321] +2[3322] + [3330] +2[3331] +2[3332] + [3333] + [4100] + [4110]
+ [4111] + [4200] +2[4210] +2[4211] + [4220] +2[4221] + [4222]+ [4300] +2[4310]
+2[4311] +2[4320] +3[4321] +2[4322] + [4330] +2[4331] +2[4332] + [4333] + [4410]
+ [4411] + [4420] +2[4421] + [4422] + [4430] +2[4431] +2[4432] + [4433] + [4441]
+ [4442] + [4443]

[2222] x [2222] = [0000] + [1000] + [1100] + [1110] + [1111] + [2000] + [2100] + [2110] + [2111]
+ [2200] + [2210] + [2211] + [2220] + [2221] + [2222] + [3000] + [3100] + [3110]
+ [3111] + [3200] + [3210] + [3211] + [3220] + [3221] + [3222] + [3300] + [3310]
+ [3311] + [3320] + [3321] + [3322] + [3330] + [3331] + [3332] + [3333] + [4000]
+ [4100] + [4110] + [4111] + [4200] + [4210] + [4211] + [4220] + [4221] + [4222]
+ [4300] + [4310] + [4311] + [4320] + [4321] + [4322]+ [4330] + [4331] + [4332]
+ [4333] + [4400] + [4410] + [4411] + [4420] + [4421] + [4422] + [4430] + [4431]
+ [4432] + [4433] + [4440] + [4441] + [4442] + [4443] + [4444]

D-7 Reduction of the Kronecker Products for R_{11}.

[00000] x [00000] = [00000]

[10000] x [00000] = [10000]

[10000] x [10000] = [00000] + [11000] + [20000]

[11000] x [00000] = [11000]

[11000] x [10000] = [10000] + [11100] + [21000]

[11000] x [11000] = [00000] + [11000] + [11110] + [20000] + [21100] + [22000]

[11100] x [00000] = [11100]

[11100] x [10000] = [11000] + [11110] + [21100]

[11100] x [11000] = [10000] + [11100] + [11111] + [21000] + [21110] + [22100]

[11100] x [11100] = [00000] + [11000] + [11110] + [11111] + [20000] + [21100] + [21111] + [22000]
+ [22110] + [22200]

[11110] x [00000] = [11110]

[11110] x [10000] = [11100] + [11111] + [21110]

[11110] x [11000] = [11000] + [11110] + [11111] + [21100] + [21111] + [22110]

[11110] x [11100] = [10000] + [11100] + [11110] + [11111] + [21000] + [21110] + [21111] + [22100]
+ [22111] + [22210]

[11110] x [11110] = [00000] + [11000] + [11100] + [11110] + [11111] + [20000] + [21100] + [21110]
+ [21111] + [22000] + [22110] + [22111] + [22200] + [22211] + [22220]

[11111] x [00000] = [11111]

[11111] x [10000] = [11110] + [11111] + [21111]

[11111] x [11000] = [11100] + [11110] + [11111] + [21110] + [21111] + [22111]

[11111] x [11100] = [11000] + [11100] + [11110] + [11111] + [21100] + [21110] + [21111] + [22110]
+ [22111] + [22211]

[11111] x [11110] = [10000] + [11000] + [11100] + [11110] + [11111] + [21000] + [21100] + [21110]
+ [21111] + [22100] + [22110] + [22111] + [22210] + [22211] + [22221]

[11111] x [11111] = [00000] + [10000] + [11000] + [11100] + [11110] + [11111] + [20000] + [21000]
+ [21100] + [21110] + [21111] + [22000] + [22100] + [22110] + [22111] + [22200]
+ [22210] + [22211] + [22220] + [22221] + [22222]

Reduction of the Kronecker Products for Sp_4.

```
<00> x <00> = <00>

<10> x <00> = <10>
<10> x <10> = <00> + <11> + <20>
<11> x <00> = <11>
<11> x <10> = <10> + <21>
<11> x <11> = <00> + <20> + <22>

<20> x <00> = <20>
<20> x <10> = <10> + <21> + <30>
<20> x <11> = <11> + <20> + <31>
<20> x <20> = <00> + <11> + <20> + <22> + <31> + <40>
<21> x <00> = <21>
<21> x <10> = <11> + <20> + <22> + <31>
<21> x <11> = <10> + <21> + <30> + <32>
<21> x <20> = <10> +2<21> + <30> + <32> + <41>
<21> x <21> = <00> + <11> +2<20> + <22> +2<31> + <33> + <40> + <42>
<22> x <00> = <22>
<22> x <10> = <21> + <32>
<22> x <11> = <11> + <31> + <33>
<22> x <20> = <20> + <22> + <31> + <42>
<22> x <21> = <10> + <21> + <30> + <32> + <41> + <43>
<22> x <22> = <00> + <20> + <22> + <40> + <42> + <44>

<30> x <00> = <30>
<30> x <10> = <20> + <31> + <40>
<30> x <11> = <21> + <30> + <41>
<30> x <20> = <10> + <21> + <30> + <32> + <41> + <50>
<30> x <21> = <11> + <20> + <22> +2<31> + <40> + <42> + <51>
<30> x <22> = <21> + <30> + <32> + <41> + <52>
<30> x <30> = <00> + <11> + <20> + <22> + <31> + <33> + <40> + <42> + <51> + <60>
<31> x <00> = <31>
<31> x <10> = <21> + <30> + <32> + <41>
<31> x <11> = <20> + <22> + <31> + <40> + <42>
<31> x <20> = <11> + <20> + <22> +2<31> + <33> + <40> + <42> + <51>
<31> x <21> = <10> +2<21> +2<30> +2<32> +2<41> + <43> + <50> + <52>
<31> x <22> = <11> + <20> +2<31> + <33> + <40> + <42> + <51> + <53>
<31> x <30> = <10> +2<21> + <30> +2<32> +2<41> + <43> + <50> + <52> + <61>
<31> x <31> = <00> + <11> +2<20> +2<22> +3<31> + <33> +2<40> +3<42> + <44> +2<51> + <53> + <60>
            + <62>
<32> x <00> = <32>
<32> x <10> = <22> + <31> + <33> + <42>
<32> x <11> = <21> + <32> + <41> + <43>
<32> x <20> = <21> + <30> +2<32> + <41> + <43> + <52>
<32> x <21> = <11> + <20> + <22> +2<31> + <33> + <40> +2<42> + <44> + <51> + <53>
<32> x <22> = <10> + <21> + <30> + <32> + <41> + <43> + <50> + <52> + <54>
<32> x <30> = <20> + <22> +2<31> + <33> + <40> +2<42> + <51> + <53> + <62>
<32> x <31> = <10> +2<21> +2<30> +2<32> +3<41> +2<43> + <50> +2<52> + <54> + <61> + <63>
<32> x <32> = <00> + <11> +2<20> + <22> +2<31> + <33> +2<40> +2<42> + <44> +2<51> +2<53> + <55>
            + <60> + <62> + <64>
<33> x <00> = <33>
<33> x <10> = <32> + <43>
<33> x <11> = <22> + <42> + <44>
<33> x <20> = <31> + <33> + <42> + <53>
<33> x <21> = <21> + <32> + <41> + <43> + <52> + <54>
<33> x <22> = <11> + <31> + <33> + <51> + <53> + <55>
<33> x <30> = <30> + <32> + <41> + <43> + <52> + <63>
<33> x <31> = <20> + <22> + <31> + <40> +2<42> + <44> + <51> + <53> + <62> + <64>
<33> x <32> = <10> + <21> + <30> + <32> + <41> + <43> + <50> + <52> + <54> + <61> + <63> + <65>
<33> x <33> = <00> + <20> + <22> + <40> + <42> + <44> + <60> + <62> + <64> + <66>

<40> x <00> = <40>
<40> x <10> = <30> + <41> + <50>
<40> x <11> = <31> + <40> + <51>
<40> x <20> = <20> + <31> + <40> + <42> + <51> + <60>
<40> x <21> = <21> + <30> + <32> +2<41> + <50> + <52> + <61>
<40> x <22> = <22> + <31> + <40> + <42> + <51> + <62>
<40> x <30> = <10> + <21> + <30> + <32> + <41> + <43> + <50> + <52> + <61> + <70>
<40> x <31> = <11> + <20> + <22> +2<31> + <33> + <40> +2<42> +2<51> + <53> + <60> + <62> + <71>
<40> x <32> = <21> + <30> +2<32> +2<41> + <43> + <50> +2<52> + <61> + <63> + <72>
```

Reduction of the Kronecker Products for Sp. (contd).

```
<40> x <33> = <31> + <33> + <40> + <42> + <51> + <53> + <62> + <73>
<40> x <40> = <00> + <11> + <20> + <22> + <31> + <33> + <40> + <42> + <44> + <51> + <53> + <60>
            + <62> + <71> + <80>
<41> x <00> = <41>
<41> x <10> = <31> + <40> + <42> + <51>
<41> x <11> = <30> + <32> + <41> + <50> + <52>
<41> x <20> = <21> + <30> + <32> +2<41> + <43> + <50> + <52> + <61>
<41> x <21> = <20> + <22> +2<31> + <33> +2<40> +2<42> +2<51> + <53> + <60> + <62>
<41> x <22> = <21> + <30> + <32> +2<41> + <43> + <50> + <52> + <61> + <63>
<41> x <30> = <11> + <20> + <22> +2<31> + <33> + <40> +2<42> + <44> +2<51> + <53> + <60> + <62>
            + <71>
<41> x <31> = <10> +2<21> +2<30> +3<32> +3<41> +2<43> +2<50> +3<52> + <54> +2<61> + <63> + <70>
            + <72>
<41> x <32> = <11> + <20> + <22> +3<31> + <33> +2<40> +3<42> + <44> +3<51> +2<53> + <60> +2<62>
            + <64> + <71> + <73>
<41> x <33> = <21> + <30> + <32> +2<41> + <43> + <50> +2<52> + <54> + <61> + <63> + <72> + <74>
<41> x <40> = <10> +2<21> + <30> +2<32> +2<41> +2<43> + <50> +2<52> + <54> +2<61> + <63> + <70>
            + <72> + <81>
<41> x <41> = <00> + <11> +2<20> +2<22> +3<31> +2<33> +2<40> +4<42> + <44> +3<51> +3<53> + <55>
            +2<60> +3<62> + <64> +2<71> + <73> + <80> + <82>
<42> x <00> = <42>
<42> x <10> = <32> + <41> + <43> + <52>
<42> x <11> = <31> + <33> + <42> + <51> + <53>
<42> x <20> = <22> + <31> + <33> + <40> +2<42> + <44> + <51> + <53> + <62>
<42> x <21> = <21> + <30> +2<32> +2<41> +2<43> + <50> +2<52> + <54> + <61> + <63>
<42> x <22> = <20> + <22> + <31> + <40> +2<42> + <44> + <51> + <53> + <60> + <62> + <64>
<42> x <30> = <21> + <30> +2<32> +2<41> +2<43> + <50> +2<52> + <54> + <61> + <63> + <72>
<42> x <31> = <11> + <20> + <22> +3<31> +2<33> +2<40> +3<42> + <44> +3<51> +3<53> + <55> + <60>
            +2<62> + <64> + <71> + <73>
<42> x <32> = <10> +2<21> +2<30> +2<32> +3<41> +2<43> +2<50> +3<52> +2<54> +2<61> +2<63> + <65>
            + <70> + <72> + <74>
<42> x <33> = <11> + <20> +2<31> + <33> + <40> + <42> +2<51> +2<53> + <55> + <60> + <62> + <64>
            + <71> + <73> + <75>
<42> x <40> = <20> + <22> +2<31> + <33> + <40> +3<42> + <44> +2<51> +2<53> + <60> +2<62> + <64>
            + <71> + <73> + <82>
<42> x <41> = <10> +2<21> +2<30> +3<32> +4<41> +3<43> +2<50> +4<52> +2<54> +3<61> +3<63> + <65>
            + <70> +2<72> + <74> + <81> + <83>
<42> x <42> = <00> + <11> +2<20> +2<22> +3<31> + <33> +3<40> +4<42> +2<44> +4<51> +3<53> + <55>
            +2<60> +4<62> +3<64> + <66> +2<71> +2<73> + <75> + <80> + <82> + <84>
<43> x <00> = <43>
<43> x <10> = <33> + <42> + <44> + <53>
<43> x <11> = <32> + <43> + <52> + <54>
<43> x <20> = <32> + <41> +2<43> + <52> + <54> + <63>
<43> x <21> = <22> + <31> + <33> +2<42> + <44> + <51> +2<53> + <55> + <62> + <64>
<43> x <22> = <21> + <32> + <41> + <43> + <52> + <54> + <61> + <63> + <65>
<43> x <30> = <31> + <33> + <40> +2<42> + <44> + <51> +2<53> + <62> + <64> + <73>
<43> x <31> = <21> + <30> +2<32> +2<41> +2<43> + <50> +3<52> +2<54> + <61> +2<63> + <65> + <72>
            + <74>
<43> x <32> = <11> + <20> + <22> +2<31> + <33> + <40> +2<42> + <44> +2<51> +2<53> + <55> + <60>
            +2<62> +2<64> + <66> + <71> + <73> + <75>
<43> x <33> = <10> + <21> + <30> + <32> + <41> + <43> + <50> + <52> + <54> + <61> + <63> + <65>
            + <70> + <72> + <74> + <76>
<43> x <40> = <30> + <32> +2<41> +2<43> + <50> +2<52> + <54> + <61> +2<63> + <72> + <74> + <83>
<43> x <41> = <20> + <22> +2<31> + <33> +2<40> +3<42> + <44> +3<51> +3<53> + <55> + <60> +3<62>
            +2<64> + <71> +2<73> + <75> + <82> + <84>
<43> x <42> = <10> +2<21> +2<30> +2<32> +3<41> +2<43> +2<50> +3<52> +2<54> +3<61> +3<63> +2<65>
            + <70> +2<72> +2<74> + <76> + <81> + <83> + <85>
<43> x <43> = <00> + <11> +2<20> + <22> +2<31> + <33> +2<40> +2<42> + <44> +2<51> +2<53> + <55>
            +2<60> +2<62> +2<64> + <66> +2<71> +2<73> +2<75> + <77> + <80> + <82> + <84> + <86>
<44> x <00> = <44>
<44> x <10> = <43> + <54>
<44> x <11> = <33> + <53> + <55>
<44> x <20> = <42> + <44> + <53> + <64>
<44> x <21> = <32> + <43> + <52> + <54> + <63> + <65>
<44> x <22> = <22> + <42> + <44> + <62> + <64> + <66>
<44> x <30> = <41> + <43> + <52> + <54> + <63> + <74>
<44> x <31> = <31> + <33> + <42> + <51> +2<53> + <55> + <62> + <64> + <73> + <75>
<44> x <32> = <21> + <32> + <41> + <43> + <52> + <54> + <61> + <63> + <65> + <72> + <74> + <76>
<44> x <33> = <11> + <31> + <33> + <51> + <53> + <55> + <71> + <73> + <75> + <77>
<44> x <40> = <40> + <42> + <44> + <51> + <53> + <62> + <64> + <73> + <84>
<44> x <41> = <30> + <32> + <41> + <43> + <50> +2<52> + <54> + <61> +2<63> + <65> + <72> + <74>
            + <83> + <85>
<44> x <42> = <20> + <22> + <31> + <40> +2<42> + <44> + <51> + <53> + <60> +2<62> +2<64> + <66>
            + <71> + <73> + <75> + <82> + <84> + <86>
<44> x <43> = <10> + <21> + <30> + <32> + <41> + <43> + <50> + <52> + <54> + <61> + <63> + <65>
            + <70> + <72> + <74> + <76> + <81> + <83> + <85> + <87>
<44> x <44> = <00> + <20> + <22> + <40> + <42> + <44> + <60> + <62> + <64> + <66> + <80> + <82>
            + <84> + <86> + <88>
```

D-9 Reduction of the Kronecker Products for Sp_6.

<000> x <000> = <000>

<100> x <000> = <100>
<100> x <100> = <000> + <110> + <200>
<110> x <000> = <110>
<110> x <100> = <100> + <111> + <210>
<110> x <110> = <000> + <110> + <200> + <211> + <220>
<111> x <000> = <111>
<111> x <100> = <110> + <211>
<111> x <110> = <100> + <210> + <221>
<111> x <111> = <000> + <200> + <220> + <222>

<200> x <000> = <200>
<200> x <100> = <100> + <210> + <300>
<200> x <110> = <110> + <200> + <211> + <310>
<200> x <111> = <111> + <210> + <311>
<200> x <200> = <000> + <110> + <200> + <220> + <310> + <400>
<210> x <000> = <210>
<210> x <100> = <110> + <200> + <211> + <220> + <310>
<210> x <110> = <100> + <111> +2<210> + <221> + <300> + <311> + <320>
<210> x <111> = <110> + <200> + <211> + <220> + <310> + <321>
<210> x <200> = <100> + <111> +2<210> + <221> + <300> + <311> + <320> + <410>
<210> x <210> = <000> +2<110> +2<200> +3<211> +2<220> + <222> +3<310> +2<321> + <330> + <400>
+ <411> + <420>
<211> x <000> = <211>
<211> x <100> = <111> + <210> + <221> + <311>
<211> x <110> = <110> + <200> + <211> + <220> + <222> + <310> + <321>
<211> x <111> = <100> + <210> + <221> + <300> + <320> + <322>
<211> x <200> = <110> +2<211> + <220> + <310> + <321> + <411>
<211> x <210> = <100> + <111> +3<210> +2<221> + <300> +2<311> +2<320> + <322> + <331> + <410>
+ <421>
<211> x <211> = <000> + <110> +2<200> + <211> +2<220> + <222> +2<310> +2<321> + <330> + <332>
+ <400> + <420> + <422>
<220> x <000> = <220>
<220> x <100> = <210> + <221> + <320>
<220> x <110> = <110> + <211> + <220> + <310> + <321> + <330>
<220> x <111> = <111> + <210> + <311> + <320> + <331>
<220> x <200> = <200> + <211> + <220> + <222> + <310> + <321> + <420>
<220> x <210> = <100> + <111> +2<210> +2<221> + <300> +2<311> +2<320> + <322> + <331> + <410>
+ <421> + <430>
<220> x <211> = <110> + <200> +2<211> + <220> +2<310> +2<321> + <330> + <332> + <411> + <420>
+ <431>
<220> x <220> = <000> + <110> + <200> + <211> +2<220> + <222> + <310> +2<321> + <330> + <400>
+ <411> + <420> + <422> + <431> + <440>
<221> x <000> = <221>
<221> x <100> = <211> + <220> + <222> + <321>
<221> x <110> = <111> + <210> + <221> + <311> + <320> + <322> + <331>
<221> x <111> = <110> + <211> + <310> + <321> + <330> + <332>
<221> x <200> = <210> +2<221> + <311> + <320> + <322> + <421>
<221> x <210> = <110> + <200> +2<211> +2<220> + <222> +2<310> +3<321> + <330> + <332> + <411>
+ <420> + <422> + <431>
<221> x <211> = <100> + <111> +2<210> + <221> + <300> +2<311> +2<320> + <322> +2<331> + <333>
+ <410> + <421> + <430> + <432>
<221> x <220> = <100> +2<210> +2<221> + <300> + <311> +2<320> + <322> + <331> + <410> +2<421>
+ <430> + <432> + <441>
<221> x <221> = <000> + <110> +2<200> + <211> +2<220> + <222> +2<310> +2<321> + <330> + <332>
+ <400> + <411> +2<420> + <422> +2<431> + <433> + <440> + <442>
<222> x <000> = <222>
<222> x <100> = <221> + <322>
<222> x <110> = <211> + <321> + <332>
<222> x <111> = <111> + <311> + <331> + <333>
<222> x <200> = <220> + <222> + <321> + <422>
<222> x <210> = <210> + <221> + <311> + <320> + <322> + <331> + <421> + <432>
<222> x <211> = <110> + <211> + <310> + <321> + <330> + <332> + <411> + <431> + <433>
<222> x <220> = <200> + <220> + <222> + <310> + <321> + <420> + <422> + <431> + <442>
<222> x <221> = <100> + <210> + <221> + <300> + <320> + <322> + <410> + <421> + <430> + <432>
+ <441> + <443>
<222> x <222> = <000> + <200> + <220> + <222> + <400> + <420> + <422> + <440> + <442> + <444>

<0000> x <0000> = <0000>

<1000> x <0000> = <1000>
<1000> x <1000> = <0000> + <1100> + <2000>
<1100> x <0000> = <1100>
<1100> x <1000> = <1000> + <1110> + <2100>
<1100> x <1100> = <0000> + <1100> + <1111> + <2000> + <2110> + <2200>
<1110> x <0000> = <1110>
<1110> x <1000> = <1100> + <1111> + <2110>
<1110> x <1100> = <1000> + <1110> + <2100> + <2111> + <2210>
<1110> x <1110> = <0000> + <1100> + <2000> + <2110> + <2200> + <2211> + <2220>
<1111> x <0000> = <1111>
<1111> x <1000> = <1110> + <2111>
<1111> x <1100> = <1100> + <2110> + <2211>
<1111> x <1110> = <1000> + <2100> + <2210> + <2221>
<1111> x <1111> = <0000> + <2000> + <2200> + <2220> + <2222>

<2000> x <0000> = <2000>
<2000> x <1000> = <1000> + <2100> + <3000>
<2000> x <1100> = <1100> + <2000> + <2110> + <3100>
<2000> x <1110> = <1110> + <2100> + <2111> + <3110>
<2000> x <1111> = <1111> + <2110> + <3111>
<2000> x <2000> = <0000> + <1100> + <2000> + <2200> + <3100> + <4000>
<2100> x <0000> = <2100>
<2100> x <1000> = <1100> + <2000> + <2110> + <2200> + <3100>
<2100> x <1100> = <1000> + <1110> +2<2100> + <2111> + <2210> + <3000> + <3110> + <3200>
<2100> x <1110> = <1100> + <1111> + <2000> +2<2110> + <2200> + <2211> + <3100> + <3111> + <3210>
<2100> x <1111> = <1110> + <2100> + <2111> + <2210> + <3110> + <3211>
<2100> x <2000> = <1000> + <1110> +2<2100> + <2210> + <3000> + <3110> + <3200> + <4100>
<2100> x <2100> = <0000> +2<1100> + <1111> +2<2000> +3<2110> +2<2200> + <2211> + <2220> +3<3100>
 + <3111> +2<3210> + <3300> + <4000> + <4110> + <4200>
<2110> x <0000> = <2110>
<2110> x <1000> = <1110> + <2100> + <2111> + <2210> + <3110>
<2110> x <1100> = <1100> + <1111> + <2000> +2<2110> + <2200> + <2211> + <2220> + <3100> + <3111>
 + <3210>
<2110> x <1110> = <1000> + <1110> +2<2100> + <2111> +2<2210> + <2221> + <3000> + <3110> + <3200>
 + <3211> + <3220>
<2110> x <1111> = <1100> + <2000> + <2110> + <2200> + <2211> + <2220> + <3100> + <3210> + <3221>
<2110> x <2000> = <1100> + <1111> +2<2110> + <2200> + <2211> + <3100> + <3111> + <3210> + <4110>
<2110> x <2100> = <1000> +2<1110> +3<2100> +3<2111> +3<2210> + <2221> + <3000> +3<3110> +2<3200>
 +2<3211> + <3220> + <3310> + <4100> + <4111> + <4210>
<2110> x <2110> = <0000> +2<1100> + <1111> +2<2000> +4<2110> +3<2200> +3<2211> +2<2220> + <2222>
 +3<3100> +2<3111> +4<3210> +2<3221> + <3300> + <3311> + <3320> + <4000> + <4110>
 + <4200> + <4211> + <4220>
<2111> x <0000> = <2111>
<2111> x <1000> = <1111> + <2110> + <2211> + <3111>
<2111> x <1100> = <1110> + <2100> + <2111> + <2210> + <2221> + <3110> + <3211>
<2111> x <1110> = <1100> + <2000> + <2110> + <2200> + <2211> + <2220> + <2222> + <3100> + <3210>
 + <3221>
<2111> x <1111> = <1000> + <2100> + <2210> + <2221> + <3000> + <3200> + <3220> + <3222>
<2111> x <2000> = <1110> +2<2111> + <2210> + <3110> + <3211> + <4111>
<2111> x <2100> = <1100> + <1111> +3<2110> + <2200> +2<2211> + <2220> + <3100> +2<3111> +2<3210>
 + <3221> + <3311> + <4110> + <4211>
<2111> x <2110> = <1000> + <1110> +3<2100> + <2111> +3<2210> +2<2221> + <3000> +2<3110> +2<3200>
 +2<3211> +2<3220> + <3222> + <3310> + <3321> + <4100> + <4210> + <4221>
<2111> x <2111> = <0000> + <1100> +2<2000> + <2110> +2<2200> + <2211> +2<2220> + <2222> +2<3100>
 +2<3210> +2<3221> + <3300> + <3320> + <3322> + <4000> + <4200> + <4220> + <4222>
<2200> x <0000> = <2200>
<2200> x <1000> = <2100> + <2210> + <3200>
<2200> x <1100> = <1100> + <2110> + <2200> + <2211> + <3100> + <3210> + <3300>
<2200> x <1110> = <1110> + <2100> + <2111> + <2210> + <3110> + <3200> + <3211> + <3310>
<2200> x <1111> = <1111> + <2110> + <2200> + <3111> + <3210> + <3311>
<2200> x <2000> = <2000> + <2110> + <2200> + <2220> + <3100> + <3210> + <4200>
<2200> x <2100> = <1000> + <1110> +2<2100> + <2111> +2<2210> + <2221> + <3000> +2<3110> +2<3200>
 + <3211> + <3220> + <3310> + <4100> + <4210> + <4300>
<2200> x <2110> = <1100> + <1111> + <2000> +3<2110> + <2200> +2<2211> + <2220> +2<3100> +2<3111>
 +3<3210> + <3221> + <3300> + <3311> + <3320> + <4110> + <4200> + <4211> + <4310>
<2200> x <2111> = <1110> + <2100> +2<2111> +2<2210> +2<3110> + <3200> +2<3211> + <3220> + <3310>
 + <3321> + <4111> + <4210> + <4311>
<2200> x <2200> = <0000> + <1100> + <1111> + <2000> + <2110> +2<2200> + <2211> + <2220> + <2222>
 + <3100> + <3111> +2<3210> + <3221> + <3300> + <3311> + <4000> + <4110> + <4200>
 + <4220> + <4310> + <4400>

```
<2210> x <0000> = <2210>
<2210> x <1000> = <2110> + <2200> + <2211> + <2220> + <3210>
<2210> x <1100> = <1110> + <2100> + <2111> +2<2210> + <2221> + <3110> + <3200> + <3211> + <3220>
                + <3310>
<2210> x <1110> = <1100> + <1111> +2<2110> + <2200> + <2211> + <2220> + <3100> + <3111> +2<3210>
                + <3221> + <3300> + <3311> + <3320>
<2210> x <1111> = <1110> + <2100> + <2111> + <2210> + <3110> + <3200> + <3211> + <3220> + <3310>
                + <3321>
<2210> x <2000> = <2100> + <2111> +2<2210> + <2221> + <3110> + <3200> + <3211> + <3220> + <4210>
<2210> x <2100> = <1100> + <1111> + <2000> +3<2110> +2<2200> +3<2211> +2<2220> + <2222> +2<3100>
                +2<3111> +4<3210> +2<3221> + <3300> + <3311> + <3320> + <4110> + <4200> + <4211>
                + <4220> + <4310>
<2210> x <2110> = <1000> +2<1110> +3<2100> +3<2111> +4<2210> +2<2221> + <3000> +4<3110> +3<3200>
                +4<3211> +3<3220> + <3222> +3<3310> +2<3321> + <3330> + <4100> + <4111> +2<4210>
                + <4221> + <4300> + <4311> + <4320>
<2210> x <2111> = <1100> + <1111> + <2000> +3<2110> +2<2200> +2<2211> + <2220> +2<3100> +2<3111>
                +4<3210> +2<3221> + <3300> +2<3311> +2<3320> + <3322> + <3331> + <4110> + <4200>
                + <4211> + <4220> + <4310> + <4321>
<2210> x <2200> = <1000> + <1110> +2<2100> +2<2111> +3<2210> +2<2221> + <3000> +2<3110> +2<3200>
                +3<3211> +2<3220> + <3222> +2<3310> + <3321> + <4100> + <4111> +2<4210> + <4221>
                + <4300> + <4311> + <4320> + <4410>
<2210> x <2210> = <0000> +2<1100> + <1111> +2<2000> +4<2110> +3<2200> +4<2211> +3<2220> + <2222>
                +3<3100> +3<3111> +6<3210> +4<3221> +2<3300> +3<3311> +3<3320> + <3322> + <3331>
                + <4000> +2<4110> +2<4200> +3<4211> +2<4220> + <4222> +3<4310> +2<4321> + <4330>
                + <4400> + <4411> + <4420>
<2211> x <0000> = <2211>
<2211> x <1000> = <2111> + <2210> + <2221> + <3211>
<2211> x <1100> = <1111> + <2110> + <2200> + <2211> + <2220> + <2222> + <3111> + <3210> + <3221>
                + <3311>
<2211> x <1110> = <1110> + <2100> + <2111> + <2210> + <2221> + <3110> + <3200> + <3211> + <3220>
                + <3222> + <3310> + <3321>
<2211> x <1111> = <1100> + <2110> + <2211> + <3100> + <3210> + <3221> + <3300> + <3320> + <3322>
<2211> x <2000> = <2110> +2<2211> + <2220> + <3111> + <3210> + <3221> + <4211>
<2211> x <2100> = <1110> + <2100> +2<2111> +3<2210> +2<2221> +2<3110> + <3200> +3<3211> +2<3220>
                + <3222> + <3310> + <3321> + <4111> + <4210> + <4221> + <4311>
<2211> x <2110> = <1100> + <1111> + <2000> +3<2110> +2<2200> +2<2211> +2<2220> + <2222> +2<3100>
                +2<3111> +4<3210> +3<3221> + <3300> +2<3311> +2<3320> + <3322> + <3331> + <4110>
                + <4200> + <4211> + <4220> + <4222> + <4310> + <4321>
<2211> x <2111> = <1000> + <1110> +2<2100> + <2111> +2<2210> + <2221> + <3000> +2<3110> +2<3200>
                +2<3211> +2<3220> + <3222> +2<3310> +2<3321> + <3330> + <3332> + <4100> + <4210>
                + <4221> + <4300> + <4320> + <4322>
<2211> x <2200> = <1100> +2<2110> + <2200> +3<2211> + <2220> + <3100> + <3111> +3<3210> +2<3221>
                + <3300> + <3311> + <3320> + <3322> + <4110> +2<4211> + <4220> + <4310> + <4321>
                + <4411>
<2211> x <2210> = <1000> + <1110> +3<2100> +2<2111> +4<2210> +2<2221> + <3000> +3<3110> +3<3200>
                +4<3211> +3<3220> + <3222> +3<3310> +3<3321> + <3330> + <3332> + <4100> + <4111>
                +3<4210> +2<4221> + <4300> +2<4311> +2<4320> + <4322> + <4331> + <4410> + <4421>
<2211> x <2211> = <0000> + <1100> + <1111> +2<2000> +2<2110> +3<2200> + <2211> +2<2220> + <2222>
                +2<3100> +2<3111> +4<3210> +2<3221> + <3300> +3<3311> +2<3320> + <3322> +2<3331>
                + <3333> + <4000> + <4110> +2<4200> + <4211> +2<4220> + <4222> +2<4310> +2<4321>
                + <4330> + <4332> + <4400> + <4420> + <4422>
<2220> x <0000> = <2220>
<2220> x <1000> = <2210> + <2221> + <3220>
<2220> x <1100> = <2110> + <2211> + <2220> + <3210> + <3221> + <3320>
<2220> x <1110> = <1110> + <2111> + <2210> + <3110> + <3211> + <3220> + <3310> + <3321> + <3330>
<2220> x <1111> = <1111> + <2110> + <3111> + <3210> + <3311> + <3320> + <3331>
<2220> x <2000> = <2200> + <2211> + <2220> + <2222> + <3210> + <3221> + <4220>
<2220> x <2100> = <2100> + <2111> +2<2210> +2<2221> + <3110> + <3200> +2<3211> +2<3220> + <3222>
                + <3310> + <3321> + <4210> + <4221> + <4320>
<2220> x <2110> = <1100> + <1111> +2<2110> + <2200> +2<2211> + <2220> + <3100> +2<3111> +3<3210>
                +2<3221> + <3300> +2<3311> +2<3320> + <3322> + <3331> + <4110> + <4211> + <4220>
                + <4310> + <4321> + <4330>
<2220> x <2111> = <1110> + <2100> +2<2111> + <2210> +2<3110> + <3200> +2<3211> + <3220> +2<3310>
                +2<3321> + <3330> + <3332> + <4111> + <4210> + <4311> + <4320> + <4331>
<2220> x <2200> = <2000> + <2110> + <2200> + <2211> +2<2220> + <2222> + <3100> + <3111> +2<3210>
                +2<3221> + <3311> + <3320> + <4200> + <4211> + <4220> + <4222> + <4310> + <4321>
                + <4420>
<2220> x <2210> = <1000> + <1110> +2<2100> + <2111> +3<2210> +2<2221> + <3000> +2<3110> +2<3200>
                +3<3211> +3<3220> + <3222> +2<3310> +2<3321> + <3330> + <4100> + <4111> +2<4210>
                +2<4221> + <4300> +2<4311> +2<4320> + <4322> + <4331> + <4410> + <4421> + <4430>
```

<2220> x <2211> = <1100> + <2000> +2<2110> + <2200> +2<2211> + <2220> +2<3100> + <3111> +3<3210>
+2<3221> + <3300> + <3311> +2<3320> + <3322> + <3331> + <4110> + <4200> +2<4211>
+ <4220> +2<4310> +2<4321> + <4330> + <4332> + <4411> + <4420> + <4431>

<2220> x <2220> = <0000> + <1100> + <2000> + <2110> +2<2200> + <2211> +2<2220> + <2222> + <3100>
+2<3210> +2<3221> + <3300> + <3311> + <3320> + <4000> + <4110> + <4200> + <4211>
+2<4220> + <4222> + <4310> +2<4321> + <4330> + <4400> + <4411> + <4420> + <4422>
+ <4431> + <4440>

<2221> x <0000> = <2221>

<2221> x <1000> = <2211> + <2220> + <2222> + <3221>

<2221> x <1100> = <2111> + <2210> + <2221> + <3211> + <3220> + <3222> + <3321>

<2221> x <1110> = <1111> + <2110> + <2211> + <3111> + <3210> + <3221> + <3311> + <3320> + <3322>
+ <3331>

<2221> x <1111> = <1110> + <2111> + <3110> + <3211> + <3310> + <3321> + <3330> + <3332>

<2221> x <2000> = <2210> +2<2221> + <3211> + <3220> + <3222> + <4221>

<2221> x <2100> = <2110> + <2200> +2<2211> +2<2220> + <2222> + <3111> +2<3210> +3<3221> + <3311>
+ <3320> + <3322> + <4211> + <4220> + <4222> + <4321>

<2221> x <2110> = <1110> + <2100> +2<2111> +2<2210> + <2221> +2<3110> + <3200> +3<3211> +2<3220>
+ <3222> +2<3310> +3<3321> + <3330> + <3332> + <4111> + <4210> + <4221> + <4311>
+ <4320> + <4322> + <4331>

<2221> x <2111> = <1100> + <1111> +2<2110> + <2211> + <3100> +2<3111> +2<3210> + <3221> + <3300>
+2<3311> +2<3320> + <3322> +2<3331> + <3333> + <4110> + <4211> + <4310> + <4321>
+ <4330> + <4332>

<2221> x <2200> = <2100> +2<2210> +2<2221> + <3110> + <3200> +2<3211> +2<3220> + <3222> + <3310>
+ <3321> + <4210> +2<4221> + <4311> + <4320> + <4322> + <4421>

<2221> x <2210> = <1100> + <2000> +2<2110> +2<2200> +2<2211> +2<2220> + <2222> +2<3100> + <3111>
+4<3210> +3<3221> + <3300> +2<3311> +2<3320> + <3322> + <3331> + <4110> + <4200>
+2<4211> +2<4220> + <4222> +2<4310> +3<4321> + <4330> + <4332> + <4411> + <4420>
+ <4422> + <4431>

<2221> x <2211> = <1000> + <1110> +2<2100> + <2111> +2<2210> + <2221> + <3000> +2<3110> +2<3200>
+2<3211> +2<3220> + <3222> +2<3310> +2<3321> + <3330> + <3332> + <4100> + <4111>
+2<4210> + <4221> + <4300> +2<4311> +2<4320> + <4322> +2<4331> + <4333> + <4410>
+ <4421> + <4430> + <4432>

<2221> x <2220> = <1000> +2<2100> +2<2210> +2<2221> + <3000> + <3110> +2<3200> + <3211> +2<3220>
+ <3222> + <3310> + <3321> + <4100> +2<4210> +2<4221> + <4300> + <4311> +2<4320>
+ <4322> + <4331> + <4410> +2<4421> + <4430> + <4432> + <4441>

<2221> x <2221> = <0000> + <1100> +2<2000> + <2110> +2<2200> + <2211> +2<2220> + <2222> +2<3100>
+2<3210> +2<3221> + <3300> + <3320> + <3322> + <4000> + <4110> +2<4200> + <4211>
+2<4220> + <4222> +2<4310> +2<4321> + <4330> + <4332> + <4400> + <4411> +2<4420>
+ <4422> +2<4431> + <4433> + <4440> + <4442>

<2222> x <0000> = <2222>

<2222> x <1000> = <2221> + <3222>

<2222> x <1100> = <2211> + <3221> + <3322>

<2222> x <1110> = <2111> + <3211> + <3321> + <3332>

<2222> x <1111> = <1111> + <3111> + <3311> + <3331> + <3333>

<2222> x <2000> = <2220> + <2222> + <3221> + <4222>

<2222> x <2100> = <2210> + <2221> + <3211> + <3220> + <3222> + <3321> + <4221> + <4322>

<2222> x <2110> = <2110> + <2211> + <3111> + <3210> + <3221> + <3311> + <3320> + <3322> + <3331>
+ <4211> + <4321> + <4332>

<2222> x <2111> = <1110> + <2111> + <3110> + <3211> + <3310> + <3321> + <3330> + <3332> + <4111>
+ <4311> + <4331> + <4333>

<2222> x <2200> = <2200> + <2220> + <2222> + <3210> + <3221> + <3311> + <4220> + <4222> + <4321>
+ <4422>

<2222> x <2210> = <2100> + <2210> + <2221> + <3110> + <3200> + <3211> + <3220> + <3222> + <3310>
+ <3321> + <4210> + <4221> + <4311> + <4320> + <4322> + <4331> + <4421> + <4432>

<2222> x <2211> = <1100> + <2110> + <2211> + <3100> + <3210> + <3221> + <3300> + <3320> + <3322>
+ <4110> + <4211> + <4310> + <4321> + <4330> + <4332> + <4411> + <4431> + <4433>

<2222> x <2220> = <2000> + <2200> + <2220> + <2222> + <3100> + <3210> + <3221> + <4200> + <4220>
+ <4222> + <4310> + <4321> + <4420> + <4422> + <4431> + <4442>

<2222> x <2221> = <1000> + <2100> + <2210> + <2221> + <3000> + <3200> + <3220> + <3222> + <4100>
+ <4210> + <4221> + <4300> + <4320> + <4322> + <4410> + <4421> + <4430> + <4432>
+ <4441> + <4443>

<2222> x <2222> = <0000> + <2000> + <2200> + <2220> + <2222> + <4000> + <4200> + <4220> + <4222>
+ <4400> + <4420> + <4422> + <4440> + <4442> + <4444>

D-11 Reduction of the Kronecker Products for Sp_{10}.

```
<00000> x <00000> = <00000>

<10000> x <00000> = <10000>
<10000> x <10000> = <00000> + <11000> + <20000>
<11000> x <00000> = <11000>
<11000> x <10000> = <10000> + <11100> + <21000>
<11000> x <11000> = <00000> + <11000> + <11110> + <20000> + <21100> + <22000>
<11100> x <00000> = <11100>
<11100> x <10000> = <11000> + <11110> + <21100>
<11100> x <11000> = <10000> + <11100> + <11111> + <21000> + <21110> + <22100>
<11100> x <11100> = <00000> + <11000> + <11110> + <20000> + <21100> + <21111> + <22000> + <22110>
                  + <22200>
<11110> x <00000> = <11110>
<11110> x <10000> = <11100> + <11111> + <21110>
<11110> x <11000> = <11000> + <11110> + <21100> + <21111> + <22110>
<11110> x <11100> = <10000> + <11100> + <21000> + <21110> + <22100> + <22111> + <22210>
<11110> x <11110> = <00000> + <11000> + <20000> + <21100> + <22000> + <22110> + <22200> + <22211>
                  + <22220>
<11111> x <00000> = <11111>
<11111> x <10000> = <11110> + <21111>
<11111> x <11000> = <11100> + <21110> + <22111>
<11111> x <11100> = <11000> + <21100> + <22110> + <22211>
<11111> x <11110> = <10000> + <21000> + <22100> + <22210> + <22221>
<11111> x <11111> = <00000> + <20000> + <22000> + <22200> + <22220> + <22222>
```

D-12 Reduction of the Kronecker Products for Sp_{12}.

```
<000000> x <000000> = <000000>

<100000> x <000000> = <100000>
<100000> x <100000> = <000000> + <110000> + <200000>
<110000> x <000000> = <110000>
<110000> x <100000> = <100000> + <111000> + <210000>
<110000> x <110000> = <000000> + <110000> + <111100> + <200000> + <211000> + <220000>
<111000> x <000000> = <111000>
<111000> x <100000> = <110000> + <111100> + <211000>
<111000> x <110000> = <100000> + <111000> + <111110> + <210000> + <211100> + <221000>
<111000> x <111000> = <000000> + <110000> + <111100> + <111111> + <200000> + <211000> + <211110>
                    + <220000> + <221100> + <222000>
<111100> x <000000> = <111100>
<111100> x <100000> = <111000> + <111110> + <211100>
<111100> x <110000> = <110000> + <111100> + <111111> + <211000> + <211110> + <221100>
<111100> x <111000> = <100000> + <111000> + <111110> + <210000> + <211100> + <211111> + <221000>
                    + <221110> + <222100>
<111100> x <111100> = <000000> + <110000> + <111100> + <200000> + <211000> + <211110> + <220000>
                    + <221100> + <221111> + <222000> + <222110> + <222200>
<111110> x <000000> = <111110>
<111110> x <100000> = <111100> + <111111> + <211110>
<111110> x <110000> = <111000> + <111110> + <211100> + <211111> + <221110>
<111110> x <111000> = <110000> + <111100> + <211000> + <211110> + <221100> + <221111> + <222110>
<111110> x <111100> = <100000> + <111000> + <210000> + <211100> + <221000> + <221110> + <222100>
                    + <222111> + <222210>
<111110> x <111110> = <000000> + <110000> + <200000> + <211000> + <220000> + <221100> + <222000>
                    + <222110> + <222200> + <222211> + <222220>
<111111> x <000000> = <111111>
<111111> x <100000> = <111110> + <211111>
<111111> x <110000> = <111100> + <211110> + <221111>
<111111> x <111000> = <111000> + <211100> + <221110> + <222111>
<111111> x <111100> = <110000> + <211000> + <221100> + <222110> + <222211>
<111111> x <111110> = <100000> + <210000> + <221000> + <222100> + <222210> + <222221>
<111111> x <111111> = <000000> + <200000> + <220000> + <222000> + <222200> + <222220> + <222222>
```

D-13 Reduction of the Kronecker Products for Sp_{14}.

```
<0000000> x <0000000> = <0000000>

<1000000> x <0000000> = <1000000>
<1000000> x <1000000> = <0000000> + <1100000> + <2000000>
<1100000> x <0000000> = <1100000>
<1100000> x <1000000> = <1000000> + <1110000> + <2100000>
<1100000> x <1100000> = <0000000> + <1100000> + <1111000> + <2000000> + <2110000> + <2200000>
<1110000> x <0000000> = <1110000>
<1110000> x <1000000> = <1100000> + <1111000> + <2110000>
<1110000> x <1100000> = <1000000> + <1110000> + <1111100> + <2100000> + <2111000> + <2210000>
<1110000> x <1110000> = <0000000> + <1100000> + <1111000> + <1111110> + <2000000> + <2110000>
                      + <2111100> + <2200000> + <2211000> + <2220000>
<1111000> x <0000000> = <1111000>
<1111000> x <1000000> = <1110000> + <1111100> + <2111000>
<1111000> x <1100000> = <1100000> + <1111000> + <1111110> + <2110000> + <2111100> + <2211000>
<1111000> x <1110000> = <1000000> + <1110000> + <1111100> + <1111111> + <2100000> + <2111000>
                      + <2111110> + <2210000> + <2211100> + <2221000>
<1111000> x <1111000> = <0000000> + <1100000> + <1111000> + <1111110> + <2000000> + <2110000>
                      + <2111100> + <2111111> + <2200000> + <2211000> + <2211110> + <2220000>
                      + <2221100> + <2222000>
<1111100> x <0000000> = <1111100>
<1111100> x <1000000> = <1111000> + <1111110> + <2111100>
<1111100> x <1100000> = <1110000> + <1111100> + <1111111> + <2111000> + <2111110> + <2211100>
<1111100> x <1110000> = <1100000> + <1111000> + <1111110> + <2110000> + <2111100> + <2111111>
                      + <2211000> + <2211110> + <2221100>
<1111100> x <1111000> = <1000000> + <1110000> + <1111100> + <2100000> + <2111000> + <2111110>
                      + <2210000> + <2211100> + <2211111> + <2221000> + <2221110> + <2222100>
<1111100> x <1111100> = <0000000> + <1100000> + <1111000> + <2000000> + <2110000> + <2111100>
                      + <2200000> + <2211000> + <2211110> + <2220000> + <2221100> + <2221111>
                      + <2222000> + <2222110> + <2222200>
<1111110> x <0000000> = <1111110>
<1111110> x <1000000> = <1111100> + <1111111> + <2111110>
<1111110> x <1100000> = <1111000> + <1111110> + <2111100> + <2111111> + <2211110>
<1111110> x <1110000> = <1110000> + <1111100> + <2111000> + <2111110> + <2211100> + <2211111>
                      + <2221110>
<1111110> x <1111000> = <1100000> + <1111000> + <2110000> + <2111100> + <2211000> + <2211110>
                      + <2221100> + <2221111> + <2222110>
<1111110> x <1111100> = <1000000> + <1110000> + <2100000> + <2111000> + <2210000> + <2211100>
                      + <2221000> + <2221110> + <2222100> + <2222111> + <2222210>
<1111110> x <1111110> = <0000000> + <1100000> + <2000000> + <2110000> + <2200000> + <2211000>
                      + <2220000> + <2221100> + <2222000> + <2222110> + <2222200> + <2222211>
                      + <2222220>
<1111111> x <0000000> = <1111111>
<1111111> x <1000000> = <1111110> + <2111111>
<1111111> x <1100000> = <1111100> + <2111110> + <2211111>
<1111111> x <1110000> = <1111000> + <2111100> + <2211110> + <2221111>
<1111111> x <1111000> = <1110000> + <2111000> + <2211100> + <2221110> + <2222111>
<1111111> x <1111100> = <1100000> + <2110000> + <2211000> + <2221100> + <2222110> + <2222211>
<1111111> x <1111110> = <1000000> + <2100000> + <2210000> + <2221000> + <2222100> + <2222210>
                      + <2222221>
<1111111> x <1111111> = <0000000> + <2000000> + <2200000> + <2220000> + <2222000> + <2222200>
                      + <2222220> + <2222222>
```

Reduction of the Kronecker Products for Sp_{16}.

<00000000> x <00000000> = <00000000>

<10000000> x <00000000> = <10000000>

<10000000> x <10000000> = <00000000> + <11000000> + <20000000>

<11000000> x <00000000> = <11000000>

<11000000> x <10000000> = <10000000> + <11100000> + <21000000>

<11000000> x <11000000> = <00000000> + <11000000> + <11110000> + <20000000> + <21100000> + <22000000>

<11100000> x <00000000> = <11100000>

<11100000> x <10000000> = <11000000> + <11110000> + <21100000>

<11100000> x <11000000> = <10000000> + <11100000> + <11111000> + <21000000> + <21110000> + <22100000>

<11100000> x <11100000> = <00000000> + <11000000> + <11110000> + <11111100> + <20000000> + <21100000> + <21111000> + <22000000> + <22110000> + <22200000>

<11110000> x <00000000> = <11110000>

<11110000> x <10000000> = <11100000> + <11111000> + <21110000>

<11110000> x <11000000> = <11000000> + <11110000> + <11111100> + <21100000> + <21111000> + <22110000>

<11110000> x <11100000> = <10000000> + <11100000> + <11111000> + <11111110> + <21000000> + <21110000> + <21111100> + <22100000> + <22111000> + <22210000>

<11110000> x <11110000> = <00000000> + <11000000> + <11110000> + <11111100> + <11111111> + <20000000> + <21100000> + <21111000> + <21111110> + <22000000> + <22110000> + <22111100> + <22200000> + <22211000> + <22220000>

<11111000> x <00000000> = <11111000>

<11111000> x <10000000> = <11110000> + <11111100> + <21111000>

<11111000> x <11000000> = <11100000> + <11111000> + <11111110> + <21110000> + <21111100> + <22111000>

<11111000> x <11100000> = <11000000> + <11110000> + <11111100> + <11111111> + <21100000> + <21111000> + <21111110> + <22110000> + <22111100> + <22211000>

<11111000> x <11110000> = <10000000> + <11100000> + <11111000> + <11111110> + <21000000> + <21110000> + <21111100> + <21111111> + <22100000> + <22111000> + <22111110> + <22210000> + <22211100> + <22221000>

<11111000> x <11111000> = <00000000> + <11000000> + <11110000> + <11111100> + <20000000> + <21100000> + <21111000> + <21111110> + <22000000> + <22110000> + <22111100> + <22111111> + <22200000> + <22211000> + <22211110> + <22220000> + <22221100> + <22222000>

<11111100> x <00000000> = <11111100>

<11111100> x <10000000> = <11111000> + <11111110> + <21111100>

<11111100> x <11000000> = <11110000> + <11111100> + <11111111> + <21111000> + <21111110> + <22111100>

<11111100> x <11100000> = <11100000> + <11111000> + <11111110> + <21110000> + <21111100> + <21111111> + <22111000> + <22111110> + <22211100>

<11111100> x <11110000> = <11000000> + <11110000> + <11111100> + <21100000> + <21111000> + <21111110> + <22110000> + <22111100> + <22111111> + <22211000> + <22211110> + <22221100>

<11111100> x <11111000> = <10000000> + <11100000> + <11111000> + <21000000> + <21110000> + <21111100> + <22100000> + <22111000> + <22111110> + <22210000> + <22211100> + <22211111> + <22221000> + <22221110> + <22222100>

<11111100> x <11111100> = <00000000> + <11000000> + <11110000> + <20000000> + <21100000> + <21111000> + <22000000> + <22110000> + <22111100> + <22200000> + <22211000> + <22211110> + <22220000> + <22221100> + <22221111> + <22222000> + <22222110> + <22222200>

<11111110> x <00000000> = <11111110>

<11111110> x <10000000> = <11111100> + <11111111> + <21111110>

<11111110> x <11000000> = <11111000> + <11111110> + <21111100> + <21111111> + <22111110>

<11111110> x <11100000> = <11110000> + <11111100> + <21111000> + <21111110> + <22111100> + <22111111> + <22211110>

<11111110> x <11110000> = <11100000> + <11111000> + <21110000> + <21111100> + <22111000> + <22111110> + <22211100> + <22211111> + <22221110>

<11111110> x <11111000> = <11000000> + <11110000> + <21100000> + <21111000> + <22110000> + <22111100> + <22211000> + <22211110> + <22221100> + <22221111> + <22222110>

<11111110> x <11111100> = <10000000> + <11100000> + <21000000> + <21110000> + <22100000> + <22111000> + <22210000> + <22211100> + <22221000> + <22221110> + <22222100> + <22222111> + <22222210>

<11111110> x <11111110> = <00000000> + <11000000> + <20000000> + <21100000> + <22000000> + <22110000> + <22200000> + <22211000> + <22220000> + <22221100> + <22222000> + <22222110> + <22222200> + <22222211> + <22222220>

<11111111> x <00000000> = <11111111>

<11111111> x <10000000> = <11111110> + <21111111>

<11111111> x <11000000> = <11111100> + <21111110> + <22111111>

<11111111> x <11100000> = <11111000> + <21111100> + <22111110> + <22211111>

<11111111> x <11110000> = <11110000> + <21111000> + <22111100> + <22211110> + <22221111>

<11111111> x <11111000> = <11100000> + <21110000> + <22111000> + <22211100> + <22221110> + <22222111>

<11111111> x <11111100> = <11000000> + <21100000> + <22110000> + <22211000> + <22221100> + <22222110> + <22222211>

<11111111> x <11111110> = <10000000> + <21000000> + <22100000> + <22210000> + <22221000> + <22222100> + <22222210> + <22222221>

<11111111> x <11111111> = <00000000> + <20000000> + <22000000> + <22200000> + <22220000> + <22222000> + <22222200> + <22222220> + <22222222>

D-15 Reduction of the Kronecker Products for Sp_{18}.

```
<000000000> x <000000000> = <000000000>

<100000000> x <000000000> = <100000000>
<100000000> x <100000000> = <000000000> + <110000000> + <200000000>
<110000000> x <000000000> = <110000000>
<110000000> x <100000000> = <100000000> + <111000000> + <210000000>
<110000000> x <110000000> = <000000000> + <110000000> + <111100000> + <200000000> + <211000000>
                          + <220000000>
<111000000> x <000000000> = <111000000>
<111000000> x <100000000> = <110000000> + <111100000> + <211000000>
<111000000> x <110000000> = <100000000> + <111000000> + <111110000> + <210000000> + <211100000>
                          + <221000000>
<111000000> x <111000000> = <000000000> + <110000000> + <111100000> + <111111000> + <200000000>
                          + <211000000> + <211110000> + <220000000> + <221100000> + <222000000>
<111100000> x <000000000> = <111100000>
<111100000> x <100000000> = <111000000> + <111110000> + <211100000>
<111100000> x <110000000> = <110000000> + <111100000> + <111111000> + <211000000> + <211110000>
                          + <221100000>
<111100000> x <111000000> = <100000000> + <111000000> + <111110000> + <111111100> + <210000000>
                          + <211100000> + <211111000> + <221000000> + <221110000> + <222100000>
<111100000> x <111100000> = <000000000> + <110000000> + <111100000> + <111111000> + <111111110>
                          + <200000000> + <211000000> + <211110000> + <211111100> + <220000000>
                          + <221100000> + <221111000> + <222000000> + <222110000> + <222200000>
<111110000> x <000000000> = <111110000>
<111110000> x <100000000> = <111100000> + <111111000> + <211110000>
<111110000> x <110000000> = <111000000> + <111110000> + <111111100> + <211100000> + <211111000>
                          + <221110000>
<111110000> x <111000000> = <110000000> + <111100000> + <111111000> + <111111110> + <211000000>
                          + <211110000> + <211111100> + <221100000> + <221111000> + <222110000>
<111110000> x <111100000> = <100000000> + <111000000> + <111110000> + <111111100> + <111111111>
                          + <210000000> + <211100000> + <211111000> + <211111110> + <221000000>
                          + <221110000> + <221111100> + <222100000> + <222111000> + <222210000>
<111110000> x <111110000> = <000000000> + <110000000> + <111100000> + <111111000> + <111111110>
                          + <200000000> + <211000000> + <211110000> + <211111100> + <211111111>
                          + <220000000> + <221100000> + <221111000> + <221111110> + <222000000>
                          + <222110000> + <222111100> + <222200000> + <222211000> + <222220000>
<111111000> x <000000000> = <111111000>
<111111000> x <100000000> = <111110000> + <111111100> + <211111000>
<111111000> x <110000000> = <111100000> + <111111000> + <111111110> + <211110000> + <211111100>
                          + <221111000>
<111111000> x <111000000> = <111000000> + <111110000> + <111111100> + <111111111> + <211100000>
                          + <211111000> + <211111110> + <221110000> + <221111100> + <222111000>
<111111000> x <111100000> = <110000000> + <111100000> + <111111000> + <111111110> + <211000000>
                          + <211110000> + <211111100> + <211111111> + <221100000> + <221111000>
                          + <221111110> + <222110000> + <222111100> + <222211000>
<111111000> x <111110000> = <100000000> + <111000000> + <111110000> + <111111100> + <210000000>
                          + <211100000> + <211111000> + <211111110> + <221000000> + <221110000>
                          + <221111100> + <221111111> + <222100000> + <222111000> + <222111110>
                          + <222210000> + <222211100> + <222221000>
<111111000> x <111111000> = <000000000> + <110000000> + <111100000> + <111111000> + <200000000>
                          + <211000000> + <211110000> + <211111100> + <220000000> + <221100000>
                          + <221111000> + <221111110> + <222000000> + <222110000> + <222111100>
                          + <222111111> + <222200000> + <222211000> + <222211110> + <222220000>
                          + <222221100> + <222222000>
<111111100> x <000000000> = <111111100>
<111111100> x <100000000> = <111111000> + <111111110> + <211111100>
<111111100> x <110000000> = <111110000> + <111111100> + <111111111> + <211111000> + <211111110>
                          + <221111100>
<111111100> x <111000000> = <111100000> + <111111000> + <111111110> + <211110000> + <211111100>
                          + <211111111> + <221111000> + <221111110> + <222111100>
<111111100> x <111100000> = <111000000> + <111110000> + <111111100> + <211100000> + <211111000>
                          + <211111110> + <221110000> + <221111100> + <221111111> + <222111000>
                          + <222111110> + <222211100>
<111111100> x <111110000> = <110000000> + <111100000> + <111111000> + <211000000> + <211110000>
                          + <211111100> + <221100000> + <221111000> + <221111110> + <222110000>
                          + <222111100> + <222111111> + <222211000> + <222211110> + <222221100>
<111111100> x <111111000> = <100000000> + <111000000> + <111110000> + <210000000> + <211100000>
                          + <211111000> + <221000000> + <221110000> + <221111100> + <222100000>
                          + <222111000> + <222111110> + <222210000> + <222211100> + <222211111>
                          + <222221000> + <222221110> + <222222100>
```

D-15 Reduction of the Kronecker Products for Sp_{18} (contd).

<111111100> x <111111100> = <000000000> + <110000000> + <111100000> + <200000000> + <211000000>
+ <211110000> + <220000000> + <221100000> + <221111000> + <222000000>
+ <222110000> + <222111100> + <222200000> + <222211000> + <222211110>
+ <222220000> + <222221100> + <222221111> + <222222000> + <222222110>
+ <222222200>

<111111110> x <000000000> = <111111110>

<111111110> x <100000000> = <111111100> + <111111111> + <211111110>

<111111110> x <110000000> = <111111000> + <111111110> + <211111100> + <211111111> + <221111110>

<111111110> x <111000000> = <111110000> + <111111100> + <211111000> + <211111110> + <221111100>
+ <221111111> + <222111110>

<111111110> x <111100000> = <111100000> + <111111000> + <211110000> + <211111100> + <221111000>
+ <221111110> + <222111100> + <222111111> + <222211110>

<111111110> x <111110000> = <111000000> + <111110000> + <211100000> + <211111000> + <221110000>
+ <221111100> + <222111000> + <222111110> + <222211100> + <222211111>
+ <222221110>

<111111110> x <111111000> = <110000000> + <111100000> + <211000000> + <211110000> + <221100000>
+ <221111000> + <222110000> + <222111100> + <222211000> + <222211110>
+ <222221100> + <222221111> + <222222110>

<111111110> x <111111100> = <100000000> + <111000000> + <210000000> + <211100000> + <221000000>
+ <221110000> + <222100000> + <222111000> + <222210000> + <222211100>
+ <222221000> + <222221110> + <222222100> + <222222111> + <222222210>

<111111110> x <111111110> = <000000000> + <110000000> + <200000000> + <211000000> + <220000000>
+ <221100000> + <222000000> + <222110000> + <222200000> + <222211000>
+ <222220000> + <222221100> + <222222000> + <222222110> + <222222200>
+ <222222211> + <222222220>

<111111111> x <000000000> = <111111111>

<111111111> x <100000000> = <111111110> + <211111111>

<111111111> x <110000000> = <111111100> + <211111110> + <221111111>

<111111111> x <111000000> = <111111000> + <211111100> + <221111110> + <222111111>

<111111111> x <111100000> = <111110000> + <211111000> + <221111100> + <222111110> + <222211111>

<111111111> x <111110000> = <111100000> + <211110000> + <221111000> + <222111100> + <222211110>
+ <222221111>

<111111111> x <111111000> = <111000000> + <211100000> + <221110000> + <222111000> + <222211100>
+ <222221110> + <222222111>

<111111111> x <111111100> = <110000000> + <211000000> + <221100000> + <222110000> + <222211000>
+ <222221100> + <222222110> + <222222211>

<111111111> x <111111110> = <100000000> + <210000000> + <221000000> + <222100000> + <222210000>
+ <222221000> + <222222100> + <222222210> + <222222221>

<111111111> x <111111111> = <000000000> + <200000000> + <220000000> + <222000000> + <222200000>
+ <222220000> + <222222000> + <222222200> + <222222220> + <222222222>

E-1 Dimensions of the Irreducible Representations of G_2.

(00)	1	(64)	4914	(98)	29667	(123)	59136	(145)	187264
		(65)	4928	(99)	19096	(124)	79002	(146)	226044
(10)	7	(66)	3542			(125)	99008	(147)	262144
(11)	14			(100)	5005	(126)	117649	(148)	292740
		(70)	1254	(101)	11648	(127)	133056	(149)	314496
(20)	27	(71)	3003	(102)	19683	(128)	142968	(1410)	323532
(21)	64	(72)	5103	(103)	28672	(129)	144704	(1411)	315392
(22)	77	(73)	7293	(104)	37961	(1210)	135135	(1412)	285012
		(74)	9177	(105)	46656	(1211)	110656	(1413)	226688
(30)	77	(75)	10206	(106)	53599	(1212)	67158	(1414)	134044
(31)	189	(76)	9660	(107)	57344				
(32)	286	(77)	6630	(108)	56133	(130)	14756	(150)	27132
(33)	273			(109)	47872	(131)	33592	(151)	61047
		(80)	2079	(1010)	30107	(132)	56133	(152)	101269
(40)	182	(81)	4928			(133)	81719	(153)	146965
(41)	448	(82)	8372	(110)	7371	(134)	109375	(154)	196911
(42)	729	(83)	12096	(111)	17017	(135)	137781	(155)	249458
(43)	896	(84)	15625	(112)	28652	(136)	165242	(156)	302498
(44)	748	(85)	18304	(113)	41769	(137)	189658	(157)	353430
		(86)	19278	(114)	55614	(138)	208494	(158)	399126
(50)	378	(87)	17472	(115)	69160	(139)	218750	(159)	435897
(51)	924	(88)	11571	(116)	81081	(1310)	216931	(1510)	459459
(52)	1547			(117)	89726	(1311)	199017	(1511)	464899
(53)	2079	(90)	3289	(118)	93093	(1312)	160433	(1512)	446641
(54)	2261	(91)	7722	(119)	88803	(1313)	96019	(1513)	398412
(55)	1729	(92)	13090	(1110)	74074			(1514)	313208
		(93)	19019	(1111)	45695	(140)	20196	(1515)	183260
(60)	714	(94)	24948			(141)	45696		
(61)	1728	(95)	30107	(120)	10556	(142)	76076		
(62)	2926	(96)	33495	(121)	24192	(143)	110592		
(63)	4096	(97)	33858	(122)	40579	(144)	148148		

E-2 Branching Rules for the Reduction $R_7 \to G_2$.

[000] = (00)

[100] = (10)
[110] = (10) + (11)
[111] = (00) + (10) + (20)

[200] = (20)
[210] = (11) + (20) + (21)
[211] = (10) + (11) + (20) + (21) + (30)
[220] = (20) + (21) + (22)
[221] = (10) + (11) + (20) + (21) + (30) + (31)
[222] = (00) + (10) + (20) + (30) + (40)

[300] = (30)
[310] = (21) + (30) + (31)
[311] = (20) + (21) + (22) + (30) + (31) + (40)
[320] = (21) + (22) + (30) + (31) + (32)
[321] = (11) + (20) +2(21) + (22) + (30) +2(31) + (32) + (40) + (41)
[322] = (10) + (11) + (20) + (21) + (30) + (31) + (40) + (41) + (50)
[330] = (30) + (31) + (32) + (33)
[331] = (20) + (21) + (22) + (30) + (31) + (32) + (40) + (41) + (42)
[332] = (10) + (11) + (20) + (21) + (30) + (31) + (40) + (41) + (50) + (51)
[333] = (00) + (10) + (20) + (30) + (40) + (50) + (60)

[400] = (40)
[410] = (31) + (40) + (41)
[411] = (30) + (31) + (32) + (40) + (41) + (50)
[420] = (22) + (31) + (32) + (40) + (41) + (42)
[421] = (21) + (22) + (30) +2(31) +2(32) + (33) + (40) +2(41) + (42) + (50) + (51)
[422] = (20) + (21) + (22) + (30) + (31) + (32) + (40) + (41) + (42) + (50) + (51) + (60)
[430] = (31) + (32) + (33) + (40) + (41) + (42) + (43)
[431] = (21) + (22) + (30) +2(31) +2(32) + (33) + (40) +2(41) +2(42) + (43) + (50) + (51) + (52)
[432] = (11) + (20) +2(21) + (22) + (30) +2(31) + (32) + (40) +2(41) + (42) + (50) +2(51) + (52) + (60) + (61)
[433] = (10) + (11) + (20) + (21) + (30) + (31) + (40) + (41) + (50) + (51) + (60) + (61) + (70)
[440] = (40) + (41) + (42) + (43) + (44)
[441] = (30) + (31) + (32) + (33) + (40) + (41) + (42) + (43) + (50) + (51) + (52) + (53)
[442] = (20) + (21) + (22) + (30) + (31) + (32) + (40) + (41) + (42) + (50) + (51) + (52) + (60) + (61) + (62)
[443] = (10) + (11) + (20) + (21) + (30) + (31) + (40) + (41) + (50) + (51) + (60) + (61) + (70) + (71)
[444] = (00) + (10) + (20) + (30) + (40) + (50) + (60) + (70) + (80)

E-3 Branching Rules for the Reduction $G_2 \to R_3$.

```
(00) = [0]

(10) = [3]
(11) = [1] + [5]

(20) = [2] + [4] + [6]
(21) = [2] + [3] + [4] + [5] + [7] + [8]
(22) = [0] + [2] + [4] + [5] + [6] + [8] + [10]

(30) = [1] + [3] + [4] + [5] + [6] + [7] + [9]
(31) = [1] + [2] +2[3] + [4] +2[5] +2[6] +2[7] + [8] + [9] + [10] + [11]
(32) = [1] + [2] +2[3] +2[4] +2[5] +2[6] +2[7] +2[8] +2[9] + [10] + [11] + [12] + [13]
(33) = [1] +2[3] + [4] +2[5] + [6] +2[7] + [8] +2[9] + [10] + [11] + [12] + [13] + [15]

(40) = [0] + [2] + [3] +2[4] + [5] +2[6] + [7] +2[8] + [9] + [10] + [12]
(41) = [1] +2[2] +2[3] +3[4] +3[5] +3[6] +3[7] +3[8] +3[9] +2[10] +2[11] + [12] + [13] + [14]
(42) = [0] + [1] +3[2] +2[3] +4[4] +3[5] +5[6] +4[7] +4[8] +3[9] +4[10] +3[11] +3[12] + [13] +2[14]
     + [15] + [16]
(43) = [0] + [1] +2[2] +3[3] +3[4] +4[5] +4[6] +4[7] +4[8] +4[9] +4[10] +3[11] +3[12] +3[13] +2[14]
     +2[15] + [16] + [17] + [18]
(44) = [0] +2[2] + [3] +3[4] +2[5] +3[6] +2[7] +4[8] +2[9] +3[10] +2[11] +3[12] +2[13] +2[14] + [15]
     +2[16] + [17] + [18] + [20]

(50) = [1] + [2] +2[3] + [4] +3[5] +2[6] +3[7] +2[8] +2[9] +2[10] +2[11] + [12] + [13] + [15]
(51) =2[1] +2[2] +3[3] +4[4] +5[5] +4[6] +5[7] +5[8] +5[9] +4[10] +4[11] +3[12] +3[13] +2[14] + [15]
     + [16] + [17]
(52) =2[1] +3[2] +5[3] +5[4] +6[5] +6[6] +7[7] +7[8] +7[9] +6[10] +6[11] +5[12] +5[13] +4[14] +3[15]
     +2[16] +2[17] + [18] + [19]
(53) =3[1] +3[2] +5[3] +5[4] +7[5] +7[6] +8[7] +7[8] +8[9] +7[10] +8[11] +6[12] +6[13] +5[14] +5[15]
     +4[16] +3[17] +2[18] +2[19] + [20] + [21]
(54) =2[1] +3[2] +4[3] +5[4] +6[5] +6[6] +7[7] +7[8] +7[9] +7[10] +7[11] +6[12] +6[13] +6[14] +5[15]
     +4[16] +4[17] +3[18] +3[19] +2[20] + [21] + [22] + [23]
(55) =2[1] + [2] +3[3] +2[4] +5[5] +3[6] +5[7] +4[8] +5[9] +4[10] +5[11] +4[12] +5[13] +3[14] +4[15]
     +3[16] +4[17] +2[18] +2[19] +2[20] +2[21] + [22] + [23] + [25]

(60) = [0] +2[2] +2[3] +3[4] +2[5] +4[6] +3[7] +4[8] +3[9] +4[10] +2[11] +3[12] +2[13] +2[14] + [15]
     + [16] + [18]
(61) = [0] +2[1] +3[2] +4[3] +5[4] +6[5] +7[6] +7[7] +7[8] +7[9] +7[10] +7[11] +6[12] +5[13] +4[14]
     +4[15] +3[16] +2[17] + [18] + [19] + [20]
(62) =2[0] +2[1] +5[2] +5[3] +8[4] +8[5] +10[6] +9[7] +11[8] +10[9] +11[10] +9[11] +10[12] +8[13]
     +8[14] +6[15] +6[16] +4[17] +4[18] +2[19] +2[20] + [21] + [22]
(63) = [0] +3[1] +5[2] +7[3] +9[4] +9[5] +11[6] +12[7] +12[8] +13[9] +12[10] +12[11] +12[12] +11[13]
     +10[14] +9[15] +8[16] +7[17] +6[18] +5[19] +3[20] +3[21] +2[22] + [23] + [24]
(64) =2[0] +2[1] +6[2] +6[3] +9[4] +9[5] +12[6] +11[7] +13[8] +12[9] +14[10]+12[11] +13[12] +11[13]
     +12[14] +10[15] +10[16] +8[17] +8[18] +6[19] +6[20] +4[21] +4[22] +2[23] +2[24] + [25] + [26]
(65) = [0] +2[1] +4[2] +6[3] +7[4] +8[5] +9[6] +10[7] +11[8] +11[9] +11[10] +11[11] +11[12] +11[13]
     +10[14] +10[15] +9[16] +8[17] +8[18] +7[19] +6[20] +5[21] +4[22] +4[23] +3[24] +2[25] + [26]
     + [27] + [28]
(66) =2[0] + [1] +3[2] +2[3] +5[4] +4[5] +7[6] +5[7] +7[8] +6[9] +8[10] +6[11] +8[12] +6[13] +7[14]
     +6[15] +7[16] +5[17] +6[18] +4[19] +5[20] +4[21] +4[22] +2[23] +3[24] +2[25] +2[26] + [27]
     + [28] + [30]

(70) =2[1] + [2] +3[3] +3[4] +4[5] +4[6] +5[7] +4[8] +6[9] +4[10] +5[11] +4[12] +4[13] +3[14] +3[15]
     +2[16] +2[17] + [18] + [19] + [21]
(71) =2[1] +4[2] +6[3] +6[4] +8[5] +8[6] +10[7] +10[8] +10[9] +10[10] +10[11] +9[12] +9[13] +8[14]
     +7[15] +5[16] +5[17] +4[18] +3[19] +2[20] + [21] + [22] + [23]
(72) = [0] +4[1] +5[2] +8[3] +9[4] +12[5] +12[6] +14[7] +14[8] +15[9] +15[10] +15[11] +14[12]
     +14[13] +12[14] +12[15] +10[16] +9[17] +7[18] +6[19] +5[20] +4[21] +2[22] +2[23] + [24] + [25]
(73) = [0] +5[1] +6[2] +10[3] +11[4] +14[5] +15[6] +17[7] +17[8] +19[9] +18[10] +19[11] +18[12]
     +18[13] +16[14] +16[15] +14[16] +13[17] +11[18] +10[19] +8[20] +7[21] +5[22] +4[23] +3[24]
     +2[25] + [26] + [27]
(74) = [0] +5[1] +7[2] +10[3] +12[4] +15[5] +16[6] +18[7] +19[8] +20[9] +20[10] +21[11] +20[12]
     +20[13] +19[14] +18[15] +17[16] +16[17] +14[18] +13[19] +11[20] +10[21] +8[22] +7[23] +5[24]
     +4[25] +3[26] +2[27] + [28] + [29]
(75) =5[1] +6[2] +10[3] +11[4] +14[5] +15[6] +18[7] +17[8] +20[9] +19[10] +21[11] +19[12] +20[13]
     +19[14] +19[15] +17[16] +17[17] +15[18] +15[19] +12[20] +12[21] +10[22] +9[23] +7[24] +6[25]
     +5[26] +4[27] +2[28] +2[29] + [30] + [31]
(76) = [0] +3[1] +5[2] +8[3] +9[4] +11[5] +12[6] +14[7] +15[8] +16[9] +16[10] +16[11] +17[12]
     +17[13] +16[14] +16[15] +15[16] +15[17] +14[18] +13[19] +12[20] +11[21] +10[22] +9[23] +8[24]
     +7[25] +5[26] +5[27] +4[28] +3[29] +2[30] + [31] + [32] + [33]
```

Branching Rules for the Reduction $G_2 \to R_3$ (contd).

(77) =3[1] +2[2] +5[3] +4[4] +8[5] +6[6] +9[7] +8[8] +10[9] +9[10] +11[11] +9[12] +11[13] +9[14]
+11[15] +9[16] +10[17] +8[18] +9[19] +8[20] +8[21] +6[22] +7[23] +5[24] +6[25] +4[26] +4[27]
+3[28] +3[29] +2[30] +2[31] + [32] + [33] + [35]

(80) = [0] + [1] +3[2] +2[3] +5[4] +4[5] +6[6] +5[7] +7[8] +6[9] +7[10] +6[11] +7[12] +5[13] +6[14]
+4[15] +5[16] +3[17] +3[18] +2[19] +2[20] + [21] + [22] + [24]

(81) = [0] +3[1] +5[2] +6[3] +8[4] +10[5] +11[6] +12[7] +13[8] +13[9] +14[10] +14[11] +13[12]
+13[13] +12[14] +11[15] +10[16] +9[17] +7[18] +6[19] +5[20] +4[21] +3[22] +2[23] + [24] + [25]
+ [26]

(82) =2[0] +3[1] +8[2] +9[3] +13[4] +13[5] +17[6] +17[7] +20[8] +19[9] +21[10] +20[11] +21[12]
+19[13] +20[14] +17[15] +17[16] +14[17] +14[18] +11[19] +10[20] +7[21] +7[22] +5[23] +4[24]
+2[25] +2[26] + [27] + [28]

(83) =2[0] +5[1] +9[2] +12[3] +15[4] +18[5] +20[6] +22[7] +24[8] +25[9] +26[10] +26[11] +26[12]
+26[13] +25[14] +24[15] +22[16] +21[17] +19[18] +17[19] +15[20] +13[21] +11[22] +9[23] +7[24]
+6[25] +4[26] +3[27] +2[28] + [29] + [30]

(84) =3[0] +5[1] +11[2] +12[3] +18[4] +19[5] +23[6] +24[7] +28[8] +27[9] +30[10] +29[11] +31[12]
+29[13] +30[14] +27[15] +28[16] +25[17] +24[18] +21[19] +21[20] +17[21] +16[22] +13[23] +12[24]
+9[25] +8[26] +5[27] +5[28] +3[29] +2[30] + [31] + [32]

(85) =2[0] +6[1] +10[2] +13[3] +17[4] +20[5] +23[6] +25[7] +27[8] +29[9] +30[10] +31[11] +31[12]
+31[13] +31[14] +30[15] +29[16] +28[17] +26[18] +25[19] +23[20] +21[21] +19[22] +17[23] +15[24]
+13[25] +11[26] +9[27] +7[28] +6[29] +4[30] +3[31] +2[32] + [33] + [34]

(86) =3[0] +4[1] +10[2] +11[3] +16[4] +17[5] +22[6] +22[7] +26[8] +25[9] +29[10] +28[11] +30[12]
+28[13] +30[14] +28[15] +29[16] +26[17] +27[18] +24[19] +24[20] +21[21] +21[22] +18[23] +17[24]
+14[25] +14[26] +11[27] +10[28] +7[29] +7[30] +5[31] +4[32] +2[33] +2[34] + [35] + [36]

(87) = [0] +4[1] +7[2] +9[3] +12[4] +14[5] +16[6] +18[7] +20[8] +21[9] +22[10] +23[11] +23[12]
+24[13] +24[14] +23[15] +23[16] +23[17] +22[18] +21[19] +20[20] +19[21] +18[22] +17[23] +15[24]
+14[25] +13[26] +11[27] +10[28] +9[29] +7[30] +6[31] +5[32] +4[33] +3[34] +2[35] + [36] + [37]
+ [38]

(88) =2[0] + [1] +5[2] +4[3] +8[4] +7[5] +10[6] +9[7] +13[8] +11[9] +14[10] +12[11] +15[12] +13[13]
+15[14] +13[15] +15[16] +13[17] +14[18] +12[19] +14[20] +11[21] +12[22] +10[23] +11[24] +9[25]
+9[26] +7[27] +8[28] +6[29] +6[30] +4[31] +5[32] +3[33] +3[34] +2[35] +2[36] + [37] + [38]
+ [40]

E-4 Reduction of the Kronecker Products for G_2.

```
(00) x (00) = (00)

(10) x (00) = (10)
(10) x (10) = (00) + (10) + (11) + (20)
(11) x (00) = (11)
(11) x (10) = (10) + (20) + (21)
(11) x (11) = (00) + (11) + (20) + (22) + (30)

(20) x (00) = (20)
(20) x (10) = (10) + (11) + (20) + (21) + (30)
(20) x (11) = (10) + (11) + (20) + (21) + (30) + (31)
(20) x (20) = (00) + (10) + (11) +2(20) +2(21) + (22) + (30) + (31) + (40)
(21) x (00) = (21)
(21) x (10) = (11) + (20) + (21) + (22) + (30) + (31)
(21) x (11) = (10) + (20) +2(21) + (30) + (31) + (32) + (40)
(21) x (20) = (10) + (11) +2(20) +2(21) + (22) +2(30) +2(31) + (32) + (40) + (41)
(21) x (21) = (00) + (10) +2(11) +2(20) +2(21) +2(22) +3(30) +3(31) + (32) + (33) +2(40) +2(41)
            + (42) + (50)
(22) x (00) = (22)
(22) x (10) = (21) + (31) + (32)
(22) x (11) = (11) + (22) + (30) + (31) + (33) + (41)
(22) x (20) = (20) + (21) + (22) + (30) + (31) + (32) + (40) + (41) + (42)
(22) x (21) = (10) + (20) +2(21) + (30) +2(31) +2(32) +2(40) + (41) + (42) + (43) + (50) + (51)
(22) x (22) = (00) + (11) + (20) +2(22) + (30) + (31) + (33) + (40) +2(41) + (42) + (44) + (50)
            + (52) + (60)

(30) x (00) = (30)
(30) x (10) = (20) + (21) + (30) + (31) + (40)
(30) x (11) = (11) + (20) + (21) + (22) + (30) + (31) + (40) + (41)
(30) x (20) = (10) + (11) + (20) +2(21) + (22) +2(30) +2(31) + (32) + (40) + (41) + (50)
(30) x (21) = (10) + (11) +2(20) +3(21) + (22) +2(30) +3(31) +2(32) +2(40) +2(41) + (42) + (50)
            + (51)
(30) x (22) = (11) + (20) + (21) + (22) +2(30) +2(31) + (32) + (33) + (40) +2(41) + (42) + (50)
            + (51) + (52)
(30) x (30) = (00) + (10) + (11) +2(20) +2(21) +2(22) +2(30) +3(31) +2(32) + (33) +2(40) +2(41)
            + (42) + (50) + (51) + (60)
(31) x (00) = (31)
(31) x (10) = (21) + (22) + (30) + (31) + (32) + (40) + (41)
(31) x (11) = (20) + (21) + (22) + (30) +2(31) + (32) + (40) + (41) + (42) + (50)
(31) x (20) = (11) + (20) +2(21) + (22) +2(30) +3(31) +2(32) + (33) +2(40) +2(41) + (42) + (50)
            + (51)
(31) x (21) = (10) + (11) +2(20) +3(21) +2(22) +3(30) +4(31) +3(32) + (33) +3(40) +4(41) +2(42)
            + (43) +2(50) +2(51) + (52) + (60)
(31) x (22) = (10) + (11) + (20) +2(21) + (22) +2(30) +3(31) +2(32) + (33) +2(40) +3(41) +2(42)
            + (43) +2(50) +2(51) + (52) + (53) + (60) + (61)
(31) x (30) = (10) + (11) +2(20) +3(21) +2(22) +3(30) +4(31) +3(32) + (33) +3(40) +4(41) +3(42)
            + (43) +2(50) +2(51) + (52) + (60) + (61)
(31) x (31) = (00) + (10) +2(11) +3(20) +4(21) +3(22) +4(30) +6(31) +5(32) +2(33) +5(40) +6(41)
            +5(42) +2(43) + (44) +4(50) +5(51) +3(52) + (53) +2(60) +2(61) + (62) + (70)
(32) x (00) = (32)
(32) x (10) = (22) + (31) + (32) + (33) + (41) + (42)
(32) x (11) = (21) + (31) +2(32) + (40) + (41) + (42) + (43) + (51)
(32) x (20) = (21) + (22) + (30) +2(31) +2(32) + (33) + (40) +2(41) +2(42) + (43) + (50) + (51)
            + (52)
(32) x (21) = (11) + (20) + (21) +2(22) +2(30) +3(31) +2(32) +2(33) +2(40) +4(41) +3(42) + (43)
            + (44) +2(50) +2(51) +2(52) + (53) + (60) + (61)
(32) x (22) = (10) + (20) +2(21) + (30) +2(31) +3(32) +2(40) +2(41) +2(42) +2(43) +2(50) +3(51)
            + (52) + (53) + (54) + (60) + (61) + (62) + (70)
(32) x (30) = (20) +2(21) + (22) +2(30) +3(31) +3(32) + (33) +3(40) +3(41) +3(42) +2(43) +2(50)
            +3(51) +2(52) + (53) + (60) + (61) + (62)
(32) x (31) = (10) + (11) +2(20) +3(21) +2(22) +3(30) +5(31) +4(32) +2(33) +4(40) +6(41) +5(42)
            +3(43) + (44) +4(50) +5(51) +4(52) +2(53) + (54) +3(60) +3(61) +2(62) + (63) + (70)
            + (71)
(32) x (32) = (00) + (10) +2(11) +2(20) +2(21) +3(22) +3(30) +4(31) +3(32) +3(33) +3(40) +6(41)
            +5(42) +2(43) +2(44) +4(50) +5(51) +5(52) +3(53) + (54) + (55) +4(60) +4(61) +2(62)
            +2(63) + (64) +2(70) +2(71) + (72) + (80)
(33) x (00) = (33)
(33) x (10) = (32) + (42) + (43)
(33) x (11) = (22) + (33) + (41) + (42) + (44) + (52)
(33) x (20) = (31) + (32) + (33) + (41) + (42) + (43) + (51) + (52) + (53)
```

Reduction of the Kronecker Products for G_2 (contd).

(33) x (21) = (21) + (31) +2(32) + (40) + (41) +2(42) +2(43) + (50) +2(51) + (52) + (53) + (54)
+ (61) + (62)
(33) x (22) = (11) + (22) + (30) + (31) +2(33) +2(41) + (42) + (44) + (50) + (51) +2(52) + (53)
+ (55) + (60) + (61) + (63) + (71)
(33) x (30) = (22) + (30) + (31) + (32) + (33) + (40) +2(41) +2(42) + (43) + (44) + (50) + (51)
+2(52) + (53) + (60) + (61) + (62) + (63)
(33) x (31) = (20) + (21) + (22) + (30) +2(31) +2(32) + (33) +2(40) +3(41) +3(42) +2(43) + (44)
+2(50) +3(51) +3(52) +2(53) + (54) +2(60) +2(61) +2(62) + (63) + (64) + (70) + (71)
+ (72)
(33) x (32) = (10) + (20) +2(21) + (30) +2(31) +3(32) +2(40) +2(41) +3(42) +3(43) +2(50) +4(51)
+2(52) +2(53) +2(54) +2(60) +3(61) +3(62) + (63) + (64) + (65) +2(70) + (71) + (72)
+ (73) + (80) + (81)
(33) x (33) = (00) + (11) + (20) +2(22) + (30) + (31) +2(33) + (40) +2(41) +2(42) +2(44) + (50)
+ (51) +3(52) + (53) + (55) +2(60) +2(61) + (62) +2(63) + (64) + (66) + (70) +2(71)
+ (72) + (74) + (80) + (82) + (90)

(40) x (00) = (40)
(40) x (10) = (30) + (31) + (40) + (41) + (50)
(40) x (11) = (21) + (30) + (31) + (32) + (40) + (41) + (50) + (51)
(40) x (20) = (20) + (21) + (22) + (30) +2(31) + (32) +2(40) +2(41) + (42) + (50) + (51) + (60)
(40) x (21) = (11) + (20) +2(21) +2(22) +2(30) +3(31) +2(32) + (33) +2(40) +3(41) +2(42) +2(50)
+2(51) + (52) + (60) + (61)
(40) x (22) = (20) +2(21) + (22) + (30) +2(31) +2(32) +2(40) +2(41) +2(42) + (43) + (50) +2(51)
+ (52) + (60) + (61) + (62)
(40) x (30) = (10) + (11) + (20) +2(21) + (22) +2(30) +3(31) +3(32) + (33) +2(40) +3(41) +2(42)
+ (43) +2(50) +2(51) + (52) + (60) + (61) + (70)
(40) x (31) = (10) + (11) +2(20) +3(21) +2(22) +3(30) +5(31) +4(32) +2(33) +3(40) +5(41) +4(42)
+2(43) +3(50) +4(51) +3(52) + (53) +2(60) +2(61) + (62) + (70) + (71)
(40) x (32) = (11) + (20) +2(21) +2(22) +3(30) +4(31) +3(32) +2(33) +3(40) +5(41) +4(42) +2(43)
+ (44) +3(50) +4(51) +4(52) +2(53) +2(60) +3(61) +2(62) + (63) + (70) + (71) + (72)
(40) x (33) = (21) + (30) +2(31) +2(32) + (33) +2(40) +2(41) +2(42) +2(43) +2(50) +3(51) +2(52)
+2(53) + (54) + (60) +2(61) +2(62) + (63) + (70) + (71) + (72) + (73)
(40) x (40) = (00) + (10) + (11) +2(20) +2(21) +2(22) +2(30) +3(31) +3(32) +2(33) +3(40) +4(41)
+4(42) +2(43) + (44) +2(50) +3(51) +2(52) + (53) +2(60) +2(61) + (62) + (70) + (71)
+ (80)
(41) x (00) = (41)
(41) x (10) = (31) + (32) + (40) + (41) + (42) + (50) + (51)
(41) x (11) = (22) + (30) + (31) + (32) + (33) + (40) +2(41) + (42) + (50) + (51) + (52) + (60)
(41) x (20) = (21) + (22) + (30) +2(31) +2(32) + (33) +2(40) +3(41) +2(42) + (43) +2(50) +2(51)
+ (52) + (60) + (61)
(41) x (21) = (20) +2(21) + (22) +2(30) +4(31) +4(32) + (33) +3(40) +4(41) +4(42) +2(43) +3(50)
+4(51) +2(52) + (53) +2(60) +2(61) + (62) + (70)
(41) x (22) = (11) + (20) + (21) +2(22) +2(30) +3(31) +2(32) +2(33) +2(40) +4(41) +3(42) + (43)
+ (44) +2(50) +3(51) +3(52) + (53) +2(60) +2(61) + (62) + (63) + (70) + (71)
(41) x (30) = (11) + (20) +2(21) +2(22) +2(30) +4(31) +3(32) +2(33) +3(40) +5(41) +4(42) +2(43)
+ (44) +3(50) +4(51) +3(52) + (53) +2(60) +2(61) + (62) + (70) + (71)
(41) x (31) = (10) + (11) +2(20) +4(21) +3(22) +4(30) +6(31) +6(32) +3(33) +5(40) +8(41) +7(42)
+4(43) + (44) +5(50) +7(51) +6(52) +3(53) + (54) +4(60) +5(61) +3(62) + (63) +2(70)
+2(71) + (72) + (80)
(41) x (32) = (10) + (11) +2(20) +4(21) +2(22) +3(30) +6(31) +6(32) +2(33) +5(40) +7(41) +7(42)
+5(43) + (44) +5(50) +8(51) +6(52) +4(53) +2(54) +4(60) +6(61) +5(62) +2(63) + (64)
+3(70) +3(71) +2(72) + (73) + (80) + (81)
(41) x (33) = (11) + (20) + (21) +2(22) +2(30) +3(31) +2(32) +2(33) +2(40) +5(41) +4(42) +2(43)
+2(44) +3(50) +4(51) +5(52) +3(53) + (54) + (55) +3(60) +4(61) +3(62) +3(63) + (64)
+2(70) +3(71) +2(72) + (73) + (74) + (80) + (81) + (82)
(41) x (40) = (10) + (11) +2(20) +3(21) +2(22) +3(30) +5(31) +5(32) +2(33) +4(40) +6(41) +6(42)
+4(43) + (44) +4(50) +6(51) +5(52) +3(53) + (54) +3(60) +4(61) +3(62) + (63) +2(70)
+2(71) + (72) + (80) + (81)
(41) x (41) = (00) + (10) +2(11) +3(20) +4(21) +4(22) +5(30) +8(31) +7(32) +5(33) +6(40) +10(41)
+10(42) +6(43) +3(44) +7(50) +10(51) +10(52) +6(53) +2(54) + (55) +6(60) +8(61) +7(62)
+4(63) + (64) +4(70) +5(71) +3(72) + (73) +2(80) +2(81) + (82) + (90)
(42) x (00) = (42)
(42) x (10) = (32) + (33) + (41) + (42) + (43) + (51) + (52)
(42) x (11) = (31) + (32) + (33) + (41) +2(42) + (43) + (50) + (51) + (52) + (53) + (61)
(42) x (20) = (22) + (31) +2(32) + (33) + (40) +2(41) +3(42) +2(43) + (44) + (50) +2(51) +2(52)
+ (53) + (60) + (61) + (62)
(42) x (21) = (21) + (22) + (30) +2(31) +3(32) +2(33) +2(40) +4(41) +4(42) +3(43) + (44) +2(50)
+4(51) +4(52) +2(53) + (54) +2(60) +2(61) +2(62) + (63) + (70) + (71)
(42) x (22) = (20) + (21) + (22) + (30) +2(31) +2(32) + (33) +2(40) +3(41) +4(42) +2(43) + (44)
+2(50) +3(51) +3(52) +2(53) + (54) +2(60) +3(61) +2(62) + (63) + (64) + (70) + (71)
+ (72) + (80)

Reduction of the Kronecker Products for G_2 (contd).

(42) x (30) = (21) + (22) + (30) +3(31) +3(32) +2(33) +2(40) +4(41) +4(42) +3(43) + (44) +3(50)
+4(51) +4(52) +3(53) + (54) +2(60) +3(61) +2(62) + (63) + (70) + (71) + (72)
(42) x (31) = (11) + (20) +2(21) +2(22) +3(30) +5(31) +5(32) +3(33) +4(40) +7(41) +7(42) +5(43)
+2(44) +5(50) +8(51) +7(52) +5(53) +2(54) + (55) +4(60) +6(61) +5(62) +3(63) + (64)
+3(70) +3(71) +2(72) + (73) + (80) + (81)
(42) x (32) = (10) + (11) +2(20) +3(21) +2(22) +3(30) +5(31) +5(32) +3(33) +4(40) +7(41) +7(42)
+5(43) +2(44) +5(50) +8(51) +8(52) +5(53) +3(54) + (55) +5(60) +7(61) +6(62) +4(63)
+2(64) + (65) +4(70) +5(71) +3(72) +2(73) + (74) +2(80) +2(81) + (82) + (90)
(42) x (33) = (10) + (11) + (20) +2(21) + (22) +2(30) +3(31) +3(32) +2(33) +2(40) +4(41) +4(42)
+3(43) + (44) +3(50) +5(51) +5(52) +4(53) +2(54) + (55) +3(60) +5(61) +4(62) +3(63)
+2(64) + (65) +3(70) +4(71) +3(72) +2(73) + (74) + (75) +2(80) +2(81) + (82) + (83)
+ (90) + (91)
(42) x (40) = (20) +2(21) +2(22) +2(30) +4(31) +4(32) +2(33) +4(40) +6(41) +6(42) +4(43) +2(44)
+4(50) +6(51) +6(52) +4(53) +2(54) +4(60) +5(61) +5(62) +3(63) + (64) +2(70) +3(71)
+2(72) + (73) + (80) + (81) + (82)
(42) x (41) = (10) + (11) +2(20) +4(21) +3(22) +4(30) +7(31) +7(32) +4(33) +6(40) +10(41) +10(42)
+7(43) +3(44) +7(50) +11(51) +11(52) +8(53) +4(54) + (55) +7(60) +10(61) +9(62) +6(63)
+3(64) + (65) +5(70) +7(71) +6(72) +3(73) + (74) +3(80) +3(81) +2(82) + (83) + (90)
+ (91)
(42) x (42) = (00) + (10) +2(11) +3(20) +4(21) +4(22) +4(30) +7(31) +7(32) +4(33) +6(40) +10(41)
+11(42) +8(43) +4(44) +7(50) +12(51) +12(52) +9(53) +5(54) +2(55) +8(60) +12(61)
+12(62) +8(63) +5(64) +2(65) + (66) +7(70) +9(71) +8(72) +5(73) +3(74) + (75) +5(80)
+6(81) +4(82) +2(83) + (84) +2(90) +2(91) + (92) + (100)
(43) x (00) = (43)
(43) x (10) = (33) + (42) + (43) + (44) + (52) + (53)
(43) x (11) = (32) + (42) +2(43) + (51) + (52) + (53) + (54) + (62)
(43) x (20) = (32) + (33) + (41) +2(42) +2(43) + (44) + (51) +2(52) +2(53) + (54) + (61) + (62)
+ (63)
(43) x (21) = (22) + (31) + (32) +2(33) +2(41) +3(42) +2(43) +2(44) + (50) +2(51) +4(52) +3(53)
+ (54) + (55) + (60) +2(61) +2(62) +2(63) + (64) + (71) + (72)
(43) x (22) = (21) + (31) +2(32) + (40) + (41) +2(42) +3(43) + (50) +3(51) +2(52) +2(53) +2(54)
+ (60) +2(61) +3(62) + (63) + (64) + (65) + (70) + (71) + (72) + (73) + (81)
(43) x (30) = (31) +2(32) + (33) + (40) +2(41) +3(42) +3(43) + (44) + (50) +3(51) +3(52) +3(53)
+2(54) + (60) +2(61) +3(62) +2(63) + (64) + (70) + (71) + (72) + (73)
(43) x (31) = (21) + (22) + (30) +2(31) +3(32) +2(33) +2(40) +4(41) +5(42) +4(43) +2(44) +3(50)
+5(51) +6(52) +5(53) +3(54) + (55) +3(60) +5(61) +5(62) +4(63) +2(64) + (65) +2(70)
+3(71) +3(72) +2(73) + (74) + (80) + (81) + (82)
(43) x (32) = (11) + (20) + (21) +2(22) +2(30) +3(31) +2(32) +3(33) +2(40) +5(41) +5(42) +3(43)
+3(44) +3(50) +5(51) +7(52) +5(53) +2(54) +2(55) +4(60) +6(61) +5(62) +5(63) +3(64)
+ (65) + (66) +3(70) +5(71) +4(72) +2(73) +2(74) + (75) +2(80) +2(81) +2(82) + (83)
+ (90) + (91)
(43) x (33) = (10) + (20) +2(21) + (30) +2(31) +3(32) +2(40) +2(41) +3(42) +4(43) +2(50) +4(51)
+3(52) +3(53) +3(54) +2(60) +4(61) +5(62) +2(63) +2(64) +2(65) +3(70) +3(71) +3(72)
+3(73) + (74) + (75) + (76) +2(80) +3(81) + (82) + (83) + (84) + (90) + (91) + (92)
+ (100)
(43) x (40) = (22) + (30) +2(31) +2(32) +2(33) +2(40) +4(41) +4(42) +3(43) +2(44) +3(50) +4(51)
+5(52) +4(53) +2(54) + (55) +3(60) +4(61) +4(62) +4(63) +2(64) +2(70) +3(71) +3(72)
+2(73) + (74) + (80) + (81) + (82) + (83)
(43) x (41) = (20) +2(21) + (22) +2(30) +4(31) +5(32) +2(33) +4(40) +6(41) +7(42) +6(43) +2(44)
+5(50) +9(51) +8(52) +7(53) +5(54) + (55) +5(60) +8(61) +9(62) +6(63) +4(64) +2(65)
+5(70) +6(71) +6(72) +5(73) +2(74) + (75) +3(80) +4(81) +3(82) +2(83) + (84) + (90)
+ (91) + (92)
(43) x (42) = (10) + (11) +2(20) +3(21) +2(22) +3(30) +5(31) +5(32) +3(33) +4(40) +7(41) +8(42)
+6(43) +3(44) +5(50) +9(51) +10(52) +8(53) +5(54) +2(55) +6(60) +10(61) +10(62) +8(63)
+5(64) +3(65) + (66) +6(70) +9(71) +8(72) +6(73) +4(74) +2(75) + (76) +5(80) +6(81)
+5(82) +3(83) +2(84) + (85) +3(90) +3(91) +2(92) + (93) + (100) + (101)
(43) x (43) = (00) + (10) +2(11) +2(20) +2(21) +3(22) +3(30) +4(31) +3(32) +4(33) +3(40) +6(41)
+6(42) +4(43) +4(44) +4(50) +6(51) +9(52) +7(53) +3(54) +3(55) +5(60) +8(61) +8(62)
+8(63) +5(64) +2(65) +2(66) +5(70) +9(71) +8(72) +5(73) +5(74) +3(75) + (76) + (77)
+5(80) +6(81) +6(82) +4(83) +2(84) +2(85) + (86) +4(90) +4(91) +2(92) +2(93) + (94)
+2(100) +2(101) + (102) + (110)
(44) x (00) = (44)
(44) x (10) = (43) + (53) + (54)
(44) x (11) = (33) + (44) + (52) + (53) + (55) + (63)
(44) x (20) = (42) + (43) + (44) + (52) + (53) + (54) + (62) + (63) + (64)
(44) x (21) = (32) + (42) +2(43) + (51) + (52) +2(53) +2(54) + (61) +2(62) + (63) + (64) + (65)
+ (72) + (73)
(44) x (22) = (22) + (33) + (41) + (42) +2(44) +2(52) + (53) + (55) + (60) + (61) + (62) +2(63)
+ (64) + (66) + (71) + (72) + (74) + (82)
(44) x (30) = (33) + (41) + (42) + (43) + (44) + (51) +2(52) +2(53) + (54) + (55) + (61) + (62)
+2(63) + (64) + (71) + (72) + (73) + (74)

Reduction of the Kronecker Products for G_2 (contd).

```
(44) x (31) = (31) + (32) + (33) + (41) +2(42) +2(43) + (44) + (50) +2(51) +3(52) +3(53) +2(54)
              + (55) + (60) +2(61) +3(62) +3(63) +2(64) + (65) + (70) +2(71) +2(72) +2(73) + (74)
              + (75) + (81) + (82) + (83)
(44) x (32) = (21) + (31) +2(32) + (40) + (41) +2(42) +3(43) + (50) +3(51) +2(52) +3(53) +3(54)
              + (60) +3(61) +4(62) +2(63) +2(64) +2(65) +2(70) +2(71) +3(72) +3(73) + (74) + (75)
              + (76) + (80) +2(81) + (82) + (83) + (84) + (91) + (92)
(44) x (33) = (11) + (22) + (30) + (31) +2(33) +2(41) + (42) +2(44) + (50) + (51) +3(52) +2(53)
              +2(55) + (60) +2(61) + (62) +3(63) + (64) + (66) + (70) +3(71) +2(72) + (73) +2(74)
              + (75) + (77) + (80) + (81) +2(82) + (83) + (85) + (90) + (91) + (93) + (101)
(44) x (40) = (32) + (40) + (41) +2(42) +2(43) + (44) + (50) +2(51) +2(52) +2(53) +2(54) + (60)
              +2(61) +3(62) +2(63) +2(64) + (65) + (70) + (71) +2(72) +2(73) + (74) + (80) + (81)
              + (82) + (83) + (84)
(44) x (41) = (22) + (30) + (31) + (32) +2(33) + (40) +3(41) +3(42) +2(43) +2(44) +2(50) +3(51)
              +5(52) +4(53) +2(54) +2(55) +3(60) +4(61) +4(62) +5(63) +3(64) + (65) + (66) +2(70)
              +4(71) +4(72) +3(73) +3(74) + (75) +2(80) +2(81) +3(82) +2(83) + (84) + (85) + (90)
              + (91) + (92) + (93)
(44) x (42) = (20) + (21) + (22) + (30) +2(31) +2(32) + (33) +2(40) +3(41) +4(42) +3(43) +2(44)
              +2(50) +4(51) +5(52) +4(53) +3(54) + (55) +3(60) +5(61) +6(62) +5(63) +4(64) +2(65)
              + (66) +3(70) +5(71) +5(72) +4(73) +3(74) +2(75) + (76) +3(80) +4(81) +4(82) +3(83)
              +2(84) + (85) + (86) +2(90) +2(91) +2(92) + (93) + (94) + (100) + (101) + (102)
(44) x (43) = (10) + (20) +2(21) + (30) +2(31) +3(32) +2(40) +2(41) +3(42) +4(43) +2(50) +4(51)
              +3(52) +4(53) +4(54) +2(60) +4(61) +6(62) +3(63) +3(64) +3(65) +3(70) +4(71) +5(72)
              +5(73) +2(74) +2(75) +2(76) +3(80) +5(81) +3(82) +3(83) +3(84) + (85) + (86) + (87)
              +2(90) +3(91) +3(92) + (93) + (94) + (95) +2(100) + (101) + (102) + (103) + (110)
              + (111)
(44) x (44) = (00) + (11) + (20) +2(22) + (30) + (31) +2(33) + (40) +2(41) +2(42) +3(44) + (50)
              + (51) +3(52) +2(53) +2(55) +2(60) +2(61) +2(62) +4(63) +2(64) +2(66) + (70) +3(71)
              +3(72) + (73) +3(74) + (75) + (77) +2(80) +2(81) +4(82) +2(83) + (84) +2(85) + (86)
              + (88) +2(90) +2(91) + (92) +2(93) + (94) + (96) + (100) +2(101) + (102) + (104)
              + (110) + (112) + (120)
```

E-5 Characters of G_2 expressed in terms of R_7 Characters.

(00) = [000]

(10) = [100]
(11)=- [100] + [110]

(20) = [200]
(21) = [100] - [110] - [200] + [210]
(22)=- [100] + [110] - [210] + [220]

(30) = [300]
(31)=- [100] + [110] + [200] - [210] - [300] + [310]
(32) = [100] - [110] + [210] - [220] - [310] + [320]
(33)=- [200] + [220] - [320] + [330]

(40) = [400]
(41) = [100] - [110] - [200] + [210] + [300] - [310] - [400] + [410]
(42) = [310] - [320] - [410] + [420]
(43)=- [100] + [110] + [200] - [210] + [320] - [330] - [420] + [430]
(44)=- [300] + [330] - [430] + [440]

(50) = [500]
(51)=- [100] + [110] + [200] - [210] - [300] + [310] + [400] - [410] - [500] + [510]
(52)=- [100] + [110] - [210] + [220] + [410] - [420] - [510] + [520]
(53) = [100] - [110] + [210] - [220] + [420] - [430] - [520] + [530]
(54) = [100] - [110] - [200] + [210] + [300] - [310] + [430] - [440] - [530] + [540]
(55)=- [400] + [440] - [540] + [550]

(60) = [600]
(61) = [100] - [110] - [200] + [210] + [300] - [310] - [400] + [410] + [500] - [510] - [600] + [610]
(62) = [100] - [110] + [210] - [220] - [310] + [320] + [510] - [520] - [610] + [620]
(63) = [520] - [530] - [620] + [630]
(64)=- [100] + [110] - [210] + [220] + [310] - [320] + [530] - [540] - [630] + [640]
(65)=- [100] + [110] + [200] - [210] - [300] + [310] + [400] - [410] + [540] - [550] - [640] + [650]
(66)=- [500] + [550] - [650] + [660]

(70) = [700]
(71)=- [100] + [110] + [200] - [210] - [300] + [310] + [400] - [410] - [500] + [510] + [600] - [610]
- [700] + [710]
(72) = [310] - [320] - [410] + [420] + [610] - [620] - [710] + [720]
(73)=- [200] + [220] - [320] + [330] + [620] - [630] - [720] + [730]
(74) = [200] - [220] + [320] - [330] + [630] - [640] - [730] + [740]
(75)=- [310] + [320] + [410] - [420] + [640] - [650] - [740] + [750]
(76) = [100] - [110] - [200] + [210] + [300] - [310] - [400] + [410] + [500] - [510] + [650] - [660]
- [750] + [760]
(77)=- [600] + [660] - [760] + [770]

(80) = [800]
(81) = [100] - [110] - [200] + [210] + [300] - [310] - [400] + [410] + [500] - [510] - [600] + [610]
+ [700] - [710] - [800] + [810]
(82)=- [100] + [110] - [210] + [220] + [410] - [420] - [510] + [520] + [710] - [720] - [810] + [820]
(83)=- [100] + [110] + [200] - [210] + [320] - [330] - [420] + [430] + [720] - [730] - [820] + [830]
(84) = [730] - [740] - [830] + [840]
(85) = [100] - [110] - [200] + [210] - [320] + [330] + [420] - [430] + [740] - [750] - [840] + [850]
(86) = [100] - [110] + [210] - [220] - [410] + [420] + [510] - [520] + [750] - [760] - [850] + [860]
(87)=- [100] + [110] + [200] - [210] - [300] + [310] + [400] - [410] - [500] + [510] + [600] - [610]
+ [760] - [770] - [860] + [870]
(88)=- [700] + [770] - [870] + [880]

Author Index

Numbers in italics refer to pages on which the complete references are listed.

Aitken, A. C., 12, *126*
Armstrong, L., 103, 105, 121, *130*, *131*, 133, 137

Bethe, H. A., 102, *130*
Biedenharn, L. C., 84, *129*
Bivins, R. L., 17, 78, *126*, *129*
Boerner, H., 8, 10, *125*
Brink, D. M., 84, *129*
Butler, P. H., 52, 77, 85, *128*, *129*, *131*, 133, 135, 136, 137, 140

Cartan, E., 1, *125*
Condon, E. U., 97, *123*
Crosswhite, H., 105, 106, *130*
Crosswhite, H. M., 105, 106, *130*
Cunningham, M. J., *131*, 133, 137

Dagis, R., 103, *130*
Duncan, D. G., 52, *128*

Eckart, C., 90, *129*
Edmonds, A. R., 84, *129*
Elliott, J. P., 1, 49, 77, *125*

Fano, U., 84, *129*
Feneuille, S., 77, 88, 103, 105, 112, 116, 121, 124, *128*, *129*, *130*, *131*, 133
Flowers, B. H., 1, 87, *125*, *129*
Foulkes, H. O., 52, *127*, *128*
Fraga, S., 105, *130*
Frame, J. S., 9, *125*
Fried, B., 98, *130*
Frobenius, G., 21, 37, 39, 41, 43, 44, *126*

Gamba, A., 21, 25, *126*
Gibbs, R. L., 84

Hamermesh, M., 5, 10, 22, *125*
Heintz, W. H., 84, *129*

Ibrahim, E. M., 42, 52, 53, *127*, *128*
Innes, F. R., 84, *129*

Jahn, H. A., 1, 44, *125*
Jucys, A., 84, 92, 96, 103, *129*, *130*
Judd, B. R., 1, 2, 46, 64, 69, 75, 76, 84, 85, 87, 88, 89, 92, 96, 97, 99, 100, 102, 104, 105, 106, 108, 111, 112, 121, *125*, *128*, *129*, *130*, *131*, 133, 137

Karaziya, R. I., 92, 96, *129*
Katiljus, R., 103, *130*
Klapisch, M., 112, *130*
Kondo, K., 17, *126*
Koster, G. F., 96, 101, 104, *130*
Kretzschmar, M., 49, *127*
Kuang Zhi-Quan, 46, 71, *127*

Lawson, R. D., 87, *129*
Levinsonas, J., 84, *129*
Lie, S., 1, *125*

Littlewood, D. E., 1, 2, 10, 12, 13, 17, 18, 21, 22, 24, 25, 26, 28, 32, 36, 37, 38, 39, 40, 41, 42, 44, 45, 47, 49, 52, 54, 58, 59, 61, 63, 64, 69, 76, *125*, *126*, *127*, *128*, 133, 135, 137, 140, 141, 142

Macfarlane, M. H., 87, *129*
MacMahon, P. A., 18, 20, *126*
McLellan, A. G., 92, *129*
Makar, R. H., 52, *128*
Malli, G., 105, *130*
Marvin, H. H., 103, 106, *130*
Metropolis, N., 17, 78, *126*, *129*
Missiha, S. A., 52, *128*
Muir, T., 12, 13, *125*
Murnaghan, F. D., 21, 25, 43, 45, 52, 66, 68, *126*, *127*, *128*

Newell, M. J., 43, 44, 45, 52, *127*
Nielson, C. W., 96, 101, 104, *130*
Nutter, P. B., 92, *129*

Racah, G., 1, 46, 71, 84, 92, 99, 100, 104, 111, *125*, *129*, *130*
Radicati, L. A., 25, *126*
Rajnak, K., 111, 112, *130*
Richardson, A. R., 1, 12, 13, 17, 21, 22, 24, 32, *125*, 141
Robinson, G. de B., 1, 8, 9, 10, 21, 22, 24, 25, 28, 51, 52, 54, 55, *125*, *126*, *127*
Rose, M. E., 84, *129*
Rotenberg, M., 78, *129*
Rudzikas, Z. B., 92, 96, 103, *129*, *130*
Rutherford, D. E., 1, 8, 21, *125*

Salpeter, E. E., 102, *130*
Satchler, G. R., 84, *130*
Scheffers, G., 1, *125*
Schur, I., 1, 20, 32, *125*
Shalit, Aide, 75, *128*
Shi Sheng-Ming, 46, 71, *127*
Shortley, G. H., 97, 98, *130*
Shudeman, C. L. B., 75, *128*
Smith, P. R., 27, 73, 92, 112, 116, *126*, *129*
Stein, J., 111, *130*
Stein, P. R., 17, *126*
Stone, A. P., 90, *129*
Szpikowski, S., 87, *129*

Talmi, I., 75, *128*
Taulbee, O. E., 25, *126*
Thrall, R. M., 9, *125*
Todd, J. A., 52, *128*

Ufford, C. W., 84, *129*

Vanagas, V., 84, *129*
Van Dam, H., 84, *129*
Vizbaraite, Ya. I., 92, 96, 103, *129*, *130*

Wadzinski, H. T., 92, 106, *129*
Wells, M. B., 17, *126*
Weyl, H., 1, 10, 22, 28, *125*
Wigner, E. P., 90, *129*
Wybourne, B. G., 27, 52, 73, 77, 85, 92, 102, 104, 111, 112, *126*, *128*, *129*, *130*, *131*, 133, 135, 136, 137

Yanagawa, S., 103, *130*
Yan Zhi-Da, 46, *127*
Young, A., 1, 7, 9, 33, 38, 56

Zhang Qing-Yu, 46, 71, *127*
Zia-ud-Din, M., 17, 52, *126*

Subject Index

Alternating group, 4

Branching rules, examples, 64
  $R(7) \longrightarrow G_2$, 46
  $R(n) \longrightarrow R(3)$, 137
  Spin representations, 137
  $U(n) \longrightarrow O(n)$, 39
  $U(n) \longrightarrow Sp(n)$, 42

Cayley's theorem, 4
Characters, 10
  compound, 11
  difference, 134
  Frobenius' formula, 21
  group G_2, 46
  group $GL(n)$, 32
  group $O(n)$, 37
  group $R(n)$, 40, 133
  group $Sp(n)$, 41
  group $U(n)$, 36
  immanants and, 13
  orthogonality of, 11
  tables of, 11
Classes, 5
  number in Sn, 6
  number of permutations in, 6
Compound matrices, 29
Coulomb interaction, 96
  in $(d + s)^n$, 119
Cycles, 4

Difference characters, 134
Dimensions of representations, 8
  $GL(n)$, 39
  $O(n)$, 39
  S_n, 8
  $Sp(n)$, 41
  $U(n)$, 36
Direct products for G_2,
  $GL(n)$, 32
  $O(n)$, 46
  $R(4)$, 62
  $R(6)$, 63
  $R_{2\mu}$, 136
  $Sp(n)$, 46
  $U(n)$, 36

Effective operators, 111
  mixed configurations and, 121
  three-particle, 112
  two particle, 111
Elementary symmetric functions, 18

Frobenius' formula, 21
Frobenius' notation, 38
Full linear group, 28

Hermite's reciprocity principle, 66
Homogeneous product sums, 19
Hook graphs, 9
Hook length, 9

Immanants, 12
Induced matrices, 31
Inner products, 22
  degree of, 26
  evaluation of, 25
Intransitive groups, 59
Invariant matrices, 32
Isoscalar factors, 93

jj-coupling, 74

LL-coupling, 75
LS-coupling, 70

Marvin integrals, 103
Matrices, compound, 29
  immanants of, 12
  induced, 31
  invariant, 32
  permanents, 12
Mixed configurations, 77
  $(d + s)^n$, 80
  $(d + s + s')^n$, 79
  $(l_1 + l_2 + l_3)^n$, 77
  parity and, 81
  symmetry properties of interactions, 117
Modification rules, 42

Nonstandard symbols, 42

Orbit-orbit interaction, 102
Orthogonal group, 36
Outer products, 22
  degree of, 25
  rule for S_n, 24

Partitions, 5
  characters and, 13
  cycle structure of, 6
  Frobenius' notation, 38
Permanents, 12
Permutations, 3
  conjugate, 5
  decrement of, 4
  even and odd, 4
  number in class, 6
  regular, 4
Plethysm, 49
  degree of, 55
  evaluation of, 52
  examples of, 55
  restricted groups and, 57
  rules of, 52
  Schur functions and, 51
Power sums, 19
Principal part, 53

Quarternary orthogonal group, 61

Quasi-spin, 86

Restricted groups, 35
Rotation group, 37

Schur functions, 20
  equivalences of, 44
  full linear group and, 28
  inner products of, 22
  multiplication rule for, 24
  outer products of, 22
  reduction of number of parts, 45
Selection rules, 90
Seniority classification, 70
Spin-other-orbit interaction, 106
Spin representations, 59
  difference characters and, 135
  Kronecker products of, 133
  Kronecker squares of, 136
Spin-spin interaction, 105
Symmetrization of interactions, 84
  Coulomb, 96
  generalized two-particle, 108
  orbit-orbit, 102
  spin-other-orbit, 106
  spin-spin, 105
Symplectic group, 40
Standard Young tableaux, 8
Symmetric functions, 18
  elementary, 18
  homogeneous product sums, 19
  power sums, 19
Symmetric group, 3

Tables of group properties, 140
Tensor operators, 84
  quasi-spin classification, 86
  transformation properties, 85
Ternary orthogonal group, 60
Transposition, 4

Unitary group, 35

Wigner-Eckart theorem, 89

Young diagrams, 7
  conjugate, 8
  hook graphs of, 9